Symbol	Meaning
i	imaginary unit [6.3]
$\sqrt{-b}$	$i\sqrt{b}$, $b > 0$ [6.3]
$a + bi$	complex number [6.3]
$\pm$	plus or minus [6.4]
(x, y)	ordered pair of numbers whose first component is x and whose second component is y [7.1]
$R \times R$	Cartesian product of R and R [7.1]
f, g, etc.	names of functions [7.3]
$f(x)$, etc.	f of x or value of f at x [7.3]
F^{-1}	F inverse or the inverse of F [8.6]
$\log_b x$	logarithm to the base b of x [9.2]
$\text{antilog}_b\, x$	antilogarithm to the base b of x [9.4]
s_n	nth term of a sequence [11.1]
S_n	sum of n terms of a sequence [11.1]
$\sum$	the sum [11.1]
S_∞	infinite sum [11.1]
$n!$	n factorial or factorial n [11.5]
$P(n, n)$	permutations of n things taken n at a time [11.6]
$P(n, r)$	permutations of n things taken r at a time [11.6]
$\binom{n}{r}$	combinations of n things taken r at a time [11.7]
$\begin{bmatrix} a_1 & b_1 \\ a_2 & b_2 \end{bmatrix}$	second-order matrix [B.1]
$\begin{bmatrix} a_1 & b_1 & c_1 \\ a_2 & b_2 & c_2 \\ a_3 & b_3 & c_3 \end{bmatrix}$	third-order matrix [B.1]
$\begin{vmatrix} a_1 & b_1 \\ a_2 & b_2 \end{vmatrix}$	second-order determinant [B.2]
$\begin{vmatrix} a_1 & b_1 & c_1 \\ a_2 & b_2 & c_2 \\ a_3 & b_3 & c_3 \end{vmatrix}$	third-order determinant [B.3]
δA	determinant of A [B.2]

INTERMEDIATE ALGEBRA 4TH EDITION

INTERMEDIATE ALGEBRA 4TH EDITION

WILLIAM WOOTON
IRVING DROOYAN

Los Angeles Pierce College

Wadsworth Publishing Company, Inc.,
Belmont, California

TO THE STUDENT

The self-help study guide that accompanies this text—*Study Guide: Intermediate Algebra*, 4th edition by Bernard Feldman—is available from your local bookstore.

Designer: Ann Wilkinson
Mathematics Editor: Don Dellen
Production Editor: Phyllis Niklas
Technical Illustrator: Mark Schroeder

Intermediate Algebra, 4th Edition originally published under the title *Intermediate Algebra, Third Alternate Edition*

ISBN-0-534-00409-1
L. C. Cat. Card No. 75-5397
Printed in the United States of America
3 4 5 6 7 8 9 10—80 79 78 77

CONTENTS

10

SYSTEMS OF EQUATIONS 302

11

NATURAL NUMBER FUNCTIONS 328

APPENDIX A

SYNTHETIC DIVISION; POLYNOMIAL FUNCTIONS 362

APPENDIX B

MATRICES AND DETERMINANTS 372

TABLES 389

ODD-NUMBERED ANSWERS 391

INDEX 451

SYMBOLS inside front cover

COMMON LOGARITHMS inside back cover

PREFACE

This edition of *Intermediate Algebra* is the result of twelve years of extensive classroom experience with earlier editions. Significant alterations—beyond the expected minor revisions of text and problem sets to improve clarity—include additions, deletions, and changes in sequence designed to improve the classroom effectiveness of the presentation.

New material in this edition includes a section on the remainder theorem, the factor theorem, and graphing polynomial functions in which synthetic division is applied. Also new is an introduction to matrices, and the solution of linear systems using row-equivalent matrices. These new topics are included in Appendices A and B, and are provided for use in courses in which more than a basic treatment of intermediate algebra is desired.

Changes in sequence include the transfer of the treatment of synthetic division from Chapter 3 to Appendix A, and the transfer of the sections dealing with determinants and the solution of linear systems by Cramer's rule from Chapter 9 to Appendix B. The third edition chapters on systems of equations and logarithms have been interchanged. Many of the word problems included in the Appendix of the third edition now appear in Chapter 5.

As in earlier editions, the axioms for a field are stated in the first chapter and used consistently throughout the course. Our axiom system is far from minimal; we have deliberately added some assumptions that are not properly axiomatic in order to smooth the presentation. Although we do not emphasize formal proofs, the statement–reason form is introduced in Chapter 1 to give the student at least a general sense of what might constitute a formal argument. In general, informal deductive or inductive arguments are used to lend plausibility to formally stated conclusions. This decision to rely on logical implication, rather than logical formalism, results from our association with students at this level, to whom the notion of a formal proof is generally profound and difficult.

The first six chapters review material normally encountered in some form in a beginning algebra course. The review, conducted from an axiomatic standpoint, presents the material as a unified, related structure. The axioms for a field which are stated in Chapter 1 are used in Chapter 2 as a basis for discussing fundamental operations with polynomials. In Chapter 3, which deals with operations with fractions, the definitions of the operations are shown to be consistent with the assumptions made concerning real numbers. The fundamental principle of fractions is always invoked when writing fractions. Chapter 4 acquaints the student with the properties of expressions involving rational number exponents, and Chapter 5 introduces both linear equations and inequalities as sentences; their solution is approached through the concept of equivalent sentences. Chapter 6 is a treatment of quadratic equations and inequalities in one variable.

The presentation in the last five chapters centers around the function concept, with heavy emphasis on graphing. In particular, linear and quadratic relations and functions and their graphs are explored in some detail; variation is treated from the function standpoint; logarithms are developed from a consideration of the inverse of the exponential function; and sequences and series are discussed through the use of function concepts. The last three chapters are independent of each other and any one may be omitted for a short course.

In classes in which students are well-prepared, instructors may only wish to review Chapter 1 briefly (one or two hours). In this case, an assignment consisting of reading the chapter summary and completing the review exercises may provide an adequate review. The course can then start formally with Chapter 2.

To aid in making assignments, the exercise sets have been graded. Exercises labeled A provide routine practice, and are sufficient for a basic course. Exercises labeled B contain more challenging problems. In this edition, we have given less emphasis to placing restrictions on variables in exercise sets, and given more emphasis to common errors made by students at this level.

Answers to the odd-numbered problems, including graphs, are provided at the end of the book. In addition, a study guide covering topics in the text is available for student use. Additional ancillary materials are available to the instructor.

We wish to thank David Wend of Montana State University and Thomas Kearns of Northern Kentucky State College for their assistance in reading galley proofs. We also wish to thank Nancy Halloran for her help in typing the manuscript.

Los Angeles, California

William Wooton
Irving Drooyan

INTERMEDIATE ALGEBRA 4TH EDITION

1

THE SET OF REAL NUMBERS

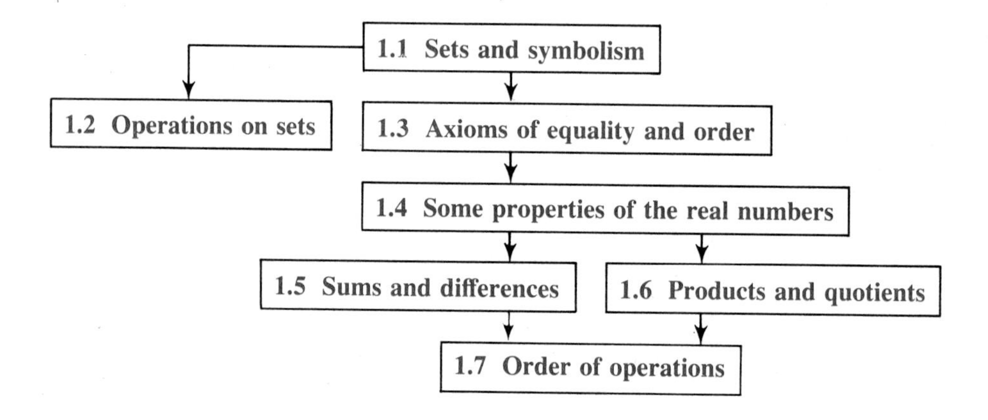

1.1

SETS AND SYMBOLISM

A **set** is a collection of some kind. It may be a collection of people, or books, or colors, or almost anything else; however, in algebra we are interested primarily in collections, or sets, of numbers. Each item in a set is called an **element** or a **member** of the set. For example, the counting numbers 1, 2, 3, ... are the elements of a set we call the set of **natural numbers**. Because there exists no last counting number, we refer to this as an **infinite set**. A set whose elements can be counted is called a **finite** set. For example, the set of numbers represented by the symbols on a die, 1, 2, 3, 4, 5, 6, is a finite set. The set containing no members is called the **empty set** or **null set,** and is considered a finite set.

Sets are designated by means of capital letters, A, I, R, etc., or by means of braces, $\{\ \}$, used in conjunction with words or symbols. Thus, when we write

$$A = \{\text{natural numbers less than } 7\}$$

or

$$A = \{1, 2, 3, 4, 5, 6\},$$

we mean that A represents the set whose elements are the numbers 1, 2, 3, 4, 5, and 6. We say that two sets are **equal** if they have the same elements; thus,

$$\{2, 3, 4\} = \{\text{natural numbers between 1 and } 5\}.$$

This example illustrates two ways in which sets are described: (1) using roster notation in which the elements are *listed*, and (2) using a *rule*. The empty set is denoted by the symbol $\varnothing$ or $\{\ \}$. Note that the symbol $\varnothing$ (read "the empty set") does not involve braces.

If every member of a given set A is also a member of another set B, we say that A is a **subset** of B. The symbol $\subset$ (read "is a subset of" or "is contained in") is used to denote this relationship. For example,

$$\{2, 4\} \subset \{1, 2, 3, 4\}$$

and

$$\{3\} \subset \{1, 2, 3, 4\}.$$

Of course, by definition, every set is a subset of itself; also by convention, $\varnothing$ is a subset of every set. Thus,

$$\{1, 2, 3, 4\} \subset \{1, 2, 3, 4\}$$

and

$$\varnothing \subset \{1, 2, 3, 4\}.$$

We generally denote an individual element in a set by means of a lowercase letter, such as a, b, c, x, y, or z. Symbols used to represent unspecified elements of a given set containing more than one member are called **variables**. If the given set, called the **replacement** set of the variable, is a set of numbers (as it will be throughout this book), then the variable represents a number. We use the symbol $\in$ to denote membership in a set and write

$$x \in N$$

(read "x is an element of N") to indicate that x represents an element of the set N. A symbol used to denote the element of a set containing only one element is called a **constant**.

The slash / is frequently used as a negation symbol in conjunction with other symbols. For example,

$\not\subset$ is read "is *not* a subset of,"

$\notin$ is read "is *not* an element of."

Variables enable us to describe sets by a rule using a convenient symbolism called **set-builder notation**. For example,

$$\{x \mid x \in A \quad \text{and} \quad x \notin B\}$$

is read "the set of all x such that x is an element of A and x is not an element of B." In particular, note that the vertical line is read "such that." Another example is:

$$\{x \mid x^2 = 9, \quad x \text{ is a natural number}\}$$

which is read "the set of all x such that $x^2 = 9$ and x is a natural number."

In this book we are going to be concerned primarily with the following sets of numbers:

1. The set N of **natural numbers** or **counting numbers**, whose elements consist of 1, 2, ..., 7, ..., 235, ...

2. The set W of **whole numbers**, whose elements consist of the natural numbers and zero.
3. The set J of **integers**, whose elements consist of the natural numbers, their negatives, and zero. Among the integers are -7, -3, 0, 5, and 11.
4. The set Q of **rational numbers**, whose elements are all the numbers that can be represented in the form a/b, where a and b represent integers and b does not represent zero. Among the rational numbers are $-\frac{3}{4}$, $\frac{18}{27}$, 3, and -6. All numbers in Q can be represented by terminating or repeating decimals. For example, $-\frac{3}{4} = -0.75$ and $\frac{2}{3} = 0.666\ldots$.
5. The set H of **irrational numbers**, whose elements are the numbers with decimal representations that are nonterminating, nonrepeating numerals. Among the irrational numbers are $\sqrt{15}$, π, and $-\sqrt{7}$. An irrational number cannot be represented in the form a/b, where a and b are integers.
6. The set R of **real numbers**, whose elements consist of all rational and all irrational numbers.

Observe from the foregoing sets of numbers that $N \subset W$, $W \subset J$, $J \subset Q$, $Q \subset R$, and $H \subset R$. These relationships are shown in Figure 1.1. It is with the real numbers or subsets of the real numbers that we shall be most concerned in this book. Many properties of the real numbers are discussed later in this chapter. However, the irrational numbers are considered in more detail in Chapter 4.

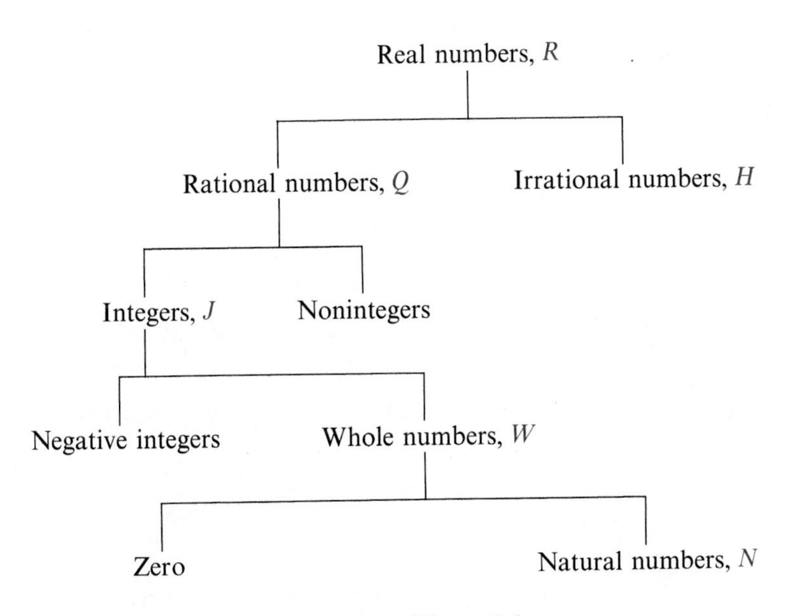

Figure 1.1

EXERCISE 1.1

A

Specify each set by listing the members.

Examples **a.** {integers between -5 and -1} **b.** {whole numbers greater than 4}

Solutions **a.** $\{-4, -3, -2\}$ **b.** $\{5, 6, \ldots\}$

1. {natural numbers between 1 and 7}
2. {first four whole numbers}
3. {first three natural numbers}
4. {integers between -2 and 2}
5. {natural number multiples of 4}
6. {odd natural numbers between 2 and 12}
7. {even integers between -3 and 3}
8. {whole number multiples of 11}

Specify each set by a rule. [*Note: There may be more than one way to specify the given set.*]

Examples **a.** $\{3, 6, 9\}$ **b.** $\{1, 3, 5, \ldots\}$

Solutions **a.** {first three natural number multiples of 3} **b.** {odd natural numbers}

9. $\{15, 30, 45, \ldots\}$
10. $\{7, 8, 9\}$
11. $\{9, 11, 13\}$
12. $\{4, 8, 12, \ldots\}$
13. $\{0, 1, 2\}$
14. $\{12, 14, 16\}$

Let $A = \{-5, -\sqrt{15}, -3.44\ldots, -\frac{2}{3}, 0, \frac{1}{5}, \frac{7}{3}, 6.1, 8\}$, *and list the members of the given set.*

Examples **a.** $\{x \mid x \in A \text{ and } x \in J\}$ **b.** $\{x \mid x \text{ is a positive number in } A\}$

Solutions **a.** We wish the set of all members of A that are also integers, $\{-5, 0, 8\}$.
b. We wish all positive members of A, $\{\frac{1}{5}, \frac{7}{3}, 6.1, 8\}$.

15. $\{x \mid x \in A \text{ and } x \in W\}$
16. $\{x \mid x \in A \text{ and } x \in N\}$
17. $\{x \mid x \in A \text{ and } x \in H\}$
18. $\{x \mid x \in A \text{ and } x \in Q\}$
19. $\{x \mid x \text{ is a negative number in } A\}$
20. $\{x \mid x \text{ is a real number in } A\}$

State whether the given set is finite or infinite.

Example {whole numbers greater than 10,000}

Solution Since there is no greatest whole number, the set is infinite.

21. {whole numbers less than 1,000,000}

22. {natural numbers less than 1,000,000}

23. {natural numbers with four-digit numerals}

24. {whole numbers with 0 as the last numeral}

25. {rational numbers between 0 and 1}

26. {real numbers between 2 and 3}

Let x represent an element of the given set. In each case, state whether x is a variable or a constant.

27. {integers}

28. {whole numbers}

29. {integers between -4 and -2}

30. {natural numbers less than 2}

31. {whole numbers less than 1}

32. {rational numbers between 0 and 1}

Replace the question mark with either $\subset$ or $\in$ to form a true statement.

33. $4 \underline{?} \{\text{integers}\}$

34. $\{0\} \underline{?} \{\text{integers}\}$

35. $\{4, 7\} \underline{?} \{4, 5, 6, 7, 8\}$

36. $\{1, 2\} \underline{?} \{1, 2\}$

37. $\varnothing \underline{?} \{1, 2\}$

38. $0 \underline{?} \{0\}$

Replace the question mark with either $\subset$ or $\not\subset$ to form a true statement.

39. $\{-1, 1\} \underline{?} N$

40. $\{3, 4\} \underline{?} J$

41. $\{\frac{1}{2}, 3\} \underline{?} R$

42. $\varnothing \underline{?} W$

Replace the question mark with either $\in$ or $\notin$ to form a true statement.

43. $2 \underline{?} Q$

44. $0 \underline{?} N$

45. $\frac{4}{3} \underline{?} R$

46. $-3 \underline{?} W$

Use the negation symbol / to write each of the statements in symbolic form.

Example Three is not an element of the set H of irrational numbers.

Solution $3 \notin H$

47. One-half is not an element in the set N of natural numbers.

48. The set J of integers is not a subset of the set N of natural numbers.

49. The set Q of rational numbers is not a subset of the set J of integers.

50. The set whose only member is 5 is not an element of $\{3, 4, 5\}$.

Use the subset symbol $\subset$ to express the relationship between the specified sets.

51. The set of natural numbers N, and the set of real numbers R.

52. The set of rational numbers Q, and the set of integers J.

53. The set of real numbers R, and the set of whole numbers W.

54. The set of irrational numbers H, and the set of real numbers R.

B

55. If $A = B$ and $4 \notin B$, can 4 be an element of A?

56. If $A \neq B$ and $4 \in B$, must 4 be an element of A?

57. If $A \neq B$ and $B = C$, can $A = C$?

58. If $A \neq B$ and $B \neq C$, must $A \neq C$?

59. If $R \subset S$ and $4 \in R$, must 4 be an element of S?

60. If $R \subset S$ and $4 \in R$, can 4 be an element of S?

61. If $R \subset S$ and $a \notin R$, can a be an element of S?

62. If $R \subset S$ and $a \in S$, must a be an element of R?

63. If $S \subset T$, can T be a subset of S?

64. If $S \subset T$, must T be a subset of S?

65. If $M \subset N$ and $N \subset M$, what other relationship exists between M and N?

66. If $M \subset N$ and $N \subset P$, what relationship always exists between M and P?

67. List all subsets of $\{1, 2, 3\}$.

68. List all subsets of $\{0, 1, 2, 3\}$.

69. In how many ways can a four-man relay team be selected from a track squad containing five members? (How many subsets containing four members are there?)

70. In how many ways can two players be selected from a tennis squad of five members?

1.2 OPERATIONS ON SETS

Let us define some operations involving sets that we shall find useful in later sections of this book. One such operation consists of forming the **union** of two sets, say A and B, and producing a third set which contains all the elements that belong *to either A or B or to both.* The symbol $\cup$ is used to denote the operation. Thus, $A \cup B$ is read "the union of A and B." For example, if $A = \{1, 2, 3, 4, 5\}$ and $B = \{5, 6, 7\}$, then

$$\begin{aligned} A \cup B &= \{x \mid x \in A \quad \text{or} \quad x \in B\} \\ &= \{1, 2, 3, 4, 5, 6, 7\}. \end{aligned}$$

Notice that each element in $A \cup B$ is listed only once. The numeral "5" denotes just one number even if we write its name twice.

A second operation involving sets A and B consists of forming the **intersection** of the sets and producing a third set which contains all the elements that belong to *both A and B.* The symbol $\cap$ is used to denote this operation. Thus, $A \cap B$

is read "the intersection of A and B." For example, if $A = \{1, 2, 3, 4, 5\}$ and $B = \{3, 4, 5, 6, 7\}$, then

$$A \cap B = \{x \mid x \in A \quad \text{and} \quad x \in B\}$$
$$= \{3, 4, 5\}.$$

Sets that do not contain any common elements are said to be **disjoint**. For example, if $A = \{2, 4\}$ and $B = \{1, 3, 5\}$, then A and B are disjoint. Note that any two sets A and B are disjoint if and only if $A \cap B = \varnothing$.

Diagrams such as those shown in Figure 1.2 offer a way of picturing set relationships. Figure 1.2a depicts a set B which is a subset of a set A, where it is understood that the region inside and on each closed curve represents the elements in the given set. Figure 1.2b shows a schematic representation of $A \cup B$, where the shading denotes the region representing $A \cup B$. Figure 1.2c pictures $A \cap B$, where again the shading covers the region depicted. In each case, note that we assume that sets A and B are both subsets of some general set U. This is frequently a useful concept in discussing sets. The set U is called the **universal set**, or simply the **universe**.

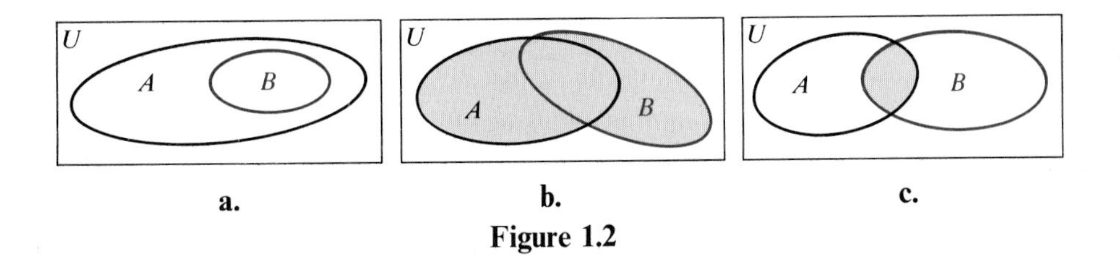

a. b. c.

Figure 1.2

EXERCISE 1.2

A

Let $A = \{2, 4, 6, 8, 10\}$, $B = \{1, 2, 3, 4, 5\}$, $C = \{1, 3, 5, 7, 9\}$, and $D = \{6, 7, 8, 9, 10\}$. List the elements of the set resulting from the operations indicated.

Examples **a.** $A \cap B$ **b.** $(A \cap B) \cup C$

Solutions **a.** $A \cap B$ contains all members that are in *both A and B*. Hence,

$$A \cap B = \{2, 4\}.$$

b. $(A \cap B) = \{2, 4\}$ and $C = \{1, 3, 5, 7, 9\}$. Hence,

$$(A \cap B) \cup C = \{1, 2, 3, 4, 5, 7, 9\}.$$

1. $B \cap C$
2. $C \cap D$
3. $A \cup D$
4. $B \cup C$
5. $A \cap D$
6. $B \cap D$
7. $A \cup C$
8. $B \cup D$
9. $A \cap A$
10. $C \cap C$
11. $B \cup B$
12. $D \cup D$
13. $A \cap \varnothing$
14. $B \cap \varnothing$
15. $C \cup \varnothing$
16. $D \cup \varnothing$
17. $(A \cap C) \cup B$
18. $(B \cap D) \cup A$
19. $(C \cup D) \cap A$
20. $(C \cup A) \cap D$
21. $(A \cup B) \cup C$
22. $(B \cap C) \cap D$

Consider any two sets G and H. For each statement, state the conditions under which it would be true.

Example $G \cap H = H$

Solution If the intersection of G and H is H, then every element of H must also be an element of G. Hence, $H \subset G$.

23. $G \cup H = \varnothing$
24. $G \cup \varnothing = \varnothing$
25. $G \cap H = G$
26. $G \cup H = G$
27. $G \cap \varnothing = G$
28. $G \cup H = G \cap H$

Use set diagrams with shading and double shading, as necessary, to represent each statement. Assume that no two sets are disjoint.

Example $(A \cap B) \cap C$

Solution The region shaded with gray *and* red represents

$$(A \cap B) \cap C.$$

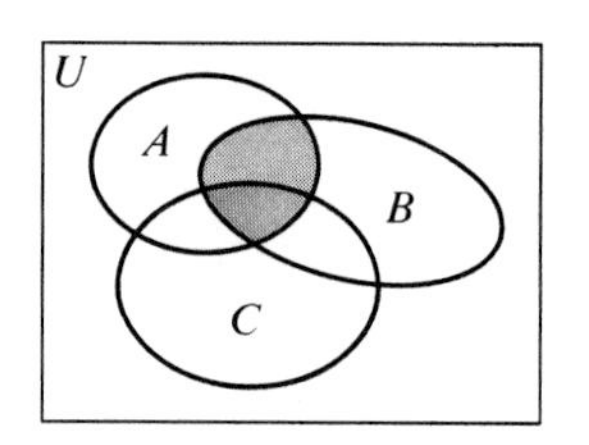

29. $(A \cup B) \cap C$
30. $(A \cap B) \cup C$
31. $A \cap (B \cup C)$
32. $A \cap (B \cap C)$

B

33. If A contains 5 members, B contains 6 members, and $A \cap B = \varnothing$, how many members does $A \cup B$ have?
34. If A contains m members, B contains n members, and $A \cap B = \varnothing$, how many members does $A \cup B$ have?
35. If A contains 3 members, B contains 8 members, and $A \cap B$ contains 1 member, how many members does $A \cup B$ have?
36. If A contains 3 members, B contains 8 members, and $A \cup B$ contains 9 members, how many members does $A \cap B$ have?

37. If $2 \in A$, must 2 be an element of $A \cup B$ for every B?

38. If $2 \in A$, must 2 be an element of $A \cap B$ for every B?

39. If $A \subset U$, $B \subset U$, and $3 \in U$, must 3 be an element of $A \cup B$?

40. If $A \subset U$, $B \subset U$, and $3 \notin A$, can 3 be an element of $A \cup B$?

1.3

AXIOMS OF EQUALITY AND ORDER

In mathematics, formal assumptions about numbers or their properties are called **axioms** or **postulates**. Such assumptions are formal statements about properties which we propose to assume as always valid. While we are free to formulate axioms in any way we please, it is clearly desirable that any axioms we adopt lead to useful consequences. A set of useful axioms must not lead to contradictory conclusions.

The words *property*, *law*, and *principle* are sometimes used to denote assumptions, although these words may also be applied to certain consequences of axioms. In this book we use, in each situation, the word we believe to be the one most frequently encountered. The first such assumptions to be considered concern equality.

An **equality**, or an "is equal to" assertion, is a mathematical statement that two symbols, or groups of symbols, are names for the same number. A number has an infinite variety of names. Thus 3, $6/2$, $4 - 1$, and $2 + 1$ are all names for the same number; hence, the equality

$$4 - 1 = 2 + 1$$

is a statement that "$4 - 1$" and "$2 + 1$" are different names for the same number.

We shall assume that the "is equal to" $(=)$ relationship has the properties listed below.

Equality Axioms for R ($a, b, c \in R$)

E-1 Reflexive property

$$a = a$$

E-2 Symmetric property

If $a = b$, then $b = a$.

E-3 Transitive property

If $a = b$ and $b = c$, then $a = c$.

E-4 Substitution property

If $a = b$, then b may be replaced by a or a by b in any statement without altering the truth or falsity of the statement.

We refer to the symbol (or the number it names) to the left of an equals sign as the left-hand member, and that to the right as the right-hand member of the equality.

Because there exists a one-to-one correspondence between the *real numbers* and the *points* on a geometric line (for each real number there corresponds one and only one point on the line, and vice versa), a geometric line can be used to visualize relationships existing between real numbers. For example, to represent 1, 3, and 5 on a line, we scale a straight line in convenient units, with increasing positive direction indicated by an arrow, and mark the required points with dots on the line. This geometric representation is called a **line graph**, or **number line** (Figure 1.3). The real number corresponding to a point on a line graph is called the **coordinate** of the point, and the point is called the **graph** of the number.

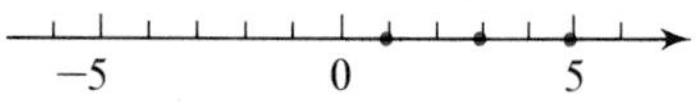

−5 0 5

Figure 1.3

A line graph can be used to separate the real numbers into three subsets—one set R_+ whose elements are associated with the points on the line to the right of a point called the **origin** and labeled 0; one set R_- whose elements are associated with the points on the line to the left of 0; and one set whose only element is 0. The elements of the set R_+ are called **positive numbers**, and in some cases we prefix their numerals with plus signs $(+1, +7, +\pi)$ to denote the fact that they are positive. The elements of the set R_- are called **negative numbers**, and numerals representing such numbers are always prefixed with minus signs $(-2, -\frac{1}{2}, -\pi)$ to identify them as such. The number 0, neither positive nor negative, serves as a point of separation for the positive and negative numbers.

We can now define the real number b to be less than the real number a if for some positive real number c,

$$b + c = a.$$

In symbols, we write $b < a$ (read "b is less than a") or, equivalently, $a > b$ (read "a is greater than b"). It is evident that, on a line graph, as shown in Figure 1.4, the graph of b will be found to the left of the graph of a. Also, if $a > 0$, then a is positive, and if $a < 0$, then a is negative. Thus,

$$R_+ = \{x \mid x \in R, \quad x > 0\},$$

$$R_- = \{x \mid x \in R, \quad x < 0\}.$$

The following symbols combine the concepts of equality and inequality and are used to write two statements at once:

$\leq$ is read "is less than or equal to,"
$\geq$ is read "is greater than or equal to."

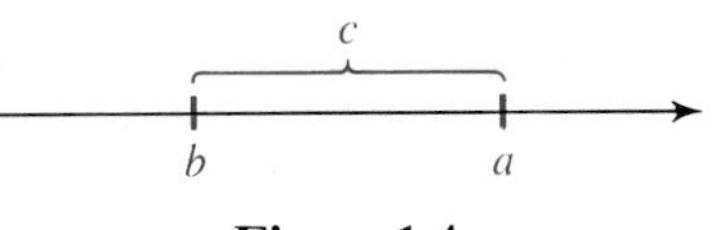

Figure 1.4

A real number that is not a negative number is called a **nonnegative** number. In symbols, we can write

$$a \geq 0$$

to indicate that a is nonnegative. Similarly, $a \leq 0$ means that a is a **nonpositive** number.

The slash is also used in conjunction with these order symbols to indicate the word "not"; thus, $\neq$ means "is *not* equal to," $\not<$ means "is *not* less than," and $\not>$ means "is *not* greater than."

Inequalities such as

$$1 < 2 \quad \text{and} \quad 3 < 5$$

are said to be of the *same sense*, because the left-hand member is less than the right-hand member in each case. Inequalities such as

$$1 < 2 \quad \text{and} \quad 5 > 3$$

are said to be of *opposite sense*, because in one case the left-hand member is less than the right-hand member and in the other case the left-hand member is greater than the right-hand member.

We assume properties listed below with respect to the order of real numbers.

Order Axioms for R ($a, b, c \in R$)

O-1 Trichotomy property

Exactly one of the following relationships holds:

$$a < b, \quad a = b, \quad \text{or} \quad a > b.$$

O-2 Transitive property

If $a < b$ and $b < c$, then $a < c$.

Line graphs can be used to display infinite sets of points as well as finite sets. For example, Figure 1.5a is the graph of the set of all integers greater than or equal to 2 *and* less than 5. The graph consists of three points. Figure 1.5b is the graph of the set of all real numbers greater than or equal to 2 and less than 5. The graph is of an infinite set of points. The closed dot on the left end of the shaded portion of the graph indicates that the end point is a part of the graph, while the open dot on the right end indicates that the end point is not in the graph.

The set of numbers whose graph is shown in Figure 1.5a can be described in set-builder notation by

$$\{x \mid 2 \le x < 5, \quad x \in J\}$$

(read "the set of all x such that x is greater than or equal to two *and* less than five, and x is an element of the set of integers"). Similarly, the set of numbers whose graph is shown in Figure 1.5b can be described by

$$\{x \mid 2 \le x < 5, \quad x \in R\}.$$

When no question exists relative to the replacement set for x, the notation $\{x \mid 2 \le x < 5, \quad x \in R\}$ is abbreviated to $\{x \mid 2 \le x < 5\}$.

If the replacement set of the variable is R, the end points of the graphs of inequalities are sometimes shown by the symbol (or) where we have used an "open" dot and [or] where we have used a closed dot. Thus, the graph of $\{x \mid 2 \le x < 5\}$ can also be represented as in Figure 1.6. In this book, however, we shall always use open and closed dots for end points.

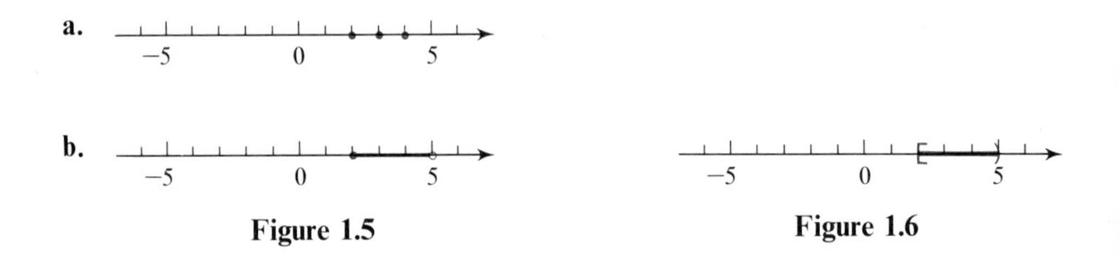

Figure 1.5

Figure 1.6

EXERCISE 1.3

A

Replace each question mark to make the given statement an application of the given property.

Example If $z = 2$ and $2 = t$, then $\underline{?} = t$; transitive property of equality.

Solution If $z = 2$ and $2 = t$, then $z = t$.

1. $3r = \underline{?}$; reflexive property of equality.

2. If $n = t + 3$, then $\underline{?} = n$; symmetric property of equality.

3. If $r = 6$ and $r - 3 = t$, then $\underline{?} - 3 = t$; substitution property of equality.

4. Either $x < \underline{?}$, $x = 3$, or $x > 3$; trichotomy property of order.

5. If $n < 3$ and $3 < t$, then $\underline{?} < \underline{?}$; transitive property of order.

6. If $a = c$ and $c = 4$, then $\underline{?} = 4$; transitive property of equality.

7. If $r = n$ and $n + 6 = 8$, then $\underline{?} + 6 = 8$; substitution property of equality.

8. If $t = 4$ and $5 \cdot t = 6s$, then $5 \cdot \underline{?} = 6s$; substitution property of equality.

9. $6 + x = \underline{?}$; reflexive property of equality.

10. If $2 + x = y$, then $y = \underline{?}$; symmetric property of equality.

Express each relation using symbols.

Examples **a.** 6 is less than 10 **b.** x is less than 3 and greater than or equal to 0

Solutions **a.** $6 < 10$ **b.** $0 \le x < 3$

11. 4 is greater than -2

12. -3 is less than 3

13. -4 is less than -2

14. -20 is greater than -29

15. $x + 3$ is positive

16. $t - 4$ is nonpositive

17. $2n - 1$ is nonnegative

18. $x + 4$ is negative

19. z is between 0 and 1

20. x is not greater than or equal to 4

21. x is greater than or equal to 1 and less than 7

22. $3t$ is greater than or equal to 0 and less than or equal to 4

Replace each question mark with an appropriate order symbol to form a true statement.

23. $-2 \ \underline{?}\ 8$ **24.** $3 \ \underline{?}\ 6$ **25.** $-7 \ \underline{?}\ -13$ **26.** $0 \ \underline{?}\ -5$

27. $-6 \ \underline{?}\ -3$ **28.** $\dfrac{-3}{2} \ \underline{?}\ \dfrac{-3}{4}$ **29.** $1\dfrac{1}{2} \ \underline{?}\ \dfrac{3}{2}$ **30.** $3 \ \underline{?}\ \dfrac{6}{2}$

31. $3 \ \underline{?}\ 5 \ \underline{?}\ 7$ **32.** $3 \ \underline{?}\ 0 \ \underline{?}\ -4$ **33.** $-7 \ \underline{?}\ 0 \ \underline{?}\ 2$ **34.** $-5 \ \underline{?}\ -2 \ \underline{?}\ 0$

Rewrite each relation without using the negation bar.

Example $7 \not> t$

Solution Since 7 is not greater than t, 7 must be either less than or equal to t. Therefore, $7 \le t$.

35. $2 \ngeq 5$ **36.** $-1 \nleq -2$ **37.** $7 \ngeq 8$

38. $-3 \ngeq 0$ **39.** $x \not< y$ **40.** $x \not> z$

Rewrite each relation in two ways—with a slash and without it.

41. x is positive

42. x is negative

43. x is nonnegative

44. x is nonpositive

Graph the members of each of the given sets. Use a separate line graph for each set $N = \{\text{natural numbers}\}$ and $J = \{\text{integers}\}$.

Example $\{x \mid -2 < x < 3, \quad x \in J\}$

Solution

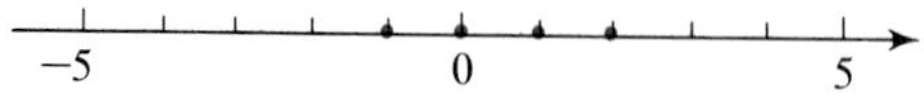

45. $\{x \mid x < 6, \quad x \in N\}$
46. $\{x \mid 4 \le x < 9, \quad x \in N\}$
47. $\{x \mid -4 < x < 4, \quad x \in J\}$
48. $\{x \mid -6 < x \le 0, \quad x \in J\}$
49. $\{x \mid 25 < x < 30, \quad x \in J\}$
50. $\{x \mid -20 \le x < 10, \quad x \in J\}$

Consider $x \in R$ and graph each given set on a number line.

Example $\{x \mid x > -3\} \cap \{x \mid x \le 2\}$

Solution $\{x|x > -3\}$

$\{x|x \le 2\}$

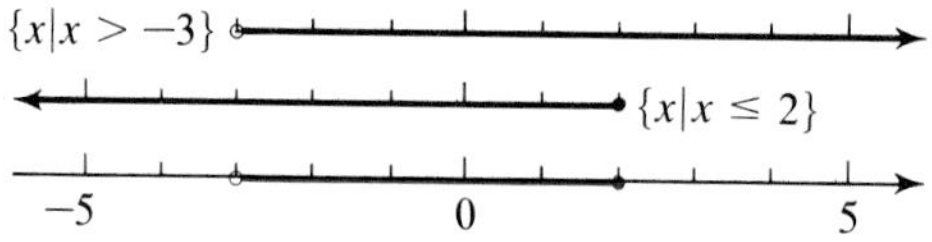

51. $\{x \mid x > 2\}$
52. $\{x \mid x \le 2\}$
53. $\{x \mid -6 < x \le -1\}$
54. $\{x \mid -2 \le x < 6\}$
55. $\{x \mid x < 7 \text{ and } x \ge 1\}$
56. $\{x \mid x \le 2 \text{ and } x > -2\}$
57. $\{x \mid x < -3 \text{ or } x > 3\}$
58. $\{x \mid x \le -5 \text{ or } x \ge 1\}$
59. $\{x \mid x < 5\} \cap \{x \mid x > 2\}$
60. $\{x \mid x < 3\} \cap \{x \mid x \ge -2\}$
61. $\{x \mid x \le 0\} \cap \{x \mid x > -5\}$
62. $\{x \mid x > 3\} \cap \{x \mid x < -2\}$
63. $\{x \mid x > 2\} \cup \{x \mid x < -5\}$
64. $\{x \mid x < 3\} \cup \{x \mid x \ge 3\}$
65. $\{x \mid x < 5\} \cup \{x \mid x \le 0\}$
66. $\{x \mid x > 1\} \cup \{x \mid x < -1\}$

B

67. $\{x \mid 1 < x < 5\} \cap \{x \mid x > 3\}$
68. $\{x \mid -2 < x < 4\} \cap \{x \mid x \le 0\}$
69. $\{x \mid x < 4\} \cup \{x \mid 2 \le x < 8\}$
70. $\{x \mid x \ge -2\} \cup \{x \mid -3 \le x < 3\}$
71. $\{x \mid -4 \le x \le 4\} \cap \{x \mid 0 \le x \le 6\}$
72. $\{x \mid -3 \le x < 3\} \cup \{x \mid -5 < x < 0\}$

1.4

SOME PROPERTIES OF THE REAL NUMBERS

In addition to the axioms of equality and order, we take as axioms the properties of the real numbers listed on page 17.

Axioms for R ($a, b, c \in R$)

R-1 Closure for addition

$a + b$ is a unique real number.

R-2 Commutative property of addition

$$a + b = b + a$$

R-3 Associative property of addition

$$(a + b) + c = a + (b + c)$$

R-4 Closure for multiplication

ab is a unique real number.

R-5 Commutative property of multiplication

$$ab = ba$$

R-6 Associative property of multiplication

$$(ab)c = a(bc)$$

R-7 Distributive property

$$a(b + c) = (ab) + (ac)$$

R-8 Identity element for addition
There exists a unique number 0 with the property

$$a + 0 = a \quad \text{and} \quad 0 + a = a.$$

R-9 Identity element for multiplication
There exists a unique number 1 with the property

$$a \cdot 1 = a \quad \text{and} \quad 1 \cdot a = a.$$

R-10 Negative or additive inverse property
For each real number a, there exists a unique real number $-a$ (called the **negative** of a) with the property

$$a + (-a) = 0 \quad \text{and} \quad (-a) + a = 0.$$

R-11 Reciprocal or multiplicative inverse property
For each real number a except 0, there exists a unique real number $1/a$ (called the **reciprocal** of a) with the property

$$a\left(\frac{1}{a}\right) = 1 \quad \text{and} \quad \left(\frac{1}{a}\right)a = 1.$$

Note that parentheses are used in some of the listed relations (and hereafter) to indicate an order of operations; that is, the operations enclosed in parentheses are performed before the other operations.

Axioms R-1 through R-11 together with axioms E-1 through E-4 imply other properties of the real numbers. Such implications are generally stated as **theorems**, which are assertions of facts that are logical consequences of the axioms and other theorems. Theorems generally consist of two parts: an "if" part, called the **hypothesis**, and a "then" part, called the **conclusion**. Proving a theorem consists of showing that if the hypothesis is true, then, because of the axioms, the conclusion must be true. Proofs are displayed in a variety of formats. We shall illustrate a few relatively formal proofs in this section and then, in general, use *informal arguments* through the remainder of the book. Let us now consider some theorems, the first of which is called the **addition property of equality**.

▸ *If a, b, and $c \in R$, and if $a = b$, then*

$$a + c = b + c \qquad \textit{and} \qquad c + a = c + b.$$

Proof

Statement	*Reason*
1. a, b, and $c \in R$ and $a = b$	1. Hypothesis
2. $a + c \in R$	2. Closure property for addition
3. $a + c = a + c$	3. Reflexive property of equality
4. $a + c = b + c$	4. Substitution in (3) from (1), by substitution property
5. $c + a = c + b$	5. Commutative property of addition

Thus, if the hypothesis "a, b, and $c \in R$ and $a = b$" is true, the conclusion "$a + c = b + c$ and $c + a = c + b$" follows logically.

The following theorem, called the **multiplication property of equality**, is closely analogous to the addition property:

▸ *If a, b, and $c \in R$, and if $a = b$, then*

$$ac = bc \qquad \textit{and} \qquad ca = cb.$$

The proof exactly parallels that of the previous theorem and is left as an exercise (Problem 43, Exercise 1.4).

Next, we have the **cancellation property for addition**.

▸ *If a, b, and $c \in R$, and if $a + c = b + c$, then*

$$a = b.$$

Proof

Statement	*Reason*
1. $a + c = b + c$	1. Hypothesis
2. $(a + c) + (-c) = (b + c) + (-c)$	2. Addition property of equality
3. $a + [c + (-c)] = b + [c + (-c)]$	3. Associative property of addition
4. $c + (-c) = 0$	4. Additive inverse property
5. $a + 0 = b + 0$	5. Substitution in (3) from (4), by substitution property
6. $a + 0 = a;\ b + 0 = b$	6. Identity element for addition
7. $a = b$	7. Substitution in (5) from (6), by substitution property

Paralleling the foregoing theorem is the **cancellation property for multiplication.**

► *If a, b, and* $c \in R$, $c \neq 0$, *and if* $ac = bc$, *then*

$$a = b.$$

The proof exactly parallels that of the cancellation property for addition and is also left as an exercise (Problem 44, Exercise 1.4).

The **zero-factor property** details the role of 0 in a product.

► *For every* $a \in R$,

$$a \cdot 0 = 0.$$

Proof

Statement	*Reason*
1. $0 + 0 = 0$	1. Identity element for addition
2. $a \cdot (0 + 0) = a \cdot 0$	2. Multiplication property of equality
3. $(a \cdot 0) + (a \cdot 0) = a \cdot 0$	3. Distributive property
4. $a \cdot 0 = 0 + (a \cdot 0)$	4. Identity element for addition
5. $(a \cdot 0) + (a \cdot 0) = 0 + (a \cdot 0)$	5. Substitution in (3) from (4) by substitution property
6. $a \cdot 0 = 0$	6. Cancellation property for addition

Since the number $-a$ is assumed to be unique, it follows that if $a + b = 0$, then $b = -a$ or $a = -b$; that is, if the sum of two numbers is zero, each is the negative of the other. Thus, the negative of a negative number is a positive number; that is,

$$(-3) + [-(-3)] = 0$$

implies that

$$-(-3) = 3.$$

This fact is formalized as the **double-negative property**.

▸ *For each $a \in R$,*

$$-(-a) = a.$$

For example,

$$-(-2) = 2, \qquad -(-3x) = 3x, \qquad \text{and} \qquad -[-(x + 3)] = x + 3.$$

We have given names to the foregoing theorems because they are used frequently throughout the book. In general, the properties we obtain as the result of our arguments are not named.

Sometimes we wish to consider only the nonnegative one of a pair of numbers a and $-a$. For example, since the graphs of $-a$ and a are each located the same distance from the origin, when we wish to refer simply to this distance and not to its direction to the left or right of 0, we can use the notation $|a|$ (read "the absolute value of a"). Thus, $|a|$ is always nonnegative. For example,

$$|3| = 3, \qquad |0| = 0, \qquad \text{and} \qquad |-3| = 3.$$

The double-negative property enables us to write a formal definition for absolute value.

▸ *For $a \in R$,*

$$|a| = \begin{cases} a, & \textit{if } a \geq 0, \\ -a, & \textit{if } a < 0. \end{cases}$$

Thus, for example, from this definition,

$$\begin{aligned} &\text{if} \quad a = 3, &&\text{then} \quad |3| = 3; \\ &\text{if} \quad a = 0, &&\text{then} \quad |0| = 0; \\ &\text{if} \quad a = -3, &&\text{then} \quad |-3| = -(-3) = 3. \end{aligned}$$

EXERCISE 1.4

A

In Problems 1–20, *replace each question mark to make the given statement an application of the given property.*

Example $2(3 + 1) = 2 \cdot 3 + \underline{?}$; distributive property

Solution $2(3 + 1) = 2 \cdot 3 + 2 \cdot 1$

1. $12 + x = x + \underline{?}$; commutative property of addition

2. $(2m)n = 2(\underline{?})$; associative property of multiplication

3. $2 \cdot 7 \in \underline{?}$; closure for multiplication

4. $(2 + z) + 3 = 2 + (\underline{?})$; associative property of addition

5. $4 \cdot t = \underline{?}$; commutative property of multiplication

6. $7 + \underline{?} = 0$; additive inverse property

7. $3 \cdot \frac{1}{3} = \underline{?}$; multiplicative inverse property

8. $m + \underline{?} = m$; identity element for addition

9. $r \cdot \underline{?} = r$; identity element for multiplication

10. $r + (s + 2) = (s + 2) + \underline{?}$; commutative property of addition

11. $3(x + y) = \underline{?} + \underline{?}$; distributive property

12. $6 \cdot \frac{1}{6} = \underline{?}$; commutative property of multiplication

13. If $z = 4$, then $z + 5 = 4 + \underline{?}$; additive property of equality.

14. $-(-3) = \underline{?}$; double-negative property

15. If $x = 6$, then $-3(x) = \underline{?} \cdot 6$; multiplicative property of equality.

16. If $2x = 2z$, then $x = \underline{?}$; cancellation property of multiplication.

17. If $n + 3 = 7 + 3$, then $n = \underline{?}$; cancellation property of addition.

18. $15 \cdot \underline{?} = 0$; zero factor property

19. If $-6x = -6y$, then $x = \underline{?}$; cancellation property of multiplication.

20. If $m = t$, then $m + 3 = \underline{?} + \underline{?}$; additive property of equality.

21. If $x < 0$, does $-x$ represent a positive number or a negative number?

22. If $x < 0$, does $-(-x)$ represent a positive number or a negative number?

Rewrite each of the expressions without using absolute value notation.

Examples **a.** $|-9|$ **b.** $|x+5|$ **c.** $|-y|$

Solutions **a.** 9

b. $x+5$ if $x+5 \geq 0$; $-(x+5)$ if $x+5<0$

c. $-y$ if $-y \geq 0$; y if $-y<0$

23. $|6|$ **24.** $|18|$ **25.** $|-10|$ **26.** $|-12|$

27. $-|-1|$ **28.** $-|-4|$ **29.** $-|\pi|$ **30.** $-\left|\frac{3}{4}\right|$

31. $|n|$ **32.** $|2z|$ **33.** $|-x|$ **34.** $-|-x|$

35. $|x-2|$ **36.** $|x-5|$ **37.** $|x+8|$ **38.** $|x+3|$

B

39. Is the set $\{0, 1\}$ closed with respect to addition? With respect to multiplication?

40. Is the set $\{-1, 0, 1\}$ closed with respect to addition? With respect to multiplication?

41. Is the set $\{1, 2\}$ closed with respect to addition? With respect to multiplication?

42. Which of the subsets of the real numbers N, W, J, and Q are closed with respect to addition? With respect to multiplication?

43. Using a proof similar to the one shown for the addition property of equality, prove the multiplication property of equality.

44. Using a proof similar to the one shown for the cancellation property for addition, prove the cancellation property for multiplication.

Let $x, y \in R$.

45. Must $|-x| = |x|$?

46. Can $|x| = -|x+3|$?

47. Can $|y| > |x| + |y|$?

48. Can $|y| < |x| + |y|$?

49. Must $|x+y| \leq |x| + |y|$?

50. Must $|x+y| > |x|$?

51. Under what conditions is $|x+y| = |x| + |y|$?

52. Under what conditions is $|y+1| = |y-1|$?

1.5

SUMS AND DIFFERENCES

Addition is an operation that associates with each pair of real numbers a, b a third real number $a+b$, called the **sum** of a and b. We assume the following property for sums of positive real numbers:

O-3 If $a, b \in R$ and if $a > 0$ and $b > 0$, then

$$a + b > 0.$$

This is simply an agreement that the sum of two positive numbers is positive. This agreement enables us to define the **difference** $a - b$ of two positive numbers a and b in terms of a sum. Thus, if a and b are real numbers and if $a > b > 0$, then $a - b$ is the number d,

$$\boldsymbol{a - b = d},$$

such that

$$\boldsymbol{b + d = a}.$$

For example,

$$7 - 2 = 5 \qquad \text{because} \qquad 2 + 5 = 7.$$

The properties of real numbers imply the familiar laws of signs for sums. For example, to argue that the sum of two negative numbers is the negative of the sum of the absolute values of the numbers, we first note that, by the additive inverse property,

$$a + (-a) = 0 \qquad \text{and} \qquad b + (-b) = 0,$$

so that

$$[a + (-a)] + [b + (-b)] = 0.$$

By means of the associative and commutative properties of addition, this can be rewritten as

$$(a + b) + [(-a) + (-b)] = 0.$$

Since by the additive inverse property, $(a + b) + [-(a + b)] = 0$, it follows, by the uniqueness of the additive inverse, that

$$(-a) + (-b) = -(a + b).$$

Now, if we assume a and b to be positive, then by O-3, $a + b$ is positive. Thus, the sum of the negative numbers $-a$ and $-b$, with absolute values a and b, is the negative of the sum of their absolute values. For example,

$$-3 + (-5) = -(|-3| + |-5|) = -8.$$

Similar arguments lead to the assertion that the absolute value of the sum of a positive and a negative number is equal to the nonnegative difference of the absolute values of the numbers, and the sum is positive or negative as the addend of greater absolute value is positive or negative. For example,

$$\begin{aligned} 8 + (-6) &= 8 - 6 \\ &= 2, \end{aligned} \qquad \begin{aligned} -3 + 5 &= 5 - |-3| \\ &= 5 - 3 = 2, \end{aligned}$$

$$\begin{aligned} -12 + 10 &= -(|-12| - 10) \\ &= -(12 - 10) = -2, \end{aligned} \qquad \begin{aligned} 5 + (-7) &= -(|7| - 5) \\ &= -(7 - 5) = -2. \end{aligned}$$

This process is easily accomplished mentally.

The definition given on page 23 for the difference of two *positive real numbers* a and b, for $a > b$, can thus be generalized to apply to *all real numbers*. We define the difference of two real numbers a and b to be the number d,

$$a - b = d,$$

such that

$$b + d = a.$$

It follows (see Problem 45, Exercise 1.5) that the difference, $a - b$, where $a, b \in R$, is equal to the sum of a and the additive inverse of b; that is,

$$\boldsymbol{a - b = a + (-b)}.$$

For example,

$$8 - (-3) = 8 + [-(-3)] = 8 + 3 = 11, \qquad (-7) - (4) = (-7) + (-4) = -(|-7| + |-4|) = -11,$$

$$(-5) - (-2) = (-5) + [-(-2)] = -(|-5| - 2) = -3.$$

Since we have seen that the difference $a - b$ is given by $a + (-b)$, we may consider the symbols $a - b$ as representing either the difference of a and b or, preferably, the sum of a and $(-b)$. Note that we now have used the sign $+$ in two ways and the sign $-$ in three ways. In Section 1.3 we used these signs to denote positive and negative numbers. In Section 1.4 we used the sign $-$ to indicate the "opposite" of a number. Here we have used $+$ and $-$ as signs of operation to indicate the sum or difference of two numbers.

In discussing the results of operations, it is convenient to use the term "basic numeral." For example, while the numeral "3 + 5" names the sum of the real numbers 3 and 5, we shall refer to "8" as the basic numeral for this number. Similarly, the basic numeral for 2 − 8 is "−6."

EXERCISE 1.5

A

Rewrite each difference as a sum.

Examples **a.** $12 - 3$ **b.** $6 - (-7t)$ **c.** $-3r - 5s$

Solutions **a.** $12 - 3 = 12 + (-3)$

b. $6 - (-7t) = 6 + [-(-7t)] = 6 + 7t$

c. $-3r - 5s = -3r + (-5s)$

1. $4 - 12$ **2.** $8 - 1$ **3.** $-6 - 2$ **4.** $-8r - 3$

5. $7x - (-2y)$ **6.** $3r - (-5s)$ **7.** $3a - 2b$ **8.** $t - (-s)$

Write each sum or difference using a basic numeral.

Examples **a.** $3 + (-7)$ **b.** $(-3) - (-7)$ **c.** $(-2) + (-3) - (4)$

Solutions

a. $3 + (-7) = -4$

b. $(-3) - (-7)$
$= (-3) + [-(-7)]$
$= (-3) + (+7)$
$= 4$

c. $(-2) + (-3) - (4)$
$= (-2) + (-3) + (-4)$
$= -9$

9. $8 + 6$ **10.** $7 + 23$ **11.** $-3 + 9$ **12.** $-7 + 2$

13. $8 + (-9)$ **14.** $6 + (-11)$ **15.** $-3 + (-12)$ **16.** $-7 + (-5)$

17. $12 - 5$ **18.** $11 - 7$ **19.** $3 - 15$ **20.** $7 - 18$

21. $-3 - 5$ **22.** $-6 - 12$ **23.** $3 - (-5)$ **24.** $7 - (-2)$

25. $-3 - (-5)$ **26.** $-11 - (-7)$ **27.** $6 + 3 - 8$ **28.** $11 - 5 + 7$

29. $3 - 8 + 5$ **30.** $4 - 3 - 11$ **31.** $-4 - 3 - (-5)$ **32.** $8 - (-2) - 3$

Examples **a.** $5 - (8 - 12)$ **b.** $(4 - 2 + 7) - (3 - 8)$

Solutions

a. $5 - (8 - 12) = 5 - (-4)$
$= 5 + [-(-4)]$
$= 5 + 4 = 9$

b. $(4 - 2 + 7) - (3 - 8) = 9 - (-5)$
$= 9 + [-(-5)]$
$= 9 + 5 = 14$

33. $7 - (5 - 2)$ **34.** $8 - (10 - 2)$ **35.** $(6 - 5) - 11$

36. $(3 - 7) - 1$ **37.** $(6 - 1 + 8) - 3$ **38.** $4 - (6 + 2 - 11)$

39. $(7 - 2) + (-3 + 1)$ **40.** $(5 - 9) + (4 - 2)$

41. $(3 - 5 + 4) - (8 - 13)$ **42.** $(38 - 25) + (13 - 17 - 2)$

43. $(5 + 24 - 29) + (12 - 8 + 4)$ **44.** $(-15 + 3 - 1) - (9 - 5 + 12)$

B

45. Show that for all $a, b \in R$, $a - b = a + (-b)$. [*Hint:* Start with the fact that $a - b = d$ implies that $b + d = a$. Then add $-b$ to each member of the latter equation.]

46. Which of the following sets are closed under subtraction?
a. N **b.** W **c.** J **d.** Q

47. If subtraction were commutative in R, we would be able to write $a - b = b - a$. Find an example to show that subtraction is not commutative.

48. If subtraction were associative in R, we would be able to write $(a - b) - c = a - (b - c)$. Find an example to show that subtraction is not associative.

1.6

PRODUCTS AND QUOTIENTS

Multiplication is an operation that associates with each pair of real numbers a and b a third real number, $a \cdot b$ or ab. The number ab is called the **product** of the **factors** a and b. In this section, we shall briefly examine some properties of products of real numbers. Let us begin by making another assumption:

O-4 If $a, b \in R$, and if $a > 0$ and $b > 0$, then

$$ab > 0.$$

This is an agreement that the product of positive real numbers is positive.

Now, what can be said about the product of a positive and a negative number? Let us assume that a and b are positive and begin by considering the equality

$$b + (-b) = 0.$$

By the multiplication property of equality,

$$a[b + (-b)] = a \cdot 0,$$

and the distributive property and the zero factor property permit us to write

$$ab + a(-b) = 0.$$

This implies that $a(-b)$ is the additive inverse of ab. We then have, as a consequence of O-4 and other properties of the real numbers, the following familiar laws of signs for products:

$$\mathbf{a(-b) = -(ab) = -ab,}$$

$$\mathbf{(-a)(b) = -(ab) = -ab,}$$

$$\mathbf{(-a)(-b) = ab,}$$

where we use the symbol $-ab$ to denote the negative of ab.

Thus, the product of two real numbers with unlike signs is negative and the product of two real numbers with like signs is positive. For example,

$(3)(-2) = -6,$ $\quad (-3)(-2) = 6,$ $\quad (-2)(3) = -6,$ $\quad$ and $\quad (3)(2) = 6.$

Note that since $-1 \cdot a = -(1 \cdot a)$ and $1 \cdot a = a$,

$$\mathbf{-1 \cdot a = -a.}$$

We define the **quotient** of two real numbers a and b to be the number q.

$$\mathbf{\frac{a}{b} = q \qquad (b \neq 0),}$$

such that

$$bq = a.$$

For example,

$$\frac{6}{2} = 3 \quad \text{because} \quad 2 \cdot 3 = 6; \qquad \frac{-6}{2} = -3 \quad \text{because} \quad 2(-3) = -6;$$

$$\frac{6}{-2} = -3 \quad \text{because} \quad (-2)(-3) = 6; \qquad \frac{-6}{-2} = 3 \quad \text{because} \quad (-2)(3) = -6.$$

Since the quotient of two numbers a/b is a number q such that $bq = a$, the sign of the quotient of two signed numbers must be consistent with the laws of signs for the product of two signed numbers. Therefore,

$$\frac{+a}{+b} = +q, \quad \frac{-a}{-b} = +q, \quad \frac{+a}{-b} = -q, \quad \text{and} \quad \frac{-a}{+b} = -q.$$

When a fraction bar is used to indicate that one algebraic expression is to be divided by another, the dividend is the numerator and the divisor is the denominator. Note that the denominator is restricted to nonzero numbers, for, if b is 0 and a is not 0, then there exists no q such that

$$0 \cdot q = a.$$

Again, if b is 0 and a is 0, then, for *any* q,

$$0 \cdot q = 0,$$

and the quotient is not unique. Thus, *division by zero in not defined.*

It follows from our definition of a quotient (see Problem 75, Exercise 1.6) that

$$\frac{a}{b} = a\left(\frac{1}{b}\right) \quad (b \neq 0).$$

For example,

$$\frac{2}{3} = 2\left(\frac{1}{3}\right), \quad \frac{a}{3b} = a\left(\frac{1}{3b}\right) \quad (b \neq 0)$$

and

$$\frac{x}{y - x} = x\left(\frac{1}{y - x}\right) \quad (y \neq x).$$

Recall that in discussing the product abc, we refer to the numbers, a, b, and c as *factors* of the product. Thus, 2 and 3 are factors of 6; 5 and 1 are factors of 5. If a natural number greater than 1 has no factors that are natural numbers other than itself and 1, it is said to be a **prime number**. Thus, 2, 3, 5, 7, 11, etc., are prime numbers. A natural number greater than 1 that is not a prime number is said to be a **composite number**. Thus, 4, 6, 8, 9, 10, etc., are composite numbers. When a composite number is exhibited as a product of prime factors only, it is said to be **completely factored**. For example, although 30 may be factored into (5)(6), (10)(3).

(15)(2), or (30)(1), if we continue the factorization, we arrive at the set of prime factors 2, 3, and 5 in each case. Although it is unnecessary to develop the argument here, it is a fact that, except for order, *each composite number has one and only one prime factorization.* This is known as the **fundamental theorem of arithmetic**.

Notice again that the words "composite" and "prime" are used in reference to natural numbers only. Integers, rational numbers, and irrational numbers are not referred to as either prime or composite. Any negative integer can, however, be expressed as the product $(-1)a$, where a is a natural number. Hence, if we refer to the completely factored form of a negative integer, we refer to the product of (-1) and the prime factors of the associated natural number.

EXERCISE 1.6

A

Write each product or quotient using a basic numeral.

Examples **a.** $(-3)(2)$ **b.** $(-7)(-5)$ **c.** $2(-3 \cdot 5)$

Solutions **a.** $(-3)(2) = -6$ **b.** $(-7)(-5) = 35$ **c.** $2(-3 \cdot 5) = 2(-15)^* = -30$

1. $(6)(-5)$ **2.** $(-8)(3)$ **3.** $(-7)(-6)$ **4.** $(-15)(-2)$

5. $(4)(2)(-1)$ **6.** $(-3)(2)(6)$ **7.** $(-5)(-2)(3)$ **8.** $(4)(-3)(-6)$

9. $(-7)(-2)(-5)$ **10.** $(-8)(-3)(-1)$ **11.** $(-7)(0)(-2)$

12. $(-6)(5)(0)$ **13.** $(3)(-1)(2)(-1)$ **14.** $(-1)(-2)(-3)(-4)$

15. $(5)(-3)(2)(1)$ **16.** $(7)(-5)(-3)(-1)$

Examples **a.** $\frac{-20}{5}$ **b.** $\frac{0}{-4}$ **c.** $\frac{-3}{0}$

Solutions **a.** $\frac{-20}{5} = -4$ **b.** $\frac{0}{-4} = 0$ **c.** $\frac{-3}{0}$; not defined

17. $\frac{-16}{4}$ **18.** $\frac{-32}{8}$ **19.** $\frac{39}{-3}$ **20.** $\frac{45}{-15}$

21. $\frac{-27}{-9}$ **22.** $\frac{-54}{-6}$ **23.** $\frac{0}{-7}$ **24.** $\frac{0}{-12}$

* *Common error:* Note that $2(-3 \cdot 5) \neq 2(-3) \cdot 2(5)$.

25. $\dfrac{-5}{0}$ **26.** $\dfrac{-27}{0}$ **27.** $-\left(\dfrac{-8}{2}\right)$ **28.** $-\left(\dfrac{-12}{-3}\right)$

From the definition of a quotient, rewrite each equation in the form $a = bq$.

Examples **a.** $\dfrac{18}{-3} = -6$ **b.** $\dfrac{-27}{-3} = 9$ **c.** $\dfrac{-12}{4} = -3$

Solutions **a.** $18 = (-3)(-6)$ **b.** $-27 = (-3)(9)$ **c.** $-12 = 4(-3)$

29. $\dfrac{15}{-3} = -5$ **30.** $\dfrac{46}{-23} = -2$ **31.** $\dfrac{-38}{19} = -2$ **32.** $\dfrac{-14}{2} = -7$

33. $\dfrac{-52}{-4} = 13$ **34.** $\dfrac{-28}{-7} = 4$ **35.** $\dfrac{0}{-7} = 0$ **36.** $\dfrac{-12}{1} = -12$

Rewrite each quotient as a product in which one factor is a natural number and the other factor is the reciprocal of a natural number.

Examples **a.** $\dfrac{3}{4}$ **b.** $\dfrac{16}{5}$ **c.** $\dfrac{7}{9}$

Solutions **a.** $\dfrac{3}{4} = 3\left(\dfrac{1}{4}\right)$ **b.** $\dfrac{16}{5} = 16\left(\dfrac{1}{5}\right)$ **c.** $\dfrac{7}{9} = 7\left(\dfrac{1}{9}\right)$

37. $\dfrac{7}{8}$ **38.** $\dfrac{27}{7}$ **39.** $\dfrac{3}{8}$ **40.** $\dfrac{9}{17}$

41. $\dfrac{82}{11}$ **42.** $\dfrac{3}{13}$ **43.** $\dfrac{7}{100}$ **44.** $\dfrac{3}{1000}$

Rewrite each product as a quotient.

45. $3\left(\dfrac{1}{2}\right)$ **46.** $8\left(\dfrac{1}{3}\right)$ **47.** $2\left(\dfrac{1}{7}\right)$ **48.** $\left(\dfrac{1}{100}\right)3$

49. $\left(\dfrac{1}{8}\right)5$ **50.** $\left(\dfrac{1}{7}\right)6$ **51.** $9\left(\dfrac{1}{2}\right)$ **52.** $7\left(\dfrac{1}{10{,}000}\right)$

Express each integer in completely factored form. If the integer is a prime number, so state.

Examples **a.** 36 **b.** 29 **c.** -51

Solutions **a.** $36 = (2)(2)(3)(3)$ **b.** Prime **c.** $-51 = -1(3)(17)$

53. 8 **54.** 26 **55.** 49 **56.** 18
57. 17 **58.** -16 **59.** -12 **60.** 23
61. 56 **62.** 65 **63.** -38 **64.** -47
65. 20 **66.** 39 **67.** 106 **68.** 117

B

If $x, y \in R$, under what conditions is each statement in Problems 69–72 a true statement?

69. $\frac{x}{y} = 0$ **70.** $\frac{x}{y} \neq 0$ **71.** $\frac{x}{y} > 0$ **72.** $\frac{x}{y} < 0$

73. What can be said about the sign on the product of an even number of negative factors? An odd number of negative factors?

74. Notice that 1 is not called a prime number. Considering the statement of the uniqueness of the prime factors of a composite number, can you tell why 1 is not included in the set of prime numbers?

75. Show that for all $a, b \in R$, $b \neq 0$, $a/b = a(1/b)$. [*Hint:* Start with the fact that $a/b = q$ implies that $bq = a$. Then multiply both members by $1/b$.]

76. Which of the following sets are closed under division as long as the divisors are not zero?

a. N **b.** W **c.** J **d.** Q **e.** R

77. If division were commutative in R, we would be able to write $a \div b = b \div a$. Find an example to show that division is not commutative.

78. If division were associative in R, we would be able to write $(a \div b) \div c = a \div (b \div c)$. Find an example to show that division is not associative.

1.7

ORDER OF OPERATIONS

On page 17, it was observed that parentheses are frequently used to show the order of performing operations. Thus,

$$a(b + c)$$

represents the product of a and the sum of b and c, while

$$(ab) + (ac)$$

represents the sum of the products ab and ac. Of course, the distributive property assures us that these expressions name the same numbers. Frequently, however, we find such expressions as $(ab) + (ac)$ written without parentheses, simply as

$$ab + ac.$$

In this case, we need an agreement about the meaning of the expression. Therefore, let us agree that operations will be performed as follows:

1. First, any expression within a symbol of inclusion (parentheses, brackets, fraction bars, etc.) is simplified, starting with the innermost inclusion symbol.
2. Next, multiplications and divisions are performed as encountered in order from left to right.
3. Last, additions and subtractions are performed in order from left to right.

EXERCISE 1.7

A

Write each expression using a basic numeral.

Examples **a.** $3 + 2 \cdot 5$ **b.** $3(2 + 7) - 3 \cdot (-4)$ **c.** $\dfrac{6 + (-15)}{-3} + \dfrac{5}{4 - (-1)}$

Solutions

a. $3 + 2 \cdot 5 = 3 + 10$*
$= 13$

b. $3(2 + 7) - 3 \cdot (-4)$
$= 3(9) + 12$
$= 27 + 12$
$= 39$

c. $\dfrac{6 + (-15)}{-3} + \dfrac{5}{4 - (-1)}$
$= \dfrac{-9}{-3} + \dfrac{5}{5}$
$= 3 + 1 = 4$

1. $4 + 4 \cdot 4$ **2.** $7 - 6 \cdot 2$ **3.** $4(-1) + 2 \cdot 2$

4. $6 \cdot 3 + (-2)(5)$ **5.** $-2 \cdot (4 + 6) + 3$ **6.** $5(6 - 12) + 7$

7. $2 - 3(6 - 1)$ **8.** $4 - 7(8 + 2)$ **9.** $\dfrac{7 + (-5)}{2} - 3$

10. $\dfrac{12}{8 - 2} - 5$ **11.** $\dfrac{3(6 - 8)}{-2} - \dfrac{6}{-2}$ **12.** $\dfrac{5(3 - 5)}{2} - \dfrac{18}{-3}$

13. $6[3 - 2(4 + 1)] - 2$ **14.** $6[5 - 3(1 - 4)] + 3$

15. $(4 - 3)[2 + 3(2 - 1)]$ **16.** $(8 - 6)[5 + 7(2 - 3)]$

17. $64 \div [8(4 - 2[3 + 1])]$ **18.** $27 \div (3[9 - 3(4 - 2)])$

* *Common error:* Note that $3 + 2 \cdot 5 \neq (3 + 2) \cdot 5$.

Evaluate each expression for the given values of the variables.

19. $\dfrac{5(F-32)}{9}$; $F = 212$

20. $\dfrac{R+r}{r}$; $R = 12$ and $r = 2$

21. $\dfrac{E-e}{R}$; $E = 18$, $e = 2$, and $R = 4$

22. $P + Prt$; $P = 1000$, $r = 0.04$, and $t = 2$

23. $\dfrac{a - rs_n}{1-r}$; $r = 2$, $s_n = 12$, and $a = 4$

24. $R_0(1 + at)$; $R_0 = 2.5$, $a = 0.05$, and $t = 20$

B

Simplify.

25. $\left[\dfrac{7-(-3)}{5-3}\right]\left[\dfrac{4+(-8)}{3-5}\right]$

26. $\left[\dfrac{12+(-2)}{3+(-8)}\right]\left[\dfrac{6+(-15)}{8-5}\right]$

27. $\left(3 - 2\left[\dfrac{5-(-4)}{2+1} - \dfrac{6}{3}\right]\right) + 1$

28. $\left(7 + 3\left[\dfrac{6+(-18)}{4+2}\right] - 5\right) + 3$

29. $\dfrac{8 - 6\left(\dfrac{5+3}{4-8}\right) - 3}{-2 + 4\left(\dfrac{6-3}{1-4}\right) + 6}$

30. $\dfrac{12 + 3\left(\dfrac{12-20}{3-1}\right) - 1}{-8 + 6\left(\dfrac{12-30}{2-5}\right) + 1}$

31. $\dfrac{3(3+2) - 3 \cdot 3 + 2}{3 \cdot 2 + 2(2-1)}$

32. $\dfrac{6 - 2\left(\dfrac{4+6}{5}\right) + 8}{3 - 3 \cdot 2 + 8}$

33. $\dfrac{4 - 2\left(\dfrac{8-2}{3}\right)}{4 - 3(6+1)} + \left[\dfrac{\dfrac{8}{2} + 4\left(\dfrac{11-1}{5}\right)}{14 - 2(5+1)}\right]\left[\dfrac{\dfrac{12-2(4-2)}{2}}{10 - 3(5-1)}\right]$

34. $\left(\left[\left(\dfrac{2-8}{3}\right)\left(\dfrac{8-2}{3}\right) + \left(\dfrac{6-1}{5} + \dfrac{1-6}{5}\right)\right] - \left[\dfrac{5-0}{5} + \dfrac{0-5}{5}\right]\right) \div 4$

CHAPTER SUMMARY

[1.1] A **set** is a collection. The items in the collection are the **members**, or **elements**, of the set. A set A is a **subset** of a set B if every member of A is a member of B. The sets of **natural numbers, whole numbers, integers, rational numbers,** and **irrational numbers** are all subsets of the set of **real numbers**.

A **variable** is a symbol used to represent an unspecified element of a given set called the **replacement set** of the variable. A symbol whose replacement set contains a single element is called a **constant**.

[1.2] The **union** and **intersection** of two sets A and B are **operations** on two sets which produce a third set. If the intersection of two sets is the **null set**, then they are **disjoint sets**. All sets are considered to be subsets of a **universal set**.

[1.3] The **equality** and **order** relations in the set of real numbers R are governed by a set of assumptions called **axioms**. Relationships between real numbers can be pictured on a **line graph**, or **number line**.

[1.4] The set of real numbers is **closed** under the operations of addition and multiplication. The operations are **commutative** and **associative**, and each operation has an **identity element** in R. Each number a in R has an **additive inverse** $-a$, and each nonzero number a in R has a **multiplicative inverse** $1/a$. Multiplication **distributes** over addition. These properties imply the following properties (theorems):

If $a = b$, then $a + c = b + c$ and $c + a = c + b$.
If $a = b$, then $ac = bc$ and $ca = cb$.
If $a + c = b + c$, then $a = b$.
If $ac = bc$ with $c \neq 0$, then $a = b$.
For every $a \in R$, $a \cdot 0 = 0$.
For every $a \in R$, $-(-a) = a$.

[1.5] The sum of two positive real numbers is positive; the sum of two negative numbers and the sum of a positive and a negative number are determined by the properties of the real numbers. The difference $a - b$ of two real numbers is equal to $a + (-b)$, the sum of a and the additive inverse of b.

[1.6] The product of two positive real numbers is positive; the product of a positive and a negative number is negative; and the product of two negative numbers is positive. The quotient of two real numbers a/b, with $b \neq 0$, is equal to $a(1/b)$, the product of a and the multiplicative inverse of b. Division by 0 is not defined.

A natural number greater than 1 that has no natural number factors other than itself and 1 is a **prime number**. A natural number greater than 1 that is not a prime number is a **composite number**, and it has one and only one prime factorization.

[1.7] In evaluating an expression, the order of operations is:

1. First, any expression within a symbol of inclusion (parentheses, brackets, fraction bars, etc.) is simplified, starting with the innermost inclusion symbol.
2. Next, multiplications and divisions are performed as encountered in order from left to right.
3. Last, additions and subtractions are performed in order from left to right.

The symbols introduced in this chapter are listed on the inside of the front cover.

REVIEW EXERCISES

[1.1] 1. Let $A = \left\{-3, -2.55, -\sqrt{2}, -\frac{3}{5}, 0, 1, 5.55\ldots, \frac{13}{2}\right\}$. List the members of

$$\{x \mid x \in A \quad \text{and} \quad x \text{ is an integer}\}.$$

2. List the subsets of $\{1, 2, 3, 4\}$ that have $\{1, 2\}$ as a subset.
3. Use subset symbols to express relationships between the sets H, J, W, N, Q, and R.

[1.2] *Let $A = \{3, 5, 7, 9\}$ and $B = \{-1, 2, 6, 9\}$.*

4. List the members of $A \cup B$.
5. List the members of $A \cap B$.
6. If $G \cap H = \varnothing$, then G and H are $\underline{?}$ sets.

[1.3] 7. If $a \leq b$ and $b < c$, then $a \underline{?} c$.

8. Use the symbols a, b, c, and $<$ to show that b is between a and c.
9. Graph $\{x \mid -2 < x \leq 9, \quad x \in J\}$.
10. Let $A = \{x \mid x \leq 3, \quad x \in R\}$ and $B = \{x \mid x > 1, \quad x \in R\}$. Graph $A \cap B$.
11. Let $A = \{x \mid x > 7, \quad x \in R\}$ and $B = \{x \mid x < -2, \quad x \in R\}$. Graph $A \cup B$.

[1.4] 12. By the commutative property of addition, $x + 2 = \underline{?} + \underline{?}$.

13. By the symmetric property of equality, if $x + 2 = 7$, then $7 = \underline{?}$.
14. By the additive property of equality, if $z = 18$, then $z + t = \underline{?}$.
15. By the multiplicative inverse property, $3 \cdot \underline{?} = 1$.
16. By the cancellation property of addition, if $n + 2 = 5 + 2$, then $n = \underline{?}$.
17. By the multiplicative identity property, $7 \cdot \underline{?} = 7$.
18. By the zero factor property, $7 \cdot 0 = \underline{?}$.

[1.5] *Write each sum or difference using a basic numeral.*

19. **a.** $2 - (-4) + 8$ **b.** $5 - (-6) - 3$
20. **a.** $3 - |6| - |-3|$ **b.** $2 + |6 - 12| - |3|$

[1.6] *Write each product or quotient using a basic numeral.*

21. **a.** $3(-2)(6)$ **b.** $-3(-6)(0)$ **c.** $7(-3)(-1)$ **d.** $-1(-2)(-3)(-4)$

22. a. $\frac{-24}{-2}$ b. $\frac{0}{-3}$ c. $\frac{-16}{2}$ d. $\frac{32}{-4}$

Express each quotient as a product.

23. a. $\frac{-7}{5}$ b. $\frac{24}{7}$

24. The completely factored form of 96 is ?.

25. The completely factored form of 138 is ?.

26. The completely factored form of 204 is ?.

[1.7] *Write each expression using a basic numeral.*

27. a. $5 - \frac{3 + 9}{6}$ b. $\frac{6 + 2 \cdot 3}{10 - 4 \cdot 2}$

28. a. $\frac{6 \cdot 8 \div 4 - 2}{7 - 5}$ b. $6 - (3 + 2)\ (-4 - 1) + \frac{-18}{6}$

2

POLYNOMIALS

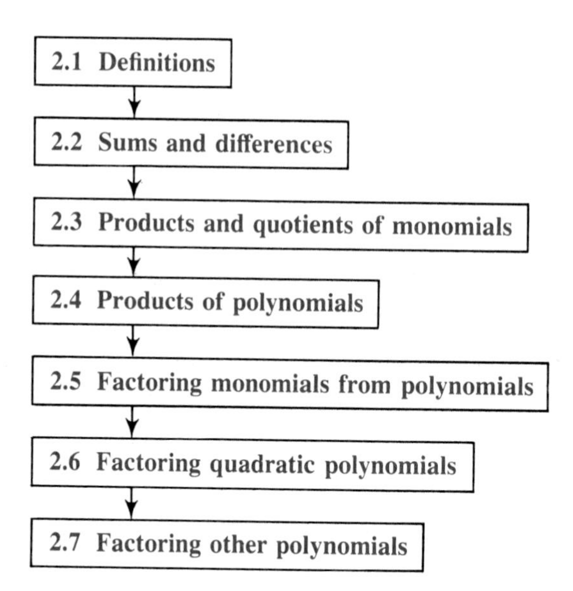

2.1

DEFINITIONS

As you know, the sum of two numbers is expressed by $a + b$ and their product is expressed by ab. In some cases, where the intent needs clarification, a centered dot is used to indicate multiplication—for example, $a \cdot b$, $x \cdot y$, $2 \cdot 3$. In other cases, parentheses are used around one or both of the symbols, as (2)(3), 2(3), $(x)(x)$. If the factors in a product are identical, the multiplicity of factors is indicated by means of a natural number **exponent**. An exponent is the number named by a small numeral written to the upper right of a given numeral to indicate how many times the number represented by the given numeral appears as a factor in a product. Thus, if a is a real number and n is a natural number,

$$\boldsymbol{a^n = a \cdot a \cdot a \cdots a \qquad (n \textit{ factors}).}$$

For example,

$$x^2 = x \cdot x, \qquad y^3 = y \cdot y \cdot y, \qquad \text{and} \qquad 3^5 = 3 \cdot 3 \cdot 3 \cdot 3 \cdot 3.$$

In the expression a^n, a is called the **base** and a^n is said to be the ***n*th power** of the base. If no exponent appears, the exponent 1 is assumed—that is, $x = x^1$.

Any meaningful collection of numerals, variables, and signs of operations is called an **expression**. In an expression of the form $A + B + C + \cdots$, A, B, and C are called **terms** of the expression. For example, $3 + x + 4y^2$ contains three terms, $2x^2y + 3xy^2$ contains two terms, and $3(x + 4y^2)$ contains only one term; however, the factor $x + 4y^2$ in the last expression contains two terms. Any factor or group of factors in a term is said to be the **coefficient** of the remaining factors in the term. Thus, in the term $3xyz$, the product $3x$ is the coefficient of yz; y is the coefficient of $3xz$; 3 is the coefficient of xyz; and so on. Hereafter, the word "coefficient" will refer to a number unless otherwise indicated. For example, the coefficient of the term $4xy$ is 4; the coefficient of $-2a^2b$ is -2.

If no coefficient appears in a term, the coefficient is understood to be 1. Thus, x is viewed as $1x$. In the expression $x^2 - y$ the coefficient of the first term is 1 and the coefficient of the second term is -1.

An expression in which the operations involved consist solely of addition, subtraction, multiplication, and division, and in which all variables occur as natural number powers only, is called a **rational expression**. Thus,

$$5, \quad \frac{x+y}{2}, \quad x - \frac{1}{x}, \quad \frac{3x+2y}{x-y}, \quad \text{and} \quad x^2 + 2x + 1$$

are rational expressions. Any rational expression in which no variable occurs in a denominator is called a **polynomial**. Thus,

$$x^2, \quad 5, \quad 3x^2 - 2x + 1, \quad \frac{1}{5}x - 2, \quad \text{and} \quad x^2y - 2x + y$$

are polynomials. A polynomial consisting of only one term is called a **monomial**. If the polynomial contains two or three terms, we refer to it as a **binomial** or **trinomial**, respectively. For example,

$3,$	$x^2y,$	and	$5x^3$	are monomials,
$x + y,$	$2x^2 - 3x,$	and	$4x + 3y$	are binomials,
$4x + 3y + 2z,$	$x^2 + 2x + 1,$	and	$5xy + 2x - 3y$	are trinomials,

and each of these examples is a polynomial.

The **degree** of a monomial in one variable is given by the exponent on the variable. Thus, $3x^4$ is of fourth degree, and $7x^5$ is of fifth degree. If the monomial contains more than one variable, the degree is given by the sum of the exponents on the variables. Thus, the monomial $3x^2y^3z$ is of sixth degree in x, y, and z. It can also be described as being of second degree in x, or third degree in y, or first degree in z. The degree of a polynomial is the same as the degree of its term of largest degree; $3x^2 + 2x + 1$ is a second-degree polynomial, $x^5 - x - 1$ is a fifth-degree polynomial, and $2x - 3$ is a first-degree polynomial.

Recall the agreement that was made in Section 1.7 concerning the order in which operations are to be performed. Because a power is a product, we should simplify any powers of ungrouped bases in an expression before performing other multiplication or division operations. For example,

$$\begin{array}{lll} 5 + 3 \cdot 4^2 & 5 + (3 \cdot 4)^2 & (5 + 3 \cdot 4)^2 \\ \quad = 5 + 3 \cdot 16 & \quad = 5 + 12^2 & \quad = (5 + 12)^2 \\ \quad = 5 + 48 & \quad = 5 + 144 & \quad = 17^2 \\ \quad = 53 & \quad = 149 & \quad = 289 \end{array}$$

Polynomials are frequently represented by symbols like

$$P(x), \quad D(y), \quad \text{and} \quad Q(z),$$

where the symbol in parentheses designates the variable. For example, we might write

$$P(x) = 2x^2 - 2x + 1,$$
$$D(y) = y^6 - 2y^2 + 3y - 2,$$
$$Q(z) = 8z^4 + 3z^3 - 2z^2 + z - 1.$$

The notation $P(x)$ can be used to denote values of the polynomial for specific values of x. Thus, $P(2)$ represents the value of the polynomial $P(x)$ when x is replaced by 2. For example, if

$$P(x) = x^2 - 2x + 1,$$

then

$$P(2) = (2)^2 - 2(2) + 1 = 1,$$
$$P(3) = (3)^2 - 2(3) + 1 = 4,$$
$$P(-4) = (-4)^2 - 2(-4) + 1 = 25.$$

EXERCISE 2.1

A

Identify each polynomial as a monomial, binomial, or trinomial and give the degree of the polynomial.

Examples **a.** $3n^2 - n$ **b.** $r^3s^2 - 2rs + rs^2$

Solutions **a.** Binomial; degree 2 **b.** Trinomial; degree 5 in r and s, degree 3 in r, and degree 2 in s

1. $4n^4 - 2n^3 + n^2$ **2.** $4x^5y^2$ **3.** $7t^3 - 3t^2$

4. $4 + 3z + 5z^2$ **5.** z^4 **6.** $3x^2y - 2xy^2 + 3xy$

How many terms are there in each expression as written.

Examples **a.** $3(t - 2)$ **b.** $4x + 4y + 2(x + y)$

Solutions **a.** One term **b.** Three terms

7. $x(y - z)$ **8.** $2x - 3x^2$ **9.** $4x - (y + 2z)$

10. $m^2n + 2mn^2 + m$ **11.** $\dfrac{2r + z}{3} - 1$ **12.** $(n - m + 3)^3$

Simplify.

Examples a. $(-4)^2 - 3^2$ b. $(3+1)^2 - \frac{4 \cdot 3^2}{6}$

Solutions a. $(-4)^2 - 3^2 = 16 - 9$*

$= 7$

b. $(3+1)^2 - \frac{4 \cdot 3^2}{6} = 4^2 - \frac{4 \cdot 9}{6}$†

$= 16 - \frac{36}{6}$

$= 16 - 6 = 10$

13. $(-6)^2$

14. -6^2

15. $5^2 - 2^2$

16. $3^2 + (-4)^2$

17. $\frac{8 \cdot 2^2}{16} + (3 \cdot 1)^2$

18. $\frac{4^2 - 3^2}{8 - 1} - (2 \cdot 1)^2$

19. $\frac{3^2 - 5}{6 - 2^2} - \frac{6^2}{3^2}$

20. $\frac{3^2 \cdot 2^2}{4 - 1} + \frac{(-3)(2)^3}{6}$

21. $\frac{(-5)^2 - 3^2}{4 - 6} + \frac{(-3)^2 \cdot 2}{5 + 1}$

22. $\frac{7^2 - 6^2}{10 + 3} - \frac{8^2 \cdot (-2)}{(-4)^2}$

23. $\frac{4^2(4 - 4)}{(-4)} - 4(-4)^2$

24. $\frac{3(3^2 - 3)}{3^2} + 3(-3)^2$

Given $x = 3$ and $y = -2$, evaluate each expression.

Example $\frac{x^2 - y}{3x + 2} + x(-y)$

Solution $\frac{(3)^2 - (-2)}{3(3) + 2} + 3[-(-2)] = \frac{9 + 2}{9 + 2} + 3(2)$

$= \frac{11}{11} + 6 = 1 + 6 = 7$

25. $3x + y$

26. $x - y^2$

27. $x^2 - 2y$

28. $(x + 2y)^2$

29. $x^2 - y^2$

30. $(3y)^2 - 3x$

31. $\frac{4x}{y} - xy$

32. $\frac{-xy^2}{6} + 2xy$

33. $\frac{(x - y)^2}{-5} + \frac{(xy)^2}{6}$

34. $(x + y)^2 + (x - y)^2$

35. $\frac{(x - y)^2}{(x + y)^2} + (x + y)^2$

36. $\frac{xy}{(x + y)^2} - xy$

* *Common error:* Note that $-3^2 \neq (-3)(-3)$ or $+9$.
† *Common error:* Note that $4 \cdot 3^2 \neq (4 \cdot 3)^2$.

Evaluate each expression for the given values of the variables.

37. $\frac{1}{2}gt^2$; $g = 32$ and $t = 2$

38. $\frac{1}{2}gt^2 - 12t$; $g = 32$ and $t = 3$

39. $\frac{Mv^2}{g}$; $M = 64$, $v = 2$, and $g = 32$

40. $\frac{32(V - v)^2}{g}$; $V = 12$, $v = 4$, and $g = 32$

41. ar^{n-1}; $a = 2$, $r = 3$, and $n = 4$

42. $\frac{a - ar^n}{1 - r}$; $a = 4$, $r = 2$, and $n = 3$

Find the values of each polynomial for the specified values of the variable.

Example If $P(x) = 2x^2 - x + 3$, find $P(-3)$, $P(0)$, and $P(a)$.

Solution

$P(-3) = 2(-3)^2 - (-3) + 3 = 24$

$P(0) = 2(0)^2 - (0) + 3 = 3$

$P(a) = 2(a)^2 - (a) + 3 = 2a^2 - a + 3$

43. If $P(x) = x^3 - 3x^2 + x + 1$, find $P(2)$, $P(-2)$, and $P(0)$.

44. If $P(x) = 2x^3 + x^2 - 3x + 4$, find $P(3)$, $P(-3)$, and $P(0)$.

45. If $D(x) = (2x - 4)^2 - x^2$, find $D(2)$, $D(-2)$, and $D(0)$.

46. If $D(x) = (3x - 1)^2 + 2x^2$, find $D(4)$, $D(-4)$, and $D(0)$.

47. If $P(x) = x^2 + 3x + 1$ and $Q(x) = x^3 - 1$, find $P(3)$ and $Q(2)$.

48. If $P(x) = -3x^2 + 1$ and $Q(x) = 2x^2 - x + 1$, find $P(0)$ and $Q(-1)$.

49. If $P(x) = x^5 + 3x - 1$ and $Q(x) = x^4 + 2x - 1$, find $P(2)$ and $Q(-2)$.

50. If $P(x) = x^6 - x^5$ and $Q(x) = x^7 - x^6$, find $P(-1)$ and $Q(-1)$.

B

Given that $P(x) = x + 1$, $Q(x) = x^2 - 1$, *and* $R(x) = x^2 + x - 1$, *find the value for each expression.*

Examples **a.** $P(0) + Q(2)$ **b.** $P[Q(3)]$

Solutions

a.

$P(0) = 0 + 1 = 1$

$Q(2) = 2^2 - 1 = 4 - 1 = 3$

$P(0) + Q(2) = 1 + 3 = 4$

b.

$Q(3) = 3^2 - 1 = 9 - 1 = 8$

$P[Q(3)] = P(8) = 8 + 1 = 9$

51. $P(2) + R(3) - Q(1)$

52. $P(1) + R(-2) \cdot Q(-1)$

53. $P(-2) - R(1) \cdot Q(2)$

54. $P(3)[R(-1) + Q(1)]$

55. $P[Q(2)]$

56. $R[P(-1)]$

57. $Q[P(2)]$

58. $Q[R(-2)]$

59. Which axiom(s) from Chapter 1 can be invoked to justify the assertion that if the variables in a monomial represent real numbers, the monomial represents a real number? That the same is true of any polynomial with real coefficients?

60. Argue that, for each real number $x > 0$, $(-x)^n = -x^n$ for n an odd natural number, and $(-x)^n = x^n$ for n an even natural number.
[*Hint:* Write $(-x)^n$ as $(-x)(-x)\cdots(-x) = (-1)(x)(-1)(x)\cdots(-1)(x)$.]

2.2

SUMS AND DIFFERENCES

By the symmetric property of equality, the distributive property can be written in the form

$$ab + ac = a(b + c),$$

and, by the commutative property, it can be written

$$ba + ca = (b + c)a.$$

This provides us with a means of simplifying certain polynomials. For example, if x and y are real numbers, then

$$\begin{aligned} 3x + 2x &= (3 + 2)x \\ &= 5x, \\ 3y + 2y + 5y &= (3 + 2 + 5)y \\ &= 10y, \\ 2x^2y + 6x^2y + x^2y &= (2 + 6 + 1)x^2y \\ &= 9x^2y. \end{aligned}$$

Terms that differ only in their numerical coefficients are commonly called **like** or **similar terms**, and the application of the distributive property in this form is referred to as *combining like terms* or *combining similar terms.* The expression obtained from combining like terms represents the same number as the original expression for all real number replacements of the variable or variables involved. Such expressions are called **equivalent expressions**.

Because we know that $a - b$ is equal to $a + (-b)$, we shall view the signs in any polynomial as indications of positive or negative coefficients and the operation involved will be understood to be addition. Thus,

$$\begin{aligned} 3x - 5x + 4x &= (3x) + (-5x) + (4x) \\ &= (3 - 5 + 4)x \\ &= 2x. \end{aligned}$$

We have been using grouping devices such as parentheses to indicate that various expressions are to be viewed as a single number. The expression

$$3x + (2x + 5x)$$

represents the sum of $3x$ and $(2x + 5x)$, while the expression

$$(3x + 2x) + 5x$$

represents the sum of $(3x + 2x)$ and $5x$. But the associative property asserts that these expressions are equivalent. Hence, the order in which we group terms in expressions of addition is immaterial. We can consider

$$3x + 2x + 5x$$

equal to either of the grouped expressions above.

An expression such as

$$a - (b + c),$$

in which a set of parentheses is preceded by a negative sign, can first be written

$$a - (b + c) = a + [-(b + c)].$$

Then, since

$$-(b + c) = -1(b + c) = -b - c,$$

we have

$$\begin{aligned} a - (b + c) &= a + [-b - c] \\ &= a - b - c. \end{aligned}$$

Thus, an expression in parentheses preceded by a negative sign can be written equivalently without parentheses by replacing each term within the parentheses with its opposite. For example,

$$\begin{aligned} (x^2 + 2x) - (2x^2 - 3x + 2) &= x^2 + 2x - 2x^2 + 3x - 2 \\ &= -x^2 + 5x - 2. \end{aligned}$$

In any expression where grouping devices are nested—that is, where groups occur within groups—a great deal of difficulty can be avoided by removing the inner devices first and working outward. Thus,

$$\begin{aligned} 3x - [2 - (3x + 1)] &= 3x - [2 - 3x - 1] \\ &= 3x - 2 + 3x + 1 \\ &= 6x - 1. \end{aligned}$$

The rewriting of a polynomial by combining like terms is a process that might be called "simplifying" the polynomial, since the result is a polynomial with fewer terms than the original. Indeed, if a polynomial is one in which no two terms contain identical variable factors and in which no grouping symbols are contained, we shall say the polynomial is in **simple form**. Thus, the polynomial in simple form that is equivalent to $3x - [2 - (3x + 1)]$ in the example above is $6x - 1$, which contains no grouping device and has all like terms combined.

EXERCISE 2.2

A

Write each expression in simple form.

Examples **a.** $(x^2 + 3x) + (4 - 2x)$ **b.** $(3y^2 - 2y + 3) - (y^2 - 1)$

Solutions **a.** $(x^2 + 3x) + (4 - 2x)$

$= x^2 + 3x + 4 - 2x$

$= x^2 + (3x - 2x) + 4$

$= x^2 + x + 4$

b. $(3y^2 - 2y + 3) - (y^2 - 1)$

$= 3y^2 - 2y + 3 - y^2 + 1$*

$= (3y^2 - y^2) + (-2y) + (3 + 1)$

$= 2y^2 - 2y + 4$

1. $4a^2 - 3a^2$ **2.** $2b - 7b$ **3.** $-6t + 12t$

4. $-8z + 8z$ **5.** $4n - 3n + 5n$ **6.** $2r - 3r + 6r$

7. $4x^2y - 3x^2y + 2x$ **8.** $5a^2b - 3ab^2 + a^2b$

9. $(n^2 - 2n) + (3n^2 - n)$ **10.** $(4t^2 + 3t) - (2t^2 + 3t)$

11. $(4x^2y - 3xy^2) - (2x^2y + xy^2)$ **12.** $(2z^3 - 3z^2) + (3z^2 - 2z^3)$

13. $(4k^2 - 3u + 2) + (u^2 - 3k + 1)$ **14.** $(2r^2 - 3r + 3) - (2r^2 - 3r + 1)$

15. $(2x^2 - 3x + 7) - (3x^2 + 3x - 5)$ **16.** $(3y^2 - 3y + 7) + (2y^2 + y + 6)$

17. $(4x^2 - 3x + 6) - (2x^2 + x - 5)$ **18.** $(2z^3 - 3z^2 + 2z) + (4z - 2z^2 - 3z^3)$

Find each sum.

Examples

a.
$$\begin{array}{r} 4x^2 - 3x + 2 \\ x^2 \qquad - 4 \\ 3x^2 + 6x - 1 \\ \hline \end{array}$$

b.
$$\begin{array}{r} 2z^3 - 3z^2 + 4z - 2 \\ z^3 \qquad\quad - 2z + 4 \\ -3z^3 + 2z^2 - \ z + 1 \\ \hline \end{array}$$

Solutions Adding by columns to combine like terms:

a.
$$\begin{array}{r} 4x^2 - 3x + 2 \\ x^2 \qquad - 4 \\ 3x^2 + 6x - 1 \\ \hline 8x^2 + 3x - 3 \end{array}$$

b.
$$\begin{array}{r} 2z^3 - 3z^2 + 4z - 2 \\ z^3 \qquad\quad - 2z + 4 \\ -3z^3 + 2z^2 - \ z + 1 \\ \hline - \ z^2 + \ z + 3 \end{array}$$

* *Common error:* Note that $-(y^2 - 1) \neq -y^2 - 1$.

19.
$$\begin{array}{r} 4a^2 + 6a - 7 \\ a^2 + 5a + 2 \\ -2a^2 - 7a + 6 \\ \hline \end{array}$$

20.
$$\begin{array}{r} 7c^2 - 10c + 8 \\ 8c + 11 \\ -6c^2 - 3c - 2 \\ \hline \end{array}$$

21.
$$\begin{array}{r} 4x^2y - 3xy + xy^2 \\ -5x^2y + xy - 2xy^2 \\ x^2y - 3xy \qquad \\ \hline \end{array}$$

22.
$$\begin{array}{r} m^2n^2 - 2mn + 7 \\ -2m^2n^2 + mn - 3 \\ 3m^2n^2 - 4mn + 2 \\ \hline \end{array}$$

23.
$$\begin{array}{r} x^3 - 2x^2 + 3x - 5 \\ 2x^3 \qquad\quad - 4x + 6 \\ 4x^2 - 3x - 2 \\ \hline \end{array}$$

24.
$$\begin{array}{r} -3y^3 + 4y^2 + 6y \qquad \\ y^3 - 2y^2 + y + 6 \\ 4y^3 - 2y^2 - 4y - 1 \\ \hline \end{array}$$

Subtract the bottom polynomial from the top polynomial.

Examples

a.
$$\begin{array}{r} 4t^2 - 3t + 6 \\ t^2 - 2t + 4 \\ \hline \end{array}$$

b.
$$\begin{array}{r} 3p^3 - 2p^2 + p - 4 \\ p^3 - 2p^2 - p + 3 \\ \hline \end{array}$$

Solutions Replace each term in the bottom polynomial with its opposite, and add.

a.
$$\begin{array}{r} 4t^2 - 3t + 6 \\ -t^2 + 2t - 4 \\ \hline 3t^2 - t + 2 \end{array}$$

b.
$$\begin{array}{r} 3p^3 - 2p^2 + p - 4 \\ -p^3 + 2p^2 + p - 3 \\ \hline 2p^3 \qquad\quad + 2p - 7 \end{array}$$

25.
$$\begin{array}{r} 2x^2 - 3x + 5 \\ 3x^2 + x - 2 \\ \hline \end{array}$$

26.
$$\begin{array}{r} 4y^2 - 3y - 7 \\ 6y^2 - y + 2 \\ \hline \end{array}$$

27.
$$\begin{array}{r} 4t^3 - 3t^2 + 2t - 1 \\ 5t^3 + t^2 + t - 2 \\ \hline \end{array}$$

28.
$$\begin{array}{r} 4s^3 - 3s^2 + 2s - 1 \\ s^3 - s^2 + 2s - 1 \\ \hline \end{array}$$

29.
$$\begin{array}{r} 4x^3 - 3x^2y + 2xy^2 - 4y^3 \\ x^3 + 2x^2y - xy^2 - 4y^3 \\ \hline \end{array}$$

30.
$$\begin{array}{r} a^3 - 3a^2b - 5ab^2 + 6b^3 \\ a^3 + a^2b - 4ab^2 - 5b^3 \\ \hline \end{array}$$

31. Subtract $4x^2 - 3x + 2$ from the sum of $x^2 - 2x + 3$ and $x^2 - 4$.

32. Subtract $2t^2 + 3t - 1$ from the sum of $2t^2 - 3t + 5$ and $t^2 + t + 2$.

33. Subtract the sum of $2b^2 - 3b + 2$ and $b^2 + b - 5$ from $4b^2 + b - 2$.

34. Subtract the sum of $7c^2 + 3c - 2$ and $3 - c - 5c^2$ from $2c^2 + 3c + 1$.

35. Subtract $x^2 - 2x$ from the sum of $x^2 + 2x$ and $2x^2 - x + 2$.

36. Subtract $2y^2 - y + 1$ from the sum of $y + 2$ and $y^2 - 4y + 3$.

Simplify.

Examples

a. $x - [2x + (3 - x)]$

b. $[x^2 - (2x + 1)] - [2x^2 - (x - 3)]$

Solutions

a. $x - [2x + (3 - x)]$
$= x - [2x + 3 - x]$
$= x - [x + 3]$
$= x - x - 3$
$= -3$

b. $[x^2 - (2x + 1)] - [2x^2 - (x - 3)]$
$= [x^2 - 2x - 1] - [2x^2 - x + 3]$
$= x^2 - 2x - 1 - 2x^2 + x - 3$
$= -x^2 - x - 4$

37. $y - [2y + (y + 1)]$

38. $3a + [2a - (a + 4)]$

39. $3 - [2x - (x + 1) + 2]$

40. $5 - [3y + (y - 4) - 1]$

41. $(3x + 2) - [x + (2 + x) + 1]$

42. $-(x - 3) + [2x - (3 + x) - 2]$

43. $[x^2 - (2x + 3)] - [2x^2 + (x - 2)]$

44. $[2y^2 - (4 - y)] + [y^2 - (2 + y)]$

45. $2x - (3y - [x - (x - y)] + x)$

46. $y - (y - [x - (2x + y)] - 2y)$

47. $3y - (2x - y) - (y - [2x - (y - 2x)] + 3y)$

48. $[x - (y + x)] - (2x - [3x - (x - y)] + y)$

49. $[x - (3x + 2)] - (2x - [x - (4 + x)] - 1)$

50. $-(2y - [2y - 4y + (y - 2)] + 1) + [2y - (4 - y) + 1]$

51. $(3y + [x - 2y + (y - x)] - 2x) - [3x + (x - y) + 2y]$

52. $[2x + (x - y) + 2y] - (3y - [x + (y - x) + y] + 2x)$

B

Given that $P(x) = x - 1$, $Q(x) = x^2 + 1$, *and* $R(x) = x^2 - x + 1$, *find the value for each expression.*

53. $P(x) + Q(x) - R(x)$

54. $P(x) - Q(x) + R(x)$

55. $R(x) - [Q(x) - P(x)]$

56. $Q(x) - [R(x) + P(x)]$

57. If $x, y \in R$ in Problems 49–52, which axioms justify the statement that the polynomials represent real numbers?

58. Which axioms justify writing $2x^2y + 5x^2y$ as $7x^2y$?

2.3

PRODUCTS AND QUOTIENTS OF MONOMIALS

Consider the product

$$a^m a^n,$$

where a is a real number and m and n are natural numbers. Since

$$a^m = aaa \cdots a \qquad (m \text{ factors}),$$

and

$$a^n = aaa \cdots a \qquad (n \text{ factors}),$$

it follows that

$$\begin{aligned} a^m a^n &= (\overbrace{aaa \cdots a}^{m \text{ factors}})(\overbrace{aaa \cdots a}^{n \text{ factors}}) \\ &= \overbrace{aaa \cdots a}^{m+n \text{ factors}}. \end{aligned}$$

Hence,

$$\boldsymbol{a^m a^n = a^{m+n}}.$$

This equation is known as the **first law of exponents**. Thus, we can simplify an expression for the product of two natural number powers of the same base simply by adding the exponents and using the sum as an exponent on the same base. For example,

$$x^2x^3 = x^5, \qquad xx^3x^4 = x^8, \qquad y^3y^4y^2 = y^9.$$

In multiplying two monomials, say $3x^2y$ and $2xy^2$, by the commutative and associative properties of multiplication, we first have

$$3 \cdot 2 \cdot x^2 \cdot x \cdot y \cdot y^2.$$

This then can be written in the simplified form

$$6x^3y^3.$$

Turning now to division, we note that in dividing x^5 by x^3 $(x \neq 0)$, we can write

$$\frac{x^5}{x^3} = \frac{xxxxx}{xxx} = xx\,\frac{xxx}{xxx} = xx \cdot 1 = xx = x^2 = x^{5-3}.$$

This example suggests that, in general,

$$\textbf{(1)} \qquad \boldsymbol{\frac{a^m}{a^n} = a^{m-n} \qquad (a \neq 0, m > n)}$$

where m and n are natural numbers.

This equation is called the **second law of exponents.** For example,

$$\begin{aligned} \frac{18x^3y^2z^4}{6x^2yz^2} &= \left(\frac{18}{6}\right)(x^{3-2}y^{2-1}z^{4-2}) \\ &= 3xyz^2 \qquad (x, y, z \neq 0). \end{aligned}$$

Note that in equation (1) if $m = n$, we have

$$\frac{x^n}{x^n} = 1 \qquad (x \neq 0)$$

for all $n \in N$.

In multiplying and dividing monomials, we do a great deal of work with symbols rather than with numbers. The fact that we can multiply and divide powers by operating on their exponents is a property of the symbolism we have adopted and not a property of the numbers these symbols represent. We are interested in multiplying or dividing numbers whose representations are x^m and x^n. We perform these operations by adding or subtracting the numbers m and n, which are simply parts of the representations. It is very important in manipulations of this kind to keep clearly in mind that the operations are performed on the numbers themselves.

EXERCISE 2.3

A

Write each product as a polynomial in simple form.

Examples **a.** $(7a^3)(4a^2b^4)$ **b.** $(-3c^2)(cd^2)(4c^3d)$

Solutions **a.** $(7a^3)(4a^2b^4) = 7 \cdot 4 \cdot a^{3+2} \cdot b^4$
$= 28a^5b^4$

b. $(-3c^2)(cd^2)(4c^3d) = -3 \cdot 4 \cdot c^{2+1+3}d^{2+1}$
$= -12c^6d^3$

1. $(7t)(-2t^2)$ **2.** $(4c^3)(3c)$ **3.** $(4a^2b)(-10ab^2c)$

4. $(-6r^2s^2)(5rs^3)$ **5.** $(11x^2yz)(4xy^3z)$ **6.** $(-8abc)(-b^2c^3)$

7. $2(3x^2y)(x^3y^4)$ **8.** $-5(ab^3)(-3a^2bc)$ **9.** $(-r^3)(-r^2s^4)(-2rt^2)$

10. $(-5mn)(2m^2n)(-n^3)$ **11.** $(y^2z)(-3x^2z^2)(-y^4z)$ **12.** $(-3xy)(2xz^4)(3x^3y^2z)$

Write each quotient as a monomial. (Assume that variables in exponents represent natural numbers.)

Examples **a.** $\dfrac{x^4b^2}{xb^2}$ **b.** $\dfrac{12a^2b^4}{-3ab^2}$

Solutions **a.** $\dfrac{x^4b^2}{xb^2} = x^{4-1} \cdot \dfrac{b^2}{b^2}$
$= x^3 \cdot 1 = x^3$

b. $\dfrac{12a^2b^4}{-3ab^2} = -4a^{2-1}b^{4-2}$
$= -4ab^2$

13. $\dfrac{6x^3y^2}{3xy}$ **14.** $\dfrac{12a^4b^2}{-4a^2b^2}$ **15.** $\dfrac{14t^3r^4}{7t^2r^2}$ **16.** $\dfrac{-22a^2bc^3}{-11ac^2}$

17. $\dfrac{14c^4d^3}{-7c^2d^3}$ 18. $\dfrac{100m^2n^3}{5m^2n^3}$ 19. $\dfrac{a^5b^7c^6}{a^4bc^3}$ 20. $\dfrac{-x^4y^8z^6}{xy^7z^5}$

21. $\dfrac{(x-4)^3}{(x-4)^2}$ 22. $\dfrac{(t+3)^4}{(t+3)^3}$ 23. $\dfrac{6m^2np^3}{-6m^2np^3}$ 24. $\dfrac{34a^6b^2c^4}{-17abc^3}$

B

Examples **a.** $x^n \cdot x^{n+3}$ **b.** $y^{2n} \cdot y^{3-n}$

Solutions **a.** $x^n \cdot x^{n+3} = x^{n+n+3}$
$= x^{2n+3}$

b. $y^{2n} \cdot y^{3-n} = y^{2n+3-n}$
$= y^{n+3}$

25. $a^{2n} \cdot a^{n-3}$ 26. $b^{-n} \cdot b^{2n+1}$ 27. $x^{n^2-n} \cdot x^{2n-n^2}$

28. $y^{2n+6} \cdot y^{4-n}$ 29. $a^{2n-2} \cdot a^{n+3}$ 30. $b^{n+2} \cdot b^{2n-1}$

Examples **a.** $\dfrac{x^{3n+1}}{x^{n+2}}$ **b.** $\dfrac{a^{n^2+2n+1}}{a^{n^2+n-3}}$

Solutions **a.** $\dfrac{x^{3n+1}}{x^{n+2}} = x^{3n+1-(n+2)}$
$= x^{3n+1-n-2}$
$= x^{2n-1}$

b. $\dfrac{a^{n^2+2n+1}}{a^{n^2+n-3}} = a^{n^2+2n+1-(n^2+n-3)}$
$= a^{n^2+2n+1-n^2-n+3}$
$= a^{n+4}$

31. $\dfrac{a^{5n}}{a^{3n}}$ 32. $\dfrac{b^{7n}}{b^n}$ 33. $\dfrac{a^{3n+1}}{a^{n-2}}$ 34. $\dfrac{x^{3n+4}}{x^{2n-1}}$

35. $\dfrac{x^{n^2+n-1}}{x^{n+1}}$ 36. $\dfrac{y^{n^2+2n-2}}{y^{n^2-2}}$ 37. $\dfrac{a^{2n}b^{n+5}}{a^nb^{n+1}}$ 38. $\dfrac{x^{n-3}b^{n+5}}{x^{n-4}b^{n+1}}$

39. Observing that

$$x^m = x^{m-n} \cdot x^n \qquad (m > n),$$

prove that

$$\frac{x^m}{x^n} = x^{m-n} \qquad (x \neq 0 \text{ and } m, n \in N \text{ with } m > n).$$

40. We have stated two laws of exponents in this chapter. Argue that, for natural numbers m and n, $(x^m)^n = (x^m)(x^m)\cdots(n \text{ factors})$, and go from here to a third law of exponents, $(x^m)^n = x^{mn}$.

41. Argue that, for any natural number n, $(xy)^n = x^ny^n$, and hence develop a fourth law of exponents.

42. Assuming we could apply the second law of exponents in dividing x^n/x^n, we would obtain x^{n-n} or x^0. What meaning must we assign to the symbol x^0 so that this result is consistent with the definition of a quotient? What restriction must be placed on x in this case?

2.4

PRODUCTS OF POLYNOMIALS

The associative property of addition can be used to extend the distributive property to cases where the right-hand factor contains more than two terms. Thus,

$$\begin{aligned} a[b + c + d] &= a[(b + c) + d] \\ &= a(b + c) + ad \\ &= ab + ac + ad, \end{aligned}$$

$$\begin{aligned} a[b + c + d + e] &= a([(b + c) + d] + e) \\ &= ab + ac + ad + ae, \qquad \text{etc.} \end{aligned}$$

We refer to this as the **generalized distributive property**. In the case of a product of a monomial and a polynomial, this property implies, for example, that

$$3x(x + y + z) = 3x(x) + 3x(y) + 3x(z) = 3x^2 + 3xy + 3xz,$$

and that

$$\begin{aligned} -2ab^2(3a^2b - ab + 2ab^2) &= (-2ab^2)(3a^2b) + (-2ab^2)(-ab) + (-2ab^2)(2ab^2) \\ &= -6a^3b^3 + 2a^2b^3 - 4a^2b^4. \end{aligned}$$

The distributive property can also be applied to simplify expressions for products of polynomials containing more than one term. For example,

$$\begin{aligned} (3x + 2y)(x - y) &= 3x(x - y) + 2y(x - y) \\ &= 3x^2 - 3xy + 2xy - 2y^2 \\ &= 3x^2 - xy - 2y^2. \end{aligned}$$

It is sometimes convenient to simplify products involving binomials or trinomials by using the familiar vertical form from arithmetic for computations. For example, the product $(2x - 3)(x^2 + 2x - 1)$ can be written as a polynomial in simple form as follows:

$$\begin{array}{rrrrl} & x^2 & + 2x & - 1 & \\ & & 2x & - 3 & \\ \hline 2x^3 & + 4x^2 & - 2x & & [2x(x^2 + 2x - 1)] \\ & - 3x^2 & - 6x & + 3 & [-3(x^2 + 2x - 1)] \\ \hline 2x^3 & + x^2 & - 8x & + 3 & \end{array}$$

Thus,

$$(2x - 3)(x^2 + 2x - 1) = 2x^3 + x^2 - 8x + 3.$$

The products listed below represent types that occur so frequently that you should learn to recognize them on sight.

(1) $$(x + a)(x + b) = x^2 + (a + b)x + ab$$

(2) $$(x + a)^2 = x^2 + 2ax + a^2 {}^*$$

(3) $$(x + a)(x - a) = x^2 - a^2$$

(4) $$(ax + by)(cx + dy) = acx^2 + (ad + bc)xy + bdy^2$$

The distributive property frequently has a role to play in the simplification of expressions involving grouping devices. For example, we can begin to simplify

$$2[x - 3y + 3(y - x)] - 2(2x + y)$$

by applying the distributive property to $3(y - x)$ in the *inner* set of parentheses and combining like terms.

$$\begin{aligned} 2[x - 3y + 3(y - x)] - 2(2x + y) &= 2[x - 3y + 3y - 3x] - 2(2x + y) \\ &= 2[-2x] - 2(2x + y) \\ &= -4x - 4x - 2y \\ &= -8x - 2y, \end{aligned}$$

so that

$$2[x - 3y + 3(y - x)] - 2(2x + y) = -8x - 2y.$$

EXERCISE 2.4

A

Write each expression as a polynomial in simple form.

Examples

a. $-(2x + y - z)$ **b.** $(x - 3)^2$ **c.** $(x - 4)(3x + 5)$

Solutions

a. $-(2x + y - z)$
$= -2x - y + z$

b. $(x - 3)^2$
$= (x - 3)(x - 3)$
$= x^2 - 6x + 9$

c. $(x - 4)(3x + 5)$
$= 3x^2 - 7x - 20$

1. $2x(x + 2y)$ **2.** $-3y(2x + y)$ **3.** $6t(t^2 - 3t + 1)$

4. $4y(2y^2 - y - 3)$ **5.** $-(3x^2 + 2x - 4)$ **6.** $-2(t^3 - 3t^2 + 2t - 1)$

7. $(x + 2)^2$ **8.** $(2x - 1)^2$ **9.** $(x - 5)(x + 1)$

* *Common error:* Note that $(x + a)^2 \neq x^2 + a^2$.

10. $(y-2)(y+3)$ **11.** $(y-6)(y-1)$ **12.** $(z-3)(z-5)$
13. $(2z+1)(z-3)$ **14.** $(3t-1)(2t+1)$ **15.** $(4r+3)(2r-1)$
16. $(2z-1)(3z+5)$ **17.** $(2x-a)(2x+a)$ **18.** $(3t-4s)(3t+4s)$

Example $(x-4)(x^2+2x-3)$

Solution **Method 1**

$$\begin{aligned}&(x-4)(x^2+2x-3)\\ &\quad = x(x^2+2x-3)-4(x^2+2x-3)\\ &\quad = x^3+2x^2-3x-4x^2-8x+12\\ &\quad = x^3-2x^2-11x+12\end{aligned}$$

Method 2

$$\begin{array}{r} x^2+2x-3\\ x-4\\ \hline x^3+2x^2-3x\\ -4x^2-8x+12\\ \hline x^3-2x^2-11x+12\end{array}$$

19. $(y+2)(y^2-2y+3)$ **20.** $(t+4)(t^2-t-1)$
21. $(x-3)(x^2+5x-6)$ **22.** $(x-7)(x^2-3x+1)$
23. $(x-2)(x-1)(x+3)$ **24.** $(y+2)(y-2)(y+4)$
25. $(z-3)(z+2)(z+1)$ **26.** $(z-5)(z+6)(z-1)$
27. $(2x+3)(3x^2-4x+2)$ **28.** $(3x-2)(4x^2+x-2)$
29. $(2a^2-3a+1)(3a^2+2a-1)$ **30.** $(b^2-3b+5)(2b^2-b+1)$
31. $(a^2-1)(a+2)(a^2-3a+1)$ **32.** $(a^2+1)(a^3+a^2-3a-2)$
33. $(a-1)(a+2)(a^2-3a+1)$ **34.** $(a+3)(a+1)(2a^2+4a-1)$
35. $(a+1)^2(a-1)^2$ **36.** $(a+1)(a-1)^3$

Examples **a.** $5+3[2a-3(a-2)]$ **b.** $-2(b-[2b+3(b-1)]+3)$

Solutions

a.
$$\begin{aligned}&5+3[2a-3(a-2)]\\ &\quad = 5+3[2a-3a+6]\\ &\quad = 5+3[-a+6]\\ &\quad = 5-3a+18\\ &\quad = 23-3a\end{aligned}$$

b.
$$\begin{aligned}&-2(b-[2b+3(b-1)]+3)\\ &\quad = -2(b-[2b+3b-3]+3)\\ &\quad = -2(b-[5b-3]+3)\\ &\quad = -2(b-5b+3+3)\\ &\quad = -2(-4b+6)\\ &\quad = 8b-12\end{aligned}$$

37. $2[a-(a-1)+2]$ **38.** $3[2a-(a+1)+3]$
39. $a[a-(2a+3)-(a-1)]$ **40.** $-2a[3a+(a-3)-(2a+1)]$
41. $-[a-3(a+1)-(2a+1)]$ **42.** $-[(a+1)-2(3a-1)+4]$
43. $2(a-[a-2(a+1)+1]+1)$ **44.** $-4(4-[3-2(a-1)+a]+a)$
45. $-x(x-3[2x-3(x+1)]+2)$ **46.** $x(4-2[3-4(x+1)]-x)$

B

In Problems 47–58, assume that all variables in exponents denote natural numbers.

Examples **a.** $2a^{2n}(3a^n - 2)$ **b.** $(2a^n + 1)(a^n - 2)$

Solutions **a.** $2a^{2n}(3a^n - 2)$
$= 3a^n(2a^{2n}) - 2(2a^{2n})$
$= 6a^{2n+n} - 4a^{2n}$
$= 6a^{3n} - 4a^{2n}$

b. $(2a^n + 1)(a^n - 2)$
$= 2a^n(a^n - 2) + 1(a^n - 2)$
$= 2a^{2n} - 4a^n + a^n - 2$
$= 2a^{2n} - 3a^n - 2$

47. $x^n(2x^n - 1)$ **48.** $3t^n(2t^n + 3)$ **49.** $a^{n+1}(a^n - 1)$
50. $b^{n-1}(b + b^n)$ **51.** $a^{2n+1}(a^n + a)$ **52.** $b^{2n+2}(b^{n-1} + b^n)$
53. $(1 + a^n)(1 - a^n)$ **54.** $(a^n - 3)(a^n + 2)$ **55.** $(a^{3n} + 2)(a^{3n} - 1)$
56. $(a^{3n} - 3)(a^{3n} + 3)$ **57.** $(2a^n - b^n)(a^n + 2b^n)$ **58.** $(a^{2n} - 2b^n)(a^{3n} + b^{2n})$

Show that the left-hand member is equivalent to the right-hand member.

59. $(x + a)(x + b) = x^2 + (a + b)x + ab$
60. $(x + a)^2 = x^2 + 2ax + a^2$
61. $(x + a)(x - a) = x^2 - a^2$
62. $(ax + by)(cx + dy) = acx^2 + (ad + bc)xy + bdy^2$
63. $(x + a)(x^2 - ax + a^2) = x^3 + a^3$
64. $(x - a)(x^2 + ax + a^2) = x^3 - a^3$

2.5

FACTORING MONOMIALS FROM POLYNOMIALS

The distributive property in the form

$$ax + bx + cx + dx = (a + b + c + d)x$$

furnishes a means of writing a polynomial as a single term comprised of two or more factors. This process is called **factoring**. Thus, by the distributive property,

$$3x^2 + 6x = 3x(x + 2).$$

Of course, we can also write

$$3x^2 + 6x = 3(x^2 + 2x)$$

or

$$3x^2 + 6x = 3x^2\left(1 + \frac{2}{x}\right) \qquad (x \neq 0)$$

or any other of an infinite number of such expressions. We are, however, primarily interested in factoring a polynomial into a unique form (except for signs and order of factors) referred to as the **completely factored form**. A polynomial with integral coefficients is in completely factored form if:

1. it is written as a product of polynomials with integral coefficients;
2. no polynomial—other than a monomial—in the factored form can be further factored into polynomials with integral coefficients.

The restriction that the factors be polynomials means that all the variables involved have exponents from $\{1, 2, 3, \ldots\}$. Restricting the coefficients to integers prohibits such factorizations as

$$x + 3 = 3\left(\frac{1}{3}x + 1\right).$$

Notice that complete factorization of monomial factors is not required. Thus, it is not necessary that the form

$$6x^2(x - 2)$$

be written

$$2 \cdot 3 \cdot x \cdot x \cdot (x - 2)$$

in order for the expression to be considered completely factored.

The earlier observation that the completely factored form is unique *except for signs and order of factors* stems from the fact that

$$-ab = (-a)(b) = a(-b)$$

or

$$ab = (-a)(-b) = (a)(b),$$

and

$$ab = ba.$$

Because the choice of signs and order of factors is arbitrary, the factored form of an expression that seems most "natural" should be used, although this is admittedly not always easy to determine. For instance, the forms

$$a(1 - x - x^2) \qquad \text{and} \qquad -a(x^2 + x - 1)$$

are equivalent, but it is difficult to affirm one as more "natural" than the other.

Common monomial factors can be factored from a polynomial by first identifying such common factors and then writing the resultant factored expression. For example, observe that the polynomial

$$6x^3 + 9x^2 - 3x$$

contains the monomial $3x$ as a factor of each term. We therefore write

$$6x^3 + 9x^2 - 3x = 3x(\qquad\quad)$$

and insert within the parentheses the appropriate polynomial factor. This factor can be determined by inspection. We ask ourselves for the monomials that multiply $3x$ to yield $6x^3$, $9x^2$, and $-3x$. The final result appears as

$$6x^3 + 9x^2 - 3x = 3x(2x^2 + 3x - 1).$$

One particularly useful factorization is of the form

$$a - b = (-1) \cdot (-a + b).$$

Since $-a + b = b - a$,

$$\boldsymbol{a - b = (-1) \cdot (b - a) = -(b - a).}$$

That is, $a - b$ and $b - a$ are negatives of each other. For example,

$$3x - y = -(y - 3x) \qquad \text{and} \qquad a - 2b = -(2b - a).$$

EXERCISE 2.5

A

Factor completely.

Examples **a.** $6x - 18$ **b.** $18x^2y - 24xy^2$ **c.** $y(x - 2) + z(x - 2)$

Solutions

a. $6x - 18$
$= 6(\underline{?} - \underline{?})$
$= 6(x - 3)$

b. $18x^2y - 24xy^2$
$= 6xy(\underline{?} - \underline{?})$
$= 6xy(3x - 4y)$

c. $y(x - 2) + z(x - 2)$
$= (x - 2)(\underline{?} + \underline{?})$
$= (x - 2)(y + z)$

1. $2x + 6$ **2.** $3x - 9$ **3.** $4x^2 + 8x$

4. $3x^2y + 6xy$ **5.** $3x^2 - 3xy + 3x$ **6.** $x^3 - x^2 + x$

7. $24a^2 + 12a - 6$ **8.** $15r^2s + 18rs^2 - 3rs$ **9.** $2x^4 - 4x^2 + 8x$

10. $ay^2 + aby + ab$ **11.** $x^2y^2z^2 + 2xyz - xz$ **12.** $3m^2n - 6mn^2 + 12mn$

13. $a(a + 3) + b(a + 3)$ **14.** $b(a - 2) + a(a - 2)$ **15.** $2x(x + 3) - y(x + 3)$

16. $y(y - 2) - 3x(y - 2)$ **17.** $2y(a + b) - x(a + b)$ **18.** $3x(2a - b) + 4y(2a - b)$

Supply the missing factors or terms.

Examples **a.** $-5x + 10 = -5(\underset{-}{?})$ **b.** $x - 3 = -(\underset{-}{?} - \underset{-}{?})$

Solutions **a.** $-5x + 10 = -5(x - 2)$ **b.** $x - 3 = -(-x + 3)$

$= -(3 - x)$

19. $7 - r = -(\underset{-}{?} - \underset{-}{?})$

20. $3m - 2n = -(\underset{-}{?} - \underset{-}{?})$

21. $2a - b = -(\underset{-}{?} - \underset{-}{?})$

22. $r^2 - s^2t^2 = -(\underset{-}{?} - \underset{-}{?})$

23. $-2x + 2 = -2(\underset{-}{?})$

24. $-6x - 9 = -3(\underset{-}{?})$

25. $-ab - ac = \underset{-}{?}(b + c)$

26. $-a^2 + ab = \underset{-}{?}(a - b)$

27. $-xy - x^2y = -xy(\underset{-}{?})$

28. $-x^3y + y^3x = -xy(\underset{-}{?})$

29. $2x - 1 = -(\underset{-}{?})$

30. $x^2 - 3x = -(\underset{-}{?})$

31. $x - y + z = -(\underset{-}{?})$

32. $3x + 3y - 2z = -(\underset{-}{?})$

33. $x - y + z = x - (\underset{-}{?})$

34. $x^2 - y^2 + y - 1 = x^2 - (\underset{-}{?})$

B

Factor completely. (Assume that variables in the exponents denote natural numbers.)

Examples **a.** $x^{3n} + x^n$ **b.** $x^{n+2} - 2x^2$ **c.** $a^{2n} + a^{n+1} - a^n$

Solutions

a. $x^{3n} + x^n$

$= x^n(\underset{-}{?} + \underset{-}{?})$

$= x^n(x^{2n} + 1)$

b. $x^{n+2} - 2x^2$

$= x^n \cdot x^2 - 2x^2$

$= x^2(\underset{-}{?} - \underset{-}{?})$

$= x^2(x^n - 2)$

c. $a^{2n} + a^{n+1} - a^n$

$= a^{2n} + a^n a - a^n$

$= a^n(\underset{-}{?} + \underset{-}{?} - \underset{-}{?})$

$= a^n(a^n + a - 1)$

35. $x^{2n} - x^n$

36. $x^{4n} - x^{2n}$

37. $a^{3n} - a^{2n} - a^n$

38. $y^{4n} + y^{3n} + y^{2n}$

39. $x^{n+2} + x^n$

40. $x^{n+2} + x^{n+1} + x^n$

41. $-x^{2n} - x^n = \underset{-}{?}(x^n + 1)$

42. $-x^{5n} + x^{2n} = \underset{-}{?}(x^{3n} - 1)$

43. $-x^{a+1} - x^a = -x^a(\underset{-}{?})$

44. $-y^{a+2} + y^2 = -y^2(\underset{-}{?})$

2.6

FACTORING QUADRATIC POLYNOMIALS

One very common type of factoring involves quadratic (second-degree) binomials or trinomials. We can rewrite the products (1)–(4) given on page 51 to obtain:

$$x^2 + (a + b)x + ab = (x + a)(x + b) \tag{1}$$

$$x^2 + 2ax + a^2 = (x + a)^2 \tag{2}$$

$$x^2 - a^2 = (x + a)(x - a) \tag{3}$$

$$acx^2 + (ad + bc)xy + bdy^2 = (ax + by)(cx + dy) \tag{4}$$

Again, we shall require integral coefficients and positive integral exponents on the variables when factoring these polynomials.

As an example of the application of form (1), consider the trinomial

$$x^2 + 6x - 16.$$

We desire, if possible, to find two binomial factors,

$$(x + a)(x + b),$$

whose product is the given trinomial. We see from form (1) that a and b are two integers such that $a + b = 6$ and $ab = -16$; that is, their sum must be the coefficient of the linear term $6x$ and their product must be -16. By inspection, or by trial and error, we determine that the two numbers are 8 and -2, so that

$$x^2 + 6x - 16 = (x + 8)(x - 2).$$

Form (2) is simply a special case of (1), the square of a binomial. Thus,

$$\begin{aligned} x^2 + 8x + 16 &= (x + 4)(x + 4) \\ &= (x + 4)^2. \end{aligned}$$

Form (3) is another special case of (1), in which the coefficient of the first-degree term in x is 0. For example,

$$x^2 - 25 = (x - 5)(x + 5).$$

In particular, form (3) states that the difference of the squares of two numbers is equal to the product of the sum and the difference of the two numbers. The factors $x - 5$ and $x + 5$ are called **conjugates** of each other. In general, any binomials of the form $a - b$ and $a + b$ are called **conjugate pairs**.

Form (4) is a generalization of (1)—that is, in (4) we are confronted with a quadratic trinomial where the coefficient of the term of second degree in x is other than 1. The factoring of such a trinomial, for example $8x^2 - 9 - 21x$, is accomplished as follows:

1. Write in decreasing powers of x.

$$8x^2 - 21x - 9$$

2. Consider possible combinations of first-degree factors of the first term.

$$\begin{array}{ll} (8x \quad)(x \quad) \\ (4x \quad)(2x \quad) \end{array}$$

3. Consider combinations of the factors ① of the last term.

$$
\begin{array}{l}
(8x \quad 9)(x \quad 1)\\
(8x \quad 1)(x \quad 9)\\
(8x \quad 3)(x \quad 3)\\
(4x \quad 9)(2x \quad 1)\\
(4x \quad 1)(2x \quad 9)\\
(4x \quad 3)(2x \quad 3)
\end{array}
$$

(① outer brackets linking the last terms of the two factors; ② product of the outer terms; ③ product of the inner terms.)

4. Select the combination(s) of products ② and ③ whose sum(s) could be the second term $(-21x)$.

$$(8x \quad 3)(x \quad 3)$$

5. Insert the proper signs.

$$(8x + 3)(x - 3)$$

Although this process can normally be done mentally, it is written in detail here for the purposes of illustration.

If a polynomial of more than one term contains a common monomial factor in each of its terms, this monomial should be factored from the polynomial before seeking other factors. Thus,

$$
\begin{aligned}
32x^2 - 84x - 36 &= 4(8x^2 - 21x - 9)\\
&= 4(8x + 3)(x - 3).
\end{aligned}
$$

EXERCISE 2.6

A

Factor completely. (Assume that variables in exponents represent natural numbers.)

Examples **a.** $x^2 - 2x - 3$ **b.** $x^2 - 9y^2$ **c.** $5x^2 - 9x - 2$

Solutions

a. $x^2 - 2x - 3$
$= (x - 3)(x + 1)$

b. $x^2 - 9y^2$
$= x^2 - (3y)^2$
$= (x - 3y)(x + 3y)$

c. $5x^2 - 9x - 2$
$= (5x + 1)(x - 2)$

1. $x^2 + 7x + 12$
2. $x^2 + 7x + 10$
3. $a^2 - a - 6$
4. $a^2 - 2a - 15$
5. $x^2 + 6xy + 5y^2$
6. $x^2 - 9xy + 20y^2$
7. $x^2 - 4$
8. $x^2 - 36.$
9. $4 - b^2$
10. $9 - a^2$
11. $(ab)^2 - 1$
12. $(a^2b)^2 - 4$
13. $x^4 - 9$
14. $y^4 - 25$
15. $x^2 - 16y^2$
16. $x^2 - 4y^2$
17. $x^2 - 36y^2$
18. $x^2 - 81y^2$
19. $2x^2 + 3x - 2$
20. $3x^2 - 7x + 2$
21. $4x^2 + 7x - 2$
22. $6x^2 - 5x + 1$
23. $3x^2 + 4x + 1$
24. $4a^2 - 5a + 1$
25. $9x^2 - 21x - 8$
26. $10x^2 - 3x - 18$
27. $3x^2 - 7ax + 2a^2$
28. $9x^2 + 9ax - 10a^2$
29. $9x^2 - y^2$
30. $4x^2 - 9y^2$
31. $4x^2 + 12x + 9$
32. $4y^2 + 4y + 1$
33. $1 - 16x^2y^2$
34. $64x^2y^2 - 1$
35. $9x^2y^2 + 6xy + 1$
36. $4x^2y^2 + 12xy + 9$

Examples **a.** $4a^3 - 5a^2 + a$ **b.** $8x^5 - 2x^3$

Solutions **a.** $4a^3 - 5a^2 + a$

$$= a(4a^2 - 5a + 1)$$
$$= a(4a - 1)(a - 1)$$

b. $8x^5 - 2x^3$

$$= 2x^3(4x^2 - 1)$$
$$= 2x^3(2x - 1)(2x + 1)$$

37. $3x^2 + 12x + 12$
38. $2x^2 + 6x - 20$
39. $2a^3 - 8a^2 - 10a$
40. $2a^3 + 15a^2 + 7a$
41. $4a^2 - 8ab + 4b^2$
42. $20a^2 + 60ab + 45b^2$
43. $4x^2y - 36y$
44. $x^2 - 4x^2y^2$
45. $12x - x^2 - x^3$
46. $x^2 - 2x^3 + x^4$
47. $x^4y^2 - x^2y^2$
48. $x^3y - xy^3$

B

Examples **a.** $x^4 + 2x^2 + 1$ **b.** $x^4 - 3x^2 - 4$

Solutions **a.** $x^4 + 2x^2 + 1 = (x^2 + 1)^2$

b. $x^4 - 3x^2 - 4 = (x^2 - 4)(x^2 + 1)$

$$= (x - 2)(x + 2)(x^2 + 1)$$

49. $y^4 + 3y^2 + 2$
50. $a^4 + 5a^2 + 6$
51. $3x^4 + 7x^2 + 2$
52. $4x^4 - 11x^2 - 3$
53. $x^4 + 3x^2 - 4$
54. $x^4 - 6x^2 - 27$
55. $x^4 - 5x^2 + 4$
56. $y^4 - 13y^2 + 36$
57. $2a^4 - a^2 - 1$

58. $3x^4 - 11x^2 - 4$ **59.** $x^4 + a^2x^2 - 2a^4$ **60.** $4x^4 - 33a^2x^2 - 27a^4$

61. $x^{4n} - 1$ **62.** $16 - y^{4n}$ **63.** $x^{4n} - y^{4n}$

64. $x^{4n} - 2x^{2n} + 1$ **65.** $3x^{4n} - 10x^{2n} + 3$ **66.** $6y^{2n} + 30y^n - 900$

2.7

FACTORING OTHER POLYNOMIALS

A few other polynomials occur frequently enough to justify a study of their factorization. In particular, the following factorizations are often encountered in mathematics:

(1) $$\boldsymbol{ax + ay + bx + by = (a + b)(x + y)}$$

(2) $$\boldsymbol{x^3 + a^3 = (x + a)(x^2 - ax + a^2)}$$

(3) $$\boldsymbol{x^3 - a^3 = (x - a)(x^2 + ax + a^2)}$$

These can be readily verified by multiplying the factors in the right-hand member of each equation. Expressions such as the left-hand member of form (1) are factorable by grouping. For example, to factor

$$3x^2y + 2y + 3xy^2 + 2x,$$

we rewrite it in the form

$$3x^2y + 2x + 3xy^2 + 2y,$$

and factor the common monomials x and y from the first group of two terms and the second group of two terms, respectively, to obtain

$$x(3xy + 2) + y(3xy + 2).$$

If we now factor the common binomial $(3xy + 2)$ from each term, we have

$$(3xy + 2)(x + y).$$

The application of forms (2) and (3) is direct. For example,

$$\begin{aligned} 8a^3 + b^3 &= (2a)^3 + b^3 \\ &= (2a + b)[(2a)^2 - 2ab + b^2] \\ &= (2a + b)[4a^2 - 2ab + b^2]. \end{aligned}$$

EXERCISE 2.7

A

Factor.

Examples **a.** $yb - ya + xb - xa$ **b.** $x^2 + xb - ax - ab$

Solutions **a.** $yb - ya + xb - xa$
$= y(b - a) + x(b - a)$
$= (b - a)(y + x)$

b. $x^2 + xb - ax - ab$
$= x(x + b) - a(x + b)$
$= (x + b)(x - a)$

1. $ax^2 + x + ax + 1$
2. $5a + ab + 5b + b^2$
3. $ax^2 + x + a^2x + a$
4. $a + ab + b + b^2$
5. $x^2 + ax + xy + ay$
6. $x^3 - x^2y + xy - y^2$
7. $3ab - cb - 3ad + cd$
8. $1 - x - y + xy$
9. $3x + y - 6x^2 - 2xy$
10. $5xz - 5yz - x + y$
11. $a^3 + 2ab^2 - 2a^2b - 4b^3$
12. $6x^3 - 4x^2 + 3x - 2$
13. $x^2 - x + 2xy - 2y$
14. $2a^2 + 3a - 2ab - 3b$
15. $2a^2b + 6a^2 - b - 3$
16. $2ab^2 + 5a - 8b^2 - 20$
17. $x^3y^2 + x^3 - 3y^2 - 3$
18. $12 - 4y^3 - 3x^2 + x^2y^3$

Examples **a.** $x^3 + 8$ **b.** $8x^3 - y^3$

Solutions **a.** $x^3 + 8$
$= x^3 + (2)^3$
$= (x + 2)(x^2 - 2x + 2^2)$
$= (x + 2)(x^2 - 2x + 4)$

b. $8x^3 - y^3$
$= (2x)^3 - y^3$
$= (2x - y)[(2x)^2 + 2xy + y^2]$
$= (2x - y)(4x^2 + 2xy + y^2)$

19. $x^3 - 1$
20. $y^3 + 27$
21. $(2x)^3 + y^3$
22. $y^3 - (3x)^3$
23. $a^3 - 8b^3$
24. $27a^3 + b^3$
25. $(xy)^3 - 1$
26. $8 + x^3y^3$
27. $27a^3 + 64b^3$
28. $a^3 - 125b^3$
29. $x^3 + (x - y)^3$
30. $(x + y)^3 - z^3$
31. $(x + 1)^3 - 1$
32. $x^6 + (x - 2y)^3$

B

33. $(x + 1)^3 - (x - 1)^3$
34. $(2y - 1)^3 + (y - 1)^3$
35. $(x + y)^3 + (x - y)^3$
36. $(x + y)^3 - (x - y)^3$

37. Show that $ac - ad + bd - bc$ can be factored as $(a - b)(c - d)$ and as $(b - a)(d - c)$.

38. Show that $a^2 - b^2 - c^2 + 2bc$ can be factored as $(a - b + c)(a + b - c)$.

39. Consider the polynomial $x^4 + x^2y^2 + y^4$. If x^2y^2 is both added to and subtracted from this expression (thus producing an equivalent expression in which the first three terms form a "perfect square"), we have

$$\begin{aligned} x^4 + x^2y^2 + y^4 &= x^4 + x^2y^2 + y^4 + x^2y^2 - x^2y^2 \\ &= x^4 + 2x^2y^2 + y^4 - x^2y^2 \\ &= (x^2 + y^2)^2 - (xy)^2 \\ &= (x^2 + y^2 - xy)(x^2 + y^2 + xy). \end{aligned}$$

By adding and subtracting an appropriate monomial, factor $x^4 + 3x^2y^2 + 4y^4$.

40. Use the method of Problem 39 to factor $x^4 - 8x^2y^2 + 4y^4$.

41. Use the method of Problem 39 to factor $a^4 + 6a^2b^2 + 25b^4$.

42. Use the method of Problem 39 to factor $4a^4 - 5a^2b^2 + b^4$.

CHAPTER SUMMARY

[2.1] Expressions of the form a^n, where

$$a^n = aaa \cdots a \qquad (n \text{ factors}),$$

are called **powers**; a is the **base** and n is the **exponent** of the power.

Any meaningful collection of numerals, variables, and signs of operation is called an **expression**. In an expression of the form $A + B + C + \cdots$, A, B, and C are called **terms**. Any factor or group of factors in a term is the **coefficient** of the remaining factors.

An expression in which the operations involved consist solely of addition, subtraction, multiplication, and division, and in which all variables occur as natural number powers only, is called a **rational expression**. Any rational expression in which no variable occurs in a denominator is called a **polynomial**. Polynomials consisting of 1, 2, or 3 terms are called **monomials**, **binomials**, or **trinomials**, respectively.

The degree of a monomial in one variable is given by the exponents on the variable. If the monomial contains more than one variable, the degree is given by the sum of the exponents on the variables. The degree of a polynomial is the degree of its term of highest degree.

Polynomials are represented by symbols such as $P(x)$, $Q(x)$, etc., and the values of these polynomials for some specific value a are represented by $P(a)$, $Q(a)$, etc.

[2.2] Terms that differ only in their numerical coefficients are called **like terms**. Two expressions that are equal for all real number replacements of any variable or variables involved are **equivalent expressions**.

[2.3] The following two laws of exponents are useful in rewriting products and quotients of powers:

(1) $$a^m \cdot a^n = a^{m+n},$$

(2) $$\frac{a^m}{a^n} = a^{m-n} \qquad (a \neq 0, \quad m > n).$$

[2.4–2.7] The distributive property can be used to rewrite products of polynomials of more than one term. The process of using the distributive property to rewrite a polynomial as a single term comprised of two or more factors is called **factoring**. Three special cases are:

(1) $$x^2 - a^2 = (x - a)(x + a)$$

(2) $$x^3 + a^3 = (x + a)(x^2 - ax + a^2)$$

(3) $$x^3 - a^3 = (x - a)(x^2 + ax + a^2)$$

The symbols introduced in this chapter are listed on the inside of the front cover.

REVIEW EXERCISES

[2.1] *In Problems 1 and 2, identify each polynomial as a monomial, binomial, or trinomial. State the degree of the polynomial.*

1. a. $2y^3 - y^2$ **b.** $3x^2 - 2x + 1$

2. a. $3x^2 - y^3$ **b.** $2x^2y + xy^3 - y^2$

3. Find the value of $\dfrac{2x^2 - 3y}{x - y}$ for $x = -2$ and $y = 3$.

4. Find the value of $(x - y^2)^2 - xy$ for $x = 3$ and $y = -1$.

5. If $P(x) = 2x^2 - 3x - 1$, find

a. $P(3)$ **b.** $P(-2)$

6. If $Q(x) = x^3 - 2x^2 - x$, find

a. $Q(-1)$ **b.** $Q(-2)$

[2.2] *Write each expression as a polynomial in simple form.*

7. a. $(2x - y) - (x - 2y + z)$

b. $(2x^2 - 3z^2) - (x - 2y) + (z^2 + y)$

Simplify each expression.

8. a. $2x - [x - 3(x + 1)]$

b. $[x^2 - (x + 1)] - [2x^2 + (x - 1)]$

[2.3] *Simplify each expression.*

9. a. $(2x^2y)(-3xy^3)$

b. $(3xy)(2xz^2)(-y^2z)$

10. a. $\dfrac{12x^4y^2}{3xy}$

b. $\dfrac{-12x^3yz^2}{-8x^2z^2}$

[2.4] *Write each expression as a polynomial in simple form.*

11. a. $2x(x^2 - 2x + 1)$

b. $2(2x - 1)(x + 3)$

12. a. $(y - 2)(y^2 - y + 1)$

b. $(z - 1)(z + 1)(z + 2)$

[2.5] *Factor each polynomial completely.*

13. a. $4y^3 - 8y^2$

b. $x^3 - 3x^2 - 10x$

14. a. $3x - y = -(?)$

b. $2x - y + z = -(?)$

15. a. $-x^2 + 2x = -x(?)$

b. $-6x^2y + 3xy - 3xy^2 = -3xy(?)$

16. a. $x(a + 2) - y(a + 2)$

b. $2a(x - y) + b(x - y)$

[2.6] *Factor each polynomial completely.*

17. a. $x^2 - 2x - 35$

b. $y^2 + 4y - 32$

18. a. $(xy)^2 - 36$

b. $a^2 - 49b^2$

19. a. $3y^2 + 11y - 4$

b. $x^3 + 3x^2 - 10x$

20. a. $9x^2 - 36$

b. $12x^2 - 3y^2$

21. a. $2x^2 + 3xy - 2y^2$

b. $6x^2 - xy - y^2$

22. a. $15a^2 + 28ab + 12b^2$

b. $12a^2 - 18ab + 6b^2$

[2.7] *Factor each polynomial.*

23. a. $2xy + 2x^2 + y + x$ b. $xy - 3x - y + 3$

24. a. $ax - 2bx + ay - 2by$ b. $2ax - 4ay + bx - 2by$

25. a. $(2x)^3 - y^3$ b. $x^3 + (4y)^3$

26. a. $27y^3 + z^3$ b. $x^3 - 8a^3$

3

FRACTIONS

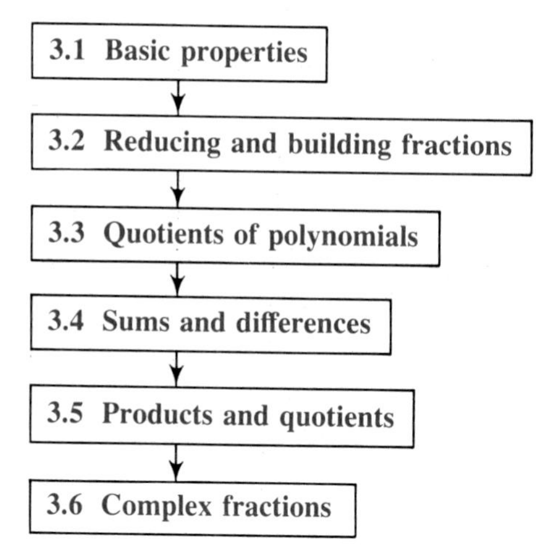

3.1 Basic properties
3.2 Reducing and building fractions
3.3 Quotients of polynomials
3.4 Sums and differences
3.5 Products and quotients
3.6 Complex fractions

3.1

BASIC PROPERTIES

Recall that a fraction is an expression denoting a quotient. If the numerator (dividend) and the denominator (divisor) are polynomials, then the fraction is a rational expression. For example,

$$\frac{y}{y+1}, \qquad \frac{x^2 - 2x + 1}{x}, \qquad \frac{1}{x^2 + 1}, \qquad \text{and} \qquad \frac{x+1}{y}$$

are rational expressions. Trivially, any polynomial can be considered a rational expression, since it is the quotient of itself and 1. Thus,

$$x^2 + 2x, \qquad 3y, \qquad \text{and} \qquad 5$$

are also rational expressions. For each replacement of the variable(s) for which the numerator and denominator of a fraction represent real numbers and for which the denominator is not zero, a rational expression represents a real number. Of course, for any value of the variable(s) for which the denominator vanishes (is equal to zero), the fraction does not represent a real number and is said to be undefined.

There are infinitely many fractions that correspond to a given quotient. Thus, for example,

$$\frac{1}{2} = \frac{2}{4} = \frac{3}{6} = \frac{4}{8} \cdots \qquad \text{and} \qquad \frac{3}{5} = \frac{6}{10} = \frac{9}{15} = \frac{12}{20} \cdots.$$

The properties of real numbers can be used to establish the following criteria for equivalent fractions:

$$\boldsymbol{\frac{a}{b} = \frac{c}{d} \qquad \textbf{if and only if} \qquad ad = bc \qquad (b, d \neq 0).}$$

This assertion establishes a means by which we can identify equivalent fractions —that is, fractions that represent the same number. For example,

$$\frac{2}{3}=\frac{4}{6} \qquad \text{because} \qquad 2\cdot 6 = 3\cdot 4,$$

$$\frac{6}{9}=\frac{8}{12} \qquad \text{because} \qquad 6\cdot 12 = 9\cdot 8,$$

$$\frac{2}{3}\neq\frac{3}{4} \qquad \text{because} \qquad 2\cdot 4 \neq 3\cdot 3.$$

There are numerous places in algebra where, for one reason or another, we wish to replace a given fraction with an equivalent fraction. We can write any number of fractions equivalent to a given fraction. For example, $4/6$, $6/9$, and $8/12$ are all equivalent to $2/3$. In general, if a, b, and c are real numbers, and b, $c \neq 0$, then

$$\frac{a}{b}=\frac{ac}{bc}.$$

This law is called the **fundamental principle of fractions** and asserts that an equivalent fraction is obtained if the numerator and the denominator of a fraction are each multiplied or divided by the same nonzero number. The validity of the fundamental principle follows from the assertion on page 67; that is,

$$\frac{a}{b}=\frac{ac}{bc} \qquad (b, c \neq 0),$$

because

$$a(bc) = b(ac).$$

There are three signs associated with a fraction: a sign for the numerator, a sign for the denominator, and a sign for the fraction itself. Although there are eight different possible symbols associated with the symbol "a/b" and the two signs "+" and "–," these symbols represent only two real numbers, a/b and its additive inverse $-(a/b)$. It follows from the definition of a quotient and the fundamental principle of fractions that, when $b \neq 0$,

$$\frac{-a}{b}=\frac{a}{-b}=-\frac{a}{b}=-\frac{-a}{-b}. \tag{1}$$

Also,

$$\frac{a}{b}=\frac{-a}{-b} \qquad \text{and} \qquad -\left(\frac{a}{-b}\right)=-\left(\frac{-a}{b}\right)=-\left(-\frac{a}{b}\right).$$

From the double-negative property,

$$-\left(\frac{-a}{b}\right) = \frac{a}{b}.$$

Therefore,

$$\frac{a}{b} = \frac{-a}{-b} = -\frac{a}{-b} = -\frac{-a}{b}. \tag{2}$$

Taken together, (1) and (2) affirm that a given fraction may be changed to an equivalent fraction by replacing any two of the fraction's three elements—the fraction itself, the numerator, and the denominator—with their negatives.

The forms a/b and $-a/b$, in which the sign of the fraction and the sign of the denominator are both positive, are generally the most convenient representations and will be referred to as **standard forms**. Thus,

$$\frac{-3}{5}, \quad \frac{3}{5}, \quad \text{and} \quad \frac{7}{10}$$

are in standard form, while

$$\frac{3}{-5}, \quad -\frac{-3}{5}, \quad \text{and} \quad -\frac{7}{-10}$$

are not.

If the numerator or denominator of a fraction is an expression containing more than one term, there are alternative standard forms. For example, since

$$a - b = -(b - a),$$

we have

$$\frac{-b}{a-b} = \frac{-b}{-(b-a)}$$
$$= \frac{b}{b-a},$$

and either

$$\frac{-b}{a-b} \quad \text{or} \quad \frac{b}{b-a}$$

may be taken as standard form, as convenience dictates. Observe that the quotient is not defined for $a - b = 0$ or $b - a = 0$; so we must make the restriction $a \neq b$.

EXERCISE 3.1

A

Write in standard form and specify any real values of the variables for which the fraction is undefined.

Examples

a. $-\dfrac{5}{-y}$

b. $-\dfrac{a}{a-2}$

c. $-\dfrac{x-1}{3}$

Solutions

a. $-\dfrac{5}{-y} = \dfrac{5}{y} \quad (y \neq 0)$

b. $-\dfrac{a}{a-2} = \dfrac{-a}{a-2}$ or $\dfrac{a}{2-a} \quad (a \neq 2)$

c. $-\dfrac{x-1}{3} = \dfrac{-(x-1)}{3}$ or $\dfrac{-x+1}{3}$*

1. $-\dfrac{2}{3}$
2. $\dfrac{3}{-4}$
3. $-\dfrac{-5}{7}$
4. $\dfrac{6}{-7}$
5. $\dfrac{-2}{-7}$
6. $-\dfrac{-3}{-5}$
7. $-\dfrac{6}{-7}$
8. $-\dfrac{-1}{2}$
9. $\dfrac{3x}{-y}$
10. $-\dfrac{2y}{x}$
11. $-\dfrac{-2x^2}{-y^2}$
12. $\dfrac{-x^3}{-y^2}$
13. $\dfrac{y+3}{-y}$
14. $-\dfrac{x+2}{x}$
15. $-\dfrac{x-y}{y}$
16. $\dfrac{2y-x}{-2x}$

Write each fraction on the left as an equal fraction in standard form with the denominator shown on the right. (Assume that no denominator equals 0.)

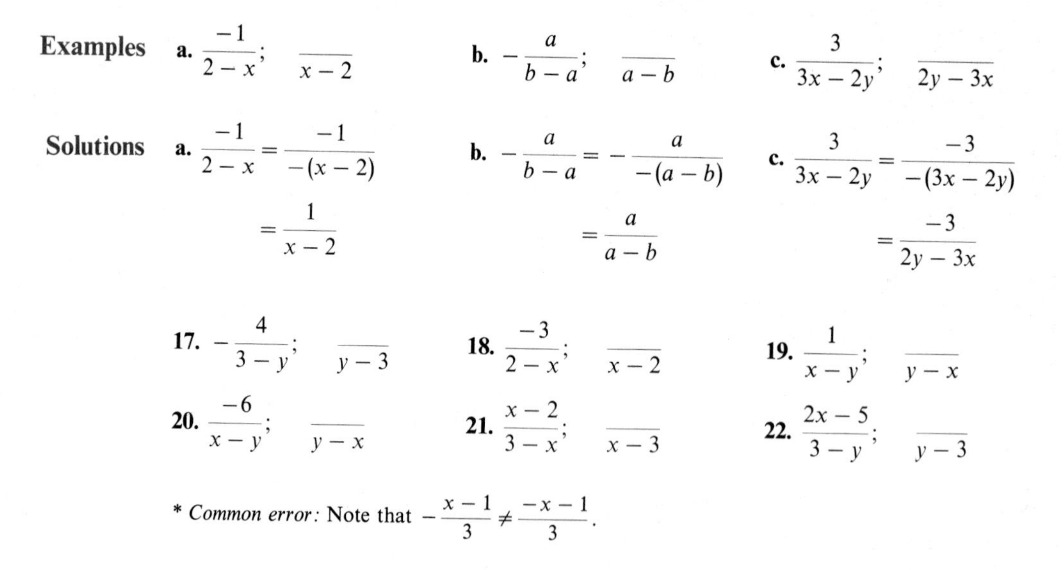

Examples

a. $\dfrac{-1}{2-x}; \quad \dfrac{\quad}{x-2}$

b. $-\dfrac{a}{b-a}; \quad \dfrac{\quad}{a-b}$

c. $\dfrac{3}{3x-2y}; \quad \dfrac{\quad}{2y-3x}$

Solutions

a. $\dfrac{-1}{2-x} = \dfrac{-1}{-(x-2)} = \dfrac{1}{x-2}$

b. $-\dfrac{a}{b-a} = -\dfrac{a}{-(a-b)} = \dfrac{a}{a-b}$

c. $\dfrac{3}{3x-2y} = \dfrac{-3}{-(3x-2y)} = \dfrac{-3}{2y-3x}$

17. $-\dfrac{4}{3-y}; \quad \dfrac{\quad}{y-3}$
18. $\dfrac{-3}{2-x}; \quad \dfrac{\quad}{x-2}$
19. $\dfrac{1}{x-y}; \quad \dfrac{\quad}{y-x}$
20. $\dfrac{-6}{x-y}; \quad \dfrac{\quad}{y-x}$
21. $\dfrac{x-2}{3-x}; \quad \dfrac{\quad}{x-3}$
22. $\dfrac{2x-5}{3-y}; \quad \dfrac{\quad}{y-3}$

* *Common error:* Note that $-\dfrac{x-1}{3} \neq \dfrac{-x-1}{3}$.

23. $\dfrac{-x}{x-y}$; $\dfrac{\quad}{y-x}$ **24.** $\dfrac{x-3}{y-x}$; $\dfrac{\quad}{x-y}$ **25.** $-\dfrac{1}{x-y}$; $\dfrac{\quad}{y-x}$

26. $-\dfrac{x}{x-2y}$; $\dfrac{\quad}{2y-x}$ **27.** $\dfrac{-a}{-3a-b}$; $\dfrac{\quad}{3a+b}$ **28.** $\dfrac{-a}{2b-3a}$; $\dfrac{\quad}{3a-2b}$

B

29. Do all fractions represent rational numbers for all real number replacements of any variables they contain? Support your answer with examples.

30. For what values of the variables are each of the fractions in Problems 23–28 undefined?

3.2

REDUCING AND BUILDING FRACTIONS

A fraction is said to be in lowest terms if the numerator and denominator do not contain common factors. The arithmetic fraction a/b, where a and b are integers and $b \neq 0$, is in lowest terms providing a and b are relatively prime—that is, providing they contain no common integral factor other than 1. If the numerator and denominator of a fraction are polynomials with integral coefficients, then the fraction is said to be in lowest terms if the numerator and denominator do not contain a common polynomial factor with integral coefficients.

To express a given fraction in lowest terms (to **reduce** the fraction), we can factor the numerator and denominator, and then apply the fundamental principle of fractions. For example,

$$\frac{8x^3y}{6x^2y^3} = \frac{4x \cdot 2x^2y}{3y^2 \cdot 2x^2y}$$

$$= \frac{4x}{3y^2} \qquad (x, y \neq 0).$$

If binomial factors are involved, the process is the same. Thus,

$$\frac{12(x+y)(x-y)^2}{15(x+y)^2(x-y)} = \frac{4(x-y)\cdot[3(x+y)(x-y)]}{5(x+y)\cdot[3(x+y)(x-y)]}$$

$$= \frac{4(x-y)}{5(x+y)} \qquad (x \neq y, -y).$$

Diagonal lines are sometimes used to abbreviate this procedure. Instead of writing

$$\frac{y}{y^2} = \frac{1 \cdot y}{y \cdot y} = \frac{1}{y} \qquad (y \neq 0),$$

we can write

$$\frac{y}{y^2} = \frac{\overset{1}{\cancel{y}}}{\underset{y}{\cancel{y^2}}} = \frac{1}{y} \qquad (y \neq 0).$$

Reducing a fraction to lowest terms should be accomplished mentally whenever possible.

Observe that the process of reducing fractions is consistent with the second law of exponents wherever this law is applicable. Thus, we can write

$$\frac{x^3}{x} = \frac{x^2 \cdot x}{1 \cdot x} = x^2 \qquad (x \neq 0)$$

from the fundamental principle of fractions, or

$$\frac{x^3}{x} = x^{3-1} = x^2 \qquad (x \neq 0)$$

directly from the second law of exponents.

The division of a polynomial containing more than one term by a monomial may also be considered a special case of changing a fraction to lowest terms, providing the monomial is contained as a factor in each term of the polynomial. For example, we can write

$$\begin{aligned} \frac{9x^3 - 6x^2 + 3x}{3x} &= \frac{3x(3x^2 - 2x + 1)}{3x} \\ &= \frac{(3x^2 - 2x + 1) \cdot (3x)}{1 \cdot (3x)} \\ &= 3x^2 - 2x + 1 \qquad (x \neq 0). \end{aligned}$$

If the divisor is contained as a factor in the dividend, then the division of one polynomial by another polynomial, where each contains more than one term, can also be considered an example of reducing a fraction to lowest terms. Thus,

$$\begin{aligned} (2x^2 + x - 15) \div (x + 3) &= \frac{2x^2 + x - 15}{x + 3} \\ &= \frac{(2x - 5)(x + 3)}{1 \cdot (x + 3)} \\ &= 2x - 5 \qquad (x \neq -3). \end{aligned}$$

Just as we change fractions to equivalent fractions in lowest terms by applying the fundamental principle in the form

$$\frac{ac}{bc} = \frac{a}{b} \qquad (b, c \neq 0),$$

we can also change fractions to equivalent fractions in higher terms by applying the fundamental principle in the form

$$\frac{a}{b} = \frac{ac}{bc} \qquad (b, c \neq 0).$$

For example, $\frac{1}{2}$ can be changed to an equivalent fraction with a denominator of 8 by multiplying the numerator by 4 and the denominator by 4. Thus,

$$\frac{1}{2} = \frac{1 \cdot 4}{2 \cdot 4} = \frac{4}{8}.$$

This process is called **building** a fraction, and the number 4 is said to be a **building factor**.

In general, when building a/b to an equivalent fraction with bc as a denominator (i.e., $a/b = ?/bc$, we can usually determine the building factor c by inspection, and then multiply the numerator and the denominator of the original fraction by this building factor. If the building factor cannot be obtained by inspection, the desired denominator (bc) can be divided by the denominator of the given fraction (b) to determine the building factor (c).

EXERCISE 3.2

A

Reduce each fraction to lowest terms. (Assume no denominator is 0.)

Examples **a.** $\dfrac{x^4y^3}{xy^5}$ **b.** $\dfrac{a-b}{b^2-a^2}$

Solutions **a.** $\dfrac{x^4y^3}{xy^5} = \dfrac{x^3 \cdot xy^3}{y^2 \cdot xy^3}$

$= \dfrac{x^3}{y^2}$

b. $\dfrac{a-b}{b^2-a^2} = \dfrac{-1(b-a)}{(b+a)(b-a)}$

$= \dfrac{-1}{b+a}$

1. $\dfrac{a^3b^2c^3}{ab^3c}$ **2.** $\dfrac{x^4y^2z}{x^3y^2}$ **3.** $\dfrac{2ab^2}{6a^3b^3c}$

4. $\dfrac{3r}{9r^2t^3}$ **5.** $\dfrac{3x+3y}{x+y}$ **6.** $\dfrac{x^3-x^2}{x-1}$

7. $\dfrac{a-b}{b-a}$ **8.** $\dfrac{x^2-1}{1-x}$ **9.** $\dfrac{(a-b)^2}{b-a}$

10. $\dfrac{(x-y)^3}{y-x}$ **11.** $\dfrac{3-y}{y^2-9}$ **12.** $\dfrac{x^2-16}{4-x}$

Examples **a.** $\dfrac{2x+4}{4}$ **b.** $\dfrac{3x+6}{3}$ **c.** $\dfrac{9x^2+3}{6x+3}$

Solutions

a. $\dfrac{2x+4}{4} = \dfrac{2(x+2)^*}{2(2)} = \dfrac{x+2}{2}$

b. $\dfrac{3x+6}{3} = \dfrac{3(x+2)^\dagger}{3} = x+2$

c. $\dfrac{9x^2+3}{6x+3} = \dfrac{3(3x^2+1)^\ddagger}{3(2x+1)} = \dfrac{3x^2+1}{2x+1}$

13. $\dfrac{4x+6}{6}$ **14.** $\dfrac{2y-8}{8}$ **15.** $\dfrac{9x-3}{9}$

16. $\dfrac{5y-10}{5}$ **17.** $\dfrac{ay-a}{a}$ **18.** $\dfrac{bx^2+b}{bx+b}$

Examples **a.** $\dfrac{2y^3-6y^2+10y}{2y}$ **b.** $\dfrac{y^2+y-6}{y-2}$ **c.** $\dfrac{x^3+8}{x^2+5x+6}$

Solutions

a. $\dfrac{2y^3-6y^2+10y}{2y} = \dfrac{2y(y^2-3y+5)}{2y} = y^2-3y+5$

b. $\dfrac{y^2+y-6}{y-2} = \dfrac{(y+3)(y-2)}{1\cdot(y-2)} = y+3$

c. $\dfrac{x^3+8}{x^2+5x+6} = \dfrac{(x+2)(x^2-2x+4)}{(x+2)(x+3)} = \dfrac{x^2-2x+4}{x+3}$

19. $\dfrac{a^3-3a^2+2a}{a}$ **20.** $\dfrac{3x^3-6x^2+3x}{-3x}$ **21.** $\dfrac{y^2+5y-14}{y-2}$

22. $\dfrac{x^2+5x+6}{x+3}$ **23.** $\dfrac{x^2-5x+4}{x^2-1}$ **24.** $\dfrac{y^2-2y-3}{y^2-9}$

25. $\dfrac{2y^2+y-6}{y^2+y-2}$ **26.** $\dfrac{6x^2-x-1}{2x^2+9x-5}$ **27.** $\dfrac{x^2+xy-2y^2}{x^2-y^2}$

28. $\dfrac{4x^2-9y^2}{2x^2+xy-6y^2}$ **29.** $\dfrac{8y^3-27}{2y-3}$ **30.** $\dfrac{(x+y)^3-8z^3}{x+y-2z}$

31. $\dfrac{x^2+ax+xy+ay}{x+a}$ **32.** $\dfrac{2x^2+6x-xy-3y}{2x-y}$

* *Common error:* Note that $\dfrac{2x+4}{4} \neq \dfrac{2x+\not{4}}{\not{4}}$.

† *Common error:* Note that $\dfrac{3x+6}{3} \neq \dfrac{\not{3}x+6}{\not{3}}$.

‡ *Common error:* Note that $\dfrac{9x^2+3}{6x+3} \neq \dfrac{9x^2+\not{3}}{6x+\not{3}}$.

Express each given fraction as an equivalent fraction with the given denominator. (Assume no denominator is 0.)

Examples **a.** $\frac{3}{4xy} = \frac{?}{8x^2y^2}$

b. $\frac{a+1}{3} = \frac{?}{6(a-3)}$

Solutions **a.** Obtain the building factor.

$$(8x^2y^2 \div 4xy = 2xy)$$

Multiply the numerator and the denominator of the given fraction by the building factor.

$$\frac{3}{4xy} = \frac{3(2xy)}{4xy(2xy)}$$

$$= \frac{6xy}{8x^2y^2}$$

b. Obtain the building factor.

$$[6(a-3) \div 3 = 2(a-3)]$$

Multiply the numerator and the denominator of the given fraction by the building factor.

$$\frac{a+1}{3} = \frac{(a+1)(2)(a-3)}{3(2)(a-3)}$$

$$= \frac{2a^2-4a-6}{6(a-3)}$$

33. $\frac{2}{3} = \frac{?}{9}$ **34.** $\frac{3}{4} = \frac{?}{8}$ **35.** $\frac{-15}{7} = \frac{?}{14}$ **36.** $\frac{-12}{5} = \frac{?}{20}$

37. $4 = \frac{?}{5}$ **38.** $6 = \frac{?}{7}$ **39.** $\frac{2}{6x} = \frac{?}{18x}$ **40.** $\frac{5}{3y} = \frac{?}{21y}$

41. $\frac{-a^2}{b^2} = \frac{?}{b^3}$ **42.** $\frac{-a}{b} = \frac{?}{ab^2}$ **43.** $y = \frac{?}{xy}$ **44.** $x = \frac{?}{xy^3}$

45. $\frac{1}{3} = \frac{?}{3(x+y)}$ **46.** $\frac{2}{5} = \frac{?}{5(x-y)}$ **47.** $\frac{a-1}{2} = \frac{?}{6(a+1)}$ **48.** $\frac{a-2}{3} = \frac{?}{6(a-3)}$

Example $\frac{3}{2a-2b} = \frac{?}{4a^2-4b^2}$

Solution Factor denominators.

$$\frac{3}{2(a-b)} = \frac{?}{4(a-b)(a+b)}$$

Obtain the building factor.

$$[4(a-b)(a+b) \div 2(a-b) = 2(a+b)]$$

Multiply the numerator and the denominator of the given fraction by the building factor $2(a+b)$

$$\frac{3}{2a-2b} = \frac{3 \cdot 2(a+b)}{2(a-b) \cdot 2(a+b)} = \frac{6(a+b)}{4a^2-4b^2}$$

49. $\dfrac{3}{a-b} = \dfrac{?}{a^2-b^2}$

50. $\dfrac{5}{2a+b} = \dfrac{?}{4a^2-b^2}$

51. $\dfrac{3x}{y+2} = \dfrac{?}{y^2-y-6}$

52. $\dfrac{5x}{y+3} = \dfrac{?}{y^2+y-6}$

53. $\dfrac{-2}{x+1} = \dfrac{?}{x^2+3x+2}$

54. $\dfrac{-3}{a+2} = \dfrac{?}{a^2+3a+2}$

55. $\dfrac{2}{a-b} = \dfrac{?}{b^2-a^2}$

56. $\dfrac{7}{x-y} = \dfrac{?}{y^2-x^2}$

57. $\dfrac{x}{2-x} = \dfrac{?}{x^2-3x+2}$

58. $\dfrac{y}{3-2y} = \dfrac{?}{2y^2-y-3}$

59. $\dfrac{3}{a+3} = \dfrac{?}{a^3+27}$

60. $\dfrac{2}{2x-3y} = \dfrac{?}{8x^3-27y^3}$

B

61. $\dfrac{3a-5b}{2x-3y} = \dfrac{?}{2ax+2bx-3ay-3by}$

62. $\dfrac{5a-3b}{3x+2y} = \dfrac{?}{6bx-3ax+4by-2ay}$

63. $\dfrac{3a+b}{4y+7x} = \dfrac{?}{-4by-7bx+12ay+21ax}$

64. $\dfrac{4a+3b}{x-8y} = \dfrac{?}{4ax-3bx-32ay+24by}$

3.3

QUOTIENTS OF POLYNOMIALS

On page 72, we saw that the quotient of polynomials $(2x^2+x-15) \div (x+3)$ could be written in fraction form as

$$\frac{2x^2+x-15}{x+3}$$

and reduced to $2x-5$. The same result can be obtained using a method similar to the long division process used in arithmetic. First write

$$x+3\,\overline{)\,2x^2+x-15}$$

and then divide $2x^2$ by x. Subtract the product of $2x$ and $x+3$ from $2x^2+x$, and "bring down" -15:

$$\begin{array}{r@{}l} & 2x \\ x+3\,\big) & \overline{2x^2+\ \ x-15} \\ & \underline{2x^2+6x} \\ & \quad\ -5x-15 \end{array}$$

Then divide $-5x$ by x. Subtract the product of -5 and $x+3$ from $-5x-15$.

$$\begin{array}{r@{}l} & 2x-5 \\ x+3\,\big) & \overline{2x^2+\ \ x-15} \\ & \underline{2x^2+6x} \\ & \quad\ -5x-15 \\ & \quad\ \underline{-5x-15} \end{array}$$

This procedure is most useful when the divisor is not a factor of the dividend. If such is the case, the division process will produce a remainder that may be expressed by a fraction. For example, using the above process for the quotient

$$\frac{x^2 + 2x + 2}{x + 1},$$

we have

$$\begin{array}{r|l} & x + 1 \\ \hline x + 1 & x^2 + 2x + 2 \\ & x^2 + x \\ \hline & \quad x + 2 \\ & \quad x + 1 \\ \hline & \qquad 1 \end{array}$$

and thus

$$\frac{x^2 + 2x + 2}{x + 1} = x + 1 + \frac{1}{x + 1} \qquad (x \neq -1).$$

We call an expression such as

$$x + 1 + \frac{1}{x + 1}$$

a **mixed expression**, just as such symbols as $3\frac{1}{2}$ denote **mixed numbers**.

If the divisor in a quotient of polynomials is a monomial, we can rewrite the quotient as the sum of two or more fractions; namely,

$$\frac{a + b + c}{g} = \frac{a}{g} + \frac{b}{g} + \frac{c}{g}.$$

As an illustration, we have

$$\begin{aligned} \frac{9x^3 - 6x^2 + 4}{3x} &= \frac{9x^3}{3x} - \frac{6x^2}{3x} + \frac{4}{3x} \\ &= \frac{3x^2 \cdot 3x}{1 \cdot 3x} - \frac{2x \cdot 3x}{1 \cdot 3x} + \frac{4}{3x} \\ &= 3x^2 - 2x + \frac{4}{3x} \qquad (x \neq 0). \end{aligned}$$

To avoid the necessity of always having to note restrictions on divisors (denominators), we shall assume in the remaining exercise sets in this chapter that no denominator is 0.

EXERCISE 3.3

A

Divide.

Examples

a. $\dfrac{4y^2 - 4y - 5}{2y + 1}$

b. $\dfrac{16y^3 - 3y + 5}{y - 2}$

Solutions

a.

$$\begin{array}{r|l} & 2y \quad - 3 \\ \hline 2y + 1 & 4y^2 - 4y - 5 \\ & 4y^2 + 2y \\ \hline & \quad -6y - 5 \\ & \quad -6y - 3 \\ \hline & \qquad\quad -2 \end{array}$$

$$\frac{4y^2 - 4y - 5}{2y + 1} = 2y - 3 + \frac{-2}{2y + 1}$$

b.

$$\begin{array}{r|l} & 16y^2 + 32y + 61 \\ \hline y - 2 & 16y^3 + 0y^2 - 3y + 5 \\ & 16y^3 - 32y^2 \\ \hline & \qquad 32y^2 - 3y \\ & \qquad 32y^2 - 64y \\ \hline & \qquad\qquad 61y + 5 \\ & \qquad\qquad 61y - 122 \\ \hline & \qquad\qquad\quad 127 \end{array}$$

$$\frac{16y^3 - 3y + 5}{y - 2} = 16y^2 + 32y + 61 + \frac{127}{y - 2}$$

1. $\dfrac{4y^2 + 12y + 5}{2y + 1}$

2. $\dfrac{2n^2 + 13n - 7}{2n - 1}$

3. $\dfrac{4t^2 - 4t - 5}{2t - 1}$

4. $\dfrac{2x^2 - 3x - 15}{2x + 5}$

5. $\dfrac{x^3 + 2x^2 + x + 1}{x - 2}$

6. $\dfrac{2x^3 - 3x^2 - 2x + 4}{x + 1}$

7. $\dfrac{a^4 - 3a^2 + 2a - 1}{a + 3}$

8. $\dfrac{2b^4 + 2b^2 + 3}{b - 4}$

9. $\dfrac{x^4 - 1}{x - 2}$

10. $\dfrac{y^5 + 1}{y - 1}$

11. $\dfrac{8z^3 + 6z^2 + z - 2}{2z + 1}$

12. $\dfrac{10t^4 - 3t^3 - 23t^2 + 7}{2t + 3}$

Examples

a. $\dfrac{2y^3 - 6y^2 + 4}{y}$

b. $\dfrac{27x^2y^2 + 18xy - 3}{9xy}$

Solutions

a. $\dfrac{2y^3 - 6y^2 + 4}{y}$

$$= \frac{2y^3}{y} - \frac{6y^2}{y} + \frac{4}{y}$$

$$= \frac{2y^2 \cdot y}{1 \cdot y} - \frac{6y \cdot y}{1 \cdot y} + \frac{4}{y}$$

$$= 2y^2 - 6y + \frac{4}{y}$$

b. $\dfrac{27x^2y^2 + 18xy - 3}{9xy}$

$$= \frac{27x^2y^2}{9xy} + \frac{18xy}{9xy} - \frac{3}{9xy}$$

$$= \frac{3xy \cdot 9xy}{1 \cdot 9xy} + \frac{2 \cdot 9xy}{9xy} - \frac{3 \cdot 1}{3 \cdot 3xy}$$

$$= 3xy + 2 - \frac{1}{3xy}$$

13. $\dfrac{8a^2 + 4a + 1}{2}$ **14.** $\dfrac{15t^3 - 12t^2 + 5t}{3t^2}$ **15.** $\dfrac{7y^4 - 14y^2 + 3}{7y^2}$

16. $\dfrac{21n^4 + 14n^2 - 7}{7n^2}$ **17.** $\dfrac{18r^2s^2 - 15rs + 6}{3rs}$ **18.** $\dfrac{12x^3 - 8x^2 + 3x}{4x}$

19. $\dfrac{8a^2x^2 - 4ax^2 + ax}{2ax}$ **20.** $\dfrac{9a^2b^2 + 3ab^2 + 4a^2b}{ab^2}$ **21.** $\dfrac{25m^6 - 15m^3 + 7}{-5m^3}$

22. $\dfrac{36t^5 + 24t^3 - 12t}{-12t^2}$ **23.** $\dfrac{40m^4 - 25m^2 + 7m}{5m^2}$ **24.** $\dfrac{15s^{10} - 21s^5 + 6}{3s^2}$

B

Example $\dfrac{z^4 - 3z^3 + 2z^2 - 3z + 1}{z^2 + 2z - 1}$

Solution

$$
\begin{array}{r|l}
 & z^2 - 5z + 13 \\
z^2 + 2z - 1 & z^4 - 3z^3 + 2z^2 - 3z + 1 \\
 & z^4 + 2z^3 - z^2 \\
 & \quad - 5z^3 + 3z^2 - 3z \\
 & \quad - 5z^3 - 10z^2 + 5z \\
 & \qquad 13z^2 - 8z + 1 \\
 & \qquad 13z^2 + 26z - 13 \\
 & \qquad\quad - 34z + 14
\end{array}
$$

$$\frac{z^4 - 3z^3 + 2z^2 - 3z + 1}{z^2 + 2z - 1} = z^2 - 5z + 13 + \frac{-34z + 14}{z^2 + 2z - 1}$$

25. $\dfrac{x^3 - 3x^2 + 2x + 5}{x^2 - 2x + 7}$ **26.** $\dfrac{2y^3 + 5y^2 - 3y + 2}{y^2 - y - 3}$

27. $\dfrac{4a^4 + 3a^3 - 2a + 1}{a^2 + 3a - 1}$ **28.** $\dfrac{2b^4 - 3b^2 + b + 2}{b^2 + b - 3}$

29. $\dfrac{t^4 - 3t^3 + 2t^2 - 2t + 1}{t^3 - 2t^2 + t + 2}$ **30.** $\dfrac{r^4 + r^3 - 2r^2 + r + 5}{r^3 + 2r + 3}$

31. Determine k so that the polynomial $x^3 - 3x + k$ has $x - 2$ as a factor.

32. Determine k so that the polynomial $x^3 + 2x^2 + k$ has $x + 3$ as a factor.

3.4

SUMS AND DIFFERENCES

Although we have defined fractions as symbols, they represent real numbers for permissible real number replacements of any variables involved. Therefore, we can use the properties of real numbers to rewrite sums involving fractions in simpler form. For example, since

$$\frac{a}{c} = a\left(\frac{1}{c}\right) \quad \text{and} \quad \frac{b}{c} = b\left(\frac{1}{c}\right),$$

for $c \neq 0$,

$$\frac{a}{c} + \frac{b}{c}$$

can be rewritten as

$$a\left(\frac{1}{c}\right) + b\left(\frac{1}{c}\right),$$

and then, by the distributive property, as

$$(a + b)\frac{1}{c}.$$

This is equal to

$$\frac{a + b}{c}.$$

Hence, if a, b, and c are real numbers, and $c \neq 0$, then

$$\boldsymbol{\frac{a}{c} + \frac{b}{c} = \frac{a + b}{c}}.$$

If the fractions in a sum have unlike denominators, we can build the fractions to equivalent fractions having common denominators and then rewrite the sum as above.

In general, for the sum of two fractions with different denominators we have

$$\frac{a}{b} + \frac{c}{d} = \frac{a \cdot d}{b \cdot d} + \frac{c \cdot b}{d \cdot b},$$

from which

$$\boldsymbol{\frac{a}{b} + \frac{c}{d} = \frac{ad + bc}{bd} \qquad (b, d \neq 0).} \tag{1}$$

A difference

$$\frac{a}{b} - \frac{c}{d} \qquad (b, d \neq 0)$$

can be viewed as the sum

$$\frac{a}{b} + \frac{-c}{d}$$

and then, using the fundamental principle of fractions, can be written as

$$\frac{a \cdot d}{b \cdot d} + \frac{-c \cdot b}{d \cdot b} = \frac{ad - cb}{bd} \qquad (b, d \neq 0),$$

from which we have

$$\textbf{(2)} \qquad \frac{a}{b} - \frac{c}{d} = \frac{ad - bc}{bd} \qquad (b, d \neq 0).$$

Note that forms (1) and (2) cannot be used directly if we want to rewrite expressions which contain three or more terms.

In rewriting sums of fractions with unlike denominators by building their terms into fractions with common denominators, any such denominator can be used. By using the **least common multiple (LCM)** of the denominators (called the **least common denominator, LCD**), however, we can obtain a simpler form for the sum than if any other common denominator is used. The LCM of two or more natural numbers is the smallest natural number that is exactly divisible by each of the given numbers. Thus, 24 is the LCM of 3 and 8, because 24 is the smallest natural number each will divide into without a remainder. To find the LCM of a set of natural numbers:

1. Express each number in completely factored form.
2. Write as factors of a product each *different* prime factor occurring in any of the numbers, including each factor the greatest number of times it occurs in any one of the given numbers.

For example, the LCM of 12, 9, and 15 can be obtained as shown below.

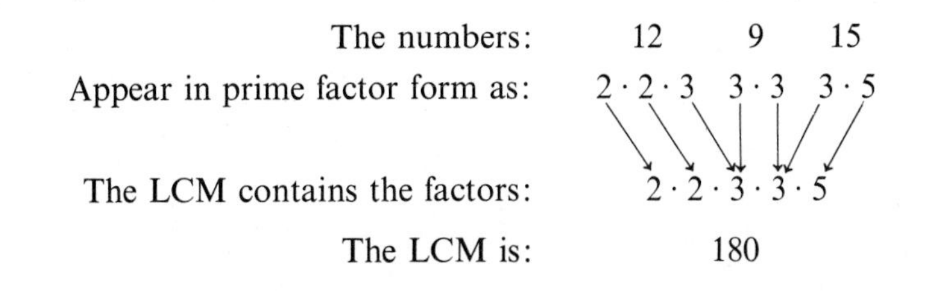

The factors 2 and 3 are each used twice because 2 appears twice as a factor of 12 and 3 appears twice as a factor of 9.

We can define the LCM of a set of polynomials in a manner analogous to that just described—namely, the polynomial of lowest degree yielding a polynomial quotient upon division by each of the given polynomials.

We can find the LCM of a set of polynomials with integral coefficients in a manner comparable to that used with a set of natural numbers. For example, the LCM of x^2, $x^2 - 9$, and $x^3 - x^2 - 6x$ can be found as follows:

Factor each polynomial.

x^2	$x^2 - 9$	$x^3 - x^2 - 6x$
$x \cdot x$	$(x - 3)(x + 3)$	$x(x - 3)(x + 2)$

Hence, the LCM is $x^2(x - 3)(x + 3)(x + 2)$. The factor x is used twice since it appears two times as a factor of the first polynomial, x^2.

Since polynomials are easier to work with in factored form, it is usually advantageous to leave the LCM of a set of polynomials in factored form rather than carry out the indicated multiplication. Sometimes it is also convenient to leave numerators or denominators of fractions in factored form.

EXERCISE 3.4

A

Find the LCM.

Examples

a. 24, 30, 20

b. $2a, 4b, 6ab^2$

Solutions

a. $24 = 2 \cdot 2 \cdot 2 \cdot 3,$
$30 = 2 \cdot 3 \cdot 5,$
$20 = 2 \cdot 2 \cdot 5$

$\text{LCM} = 2^3 \cdot 3 \cdot 5 = 120$

b. $2a = 2 \cdot a,$
$4b = 2 \cdot 2 \cdot b,$
$6ab^2 = 2 \cdot 3 \cdot a \cdot b \cdot b$

$\text{LCM} = 2^2 \cdot 3ab^2 = 12ab^2$

1. 4, 6, 10
2. 3, 4, 5
3. 6, 8, 15
4. 4, 15, 18
5. 14, 21, 36
6. 4, 11, 22
7. $2ab, 6b^2$
8. $12xy, 24x^3y^2$
9. $6xy, 8x^2, 3xy^2$
10. $7x, 8y, 6z$
11. $(a - b), a(a - b)^2$
12. $6(x + y)^2, 4xy^2$

Examples

a. $2a - 2, a - 1$

b. $x^2 - 1, 2(x - 1)^2$

Solutions

a. $2a - 2 = 2(a - 1),$
$a - 1 = (a - 1)$

$\text{LCM} = 2(a - 1)$

b. $x^2 - 1 = (x - 1)(x + 1),$
$2(x - 1)^2 = 2(x - 1)(x - 1)$

$\text{LCM} = 2(x - 1)^2(x + 1)$

13. $a^2 - b^2, a - b$
14. $x + 2, x^2 - 4$
15. $a^2 + 5a + 4, (a + 1)^2$
16. $x^2 - 3x + 2, (x - 1)^2$
17. $x^2 + 3x - 4, (x - 1)^2$
18. $x^2 - x - 2, (x - 2)^2$

19. $x^2 - x, (x-1)^3$ **20.** $y^2 + 2y, (y+2)^2$ **21.** $4a^2 - 4, (a-1)^2, 2$

22. $3x^2 - 3, (x-1)^2, 4$ **23.** $x^3, x^2 - x, (x-1)^2$ **24.** $y, y^3 - y, (y-1)^3$

Write each sum or difference as a single fraction in lowest terms.

Examples **a.** $\frac{x}{7} + \frac{2x}{7} - \frac{4}{7}$ **b.** $\frac{a+1}{b} - \frac{a-1}{b}$

Solutions **a.** First write $-\frac{4}{7}$ in standard form.

$$\frac{x}{7} + \frac{2x}{7} - \frac{4}{7} = \frac{x}{7} + \frac{2x}{7} + \frac{-4}{7}$$

$$= \frac{x + 2x - 4}{7}$$

$$= \frac{3x - 4}{7}$$

b. First write $-\frac{a-1}{b}$ in standard form.

$$\frac{a+1}{b} - \frac{a-1}{b} = \frac{a+1}{b} + \frac{-(a-1)}{b}$$

$$= \frac{a + 1 - (a-1)}{b}$$

$$= \frac{a + 1 - a + 1}{b} = \frac{2}{b}$$

25. $\frac{x}{2} - \frac{3}{2}$ **26.** $\frac{y}{7} - \frac{5}{7}$ **27.** $\frac{a}{6} + \frac{b}{6} - \frac{c}{6}$

28. $\frac{x}{3} - \frac{2y}{3} + \frac{z}{3}$ **29.** $\frac{x-1}{2y} + \frac{x}{2y}$ **30.** $\frac{y+1}{b} + \frac{y-1}{b}$

31. $\frac{3}{x+2y} - \frac{x+3}{x+2y} - \frac{x-1}{x+2y}$ **32.** $\frac{2}{a-3b} - \frac{b-2}{a-3b} + \frac{b}{a-3b}$

33. $\frac{a+1}{a^2 - 2a + 1} + \frac{5 - 3a}{a^2 - 2a + 1}$ **34.** $\frac{x+4}{x^2 - x + 2} + \frac{2x - 3}{x^2 - x + 2}$

Example $\frac{5}{a^2 - 9} - \frac{1}{a-3}$

Solution Factor denominators and write in standard form.

$$\frac{5}{a^2 - 9} - \frac{1}{a-3} = \frac{5}{(a-3)(a+3)} + \frac{-1}{a-3}$$

Build each fraction to a fraction with the denominator $(a-3)(a+3)$.

$$\frac{5}{a^2 - 9} - \frac{1}{a-3} = \frac{5}{(a-3)(a+3)} + \frac{-1(a+3)}{(a-3)(a+3)}$$

$$= \frac{5 - (a+3)}{(a-3)(a+3)} = \frac{2 - a}{(a-3)(a+3)}$$

The denominator $(a-3)(a+3)$ can also be expressed in the alternative form $a^2 - 9$.

35. $\dfrac{2}{ax} - \dfrac{2}{x}$

36. $\dfrac{3}{by} - \dfrac{2}{b}$

37. $\dfrac{a-2}{6} - \dfrac{a+1}{3}$

38. $\dfrac{x-3}{4} + \dfrac{5-x}{10}$

39. $\dfrac{2x-y}{2y} + \dfrac{x+y}{x}$

40. $\dfrac{3a+2b}{3b} - \dfrac{a+2b}{6a}$

41. $\dfrac{7}{5x-10} - \dfrac{5}{3x-6}$

42. $\dfrac{2}{y+2} - \dfrac{3}{y+3}$

43. $\dfrac{5}{2x-6} - \dfrac{3}{x+3}$

44. $\dfrac{2}{3-x} - \dfrac{1}{x-3}$

45. $\dfrac{7}{r-3} + \dfrac{3}{3-r}$

46. $\dfrac{1}{2x+1} - \dfrac{3}{x-2}$

47. $\dfrac{a}{3x+2} - \dfrac{a}{x-1}$

48. $\dfrac{a+1}{a+2} - \dfrac{a+2}{a+3}$

49. $\dfrac{5x-y}{3x+y} - \dfrac{6x-5y}{2x-y}$

50. $\dfrac{x+2y}{2x-y} - \dfrac{2x+y}{x-2y}$

51. $\dfrac{2}{x^2-x-2} - \dfrac{2}{x^2+2x+1}$

52. $\dfrac{1}{b^2-1} - \dfrac{1}{b^2+2b+1}$

53. $\dfrac{y}{y^2-16} - \dfrac{y+1}{y^2-5y+4}$

54. $\dfrac{8}{a^2-4b^2} - \dfrac{2}{a^2-5ab+6b^2}$

55. $\dfrac{y}{y^2-9} - \dfrac{1}{y^2+4y-21}$

56. $\dfrac{3a}{a^2+3a-10} - \dfrac{2a}{a^2+a-6}$

57. $x + \dfrac{1}{x-1} - \dfrac{1}{(x-1)^2}$

58. $y - \dfrac{2y}{y^2-1} + \dfrac{3}{y+1}$

B

59. $\dfrac{1}{z^2-7z+12} + \dfrac{2}{z^2-5z+6} - \dfrac{3}{z^2-6z+8}$

60. $\dfrac{4}{a^2-4b^2} + \dfrac{2}{a^2+3ab+2b^2} + \dfrac{4}{a^2-ab-2b^2}$

61. $\dfrac{1}{(a-b)(b-c)} + \dfrac{1}{(b-c)(c-a)} + \dfrac{1}{(c-a)(a-b)}$

62. $\dfrac{1}{(a+b)(b+c)} - \dfrac{1}{(a+c)(b+c)} - \dfrac{1}{(a+b)(a+c)}$

63. $\dfrac{y}{y+1} - \dfrac{1}{y^3+1}$

64. $\dfrac{2}{y^3-1} - \dfrac{y}{y-1}$

65. $\dfrac{-3y}{2y-3} - \dfrac{5}{8y^3-27}$

66. $\dfrac{1}{y^2+4y+16} + \dfrac{y}{y^3-64}$

67. For what value(s) of the variables is each of the sums or differences in Problems 47–53 undefined?

68. A set of fractions has an infinite number of common denominators. Why is it convenient to use the least common denominator in finding sums or differences of fractions?

3.5

PRODUCTS AND QUOTIENTS

To discover how to rewrite products involving fractions, let us examine the product

$$\frac{2}{3} \cdot \frac{5}{7} = p$$

and inquire into the nature of p. We have

$\frac{2}{3} \cdot \frac{5}{7} = p$	Given
$3 \cdot 7 \cdot \frac{2}{3} \cdot \frac{5}{7} = 3 \cdot 7 \cdot p$	Multiplication property of equality
$3 \cdot 7 \cdot 2 \cdot \frac{1}{3} \cdot 5 \cdot \frac{1}{7} = 3 \cdot 7 \cdot p$	$\frac{a}{b} = a \cdot \frac{1}{b}$ and $\frac{c}{d} = c \cdot \frac{1}{d}$
$2 \cdot 5 \cdot \left(3 \cdot \frac{1}{3}\right) \cdot \left(7 \cdot \frac{1}{7}\right) = 3 \cdot 7 \cdot p$	Associative and commutative properties of multiplication
$2 \cdot 5 \cdot 1 \cdot 1 = 3 \cdot 7 \cdot p$	Reciprocal property
$2 \cdot 5 = 3 \cdot 7 \cdot p$	Identity element for multiplication
$p = \frac{2 \cdot 5}{3 \cdot 7}$	Definition of a quotient
$\frac{2}{3} \cdot \frac{5}{7} = \frac{2 \cdot 5}{3 \cdot 7} = \frac{10}{21}$	Substitution for p and simplification

Using a similar argument, it follows that if a, b, c, and d are real numbers and $b, d \neq 0$, then

$$\frac{a}{b} \cdot \frac{c}{d} = \frac{ac}{bd}.$$

For example,

$$\begin{aligned} \frac{6x^2}{y} \cdot \frac{xy}{2} &= \frac{6x^3y}{2y} \\ &= \frac{3x^3 \cdot 2y}{1 \cdot 2y} \\ &= 3x^3 \qquad (y \neq 0). \end{aligned}$$

If any of the fractions, or any of the factors of the numerators or denominators of the fractions, have negative signs, it is advisable to proceed as if all the signs were positive and then attach the appropriate sign to the simplified product.

If there is an even number of negative signs involved, the result has a positive sign; if there is an odd number of negative signs involved, the result has a negative sign. For example,

$$\frac{-x}{y^2}\cdot\frac{-2y^3}{x^2}=\frac{2\cdot y\cdot xy^2}{x\cdot xy^2}=\frac{2y}{x}\qquad (x, y \neq 0)$$

and

$$\frac{-4x^2}{3y}\cdot\frac{-y}{2x}\cdot\frac{-3y}{5x}=\frac{-2y\cdot 2\cdot 3x^2y}{5\cdot 2\cdot 3x^2y}=\frac{-2y}{5}\qquad (x, y \neq 0).$$

When rewriting quotients

$$\frac{a}{b}\div\frac{c}{d}\qquad (b, c, d \neq 0),$$

we seek a quotient q such that

$$\left(\frac{c}{d}\right)q=\frac{a}{b}\qquad (b, c, d \neq 0).$$

To obtain q in terms of the other variables, we use the multiplication law of equality and multiply each member of this equality by d/c,

$$\left(\frac{d}{c}\right)\left(\frac{c}{d}\right)q=\left(\frac{d}{c}\right)\left(\frac{a}{b}\right),$$

from which

$$q=\left(\frac{d}{c}\right)\left(\frac{a}{b}\right)=\left(\frac{a}{b}\right)\left(\frac{d}{c}\right).$$

Therefore it follows that if a, b, c, and d are real numbers, and $b, c, d \neq 0$, then

$$(1)\qquad \boldsymbol{\frac{a}{b}\div\frac{c}{d}=\frac{a}{b}\cdot\frac{d}{c}}.$$

For example,

$$\begin{aligned}\frac{2x^3}{3y}\div\frac{4x}{5y^2}&=\frac{2x^3}{3y}\cdot\frac{5y^2}{4x}\\&=\frac{5x^2y\cdot 2xy}{6\cdot 2xy}\\&=\frac{5x^2y}{6}\qquad (x, y \neq 0).\end{aligned}$$

As special cases of (1), observe that

$$a\div\frac{c}{d}=\frac{a}{1}\cdot\frac{d}{c}=\frac{ad}{c},\qquad \frac{a}{b}\div c=\frac{a}{b}\cdot\frac{1}{c}=\frac{a}{bc},$$

and

$$1 \div \frac{a}{b} = 1 \cdot \frac{b}{a} = \frac{b}{a}.$$

For example,

$$2 \div \frac{3}{4} = \frac{2 \cdot 4}{3} = \frac{8}{3}, \qquad \frac{5}{7} \div 3 = \frac{5}{7} \cdot \frac{1}{3} = \frac{5}{21},$$

and

$$1 \div \frac{2}{3} = 1 \cdot \frac{3}{2} = \frac{3}{2}.$$

EXERCISE 3.5

A

Write each product as a single fraction in lowest terms.

Examples **a.** $\frac{-3c^2}{5ab} \cdot \frac{10a^2b}{9c}$ **b.** $\frac{4a^2 - 1}{a^2 - 4} \cdot \frac{a^2 + 2a}{4a + 2}$

Solutions **a.** $\frac{-3c^2}{5ab} \cdot \frac{10a^2b}{9c} = \frac{-2ac \cdot 15abc}{3 \cdot 15abc}$

$$= \frac{-2ac}{3}$$

b. $\frac{4a^2 - 1}{a^2 - 4} \cdot \frac{a^2 + 2a}{4a + 2}$

$$= \frac{(2a-1)(2a+1)}{(a-2)(a+2)} \cdot \frac{a(a+2)}{2(2a+1)}$$

$$= \frac{a(2a-1)(2a+1)(a+2)}{2(a-2)(2a+1)(a+2)}$$

$$= \frac{a(2a-1)}{2(a-2)}$$

1. $\frac{16}{38} \cdot \frac{19}{12}$ **2.** $\frac{4}{15} \cdot \frac{3}{16}$ **3.** $\frac{21}{4} \cdot \frac{2}{15}$

4. $\frac{7}{8} \cdot \frac{48}{64}$ **5.** $\frac{24}{3} \cdot \frac{20}{36} \cdot \frac{3}{4}$ **6.** $\frac{3}{10} \cdot \frac{16}{27} \cdot \frac{30}{36}$

7. $\frac{-12a^2b}{5c} \cdot \frac{10b^2c}{24a^3b}$ **8.** $\frac{a^2}{xy} \cdot \frac{3x^3y}{4a}$ **9.** $\frac{-2ab}{7c} \cdot \frac{3c^2}{4a^3} \cdot \frac{-6a}{15b^2}$

10. $\frac{10x}{12y} \cdot \frac{3x^2z}{5x^3z} \cdot \frac{6y^2x}{3yz}$ **11.** $5a^2b^2 \cdot \frac{1}{a^3b^3}$ **12.** $15x^2y \cdot \frac{3}{45xy^2}$

13. $\frac{5x+25}{2x} \cdot \frac{4x}{2x+10}$

14. $\frac{3y}{4xy-6y^2} \cdot \frac{2x-3y}{12x}$

15. $\frac{4a^2-1}{a^2-16} \cdot \frac{a^2-4a}{2a+1}$

16. $\frac{9x^2-25}{2x-2} \cdot \frac{x^2-1}{6x-10}$

17. $\frac{x^2-x-20}{x^2+7x+12} \cdot \frac{(x+3)^2}{(x-5)^2}$

18. $\frac{4x^2+8x+3}{2x^2-5x+3} \cdot \frac{6x^2-9x}{1-4x^2}$

19. $\frac{x^2-6x+5}{x^2+2x-3} \cdot \frac{x^2-4x-21}{x^2-10x+25}$

20. $\frac{x^2-x-2}{x^2+4x+3} \cdot \frac{x^2-4x-5}{x^2-3x-10}$

21. $\frac{2x^2-x-6}{3x^2-4x+1} \cdot \frac{3x^2+7x+2}{2x^2+7x+6}$

22. $\frac{3x^2-7x-6}{2x^2-x-1} \cdot \frac{2x^2-9x-5}{3x^2-13x-10}$

23. $\frac{7a+14}{14a-28} \cdot \frac{4-2a}{a+2} \cdot \frac{a-3}{a+1}$

24. $\frac{5x^2-5x}{3} \cdot \frac{x^2-9x-10}{4x-40} \cdot \frac{y^2}{2-2x^2}$

Write each quotient as a single fraction in lowest terms.

Examples a. $\frac{a^2b}{c} \div \frac{ab^3}{c^2}$ b. $\frac{3a^2-3}{2a+2} \div \frac{3a-3}{2}$

Solutions a. $\frac{a^2b}{c} \div \frac{ab^3}{c^2} = \frac{a^2b}{c} \cdot \frac{c^2}{ab^3}$

$$= \frac{ac \cdot abc}{b^2 \cdot abc}$$

$$= \frac{ac}{b^2}$$

b. $\frac{3a^2-3}{2a+2} \div \frac{3a-3}{2}$

$$= \frac{3(a-1)(a+1)}{2(a+1)} \cdot \frac{2}{3(a-1)}$$

$$= \frac{2 \cdot 3(a-1)(a+1)}{2 \cdot 3(a-1)(a+1)} = 1$$

25. $\frac{3}{4} \div \frac{9}{16}$

26. $\frac{2}{3} \div \frac{9}{15}$

27. $\frac{xy}{a^2b} \div \frac{x^3y^2}{ab}$

28. $\frac{9ab^3}{x} \div \frac{3}{2x^3}$

29. $28x^2y^3 \div \frac{21x^2y^2}{5a}$

30. $24a^3b \div \frac{3a^2b}{7x}$

31. $\frac{4x-8}{3y} \div (6x-12)$

32. $\frac{6y-27}{5x} \div (4y-18)$

33. $\frac{a^2-a-6}{a^2+2a-15} \div \frac{a^2-4}{a^2+6a+5}$

34. $\frac{a^2+2a-15}{a^2+3a-10} \div \frac{a^2-9}{a^2-9a+14}$

35. $\frac{x^2+x-2}{x^2+2x-3} \div \frac{x^2+7x+10}{x^2-2x-15}$

36. $\frac{x^2+6x-7}{x^2+x-2} \div \frac{x^2+5x-14}{x^2-3x-10}$

37. $\frac{10x^2-13x-3}{2x^2-x-3} \div \frac{5x^2-9x-2}{3x^2+2x-1}$

38. $\frac{9x^2+3x-2}{12x^2+5x-2} \div \frac{9x^2-6x+1}{8x^2+10x-3}$

B

39. $\frac{x^3 + y^3}{x} \div \frac{x + y}{3x}$

40. $\frac{8x^3 - y^3}{x + y} \div \frac{2x - y}{x^2 - y^2}$

41. $\frac{xy - 3x + y - 3}{x - 2} \div \frac{x + 1}{x^2 - 4}$

42. $\frac{2xy + 4x + 3y + 6}{2x + 3} \div \frac{y + 2}{y - 1}$

43. $\frac{a^2 - a}{a^2 - 2a - 3} \cdot \frac{a^2 + 2a + 1}{a^2 + 4a} \div \frac{a^2 - 3a - 4}{a^2 - 16}$

44. $\frac{y^2 - 4y + 3}{y^2} \cdot \frac{y^2 + y}{y^2 - 6y + 9} \div \frac{y^2 - 2y - 3}{y^2 - y - 6}$

Let $P(x) = \frac{x}{x - 1}$, $Q(x) = \frac{x^2}{x^2 - 1}$, *and* $R(x) = \frac{x^3}{(x - 1)^2}$. *Write each quotient as a single fraction in lowest terms.*

45. $P(x) \div Q(x)$ 46. $Q(x) \div R(x)$ 47. $P(x) \cdot Q(x) \div R(x)$ 48. $P(x) \cdot R(x) \div Q(x)$

49. For what value(s) of the variables is each of the products or quotients in Problems 33–44 undefined?

3.6

COMPLEX FRACTIONS

A fraction that contains a fraction or fractions in either the numerator or the denominator or in both is called a **complex fraction**. In simplifying complex fractions, it is helpful to remember that fraction bars serve as grouping devices in the same sense as parentheses or brackets. Thus, the complex fraction

$$\frac{x + \frac{3}{4}}{x - \frac{1}{2}} \quad \text{means} \quad \left(x + \frac{3}{4}\right) \div \left(x - \frac{1}{2}\right).$$

Complex fractions can be simplified in either of two ways. In simple examples, it is easier to apply the fundamental principle of fractions and multiply the numerator and the denominator of the complex fraction by the LCD of all the fractions that appear. Thus, in the preceding example,

$$\frac{4\left(x + \frac{3}{4}\right)}{4\left(x - \frac{1}{2}\right)} = \frac{4x + 3}{4x - 2} \qquad \left(x \neq \frac{1}{2}\right).$$

Alternatively, this fraction may be simplified by first representing the numerator as a single fraction and the denominator as a single fraction, and then writing the quotient as the product of the numerator and the reciprocal of the denominator:

$$\frac{x+\frac{3}{4}}{x-\frac{1}{2}} = \frac{\frac{4x+3}{4}}{\frac{2x-1}{2}} = \frac{4x+3}{4} \cdot \frac{2}{2x-1}$$

$$= \frac{(4x+3)\cdot 2}{2(2x-1)\cdot 2} = \frac{4x+3}{2(2x-1)} = \frac{4x+3}{4x-2} \qquad \left(x \neq \frac{1}{2}\right).$$

The first method is generally more convenient to use.

Sometimes it is easier to simplify parts of a complex fraction before using either of the above methods. For example, since

$$1 + \frac{2}{3} = \frac{3}{3} + \frac{2}{3} = \frac{5}{3},$$

we have

$$3 + \frac{2}{1+\frac{2}{3}} = 3 + \frac{2}{\frac{5}{3}} = 3 + 2 \cdot \frac{3}{5}$$

$$= 3 + \frac{6}{5} = \frac{15}{5} + \frac{6}{5} = \frac{21}{5}.$$

EXERCISE 3.6

A

Write each complex fraction as a single fraction in lowest terms.

Example $$\frac{\frac{3}{a} - \frac{1}{2a}}{\frac{1}{3a} + \frac{5}{6a}}$$

Solution The LCD for all fractions in the numerator and the denominator is $6a$. Multiply the numerator and the denominator by $6a$ and simplify.

$$\frac{6a\left(\frac{3}{a} - \frac{1}{2a}\right)}{6a\left(\frac{1}{3a} + \frac{5}{6a}\right)} = \frac{(6a)\frac{3}{a} - (6a)\frac{1}{2a}}{(6a)\frac{1}{3a} + (6a)\frac{5}{6a}} = \frac{18-3}{2+5} = \frac{15}{7}$$

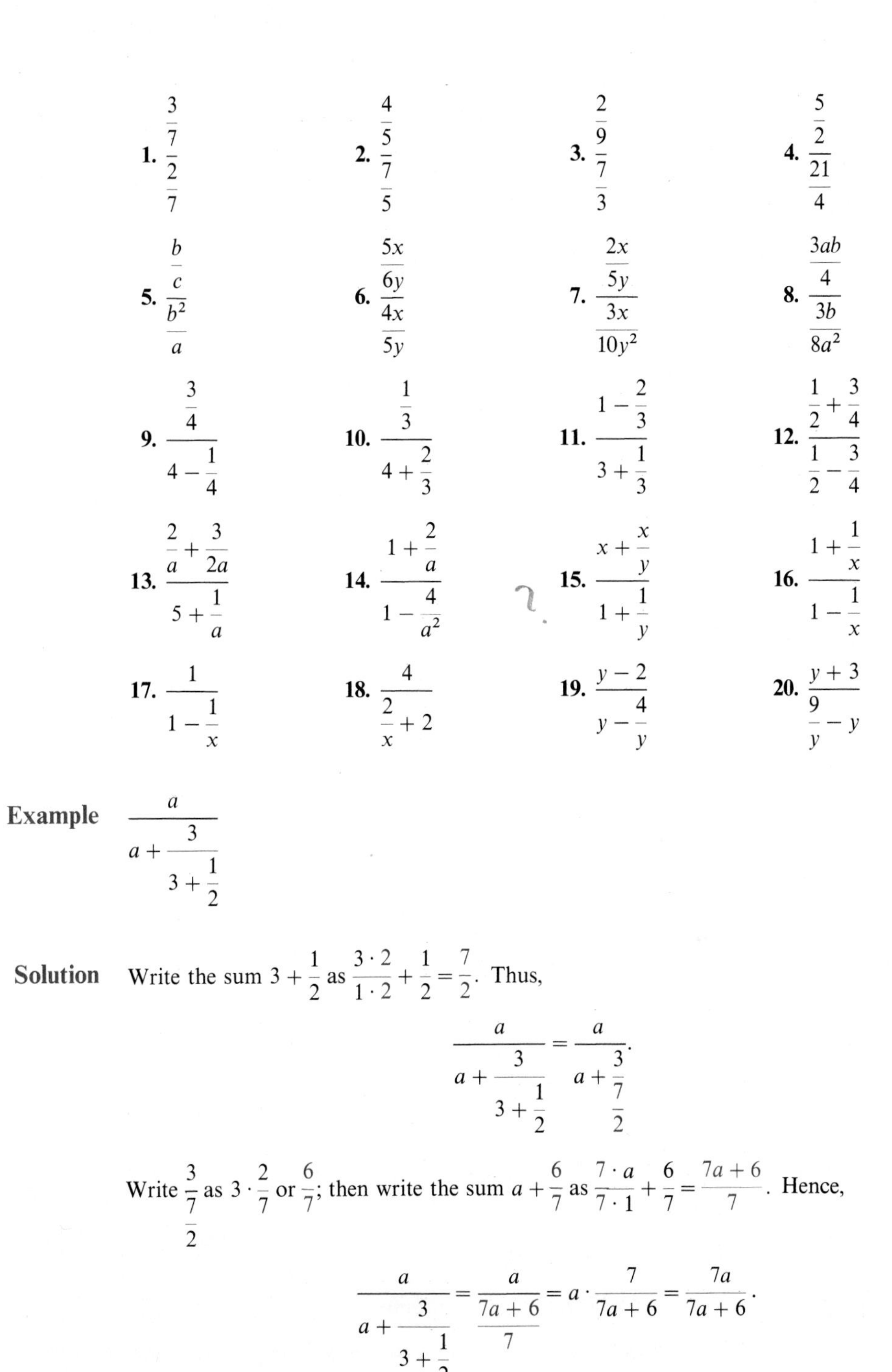

1. $\dfrac{\frac{3}{7}}{\frac{2}{7}}$ **2.** $\dfrac{\frac{4}{5}}{\frac{7}{5}}$ **3.** $\dfrac{\frac{2}{9}}{\frac{7}{3}}$ **4.** $\dfrac{\frac{5}{2}}{\frac{21}{4}}$

5. $\dfrac{\frac{b}{c}}{\frac{b^2}{a}}$ **6.** $\dfrac{\frac{5x}{6y}}{\frac{4x}{5y}}$ **7.** $\dfrac{\frac{2x}{5y}}{\frac{3x}{10y^2}}$ **8.** $\dfrac{\frac{3ab}{4}}{\frac{3b}{8a^2}}$

9. $\dfrac{\frac{3}{4}}{4-\frac{1}{4}}$ **10.** $\dfrac{\frac{1}{3}}{4+\frac{2}{3}}$ **11.** $\dfrac{1-\frac{2}{3}}{3+\frac{1}{3}}$ **12.** $\dfrac{\frac{1}{2}+\frac{3}{4}}{\frac{1}{2}-\frac{3}{4}}$

13. $\dfrac{\frac{2}{a}+\frac{3}{2a}}{5+\frac{1}{a}}$ **14.** $\dfrac{1+\frac{2}{a}}{1-\frac{4}{a^2}}$ **15.** $\dfrac{x+\frac{x}{y}}{1+\frac{1}{y}}$ **16.** $\dfrac{1+\frac{1}{x}}{1-\frac{1}{x}}$

17. $\dfrac{1}{1-\frac{1}{x}}$ **18.** $\dfrac{4}{\frac{2}{x}+2}$ **19.** $\dfrac{y-2}{y-\frac{4}{y}}$ **20.** $\dfrac{y+3}{\frac{9}{y}-y}$

Example $\dfrac{a}{a+\dfrac{3}{3+\frac{1}{2}}}$

Solution Write the sum $3+\frac{1}{2}$ as $\frac{3\cdot 2}{1\cdot 2}+\frac{1}{2}=\frac{7}{2}$. Thus,

$$\frac{a}{a+\dfrac{3}{3+\frac{1}{2}}}=\frac{a}{a+\dfrac{3}{\frac{7}{2}}}.$$

Write $\dfrac{3}{\frac{7}{2}}$ as $3\cdot\frac{2}{7}$ or $\frac{6}{7}$; then write the sum $a+\frac{6}{7}$ as $\frac{7\cdot a}{7\cdot 1}+\frac{6}{7}=\frac{7a+6}{7}$. Hence,

$$\frac{a}{a+\dfrac{3}{3+\frac{1}{2}}}=\frac{a}{\frac{7a+6}{7}}=a\cdot\frac{7}{7a+6}=\frac{7a}{7a+6}.$$

21. $2-\dfrac{2}{3+\frac{1}{2}}$ **22.** $1-\dfrac{1}{1-\frac{1}{3}}$ **23.** $a-\dfrac{a}{a+\frac{1}{4}}$

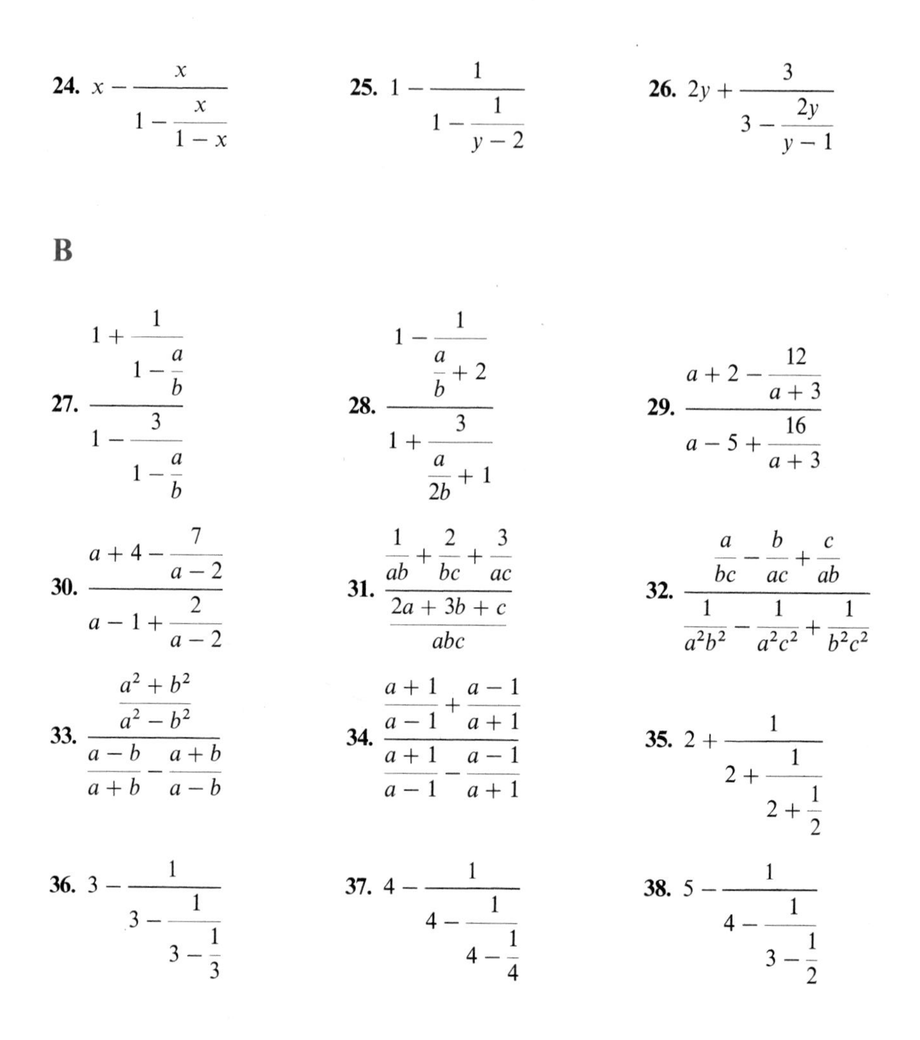

24. $x - \dfrac{x}{1 - \dfrac{x}{1 - x}}$ **25.** $1 - \dfrac{1}{1 - \dfrac{1}{y - 2}}$ **26.** $2y + \dfrac{3}{3 - \dfrac{2y}{y - 1}}$

B

27. $\dfrac{1 + \dfrac{1}{1 - \dfrac{a}{b}}}{1 - \dfrac{3}{1 - \dfrac{a}{b}}}$ **28.** $\dfrac{1 - \dfrac{1}{\dfrac{a}{b} + 2}}{1 + \dfrac{3}{\dfrac{a}{2b} + 1}}$ **29.** $\dfrac{a + 2 - \dfrac{12}{a + 3}}{a - 5 + \dfrac{16}{a + 3}}$

30. $\dfrac{a + 4 - \dfrac{7}{a - 2}}{a - 1 + \dfrac{2}{a - 2}}$ **31.** $\dfrac{\dfrac{1}{ab} + \dfrac{2}{bc} + \dfrac{3}{ac}}{\dfrac{2a + 3b + c}{abc}}$ **32.** $\dfrac{\dfrac{a}{bc} - \dfrac{b}{ac} + \dfrac{c}{ab}}{\dfrac{1}{a^2b^2} - \dfrac{1}{a^2c^2} + \dfrac{1}{b^2c^2}}$

33. $\dfrac{\dfrac{a^2 + b^2}{a^2 - b^2}}{\dfrac{a - b}{a + b} - \dfrac{a + b}{a - b}}$ **34.** $\dfrac{\dfrac{a + 1}{a - 1} + \dfrac{a - 1}{a + 1}}{\dfrac{a + 1}{a - 1} - \dfrac{a - 1}{a + 1}}$ **35.** $2 + \dfrac{1}{2 + \dfrac{1}{2 + \dfrac{1}{2}}}$

36. $3 - \dfrac{1}{3 - \dfrac{1}{3 - \dfrac{1}{3}}}$ **37.** $4 - \dfrac{1}{4 - \dfrac{1}{4 - \dfrac{1}{4}}}$ **38.** $5 - \dfrac{1}{4 - \dfrac{1}{3 - \dfrac{1}{2}}}$

CHAPTER SUMMARY

[3.1] A **fraction** is an expression denoting a quotient.

The **fundamental principle of fractions** states:

If the numerator and the denominator of a fraction are multiplied or divided by the same nonzero number, the result is a fraction **equivalent** to the given fraction:

$$\frac{a}{b} = \frac{ac}{bc} \qquad (b, c \neq 0).$$

For a given fraction, the replacement of any two of the three elements of the fraction—the fraction itself, the numerator, the denominator—by their negatives results in an equivalent fraction:

$$\frac{-a}{b} = \frac{a}{-b} = -\frac{a}{b} = -\frac{-a}{-b} \qquad (b \neq 0),$$

$$\frac{a}{b} = \frac{-a}{-b} = -\frac{a}{-b} = -\frac{-a}{b} \qquad (b \neq 0).$$

Standard forms for fractions are

$$\frac{a}{b} \quad \text{and} \quad \frac{-a}{b}.$$

[3.2] A fraction is in **lowest terms** if the numerator and the denominator do not contain factors in common. We can reduce a fraction to lowest terms by using the fundamental principle of fractions to divide the numerator and the denominator by their common nonzero factors. We can build a fraction to **higher terms** by using the fundamental principle to multiply the numerator and the denominator by the same nonzero factor.

[3.3] Quotients of polynomials can be rewritten as equivalent **mixed expressions** using a method similar to the long division process used in arithmetic.

[3.4] The operations of addition and subtraction of fractions are governed by the following properties:

$$\frac{a}{c} + \frac{b}{c} = \frac{a+b}{c} \quad (c \neq 0) \qquad \frac{a}{b} + \frac{c}{d} = \frac{ad+bc}{bd} \quad (b, d \neq 0)$$

$$\frac{a}{c} - \frac{b}{c} = \frac{a-b}{c} \quad (c \neq 0) \qquad \frac{a}{b} - \frac{c}{d} = \frac{ad-bc}{bd} \quad (b, d \neq 0)$$

The **least common denominator** (LCD) of a set of fractions with natural number denominators is the smallest natural number that is exactly divisible by each of the denominators.

[3.5] The operations of multiplication and division of fractions are governed by the following properties:

$$\frac{a}{b} \cdot \frac{c}{d} = \frac{ac}{bd} \quad (b, d \neq 0) \qquad \frac{a}{b} \div \frac{c}{d} = \frac{a}{b} \cdot \frac{d}{c} \quad (b, c, d \neq 0)$$

[3.6] A **complex fraction** is a fraction containing other fractions in its numerator or denominator or both. It is often convenient to use the fundamental principle in simplifying such fractions.

REVIEW EXERCISES

[3.1] **1. a.** Write six fractions equal to $\frac{1}{a-b}$ by changing the sign or signs of the numerator, the denominator, or the fraction itself.

b. What are the conditions on a and b for the fraction in part a to represent a positive number? A negative number?

2. Express $-\frac{1}{1-a}$ in standard form in two ways.

[3.2] *Reduce each fraction to lowest terms.*

3. a. $\frac{4x^2y^3}{10xy^4}$ **b.** $\frac{2x^2-8}{2x+4}$

4. a. $\frac{4x^2-1}{1-2x}$ **b.** $\frac{8x^3+y^3}{4x^2-y^2}$

Express each given fraction as an equivalent fraction with the given denominator.

5. a. $\frac{-3}{4}$; $\frac{?}{24}$ **b.** $\frac{x}{2y}$; $\frac{?}{2xy^2}$

6. a. $\frac{2}{x+3y}$; $\frac{?}{x^2-9y^2}$ **b.** $\frac{y}{3-y}$; $\frac{?}{y^2-4y+3}$

[3.3] *Divide.*

7. a. $\frac{12x^2-6x+3}{2x}$ **b.** $\frac{6y^3+3y^2-y}{3y^2}$

8. a. $\frac{2y^2-3y+1}{2y+3}$ **b.** $\frac{6x^4-3x^3+2x+2}{2x-1}$

[3.4] *Write each expression as a single fraction in lowest terms.*

9. a. $\frac{2x+y}{3}+\frac{x-y}{3}$ **b.** $\frac{x-2y}{4x}-\frac{2x-y}{4x}$

10. a. $\frac{2}{5x}-\frac{3}{4y}+\frac{7}{20xy}$ **b.** $\frac{y}{2y-6}+\frac{2}{3y-9}$

11. a. $\frac{2}{x^2-1}+\frac{3}{x^2-2x+1}$ **b.** $\frac{y}{y^2-16}-\frac{y+1}{y^2-5y+4}$

12. a. $3+\frac{2}{x-2}+\frac{4}{x^2-4}$ **b.** $\frac{2}{x-2y}-\frac{1}{x+2y}+\frac{3}{x^2-4y^2}$

[3.5] *Write each expression as a single fraction in lowest terms.*

13. a. $\dfrac{2x}{y^2} \cdot \dfrac{3y^3}{8x} \cdot \dfrac{4x^2}{9y}$ b. $\dfrac{4x}{2xy + 8y^2} \cdot \dfrac{x + 4y}{x}$

14. a. $\dfrac{x^2 + 2x - 3}{x^2 + 6x + 9} \cdot \dfrac{2x + 6}{2x - 2}$ b. $\dfrac{y^3 - y}{2y + 1} \cdot \dfrac{4y + 2}{y^2 + 2y + 1}$

15. a. $\dfrac{10x^2y^3}{9} \div \dfrac{2xy^2}{3}$ b. $\dfrac{2y - 6}{y + 2x} \div \dfrac{4y - 12}{2y + 4x}$

16. a. $\dfrac{y^2 + 4y + 3}{y^2 - y - 2} \div \dfrac{y^2 - 4y - 5}{y^2 - 3y - 10}$ b. $\dfrac{x^2}{y - x} \div \dfrac{x^3 - x^2}{x - y}$

[3.6] *Write each expression as a single fraction in lowest terms.*

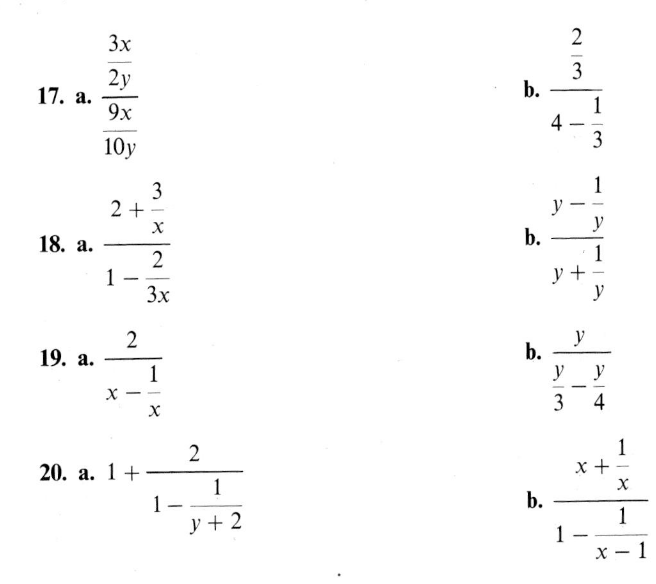

17. a. $\dfrac{\dfrac{3x}{2y}}{\dfrac{9x}{10y}}$ b. $\dfrac{\dfrac{2}{3}}{4 - \dfrac{1}{3}}$

18. a. $\dfrac{2 + \dfrac{3}{x}}{1 - \dfrac{2}{3x}}$ b. $\dfrac{y - \dfrac{1}{y}}{y + \dfrac{1}{y}}$

19. a. $\dfrac{2}{x - \dfrac{1}{x}}$ b. $\dfrac{y}{\dfrac{y}{3} - \dfrac{y}{4}}$

20. a. $1 + \dfrac{2}{1 - \dfrac{1}{y + 2}}$ b. $\dfrac{x + \dfrac{1}{x}}{1 - \dfrac{1}{x - 1}}$

4

EXPONENTS, ROOTS, AND RADICALS

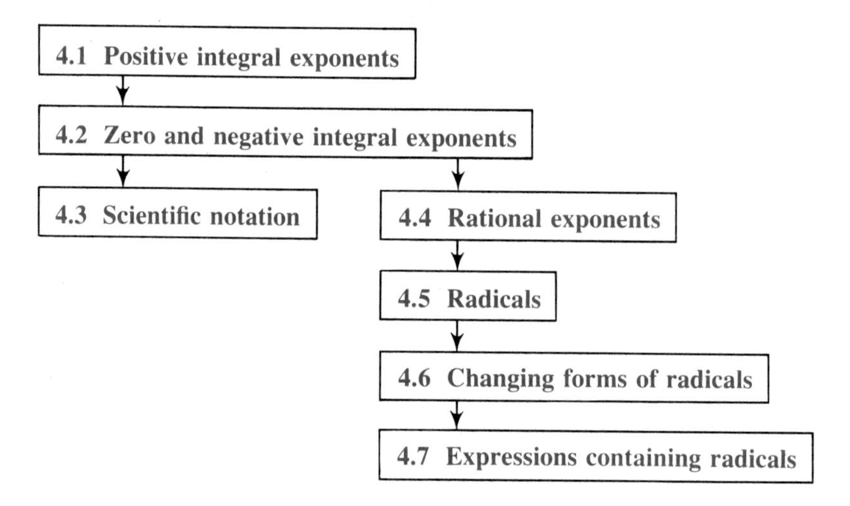

4.1

POSITIVE INTEGRAL EXPONENTS

In Chapter 2 the expression a^n, where a is any real number and n is a positive integer, was defined by

$$a^n = a \cdot a \cdot a \cdots a \qquad (n \text{ factors}).$$

The following two laws were developed from this definition:

(I) $a^m \cdot a^n = a^{m+n}$.

(II) $\dfrac{a^m}{a^n} = a^{m-n} \qquad (m > n, a \neq 0)$.

The following four laws are also very useful in simplifying expressions involving positive integral exponents:

(IIa) $\dfrac{a^m}{a^n} = \dfrac{1}{a^{n-m}} \qquad (m < n, a \neq 0)$,

since

$$\frac{a^m}{a^n} = \frac{\overbrace{a \cdot a \cdots a}^{m \text{ factors}}}{\underbrace{a \cdot a \cdots a}_{n \text{ factors}}} = \frac{\overbrace{a \cdot a \cdots a}^{m \text{ factors}}}{\underbrace{(a \cdot a \cdots a)}_{m \text{ factors}}\underbrace{(a \cdot a \cdots a)}_{n-m \text{ factors}}}$$

$$= \frac{1}{a^{n-m}}.$$

For example,

$$\frac{x^5}{x^8} = \frac{1}{x^{8-5}} = \frac{1}{x^3} \quad \text{and} \quad \frac{x^2y}{x^3y^2} = \frac{1}{x^{3-2}y^{2-1}} = \frac{1}{xy} \qquad (x, y \neq 0).$$

(III) $(\boldsymbol{a^m})^n = \boldsymbol{a^{mn}}$,

since

$$\begin{aligned}(a^m)^n &= a^m \cdot a^m \cdot a^m \cdots a^m \qquad (n \text{ factors})\\ &= a^{m+m+m+\cdots+m} \quad (n \text{ terms})\\ &= a^{mn}.\end{aligned}$$

For example,

$$(x^2)^3 = x^6 \quad \text{and} \quad (x^5)^2 = x^{10}.$$

(IV) $(\boldsymbol{ab})^n = \boldsymbol{a^n b^n}$,

since

$$\begin{aligned}(ab)^n &= (ab)(ab)(ab)\cdots(ab) \qquad (n \text{ factors})\\ &= \overbrace{(a \cdot a \cdots a)}^{n \text{ factors}}\overbrace{(b \cdot b \cdots b)}^{n \text{ factors}}\\ &= a^n b^n.\end{aligned}$$

For example,

$$(xy)^3 = x^3y^3 \quad \text{and} \quad (2x^2y^3)^3 = 2^3x^6y^9.$$

(V) $\left(\boldsymbol{\frac{a}{b}}\right)^n = \boldsymbol{\frac{a^n}{b^n}} \quad (\boldsymbol{b \neq 0})$,

since

$$\begin{aligned}\left(\frac{a}{b}\right)^n &= \frac{a}{b} \cdot \frac{a}{b} \cdot \frac{a}{b} \cdots \left(\frac{a}{b}\right) \qquad (n \text{ factors})\\ &= \frac{\overbrace{a \cdot a \cdot a \cdots a}^{n \text{ factors}}}{\underbrace{b \cdot b \cdot b \cdots b}_{n \text{ factors}}}\\ &= \frac{a^n}{b^n}.\end{aligned}$$

For example,

$$\left(\frac{x}{y}\right)^3 = \frac{x^3}{y^3} \quad \text{and} \quad \left(\frac{2x^2}{y}\right)^4 = \frac{(2x^2)^4}{y^4} \qquad (y \neq 0).$$

Ordinarily, two or more of the laws of exponents are required to simplify expressions containing exponents. In the last example, from Law IV,

$$\frac{(2x^2)^4}{y^4} = \frac{2^4(x^2)^4}{y^4} \qquad (y \neq 0),$$

and from Law III,

$$\frac{2^4(x^2)^4}{y^4} = \frac{16x^8}{y^4} \qquad (y \neq 0).$$

As in the latter part of Chapter 3, in the exercise sets in this chapter we shall assume that no denominator equals 0 unless otherwise stated.

EXERCISE 4.1

A

Using one or more of the laws of exponents, write each expression as a product or quotient in which each variable occurs but once, and all exponents are positive.

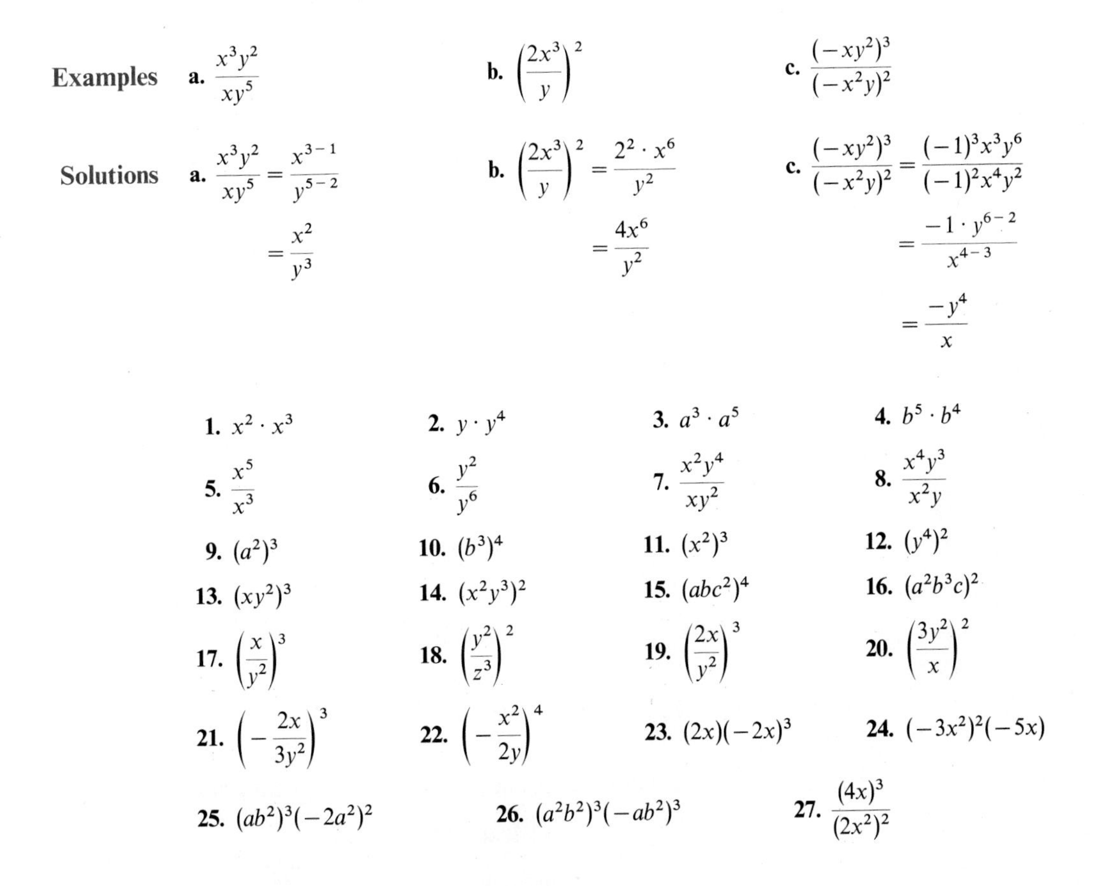

Examples **a.** $\dfrac{x^3y^2}{xy^5}$ **b.** $\left(\dfrac{2x^3}{y}\right)^2$ **c.** $\dfrac{(-xy^2)^3}{(-x^2y)^2}$

Solutions **a.** $\dfrac{x^3y^2}{xy^5} = \dfrac{x^{3-1}}{y^{5-2}} = \dfrac{x^2}{y^3}$

b. $\left(\dfrac{2x^3}{y}\right)^2 = \dfrac{2^2 \cdot x^6}{y^2} = \dfrac{4x^6}{y^2}$

c. $\dfrac{(-xy^2)^3}{(-x^2y)^2} = \dfrac{(-1)^3x^3y^6}{(-1)^2x^4y^2} = \dfrac{-1 \cdot y^{6-2}}{x^{4-3}} = \dfrac{-y^4}{x}$

1. $x^2 \cdot x^3$ **2.** $y \cdot y^4$ **3.** $a^3 \cdot a^5$ **4.** $b^5 \cdot b^4$

5. $\dfrac{x^5}{x^3}$ **6.** $\dfrac{y^2}{y^6}$ **7.** $\dfrac{x^2y^4}{xy^2}$ **8.** $\dfrac{x^4y^3}{x^2y}$

9. $(a^2)^3$ **10.** $(b^3)^4$ **11.** $(x^2)^3$ **12.** $(y^4)^2$

13. $(xy^2)^3$ **14.** $(x^2y^3)^2$ **15.** $(abc^2)^4$ **16.** $(a^2b^3c)^2$

17. $\left(\dfrac{x}{y^2}\right)^3$ **18.** $\left(\dfrac{y^2}{z^3}\right)^2$ **19.** $\left(\dfrac{2x}{y^2}\right)^3$ **20.** $\left(\dfrac{3y^2}{x}\right)^2$

21. $\left(-\dfrac{2x}{3y^2}\right)^3$ **22.** $\left(-\dfrac{x^2}{2y}\right)^4$ **23.** $(2x)(-2x)^3$ **24.** $(-3x^2)^2(-5x)$

25. $(ab^2)^3(-2a^2)^2$ **26.** $(a^2b^2)^3(-ab^2)^3$ **27.** $\dfrac{(4x)^3}{(2x^2)^2}$

28. $\dfrac{(5x)^2}{(3x^2)^3}$

29. $\dfrac{(xy^2)^3}{(x^2y)^2}$

30. $\dfrac{(-xy^2)^2}{(x^2y)^3}$

31. $\dfrac{(xy)^2(x^2y)^3}{(x^2y^2)^2}$

32. $\dfrac{(-x)^2(-x^2)^4}{(x^2)^3}$

33. $\left(\dfrac{2x}{y^2}\right)^3\left(\dfrac{y^2}{3x}\right)^2$

34. $\left(\dfrac{x^2z}{2}\right)^2\left(-\dfrac{2}{x^2z}\right)^3$

35. $\left(\dfrac{-3}{y^2}\right)^2(2y^3)^2$

36. $\left(\dfrac{y}{x}\right)^2\left(-\dfrac{3}{4xy}\right)^3$

37. $\left[\left(\dfrac{r^2s^3t}{xy}\right)^3\left(\dfrac{x^2y}{r^3st^2}\right)^2\right]^2$

38. $\left[\left(\dfrac{a^3bc}{x^2y}\right)^4\left(\dfrac{x^2yz}{ab^2c^3}\right)^2\right]^2$

39. $\left(\dfrac{x^2}{a^2b}\right)^2\left(-\dfrac{ab}{x^3}\right)^3\left(\dfrac{x}{ab}\right)^2$

40. $\left(\dfrac{m^3n^2p}{r^2s}\right)^2\left(\dfrac{rs}{mn^2p^2}\right)^3\left(-\dfrac{mnp}{rs}\right)^2$

41. $\left(\dfrac{ab^2}{x}\right)^3\left(-\dfrac{x^2}{a^2b^3}\right)^2\left(\dfrac{ab^2}{x^3}\right)^3$

42. $\left(\dfrac{2x^2y}{3z}\right)^2\left(\dfrac{2z^2}{3xy^2}\right)^3\left(-\dfrac{4xz}{3y}\right)^2$

B

Examples

a. $\dfrac{x^n \cdot x^{n+1}}{x^{n-1}}$

b. $\dfrac{(y^{n-1})^2}{y^{n-2}}$

c. $(x^{n-1} \cdot x^{2n+3})^2$

Solutions

a. $\dfrac{x^n \cdot x^{n+1}}{x^{n-1}}$

$= x^{n+(n+1)-(n-1)} = x^{n+2}$

b. $\dfrac{(y^{n-1})^2}{y^{n-2}}$

$= y^{(2n-2)-(n-2)} = y^n$

c. $(x^{n-1} \cdot x^{2n+3})^2$

$= (x^{3n+2})^2 = x^{6n+4}$

43. $x^n \cdot x^n$

44. $\dfrac{x^{2n}x^n}{x^{n+1}}$

45. $\dfrac{(x^{n+1}x^{2n-1})^2}{x^{3n}}$

46. $\left(\dfrac{y^2 \cdot y^3}{y}\right)^{2n}$

47. $\left(\dfrac{x^{3n}x^{2n}}{x^{4n}}\right)^2$

48. $\dfrac{(y^{n+1})^n}{y^n}$

49. $\dfrac{(x^ny^{n+1})^2}{x^{2n+1}y^n}$

50. $\left(\dfrac{x^{2n-1}y^{3n}}{x^ny^{2n+2}}\right)^2$

51. $\left(\dfrac{x^ny^{3n-1}}{xy}\right)^3$

52. $\dfrac{(x^{3n}y^{2n-1})^3}{(xy)^2}$

4.2

ZERO AND NEGATIVE INTEGRAL EXPONENTS

To extend the first law of exponents to include zero exponents, that is, for

$$a^0a^n = a^{0+n} = a^n$$

to hold, it is clear that a^0 *must be defined* as 1. This also follows if the second law of exponents is to hold for the case where $m = n$, since, by this law,

$$\frac{a^n}{a^n} = a^{n-n} = a^0 \qquad (a \neq 0),$$

and, by the definition of a quotient,

$$\frac{a^n}{a^n} = 1 \qquad (a \neq 0).$$

Hence, for consistency, for all real numbers a, we define

$$\mathbf{a^0 = 1 \qquad (a \neq 0).}$$

For example,

$$3^0 = 1; \qquad \left(\frac{1}{2}\right)^0 = 1; \qquad (-4)^0 = 1; \qquad x^0 = 1 \qquad (x \neq 0).$$

If $a = 0$, the expression a^0 is not defined.

We would like the first law of exponents to hold also for negative exponents. We observe that

$$a^n \cdot a^{-n} = a^{n-n} = a^0 = 1.$$

Also, by the reciprocal axiom,

$$a^n \cdot \frac{1}{a^n} = 1 \qquad (a \neq 0).$$

It is evident, then, that for consistency a^{-n} *must be defined by*

$$\mathbf{a^{-n} = \frac{1}{a^n} \qquad (a \neq 0).}$$

It follows that

$$\mathbf{\frac{1}{a^{-n}} = \frac{1}{\frac{1}{a^n}} = a^n \qquad (a \neq 0).}$$

For example,

$$3^{-2} = \frac{1}{3^2} = \frac{1}{9}; \qquad \frac{1}{2^{-3}} = 2^3 = 8; \qquad \frac{x^{-2}}{y^{-1}} = \frac{\frac{1}{x^2}}{\frac{1}{y}} = \frac{y}{x^2} \qquad (x, y \neq 0).$$

Since a^{-n} $(a \neq 0)$ was defined as $1/a^n$, and a^0 was defined as 1, it follows that

$$\frac{1}{a^{n-m}} = \frac{1}{a^{-(m-n)}} = a^{m-n}.$$

Therefore, the laws of exponents II and IIa can be generalized into one law:

$$\text{(II)} \quad \frac{a^m}{a^n} = a^{m-n} \qquad (a \neq 0),$$

where m and n are any integers. It can be shown that all the laws of exponents on pages 97–98 are valid for integral exponents.

The laws of exponents may be applied in an arbitrary order. For example, we can first apply Law II to write

$$\left(\frac{x^3}{x^2}\right)^{-3} = (x)^{-3} = x^{-3} = \frac{1}{x^3}.$$

Or we can first apply Law V and then Law IV to write

$$\left(\frac{x^3}{x^2}\right)^{-3} = \frac{(x^3)^{-3}}{(x^2)^{-3}} = \frac{x^{-9}}{x^{-6}},$$

from which by Law II, we have

$$\frac{x^{-9}}{x^{-6}} = x^{-9-(-6)} = x^{-9+6} = x^{-3} = \frac{1}{x^3}.$$

EXERCISE 4.2

A

Assume that all variables in this exercise represent positive real numbers only.

Write each expression as a basic numeral or fraction in lowest terms.

Examples **a.** $3 \cdot 5^{-2}$ **b.** $\frac{3}{2^{-3}}$ **c.** $4^2 + 4^{-2}$

Solutions

a. $3 \cdot 5^{-2} = 3 \cdot \frac{1}{5^2} = 3 \cdot \frac{1}{25} = \frac{3}{25}$

b. $\frac{3}{2^{-3}} = 3 \cdot \frac{1}{2^{-3}} = 3 \cdot 2^3 = 3 \cdot 8 = 24$

c. $4^2 + 4^{-2} = 16 + \frac{1}{16} = \frac{256}{16} + \frac{1}{16} = \frac{257}{16}$

1. 2^{-1} **2.** 3^{-2} **3.** $\frac{1}{3^{-1}}$ **4.** $\frac{3}{4^{-2}}$

5. $(-2)^{-3}$ **6.** $\frac{1}{(-3)^{-2}}$ **7.** $\frac{5^{-1}}{3^0}$ **8.** $\frac{2^0}{3^{-2}}$

9. $\left(\frac{3}{5}\right)^{-1}$ 10. $\left(\frac{1}{3}\right)^{-2}$ 11. $\frac{5^{-1}}{3^{-2}}$ 12. $\frac{3^{-3}}{6^{-2}}$

13. $3^{-2} + 3^2$ 14. $5^{-1} + 25^0$ 15. $4^{-1} - 4^{-2}$ 16. $8^{-2} - 2^0$

Write each expression as a product or quotient of powers in which each variable occurs but once, and all exponents are positive.

Examples **a.** $x^{-3} \cdot x^5$ **b.** $(x^2y^{-3})^{-1}$ **c.** $\left(\frac{x^{-1}y^2z^0}{x^3y^{-4}z^2}\right)^{-1}$

Solutions **a.** $x^{-3} \cdot x^5 = x^{-3+5} = x^2$

b. $(x^2y^{-3})^{-1} = x^{-2}y^3 = \frac{y^3}{x^2}$

c. $\left(\frac{x^{-1}y^2z^0}{x^3y^{-4}z^2}\right)^{-1} = \frac{xy^{-2}z^0}{x^{-3}y^4z^{-2}} = \frac{x^4z^2}{y^6}$

17. x^2y^{-3} 18. $\frac{x^3}{y^{-2}}$ 19. $(x^2 \cdot y)^{-3}$ 20. $(xy^3)^{-2}$

21. $\left(\frac{x}{y^3}\right)^2$ 22. $\left(\frac{2x}{y^2}\right)^3$ 23. $\frac{(xy^2)^3}{(x^2y)^2}$ 24. $\left(\frac{3x}{y^2}\right)^2\left(\frac{2y^3}{x}\right)^2$

25. $x^{-3} \cdot x^7$ 26. $\frac{x^3}{x^{-2}}$ 27. $(x^{-2}y^0)^3$ 28. $(x^{-2}y^3)^0$

29. $\frac{x^{-1}}{y^{-1}}$ 30. $\frac{x^{-3}}{y^{-2}}$ 31. $\frac{8^{-1}x^0y^{-3}}{(2xy)^{-5}}$ 32. $\left(\frac{x^{-1}y^3}{2x^0y^{-5}}\right)^{-2}$

33. $\left[\left(\frac{a^{-1}b^{-1}c}{a^2b^2c^{-2}}\right)^{-3}\right]^2$ 34. $\left[\left(\frac{ab^{-2}c^{-1}}{a^{-2}bc}\right)^4\right]^{-1}$

Write each expression as a single fraction involving positive exponents only.

Examples **a.** $x^{-1} + y^{-2}$ **b.** $(x^{-1} + x^{-2})^{-1}$ **c.** $x^{-1} + \frac{1}{x^{-1}}$

Solutions **a.**
$$x^{-1} + y^{-2} = \frac{1}{x} + \frac{1}{y^2} = \frac{(y^2)}{(y^2)}\frac{1}{x} + \frac{1(x)}{y^2(x)} = \frac{y^2 + x}{xy^2}$$

b.
$$(x^{-1} + x^{-2})^{-1} = \left(\frac{1}{x} + \frac{1}{x^2}\right)^{-1} = \left(\frac{(x)1}{(x)x} + \frac{1}{x^2}\right)^{-1} = \left(\frac{x+1}{x^2}\right)^{-1} = \frac{x^2}{x+1}$$

c.
$$x^{-1} + \frac{1}{x^{-1}} = \frac{1}{x} + x = \frac{1}{x} + \frac{x(x)}{(x)} = \frac{1}{x} + \frac{x^2}{x} = \frac{1 + x^2}{x}$$

35. $a^{-2}+b^{-2}$

36. $\dfrac{x^{-1}}{y^{-1}}+\dfrac{y}{x}$

37. $\dfrac{r}{s^{-1}}+\dfrac{r^{-1}}{s}$

38. $(x+y)^{-1}$

39. $(a-b)^{-2}$

40. $xy^{-1}+x^{-1}y$

41. $x^{-1}y-xy^{-1}$

42. $\dfrac{x^{-1}+y^{-1}}{(xy)^{-1}}$

43. $\dfrac{a}{b^{-1}}+\left(\dfrac{a}{b}\right)^{-1}$

44. $(x^{-1}-y^{-1})^{-1}$

45. $\dfrac{x^{-1}+y^{-1}}{x^{-1}-y^{-1}}$

46. $\dfrac{x^{-1}-y^{-1}}{x^{-1}+y^{-1}}$

B

Write each expression as a product free of fractions in which each variable occurs but once.

Examples

a. $x^{-2n}\cdot x^{n}$

b. $\left(\dfrac{x^{1-n}}{x^{2-n}}\right)^{-2}$

c. $\left(\dfrac{x^{n}y^{2n-1}}{y^{n}}\right)^{2}$

Solutions

a. $x^{-2n}\cdot x^{n}=x^{-2n+n}$
$=x^{-n}$

b. $\left(\dfrac{x^{1-n}}{x^{2-n}}\right)^{-2}$
$=(x^{(1-n)-(2-n)})^{-2}$
$=(x^{-1})^{-2}=x^{2}$

c. $\left(\dfrac{x^{n}y^{2n-1}}{y^{n}}\right)^{2}$
$=(x^{n}y^{2n-1-n})^{2}$
$=(x^{n}y^{n-1})^{2}=x^{2n}y^{2n-2}$

47. $a^{3-n}a^{0}$

48. $x^{-n}x^{n+1}$

49. $\left(\dfrac{a^{2n}}{a^{n+1}}\right)^{-2}$

50. $\dfrac{x^{n}y^{n+1}}{x^{2n-1}y^{n}}$

51. $\dfrac{b^{n}c^{2n-1}}{b^{n+1}c^{2n}}$

52. $\left(\dfrac{x^{n-1}y^{n}}{x^{-2}y^{-n}}\right)^{2}$

53. $\left(\dfrac{x^{n}}{x^{n-1}}\right)^{-1}$

54. $\left(\dfrac{a^{2n}b^{n-1}}{a^{n-1}b}\right)^{2}$

55. Show that $\left(\dfrac{a}{b}\right)^{-n}=\left(\dfrac{b}{a}\right)^{n}$.

56. Show that $\dfrac{a^{m}b^{-n}}{c^{-p}d^{q}}=\dfrac{a^{m}c^{p}}{b^{n}d^{q}}$.

4.3

SCIENTIFIC NOTATION

In Chapter 9 in our discussion of logarithms, we shall find it convenient to use an exponential form of notation called **scientific notation.** This form of notation is particularly useful in scientific applications of mathematics which involve very large or very small quantities. For example, the mass of the earth is approximately

$$5{,}980{,}000{,}000{,}000{,}000{,}000{,}000{,}000{,}000 = 5.98\times 10^{27}\text{ grams,}$$

and the mass of a hydrogen atom is approximately

$$0.00000000000000000000000167 = 1.67\times 10^{-24}\text{ gram.}$$

In each case we have represented a number as the product of a number between 1 and 10, and a power of 10; that is, we have factored a power of 10 from each number. The absolute value of the exponent of the power of 10 is identical to the number of places we moved the decimal point in going from the first digital form to the second.

A number written in scientific notation may be written in standard form by moving the decimal point in the numeral for the first factor the number of places indicated by the exponent on 10—to the left if the exponent is negative and to the right if it is positive. For example,

$$3.75 \times 10^4 = 37{,}500,$$
$$2.03 \times 10 = 20.3,$$
$$7.34 \times 10^{-4} = 0.000734.$$

Scientific notation can frequently be used to simplify numerical calculations. For example, if we write the quotient

$$\frac{248{,}000}{0.0124}$$

in the form

$$\frac{2.48 \times 10^5}{1.24 \times 10^{-2}},$$

we can perform the computation as follows:

$$\frac{2.48}{1.24} \times \frac{10^5}{10^{-2}} = 2 \times 10^7$$
$$= 20{,}000{,}000.$$

Sometimes it is more convenient to express a number as a product of a power of 10 and a number that is not between 1 and 10. For example, under certain circumstances, any of the following forms may be a useful representation for 6280:

$$628 \times 10, \qquad 62.8 \times 10^2, \qquad 6.28 \times 10^3, \qquad 0.628 \times 10^4.$$

EXERCISE 4.3

A

Express each number using scientific notation.

Examples **a.** 680,000 **b.** 0.000043 **c.** 0.002451

Solutions **a.** $680{,}000 = 6.8 \times 10^5$ **b.** $0.000043 = 4.3 \times 10^{-5}$ **c.** $0.002451 = 2.451 \times 10^{-3}$

1. 285 **2.** 3476 **3.** 21 **4.** 68,742

5. 8,372,000 **6.** 481,000 **7.** 0.024 **8.** 0.0063

9. 0.421 **10.** 0.000523 **11.** 0.000004 **12.** 0.0006

Express each number using standard form.

Examples **a.** 1.01×10^3 **b.** 6.3×10^{-4} **c.** 4.317×10^{-2}

Solutions **a.** $1.01 \times 10^3 = 1010$ **b.** $6.3 \times 10^{-4} = 0.00063$ **c.** $4.317 \times 10^{-2} = 0.04317$

13. 2.4×10^2 **14.** 4.8×10^3 **15.** 6.87×10^5 **16.** 8.31×10^4

17. 5.0×10^{-3} **18.** 8.0×10^{-1} **19.** 2.02×10^{-2} **20.** 4.31×10^{-3}

21. 12.27×10^3 **22.** 14.38×10^4 **23.** 23.5×10^{-4} **24.** 621.0×10^{-2}

Examples **a.** $\frac{1}{4 \times 10^3}$ **b.** $\frac{1}{5 \times 10^{-1}}$

Solutions **a.** $\frac{1}{4 \times 10^3} = \frac{1}{4} \times \frac{1}{10^3} = 0.25 \times 10^{-3}$

$= 0.00025$

b. $\frac{1}{5 \times 10^{-1}} = \frac{1}{5} \times \frac{1}{10^{-1}} = 0.2 \times 10^1$

$= 2$

25. $\frac{1}{2 \times 10^3}$ **26.** $\frac{1}{4 \times 10^4}$ **27.** $\frac{1}{8 \times 10^{-2}}$

28. $\frac{1}{5 \times 10^{-3}}$ **29.** $\frac{3}{5 \times 10^4}$ **30.** $\frac{5}{8 \times 10^2}$

Compute.

Example $\frac{4 \times 10^3 \times 6 \times 10^{-5}}{8 \times 10^{-3}}$

Solution $\frac{4 \times 10^3 \times 6 \times 10^{-5}}{8 \times 10^{-3}} = \frac{24 \times 10^{-2}}{8 \times 10^{-3}}$

$= 3 \times 10$ or 30

31. $\frac{10^3 \times 10^{-6}}{10^2}$ **32.** $\frac{10^3 \times 10^{-7} \times 10^2}{10^{-2} \times 10^4}$

33. $\frac{10^2 \times 10^5 \times 10^{-3}}{10^2 \times 10^2}$ **34.** $\frac{(4 \times 10^3) \times (6 \times 10^{-2})}{3 \times 10^{-7}}$

35. $\dfrac{(2 \times 10^2)^2 \times (3 \times 10^{-3})}{2 \times 10^4}$

36. $\dfrac{(3 \times 10)^3 \times (2 \times 10^{-1})}{2 \times 10^{-2}}$

37. $\dfrac{(2 \times 10^{-3}) \times (6 \times 10^2)^2}{(2 \times 10^{-2})^2}$

38. $\dfrac{(8 \times 10^4)^2 \times (3 \times 10)^3}{(6 \times 10^{-2})^2}$

B

Example $\dfrac{0.016 \times 3 \times 0.0028}{0.064}$

Solution

$$\frac{0.016 \times 3 \times 0.0028}{0.064} = \frac{16 \times 10^{-3} \times 3 \times 28 \times 10^{-4}}{64 \times 10^{-3}}$$

$$= \frac{16 \times 3 \times 28}{64} \times 10^{-4}$$

$$= 21 \times 10^{-4} \quad \text{or} \quad 0.0021$$

39. $\dfrac{0.6 \times 0.00084 \times 0.093}{0.00021 \times 0.00031}$

40. $\dfrac{0.065 \times 2.2 \times 50}{1.30 \times 0.011 \times 0.05}$

41. $\dfrac{28 \times 0.0006 \times 450}{1.5 \times 700 \times 0.018}$

42. $\dfrac{0.0054 \times 0.05 \times 300}{0.0015 \times 0.27 \times 80}$

43. $\dfrac{420 \times 0.0016 \times 800}{0.0028 \times 1200 \times 20}$

44. $\dfrac{0.0027 \times 0.004 \times 650}{260 \times 0.0001 \times 0.009}$

45. The speed of light is approximately 300,000,000 meters per second.

a. Express this number in scientific notation.

b. Express the speed of light in inches per second (1 inch equals 2.54 centimeters and 1 meter equals 100 centimeters).

46. One light-year is the number of miles traveled by light in 1 year and the speed of light is approximately 186,000 miles per second. Express in scientific notation the number of miles in 1 light-year.

4.4

RATIONAL EXPONENTS

If the laws of exponents developed in Section 4.1 are to hold for rational exponents, meanings consistent with these laws must be assigned to powers with rational exponents. Let us examine exponents that are the reciprocals of natural numbers, that is, exponents of the form $1/n$, where n is a natural number. We first assume that there exists a number $a^{1/n}$, where a is positive. If the third law of exponents is to hold for these new numbers, we have, for $n \in N$, $a > 0$,

$$(a^{1/n})^n = a^{n/n} = a,$$

and $a^{1/n}$ *must be defined as one of n equal factors of a.* The number $a^{1/n}$ is called an ***n*th root of *a***.

A real number a may have one, two, or no real nth roots, where $n \in N$.

1. If n is odd, then there is one real nth root of a. Thus, $(-8)^{1/3} = -2$ because $(-2)^3 = -8$, and $64^{1/3} = 4$ because $4^3 = 64$.
2. If n is even and a is positive, then there are two real nth roots. In this case, we use $a^{1/n}$ *for the positive (principal) root, and* $-a^{1/n}$ *for the negative root.* Thus, $16^{1/2} = 4$ because $4^2 = 16$, $-16^{1/2} = -4$ because $-4^2 = -16$, $81^{1/4} = 3$ because $3^4 = 81$, and $-81^{1/4} = -3$ because $-3^4 = -81$.
3. If n is even and a is negative, there are no real nth roots of a. [*Note*: Although $-4^{1/2} = -2$, $(-4)^{1/2}$ is not defined in R because there is no number a in R for which $a^2 = -4$.]
4. For n even or odd, there is one nth root of 0. Thus,

$$0^{1/7} = 0, \qquad 0^{1/4} = 0, \qquad \text{and} \qquad 0^{1/50} = 0.$$

We define the general positive rational exponent to obey the third law of exponents. Thus, for nonnegative real numbers a, the number $a^{m/n}$ is defined by

$$a^{m/n} = (a^{1/n})^m.$$

It can be shown that $(a^m)^{1/n}$ is also equal to $a^{m/n}$, so that

$$a^{m/n} = (a^{1/n})^m = (a^m)^{1/n},$$

and we can look at $a^{m/n}$ in two ways, either as the mth power of the nth root of a, or as the nth root of the mth power of a. For example,

$$8^{2/3} = (8^{1/3})^2 = (2)^2 = 4,$$

or

$$8^{2/3} = (8^2)^{1/3} = (64)^{1/3} = 4.$$

Hereafter, we shall use whichever form is most convenient for the purpose at hand. In the foregoing case the first form is preferred because it is easier to extract the root first than it is to recognize the root after the number is squared.

Because we defined $a^{1/n}$ to be the positive nth root of a for a positive and n an even natural number, and since a^m is positive for a negative and m an even natural number, we must also agree that, for m and n even natural numbers and a any real number,

$$a^{m/n} = (a^m)^{1/n} = |a|^{m/n}.$$

For example,

$$(-4)^{2/2} = |-4|^{2/2} = 4^1 = 4.$$

Note that, in the case where a is negative, despite the fact that $\frac{2}{2} = 1$, we cannot use $(-4)^1$ in place of $(-4)^{2/2}$, because $-4^1 = -4$ whereas, as was observed above, $(-4)^{2/2} = 4$.

To extend meaning to powers with negative rational exponents, let us first observe that for positive integers m and n, we have

$$-\frac{m}{n} = \frac{-m}{n}.$$

This fact, together with the definition in Section 4.2 for a^{-n}, implies that, for $a \neq 0$,

$$a^{-m/n} = (a^{1/n})^{-m} = \frac{1}{(a^{1/n})^m} = \frac{1}{a^{m/n}}.$$

Hence, it is consistent to define

$$a^{-m/n} = \frac{1}{a^{m/n}},$$

for m and n positive integers, $a \neq 0$, and $a^{1/n}$ a real number.

We assume that powers with rational exponents—positive, negative, or 0—obey all the laws of exponents set forth on pages 97–98.

Although the matter will not be discussed here, powers of positive numbers can be and are defined for all real-number exponents. We shall consider this again in Chapter 9 when we examine exponential functions.

Recall that any number that can be expressed as the quotient of two integers is called a rational number, and any real number that cannot be so expressed is called an irrational number. Some powers with rational number exponents are rational numbers and some are irrational numbers. Any expression such as $a^{1/n}$ represents a rational number if and only if a is the nth power of a rational number. For example, $4^{1/2}$, $(^{-27}/_8)^{1/3}$, and $(81)^{1/4}$ are rational numbers equal to 2, $^{-3}/_2$, and 3, respectively, while $2^{1/2}$, $5^{1/3}$, and $7^{1/4}$ are irrational. In Section 4.5 we shall consider how to obtain rational approximations for some irrational numbers.

EXERCISE 4.4

A

In Problems 1–62, assume that all bases are positive unless otherwise specified.

Write each expression using a basic numeral or fraction in lowest terms

Examples **a.** $64^{1/2}$ **b.** $(-27)^{4/3}$ **c.** $8^{-2/3}$

Solutions **a.** $64^{1/2} = 8$

b. $(-27)^{4/3} = [(-27)^{1/3}]^4$
$= [-3]^4$
$= 81$

c. $8^{-2/3} = (8^{1/3})^{-2}$
$= (2)^{-2}$
$= \frac{1}{4}$

1. $9^{1/2}$ **2.** $25^{1/2}$ **3.** $32^{1/5}$ **4.** $27^{1/3}$

5. $(-8)^{1/3}$ **6.** $(-27)^{1/3}$ **7.** $27^{2/3}$ **8.** $32^{3/5}$

9. $81^{3/4}$ **10.** $125^{2/3}$ **11.** $(-8)^{4/3}$ **12.** $(-64)^{2/3}$

13. $16^{-1/2}$ **14.** $8^{-1/3}$ **15.** $16^{-3/4}$ **16.** $27^{-2/3}$

Write each expression as a product or quotient of powers in which each variable occurs but once and all exponents are positive.

Examples **a.** $y^{3/4}y^{-1/2}$ **b.** $\dfrac{x^{5/6}}{x^{2/3}}$ **c.** $\dfrac{(x^{1/2}y^2)^2}{(x^{2/3}y)^3}$

Solutions **a.** $y^{3/4}y^{-1/2} = y^{3/4+(-1/2)}$

$= y^{1/4}$

b. $\dfrac{x^{5/6}}{x^{2/3}} = \dfrac{x^{5/6}}{x^{4/6}}$

$= x^{1/6}$

c. $\dfrac{(x^{1/2}y^2)^2}{(x^{2/3}y)^3} = \dfrac{xy^4}{x^2y^3}$

$= \dfrac{y}{x}$

17. $x^{1/3}x^{1/3}$ **18.** $y^{1/2}y^{3/2}$ **19.** $\dfrac{x^{2/3}}{x^{1/3}}$ **20.** $\dfrac{x^{3/4}}{x^{1/4}}$

21. $(a^{1/2})^3$ **22.** $(b^6)^{2/3}$ **23.** $x^{-3/4}x^{1/4}$ **24.** $y^{-2/3}y^{5/3}$

25. $(a^{2/3}b)^{1/2}$ **26.** $(a^{1/2}b^{1/3})^6$ **27.** $\left(\dfrac{a^6}{b^3}\right)^{2/3}$ **28.** $\left(\dfrac{a^{1/2}}{a^2}\right)^2$

29. $\left(\dfrac{x^{-2}y^{-1/3}z}{x^{-5/3}y^{-2/3}z^{2/3}}\right)^3$ **30.** $\left(\dfrac{x^{1/4}y^{3/4}z^{-1}}{x^{-3/4}y^{1/4}z^0}\right)^2$

31. $\left(\dfrac{a^{2/3}b^{-5/3}c^{-2}}{a^{-1}b^0c^{-7}}\right)^{-1/5}$ **32.** $\left(\dfrac{a^{1/4}b^{3/2}c^{1/2}}{a^{-7/4}b^{13/2}c^{5/2}}\right)^{-1/2}$

Write each product so that each base of a power occurs at most once in each term.

Examples **a.** $y^{1/3}(y + y^{2/3})$ **b.** $x^{-3/4}(x^{1/4} + x^{3/4})$

Solutions **a.** $y^{1/3}(y + y^{2/3}) = y^{1/3+1} + y^{1/3+2/3}$

$= y^{4/3} + y$

b. $x^{-3/4}(x^{1/4} + x^{3/4}) = x^{-3/4+1/4} + x^{-3/4+3/4}$

$= x^{-2/4} + x^0$

$= x^{-1/2} + 1$

33. $x^{1/2}(x + x^{1/2})$ **34.** $x^{1/5}(x^{2/5} + x^{4/5})$ **35.** $x^{1/3}(x^{2/3} - x^{1/3})$

36. $x^{3/8}(x^{1/4} - x^{1/2})$ **37.** $x^{-3/4}(x^{-1/4} + x^{3/4})$ **38.** $y^{-1/4}(y^{3/4} + y^{5/4})$

Factor as indicated.

Examples **a.** $y^{3/4} = y^{1/4}(\underline{?})$ **b.** $y^{-1/2} + y^{1/2} = y^{-1/2}(\underline{?})$

Solutions **a.** $y^{3/4} = y^{1/4}(y^{1/2})$ **b.** $y^{-1/2} + y^{1/2} = y^{-1/2}(1 + y)$

39. $x^{3/5} = x^{1/5}(\underline{?})$ **40.** $x^{7/8} = x^{3/8}(\underline{?})$ **41.** $x^{-1/3} = x^{-2/3}(\underline{?})$

42. $y^{-1/4} = y(\underline{?})$ **43.** $x^{1/3} = x(\underline{?})$ **44.** $y^{3/5} = y(\underline{?})$

45. $x^{3/2} + x = x(\underline{?})$ **46.** $y^{1/2} + y = y(\underline{?})$ **47.** $x - x^{2/3} = x^{1/3}(\underline{?})$

48. $a^{2/3} + a^{1/3} = a(\underline{?})$ **49.** $x^{1/2} + x^{3/2} = x^{3/2}(\underline{?})$ **50.** $b^{3/5} + b^{6/5} = b^{6/5}(\underline{?})$

51. $x^{-3/2} + x^{-1/2} = x^{-1/2}(\underline{?})$ **52.** $z^{1/2} + z^{-1/2} = z^{-1/2}(\underline{?})$

B

Simplify. Assume that m, n > 0.

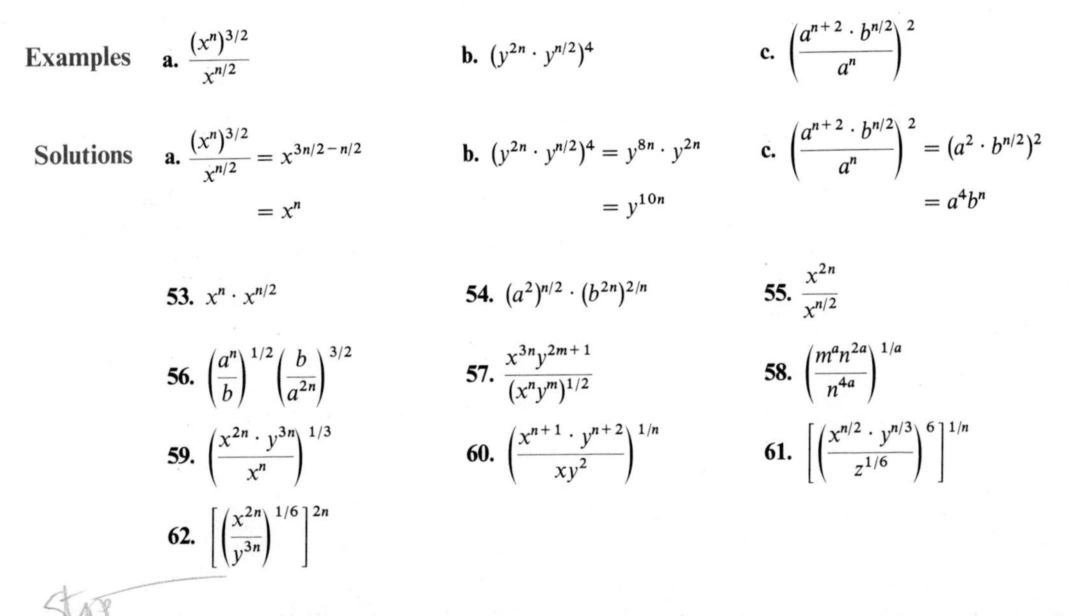

Examples **a.** $\dfrac{(x^n)^{3/2}}{x^{n/2}}$ **b.** $(y^{2n} \cdot y^{n/2})^4$ **c.** $\left(\dfrac{a^{n+2} \cdot b^{n/2}}{a^n}\right)^2$

Solutions **a.** $\dfrac{(x^n)^{3/2}}{x^{n/2}} = x^{3n/2 - n/2}$
$= x^n$

b. $(y^{2n} \cdot y^{n/2})^4 = y^{8n} \cdot y^{2n}$
$= y^{10n}$

c. $\left(\dfrac{a^{n+2} \cdot b^{n/2}}{a^n}\right)^2 = (a^2 \cdot b^{n/2})^2$
$= a^4 b^n$

53. $x^n \cdot x^{n/2}$ **54.** $(a^2)^{n/2} \cdot (b^{2n})^{2/n}$ **55.** $\dfrac{x^{2n}}{x^{n/2}}$

56. $\left(\dfrac{a^n}{b}\right)^{1/2}\left(\dfrac{b}{a^{2n}}\right)^{3/2}$ **57.** $\dfrac{x^{3n}y^{2m+1}}{(x^n y^m)^{1/2}}$ **58.** $\left(\dfrac{m^a n^{2a}}{n^{4a}}\right)^{1/a}$

59. $\left(\dfrac{x^{2n} \cdot y^{3n}}{x^n}\right)^{1/3}$ **60.** $\left(\dfrac{x^{n+1} \cdot y^{n+2}}{xy^2}\right)^{1/n}$ **61.** $\left[\left(\dfrac{x^{n/2} \cdot y^{n/3}}{z^{1/6}}\right)^6\right]^{1/n}$

62. $\left[\left(\dfrac{x^{2n}}{y^{3n}}\right)^{1/6}\right]^{2n}$

stop

In the foregoing problems, the values of the variable bases were restricted to positive real numbers. In Problems 63–68, simplify each expression. Consider variable bases to be any element of the set of real numbers, and state all necessary restrictions.

Examples **a.** $[(-3)^2]^{1/2}$ **b.** $[u^2(u + 5)]^{1/2}$

Solutions **a.** $[(-3)^2]^{1/2} = |-3| = 3$ **b.** $[u^2(u + 5)]^{1/2} = |u|(u + 5)^{1/2} \quad (u \geq -5)$

63. $[(-5)^2]^{1/2}$ **64.** $[(-3)^{12}]^{1/4}$ **65.** $(4x^2)^{1/2}$

66. $(9y^6)^{1/2}$ **67.** $\dfrac{2}{[x^2(x+5)]^{1/2}}$ **68.** $\left[\dfrac{9}{x^6(x^2+2)}\right]^{1/2}$

69. Which is greater, $16^{1/4}$ or $16^{1/2}$? $\left(\dfrac{1}{16}\right)^{1/4}$ or $\left(\dfrac{1}{16}\right)^{1/2}$? Make a conjecture about the order of $a^{1/n}$ and $a^{1/m}$ when $n > m$ and $a > 1$, and when $n > m$ and $0 < a < 1$.

70. Show by an example that $(a+b)^{1/2} \neq a^{1/2} + b^{1/2}$ for all $a, b \in R$.

4.5

RADICALS

In Section 4.4 we referred to $a^{1/n}$ (when it exists) as the nth root of a; that is, $a^{1/n}$ is one of n equal factors of a. An alternative symbol for the nth root of a, when $n \in N$ and $n \geq 2$, is defined by

$$\boldsymbol{a^{1/n} = \sqrt[n]{a}}.$$

In many cases the latter form is more convenient to use. In such a representation, the symbol $\sqrt{}$ is called a **radical**, a is called the **radicand**, n is called the **index**, and the expression is said to be a **radical of order *n***. We require that the index be a natural number greater than or equal to 2. If no index is written, the index is understood to be 2, and the expression represents the square root of the radicand.

Since, from Section 4.4, for $a > 0$,

$$a^{m/n} = (a^m)^{1/n} = (a^{1/n})^m,$$

we may write a power with a rational exponent as

$$\boldsymbol{a^{m/n} = \sqrt[n]{a^m} = \left(\sqrt[n]{a}\right)^m},$$

where the denominator of the exponent is the index of the radical and the numerator of the exponent is either the exponent of the radicand or the exponent of the root. For example,

$$x^{2/3} = \sqrt[3]{x^2} = \left(\sqrt[3]{x}\right)^2,$$

and

$$8^{2/3} = \sqrt[3]{8^2} = \left(\sqrt[3]{8}\right)^2.$$

Since we have restricted the index of a radical to be a natural number, we must always express a fractional exponent in standard form (m/n or $-m/n$) before writing the power in radical form. Thus,

$$x^{-(3/4)} = x^{(-3/4)} = \sqrt[4]{x^{-3}}.$$

In Section 4.4 we observed that, for n even, $a^{n/n} = |a|$ for all real numbers a. Therefore, we have

$$\boldsymbol{\sqrt[n]{a^n} = |a| \qquad (n \text{ even})}.$$

Thus, the symbol $\sqrt[n]{a^n}$ represents only the positive (principal) nth root of a^n when n is even. In particular,

$$\sqrt{a^2} = |a|.$$

The negative nth root of a^n is given by $-\sqrt[n]{a^n}$. In considering odd indices, there is no ambiguous interpretation possible. That is,

$$\sqrt[n]{a^n} = a \qquad (n \text{ odd})$$

for all values of a. For example,

$$\sqrt{2^2} = |2| = 2, \qquad \sqrt{(-2)^2} = |-2| = 2,$$
$$-\sqrt{2^2} = -|2| = -2, \qquad -\sqrt{(-2)^2} = -|-2| = -2;$$

$$\sqrt[3]{8} = 2, \qquad \sqrt[3]{-8} = -2, \qquad -\sqrt[3]{8} = -2, \qquad -\sqrt[3]{-8} = -(-2) = 2;$$

$$\sqrt{9x^2} = \sqrt{9}\sqrt{x^2} = 3|x|;$$

$$\sqrt{x^2 - 2x + 1} = \sqrt{(x-1)^2} = |x-1|.$$

Because, as observed in Section 4.4, $a^{1/n}$ represents a rational number if and only if a is the nth power of a rational number, the same fact is true of $\sqrt[n]{a}$. Thus, $\sqrt{4}$, $-\sqrt[3]{27}$, $\sqrt[4]{81/16}$, and $\sqrt[5]{-32}$ are rational numbers equal to 2, -3, $3/2$, and -2. Although irrational numbers such as $\sqrt{5}$, $\sqrt[3]{9}$, $\sqrt[4]{15}$, and $\sqrt[5]{61}$ do not have terminating or repeating decimal representations, we can obtain decimal approximations correct to any desired degree of accuracy. The Table of Squares, Square Roots, and Prime Factors at the end of the book provides approximations for some irrational numbers. For example,

$$\sqrt{2} \approx 1.414,$$

where the symbol $\approx$ is read "is approximately equal to."

EXERCISE 4.5

A

In Problems 1–40, assume each variable and each radicand represents a positive real number.

Write each expression in radical form.

Examples **a.** $5^{1/2}$ **b.** $(xy)^{2/3}$ **c.** $(x - y^2)^{-1/2}$

Solutions **a.** $5^{1/2} = \sqrt{5}$ **b.** $(xy)^{2/3} = \sqrt[3]{x^2y^2}$ **c.** $(x - y^2)^{-1/2} = \dfrac{1}{\sqrt{x - y^2}}$

1. $3^{1/2}$ 2. $2^{1/3}$ 3. $x^{3/2}$ 4. $y^{2/3}$
5. $3y^{1/4}$ 6. $5x^{1/3}$ 7. $xy^{1/3}$ 8. $x^2y^{1/2}$
9. $(xy)^{1/3}$ 10. $(x^2y)^{1/3}$ 11. $-3x^{3/5}$ 12. $-5y^{2/3}$
13. $(x+2y)^{1/2}$ 14. $(x-y)^{1/3}$ 15. $(2x-y)^{2/3}$ 16. $(3x+y)^{3/4}$
17. $4^{-1/3}$ 18. $6^{-2/5}$ 19. $x^{-2/3}$ 20. $y^{-2/7}$

Represent each expression with positive fractional exponents.

Examples a. $\sqrt{2^3}$ b. $\sqrt[3]{7a^2}$

Solutions a. $\sqrt{2^3} = 2^{3/2}$ b. $\sqrt[3]{7a^2} = (7a^2)^{1/3} = 7^{1/3}a^{2/3}$

21. $\sqrt{5}$ 22. $\sqrt[3]{2}$ 23. $\sqrt[3]{x^2}$ 24. $\sqrt{y^3}$
25. $\sqrt{ab}$ 26. $\sqrt[3]{ab^2}$ 27. $x\sqrt{y}$ 28. $y\sqrt[3]{x^2}$
29. $\sqrt[3]{2ab^2}$ 30. $\sqrt[4]{a^3b}$ 31. $\sqrt{a-b}$ 32. $\sqrt[4]{a+2b}$

Examples a. $\sqrt[3]{a} - 3\sqrt{b}$ b. $\dfrac{3}{\sqrt{x-1}}$

Solutions a. $\sqrt[3]{a} - 3\sqrt{b} = a^{1/3} - 3b^{1/2}$ b. $\dfrac{3}{\sqrt{x-1}} = \dfrac{3}{(x-1)^{1/2}}$

33. $\sqrt{a} - 2\sqrt{b}$ 34. $\sqrt[3]{b} + 2\sqrt{a}$ 35. $\sqrt[3]{a}\sqrt{b}$ 36. $\sqrt{a}\sqrt[3]{ab}$
37. $\dfrac{1}{\sqrt{x}}$ 38. $\dfrac{2}{\sqrt[3]{y}}$ 39. $\dfrac{2}{\sqrt{x+y}}$ 40. $\dfrac{5}{\sqrt[3]{a-b}}$

Find each root indicated.

Examples a. $\sqrt[3]{-8}$ b. $\sqrt[3]{x^6y^3}$ c. $-\sqrt[4]{81x^4}$

Solutions a. $\sqrt[3]{-8} = -2$ b. $\sqrt[3]{x^6y^3} = x^2y$ c. $-\sqrt[4]{81x^4} = -3x$

41. $\sqrt{16}$ 42. $\sqrt{144}$ 43. $-\sqrt{25}$ 44. $-\sqrt{169}$
45. $\sqrt[3]{27}$ 46. $\sqrt[3]{125}$ 47. $\sqrt[3]{-64}$ 48. $\sqrt[5]{32}$
49. $\sqrt[3]{x^3}$ 50. $\sqrt[5]{y^5}$ 51. $\sqrt{x^4}$ 52. $\sqrt{a^6}$
53. $\sqrt[3]{8y^6}$ 54. $\sqrt[3]{27y^9}$ 55. $-\sqrt{x^4y^6}$ 56. $-\sqrt{a^8b^{10}}$
57. $\sqrt{\dfrac{4}{9}x^2y^8}$ 58. $\sqrt{\dfrac{9}{16}a^2b^4}$ 59. $\sqrt[3]{\dfrac{-8}{125}x^3}$ 60. $\sqrt[3]{\dfrac{8}{27}a^3b^6}$
61. $\sqrt[4]{16x^4y^8}$ 62. $\sqrt[5]{-32x^5y^{10}}$ 63. $-\sqrt[3]{-8a^6b^9}$ 64. $-\sqrt[4]{81a^8b^{12}}$

Graph each set of real numbers on a separate line graph. (Use the Table of Squares, Square Roots, and Prime Factors to obtain rational number approximations for irrational numbers.)

Example $\sqrt{4}, -\sqrt{3}, \sqrt{17}, -\sqrt{9}$

Solution

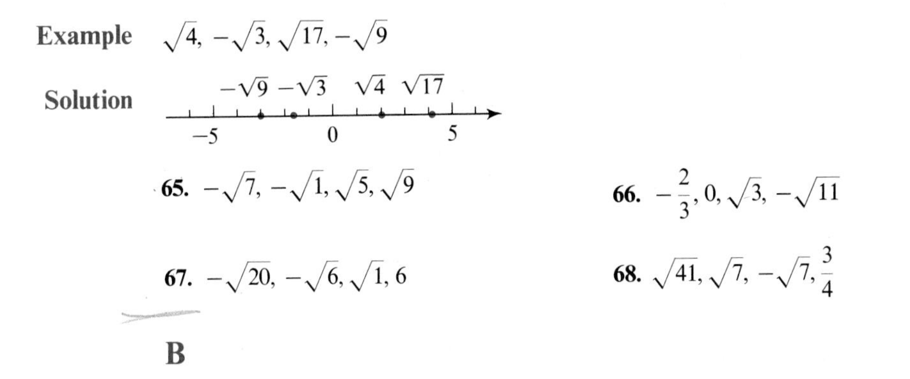

65. $-\sqrt{7}, -\sqrt{1}, \sqrt{5}, \sqrt{9}$

66. $-\frac{2}{3}, 0, \sqrt{3}, -\sqrt{11}$

67. $-\sqrt{20}, -\sqrt{6}, \sqrt{1}, 6$

68. $\sqrt{41}, \sqrt{7}, -\sqrt{7}, \frac{3}{4}$

B

In the foregoing problems, variables and radicands were restricted to represent positive numbers. In Problems 69–76, consider variables and radicands to represent elements of the set of real numbers and use absolute value notation as needed.

Examples **a.** $\sqrt{16x^2}$ **b.** $\sqrt{x^2 - 2xy + y^2}$

Solutions **a.** $\sqrt{16x^2} = 4|x|$ **b.** $\sqrt{x^2 - 2xy + y^2} = \sqrt{(x-y)^2}$
$= |x - y|$

69. $\sqrt{4x^2}$

70. $\sqrt{9x^2y^4}$

71. $\sqrt{x^2 + 2x + 1}$

72. $\sqrt{4x^2 - 4x + 1}$

73. $\dfrac{2}{\sqrt{x^2 + 2xy + y^2}}$

74. $\sqrt{x^4 + 2x^2y^2 + y^4}$

75. $\dfrac{5}{\sqrt{4x^2 - 4xy + y^2}}$

76. $\dfrac{2}{\sqrt{x^4 + 6x^2y^2 + 9y^4}}$

77. Show that $|x| = \sqrt{x^2}$ is equivalent to the definition of $|x|$ given in Section 1.4.

4.6

CHANGING FORMS OF RADICALS

From the definition of a radical and the laws of exponents, we can derive three important relationships. In each of these cases we consider $a, b > 0$ and n a natural number. We have, first,

(1) $$\sqrt[n]{ab} = \sqrt[n]{a}\sqrt[n]{b}.$$

This follows from the fact that

$$\sqrt[n]{ab} = (ab)^{1/n} = a^{1/n}b^{1/n} = \sqrt[n]{a}\sqrt[n]{b}.$$

Relationship (1) can be used to write a radical in a form in which the radicand contains no prime factor or polynomial factor raised to a power greater than or equal to the index of the radical. Thus, we can write

$$\sqrt{18} = \sqrt{3^2}\sqrt{2} = 3\sqrt{2}$$

and

$$\sqrt[3]{16x^3y^5} = \sqrt[3]{2^3x^3y^3}\sqrt[3]{2y^2} = 2xy\sqrt[3]{2y^2},$$

where in each case the radicand is first factored into two factors, one of which consists of factors raised to the same power as the index of the radical. This factor is then "removed" from the radicand.

The second important relationship is

(2) $$\sqrt[n]{\frac{\boldsymbol{a}}{\boldsymbol{b}}} = \frac{\sqrt[n]{\boldsymbol{a}}}{\sqrt[n]{\boldsymbol{b}}}.$$

This follows from the fact that

$$\sqrt[n]{\frac{a}{b}} = \left(\frac{a}{b}\right)^{1/n} = \frac{a^{1/n}}{b^{1/n}} = \frac{\sqrt[n]{a}}{\sqrt[n]{b}}.$$

We can use relation (2) to write a radical in a form in which the radicand contains no fraction. For example,

$$\sqrt{\frac{3}{4}} = \frac{\sqrt{3}}{\sqrt{4}} = \frac{\sqrt{3}}{2}.$$

If the denominator of the radicand is not the square of a monomial, we can use the fundamental principle of fractions to obtain an equivalent form that has such a denominator. Thus, if $x > 0$, then:

$\sqrt{\frac{2}{5x}} = \frac{\sqrt{2}}{\sqrt{5x}}$	Relation (2)
$= \frac{\sqrt{2}\sqrt{5x}}{\sqrt{5x}\sqrt{5x}}$	Fundamental principle of fractions
$= \frac{\sqrt{10x}}{5x}$	Relation (1) and the definition of square root

The foregoing process is called "rationalizing the denominator" of the fraction. If the radical is of order n, we must build the denominator to a radicand that is the nth power of a monomial. For example, if $x \neq 0$, then:

$$\frac{1}{\sqrt[3]{2x}} = \frac{1\sqrt[3]{2x}\sqrt[3]{2x}}{\sqrt[3]{2x}\sqrt[3]{2x}\sqrt[3]{2x}} \qquad \text{Fundamental principle of fractions}$$

$$= \frac{\sqrt[3]{(2x)^2}}{\sqrt[3]{(2x)^3}} \qquad \text{Relation (1)}$$

$$= \frac{\sqrt[3]{4x^2}}{2x} \qquad \text{Definition of a radical}$$

Consider another example. For $x \neq 0$:

$$\sqrt[5]{\frac{6}{16x^3}} = \frac{\sqrt[5]{6}}{\sqrt[5]{16x^3}} \qquad \text{Relation (2)}$$

$$= \frac{\sqrt[5]{6}\sqrt[5]{2x^2}}{\sqrt[5]{16x^3}\sqrt[5]{2x^2}} \qquad \text{Fundamental principle of fractions}$$

$$= \frac{\sqrt[5]{12x^2}}{\sqrt[5]{32x^5}} \qquad \text{Relation (1)}$$

$$= \frac{\sqrt[5]{12x^2}}{2x} \qquad \text{Definition of a radical}$$

The third important relationship is

$$(3) \qquad \sqrt[cn]{a^{cm}} = \sqrt[n]{a^m},$$

where $m \in J$; $c, n \in N$; and $n \geq 2$. This follows from the fact that

$$\sqrt[cn]{a^{cm}} = a^{cm/cn} = a^{m/n} = \sqrt[n]{a^m}.$$

If the right-hand member of (3) is to have meaning in the sense in which we have defined radicals, c must be a natural number factor of the index. Relation (3) permits us to reduce the order of a radical by dividing the index of the radical and the exponent of the radicand by the same natural number. For example,

$$\sqrt[6]{8} = \sqrt[6]{2^3} = \sqrt{2}; \qquad \sqrt[4]{x^2} = \sqrt{x} \qquad (x \geq 0).$$

We must be careful in reading equality into two radicals. For instance, in rewriting $\sqrt[4]{x^2}$ as $\sqrt{x}$, it is necessary that we restrict x to nonnegative values. If x is -3 for example,

$$\sqrt[4]{x^2} = \sqrt[4]{(-3)^2} = \sqrt[4]{9} = \sqrt{3},$$

which is a real number, but

$$\sqrt{x} = \sqrt{-3},$$

which is not a real number.

Because $\sqrt[n]{a^m} = a^{m/n}$, radical expressions similar to those above can be rewritten by first writing the expression in exponential form. For example,

$$\sqrt[4]{x^2} = x^{2/4} = x^{(1 \cdot 2)/(2 \cdot 2)} = x^{1/2} = \sqrt{x}.$$

Application of relations (1), (2), or (3) on pages 115–117, can be used to rewrite radical expressions in various ways, and, in particular, to write them in what is called **simplest form**. A radical expression is said to be in simplest form if the following conditions exist:

1. The radicand contains no polynomial factor raised to a power equal to or greater than the index of the radical.
2. The radicand contains no fractions.
3. No radical expressions are contained in denominators of fractions.
4. The index of the radical and exponents on factors in the radicand have no common factors.

Although we generally change the form of radicals to one of the forms implied above, there are times when such forms are not preferred. For example, in certain situations, $\sqrt{\frac{1}{2}}$ or $\frac{1}{\sqrt{2}}$ may be more useful than the equivalent form $\frac{\sqrt{2}}{2}$.

EXERCISE 4.6

A

Assume that all variables in radicands in this exercise denote positive real numbers only.

Change to simplest form.

Examples **a.** $\sqrt{300}$ **b.** $\sqrt[3]{2x^7y^3}$ **c.** $\sqrt{2xy}\sqrt{8x}$

Solutions

a. $\sqrt{300} = \sqrt{100}\sqrt{3}$
$= 10\sqrt{3}$

b. $\sqrt[3]{2x^7y^3} = \sqrt[3]{x^6y^3}\sqrt[3]{2x}$
$= x^2y\sqrt[3]{2x}$

c. $\sqrt{2xy}\sqrt{8x} = \sqrt{16x^2y}$
$= \sqrt{16x^2}\sqrt{y}$
$= 4x\sqrt{y}$

1. $\sqrt{18}$ **2.** $\sqrt{50}$ **3.** $\sqrt{20}$ **4.** $\sqrt{72}$

5. $\sqrt{75}$ **6.** $\sqrt{48}$ **7.** $\sqrt{160}$ **8.** $\sqrt{200}$

9. $\sqrt{x^4}$ **10.** $\sqrt{y^6}$ **11.** $\sqrt{x^3}$ **12.** $\sqrt{y^{11}}$

13. $\sqrt{9x^3}$ **14.** $\sqrt{4y^5}$ **15.** $\sqrt{8x^6}$ **16.** $\sqrt{18y^8}$

17. $\sqrt[3]{x^6}$ **18.** $\sqrt[5]{y^{10}}$ **19.** $\sqrt[4]{x^5}$ **20.** $\sqrt[3]{y^4}$

21. $\sqrt[4]{x^5y^5}$ **22.** $\sqrt[5]{x^{11}y^6}$ **23.** $\sqrt[4]{16xy^7}$ **24.** $\sqrt[3]{8x^4y^5}$

25. $\sqrt[5]{x^7y^9z^{11}}$ **26.** $\sqrt[5]{243a^{12}b^{15}}$ **27.** $\sqrt[6]{a^7b^{12}c^{15}}$ **28.** $\sqrt[6]{128m^8n^7}$

29. $\sqrt[3]{16a^{11}b^{12}c^{13}}$ **30.** $\sqrt[3]{54x^{14}y^{15}z^{16}}$ **31.** $\sqrt[7]{3^7a^8b^9c^{10}}$ **32.** $\sqrt[7]{4^8x^9y^{10}z^{14}}$

33. $\sqrt{18}\sqrt{2}$ **34.** $\sqrt{3}\sqrt{27}$ **35.** $\sqrt{xy}\sqrt{x^5y}$ **36.** $\sqrt{a}\sqrt{ab^2}$

37. $\sqrt[3]{2}\sqrt[3]{4}$ **38.** $\sqrt[4]{3}\sqrt[4]{27}$ **39.** $\sqrt[4]{x^3}\sqrt[4]{x^2}$ **40.** $\sqrt[5]{y^3}\sqrt[5]{y^4}$

41. $\sqrt{3 \times 10^2}$ **42.** $\sqrt{5 \times 10^3}$ **43.** $\sqrt{60{,}000}$ **44.** $\sqrt{800{,}000}$

Rationalize denominators.

Examples **a.** $\sqrt{\frac{1}{3}}$ **b.** $\sqrt[3]{\frac{2}{y^2}}$

Solutions **a.** $\sqrt{\frac{1}{3}} = \frac{\sqrt{1}\sqrt{3}}{\sqrt{3}\sqrt{3}}$

$= \frac{\sqrt{3}}{3}$

b. $\sqrt[3]{\frac{2}{y^2}} = \frac{\sqrt[3]{2}\sqrt[3]{y}}{\sqrt[3]{y^2}\sqrt[3]{y}}$

$= \frac{\sqrt[3]{2}\sqrt[3]{y}}{\sqrt[3]{y^3}}$

$= \frac{\sqrt[3]{2y}}{y}$

45. $\sqrt{\frac{1}{5}}$ **46.** $\sqrt{\frac{2}{3}}$ **47.** $\frac{-1}{\sqrt{2}}$ **48.** $\frac{-\sqrt{3}}{\sqrt{7}}$

49. $\sqrt{\frac{x}{2}}$ **50.** $-\sqrt{\frac{y}{3}}$ **51.** $-\sqrt{\frac{y}{x}}$ **52.** $\sqrt{\frac{2a}{b}}$

53. $\frac{x}{\sqrt{x}}$ **54.** $\frac{-x}{\sqrt{2y}}$ **55.** $\frac{-xy}{\sqrt{y}}$ **56.** $\frac{x}{\sqrt{xy}}$

57. $\sqrt[3]{\frac{y}{2x}}$ **58.** $\sqrt[3]{\frac{1}{6x^2}}$ **59.** $\frac{1}{\sqrt[4]{8}}$ **60.** $\frac{3}{\sqrt[4]{9}}$

Examples **a.** $\frac{\sqrt{a}\sqrt{ab^3}}{\sqrt{b}}$ **b.** $\sqrt[3]{\frac{16y^4}{x^7}}$

Solutions **a.** $\frac{\sqrt{a}\sqrt{ab^3}}{\sqrt{b}} = \sqrt{\frac{a^2b^3}{b}}$

$= \sqrt{a^2b^2}$

$= ab$

b. $\sqrt[3]{\frac{16y^4}{x^7}} = \sqrt[3]{\frac{16y^4 \cdot x^2}{x^7 \cdot x^2}}$

$= \sqrt[3]{\frac{8y^3}{x^9}} \cdot \sqrt[3]{2yx^2}$

$= \frac{2y\sqrt[3]{2x^2y}}{x^3}$

61. $\dfrac{\sqrt{a^5b^3}}{\sqrt{ab}}$ 62. $\dfrac{\sqrt{x}\sqrt{xy^3}}{\sqrt{y}}$ 63. $\sqrt{\dfrac{98x^2y^3}{xy^4}}$ 64. $\sqrt{\dfrac{45x^3}{5y^4}}$

65. $\sqrt[3]{\dfrac{8b^4}{a^6}}$ 66. $\sqrt[3]{\dfrac{16r^4}{4t^3}}$ 67. $\dfrac{\sqrt[5]{a}\sqrt[5]{b^2}}{\sqrt[5]{a^3b}}$ 68. $\dfrac{\sqrt[5]{x^2}\sqrt[5]{y^3}}{\sqrt[5]{xy^2}}$

Rationalize numerators.

69. $\dfrac{\sqrt{3}}{3}$ 70. $\dfrac{\sqrt{2}}{3}$ 71. $\dfrac{\sqrt{x}}{\sqrt{y}}$ 72. $\dfrac{\sqrt{xy}}{x}$

B

Reduce the order of each radical.

Examples a. $\sqrt[4]{5^2}$ b. $\sqrt[6]{9}$ c. $\sqrt[4]{x^2y^2}$

Solutions a. $\sqrt[4]{5^2} = \sqrt[2\cdot 2]{5^2} = \sqrt{5}$ b. $\sqrt[6]{9} = \sqrt[3\cdot 2]{3^2} = \sqrt[3]{3}$ c. $\sqrt[4]{x^2y^2} = \sqrt[2\cdot 2]{x^2y^2} = \sqrt{xy}$

73. $\sqrt[4]{3^2}$ 74. $\sqrt[6]{2^2}$ 75. $\sqrt[6]{3^3}$ 76. $\sqrt[8]{5^2}$

77. $\sqrt[6]{81}$ 78. $\sqrt[10]{32}$ 79. $\sqrt[6]{x^3}$ 80. $\sqrt[9]{y^3}$

81. $\sqrt[9]{8x^3}$ 82. $\sqrt[4]{16y^2}$ 83. $\sqrt[6]{x^3y^3}$ 84. $\sqrt[4]{4x^2y^2}$

Express as radicals of the same order and multiply.

Examples a. $\sqrt{3}, \sqrt[3]{5}$ b. $\sqrt[3]{xy}, \sqrt[4]{xy}$

Solutions a. (Least common index is 6)

$\sqrt{3} = \sqrt[3\cdot 2]{3^3} = \sqrt[6]{27}$ $\sqrt[3]{5} = \sqrt[2\cdot 3]{5^2} = \sqrt[6]{25}$

$\sqrt[6]{27}\cdot\sqrt[6]{25} = \sqrt[6]{675}$

b. (Least common index is 12)

$\sqrt[3]{xy} = \sqrt[4\cdot 3]{x^4y^4} = \sqrt[12]{x^4y^4}$ $\sqrt[4]{xy} = \sqrt[3\cdot 4]{x^3y^3} = \sqrt[12]{x^3y^3}$

$\sqrt[12]{x^4y^4}\cdot\sqrt[12]{x^3y^3} = \sqrt[12]{x^7y^7}$

85. $\sqrt[3]{3}, \sqrt{2}$ 86. $\sqrt[3]{2}, \sqrt[4]{2}$ 87. $\sqrt[4]{5}, \sqrt{2}$ 88. $\sqrt[3]{x}, \sqrt{x}$

89. $\sqrt[5]{y}, \sqrt{y}$ 90. $\sqrt{2x}, \sqrt[5]{3y^2}$ 91. $\sqrt[3]{2x}, \sqrt{y}, \sqrt[4]{xy}$ 92. $\sqrt[3]{x}, \sqrt[4]{y}, \sqrt{z}$

93. Show by an example that $(\sqrt{a} + \sqrt{b})^2 \neq a + b$ for all $a, b \in R$.

94. Show by an example that $\sqrt{a+b} \neq \sqrt{a} + \sqrt{b}$ for all $a, b \in R$.

95. Show by an example that $\sqrt{a^2} \neq a$ for all $a \in R$.

4.7

EXPRESSIONS CONTAINING RADICALS

The distributive property,

$$a(b + c) = ab + ac, \tag{1}$$

is assumed to hold for all real numbers. By the symmetric property of equality and the commutative property of multiplication, (1) can be written as

$$ba + ca = (b + c)a.$$

Since at this time all radical expressions have been defined so that they represent real numbers, the distributive property holds for radical expressions, and we may invoke it to write sums containing radicals of the same order as a single term. For example,

$$3\sqrt{3} + 4\sqrt{3} = (3 + 4)\sqrt{3} = 7\sqrt{3},$$

and

$$\sqrt{x} + 7\sqrt{x} = (1 + 7)\sqrt{x} = 8\sqrt{x}.$$

A direct application of (1) permits us to write certain products which contain parentheses as expressions without parentheses. For example,

$$x(\sqrt{2} + \sqrt{3}) = x\sqrt{2} + x\sqrt{3},$$

and

$$\begin{aligned}\sqrt{3}(\sqrt{2x} + \sqrt{6}) &= \sqrt{6x} + \sqrt{18}\\ &= \sqrt{6x} + \sqrt{9}\sqrt{2}\\ &= \sqrt{6x} + 3\sqrt{2}.\end{aligned}$$

In Chapter 2 we agreed to factor from each term of an expression only those common factors that are integers or positive integral powers of variables. However, we can (if we wish) consider other real numbers for factors. Thus, when the distributive property in the form

$$ab + ac = a(b + c)$$

is invoked, radicals common to each term in an expression may be factored from the expression. For example,

$$\begin{aligned}\sqrt{a} + \sqrt{ab} &= \sqrt{a} + \sqrt{a}\sqrt{b}\\ &= \sqrt{a}(1 + \sqrt{b}).\end{aligned}$$

Recall from Section 4.6 that a monomial denominator of a fraction of the form $a/\sqrt{b}$ can be rationalized by multiplying the numerator and the denominator by $\sqrt{b}$. Thus,

$$\frac{a}{\sqrt{b}} = \frac{a\sqrt{b}}{\sqrt{b}\sqrt{b}} = \frac{a\sqrt{b}}{b}.$$

The distributive property provides us with a means of rationalizing denominators of fractions in which radicals occur in one or both of the two terms of a binomial. To accomplish this, we first recall from Section 2.6 that

$$(a - b)(a + b) = a^2 - b^2,$$

where the product contains no linear term. Each of the two factors of a product exhibiting this property is said to be the **conjugate** of the other. Now consider a fraction of the form

$$\frac{a}{b + \sqrt{c}} \qquad (b + \sqrt{c} \neq 0).$$

If we multiply the numerator and the denominator of this fraction by the conjugate of the denominator, the denominator of the resulting fraction will contain no term linear in $\sqrt{c}$, and hence, it will be free of radicals. That is,

$$\frac{a(b - \sqrt{c})}{(b + \sqrt{c})(b - \sqrt{c})} = \frac{ab - a\sqrt{c}}{b^2 - c} \qquad (b^2 - c \neq 0),$$

where the denominator has been rationalized. This process is equally applicable to fractions of the form

$$\frac{a}{\sqrt{b} + \sqrt{c}},$$

since

$$\frac{a(\sqrt{b} - \sqrt{c})}{(\sqrt{b} + \sqrt{c})(\sqrt{b} - \sqrt{c})} = \frac{a\sqrt{b} - a\sqrt{c}}{b - c} \qquad (b - c \neq 0).$$

EXERCISE 4.7

Assume that all radicands and variables in this exercise are positive real numbers.

Write each sum as a single term.

Examples **a.** $3\sqrt{20} + \sqrt{45}$

b. $\sqrt{32x} + \sqrt{2x} - \sqrt{18x}$

Solutions **a.** $3\sqrt{20} + \sqrt{45}$
$= 3 \cdot 2\sqrt{5} + 3\sqrt{5}$
$= 6\sqrt{5} + 3\sqrt{5}$
$= 9\sqrt{5}$

b. $\sqrt{32x} + \sqrt{2x} - \sqrt{18x}$
$= 4\sqrt{2x} + \sqrt{2x} - 3\sqrt{2x}$
$= 2\sqrt{2x}$

1. $3\sqrt{7} + 2\sqrt{7}$
2. $5\sqrt{2} - 3\sqrt{2}$
3. $4\sqrt{3} - \sqrt{27}$
4. $\sqrt{75} + 2\sqrt{3}$
5. $\sqrt{50x} + \sqrt{32x}$
6. $\sqrt{8y} - \sqrt{18y}$
7. $3\sqrt{4xy^2} - 4\sqrt{9xy^2}$
8. $2\sqrt{8y^2z} + 3\sqrt{32y^2z}$
9. $3\sqrt{8a} + 2\sqrt{50a} - \sqrt{2a}$
10. $\sqrt{3b} - 2\sqrt{12b} + 3\sqrt{48b}$
11. $3\sqrt[3]{16} - \sqrt[3]{2}$
12. $\sqrt[3]{54} + 2\sqrt[3]{128}$

Write each expression without parentheses and then write all radicals in simple form.

Examples

a. $4(\sqrt{3} + 1)$

b. $\sqrt{x}(\sqrt{2x} - \sqrt{x})$

c. $(\sqrt{x} - \sqrt{y})(\sqrt{x} + \sqrt{y})$

Solutions

a. $4(\sqrt{3} + 1) = 4\sqrt{3} + 4$

b. $\sqrt{x}(\sqrt{2x} - \sqrt{x})$
$= x\sqrt{2} - x$

c. $(\sqrt{x} - \sqrt{y})(\sqrt{x} + \sqrt{y})$
$= x - y$

13. $2(3 - \sqrt{5})$
14. $5(2 - \sqrt{7})$
15. $\sqrt{2}(3 + \sqrt{3})$
16. $\sqrt{3}(5 - \sqrt{2})$
17. $\sqrt{2}(\sqrt{6} + \sqrt{10})$
18. $\sqrt{3}(\sqrt{12} - \sqrt{15})$
19. $\sqrt{6}(\sqrt{2} + \sqrt{3})$
20. $\sqrt{15}(\sqrt{3} - \sqrt{5})$
21. $(3 + \sqrt{5})(2 - \sqrt{5})$
22. $(1 - \sqrt{2})(2 + \sqrt{2})$
23. $(\sqrt{x} - 3)(\sqrt{x} + 3)$
24. $(2 + \sqrt{x})(2 - \sqrt{x})$
25. $(\sqrt{2} - \sqrt{3})(\sqrt{2} + 2\sqrt{3})$
26. $(\sqrt{3} - \sqrt{5})(2\sqrt{3} + \sqrt{5})$
27. $(\sqrt{5} - \sqrt{2})^2$
28. $(\sqrt{2} - 2\sqrt{3})^2$
29. $(2 - \sqrt[3]{4})(2 + \sqrt[3]{4})$
30. $(\sqrt[3]{a^2} + \sqrt[3]{b^2})(\sqrt[3]{a} - \sqrt[3]{b})$

Change each expression to the form indicated.

Examples

a. $3 + \sqrt{18} = 3(\underline{?})$

b. $\sqrt{x} + \sqrt{xy} = \sqrt{x}(\underline{?})$

Solutions

a. $3 + \sqrt{18} = 3 + 3\sqrt{2}$
$= 3(1 + \sqrt{2})$

b. $\sqrt{x} + \sqrt{xy} = \sqrt{x} + \sqrt{x}\sqrt{y}$
$= \sqrt{x}(1 + \sqrt{y})$

31. $2 + 2\sqrt{3} = 2(\underline{?})$
32. $5 + 10\sqrt{2} = 5(\underline{?})$
33. $2\sqrt{27} + 6 = 6(\underline{?})$
34. $5\sqrt{5} - \sqrt{25} = 5(\underline{?})$
35. $4 + \sqrt{16y} = 4(\underline{?})$
36. $3 + \sqrt{18x} = 3(\underline{?})$
37. $y\sqrt{3} - x\sqrt{3} = \sqrt{3}(\underline{?})$
38. $x\sqrt{2} - \sqrt{8} = \sqrt{2}(\underline{?})$
39. $\sqrt{2} - \sqrt{6} = \sqrt{2}(\underline{?})$
40. $\sqrt{12} - 2\sqrt{6} = 2\sqrt{3}(\underline{?})$
41. $\sqrt{x} + \sqrt{3x} = \sqrt{x}(\underline{?})$
42. $\sqrt{2x} - \sqrt{x} = \sqrt{x}(\underline{?})$

Reduce each fraction to lowest terms.

Examples **a.** $\dfrac{4+6\sqrt{3}}{2}$ **b.** $\dfrac{2x-\sqrt{8x^2}}{4x}$ **c.** $\dfrac{\sqrt{6}-\sqrt{8}}{\sqrt{2}}$

Solutions

a. $$\frac{4+6\sqrt{3}}{2} = \frac{2(2+3\sqrt{3})}{2} = 2+3\sqrt{3}$$

b. $$\frac{2x-\sqrt{8x^2}}{4x} = \frac{2x-2x\sqrt{2}}{4x} = \frac{2x(1-\sqrt{2})}{2\cdot 2x} = \frac{1-\sqrt{2}}{2}$$

c. $$\frac{\sqrt{6}-\sqrt{8}}{\sqrt{2}} = \frac{\sqrt{2}\sqrt{3}-2\sqrt{2}}{\sqrt{2}} = \frac{\sqrt{2}(\sqrt{3}-2)}{\sqrt{2}} = \sqrt{3}-2$$

43. $\dfrac{2+2\sqrt{3}}{2}$ **44.** $\dfrac{6+2\sqrt{5}}{2}$ **45.** $\dfrac{6+2\sqrt{18}}{6}$ **46.** $\dfrac{8-2\sqrt{12}}{4}$

47. $\dfrac{x-\sqrt{x^3}}{x}$ **48.** $\dfrac{xy-x\sqrt{xy^2}}{xy}$ **49.** $\dfrac{x\sqrt{y}-\sqrt{y^3}}{\sqrt{y}}$ **50.** $\dfrac{\sqrt{x}-y\sqrt{x^3}}{\sqrt{x}}$

Rationalize denominators.

Examples **a.** $\dfrac{3}{\sqrt{2}-1}$ **b.** $\dfrac{1}{\sqrt{x}-\sqrt{y}}$

Solutions

a. $$\frac{3}{\sqrt{2}-1} = \frac{3(\sqrt{2}+1)}{(\sqrt{2}-1)(\sqrt{2}+1)} = \frac{3\sqrt{2}+3}{2-1} = 3\sqrt{2}+3$$

b. $$\frac{1}{\sqrt{x}-\sqrt{y}} = \frac{1(\sqrt{x}+\sqrt{y})}{(\sqrt{x}-\sqrt{y})(\sqrt{x}+\sqrt{y})} = \frac{\sqrt{x}+\sqrt{y}}{x-y}$$

51. $\dfrac{4}{1+\sqrt{3}}$ **52.** $\dfrac{1}{2-\sqrt{2}}$ **53.** $\dfrac{2}{\sqrt{7}-2}$ **54.** $\dfrac{2}{4-\sqrt{5}}$

55. $\dfrac{4}{1+\sqrt{x}}$ **56.** $\dfrac{1}{2-\sqrt{y}}$ **57.** $\dfrac{x}{\sqrt{x}-3}$ **58.** $\dfrac{y}{\sqrt{3}-y}$

59. $\dfrac{\sqrt{x}}{\sqrt{x}-\sqrt{y}}$ **60.** $\dfrac{\sqrt{6}-3}{2-\sqrt{6}}$ **61.** $\dfrac{\sqrt{x}+\sqrt{y}}{\sqrt{x}-\sqrt{y}}$ **62.** $\dfrac{\sqrt{x+a}}{1-\sqrt{x+a}}$

Write each expression as a single fraction in which the denominator is rationalized.

63. $\dfrac{1}{\sqrt{2}}+\dfrac{1}{\sqrt{3}}$ **64.** $\dfrac{3}{\sqrt{6}}-\dfrac{2}{\sqrt{3}}$ **65.** $\sqrt{3}-\dfrac{2}{\sqrt{3}}$

66. $\sqrt{5}+\dfrac{1}{\sqrt{5}}$ **67.** $\sqrt{x}+\dfrac{2x}{\sqrt{x}}$ **68.** $\dfrac{3}{\sqrt{2x}}-\dfrac{1}{\sqrt{x}}$

69. $\sqrt{x+1}-\dfrac{x}{\sqrt{x+1}}$ **70.** $\sqrt{x^2-2}-\dfrac{x^2+1}{\sqrt{x^2-2}}$

71. $\dfrac{x}{\sqrt{x^2+1}}-\dfrac{\sqrt{x^2+1}}{x}$ **72.** $\dfrac{x}{\sqrt{x^2-1}}+\dfrac{\sqrt{x^2-1}}{x}$

73. $\sqrt{y^2+1}-\dfrac{x}{\sqrt{y^2+1}}$ **74.** $\dfrac{2y^2}{\sqrt{1-y^2}}+\sqrt{1-y^2}$

75. $\dfrac{x}{\sqrt{x^2+2}}+\dfrac{\sqrt{x^2+2}}{3}$ **76.** $\dfrac{2x}{\sqrt{3-x^2}}-\dfrac{\sqrt{3-x^2}}{x}$

Rationalize numerators.

Examples **a.** $\dfrac{\sqrt{3}-2}{2}$ **b.** $\dfrac{\sqrt{x}-2}{x}$

Solutions **a.**

$$\frac{\sqrt{3}-2}{2}=\frac{(\sqrt{3}-2)(\sqrt{3}+2)}{2(\sqrt{3}+2)}=\frac{3-4}{2\sqrt{3}+4}=\frac{-1}{2\sqrt{3}+4}$$

b.

$$\frac{\sqrt{x}-2}{x}=\frac{(\sqrt{x}-2)(\sqrt{x}+2)}{x(\sqrt{x}+2)}=\frac{x-4}{x\sqrt{x}+2x}$$

77. $\dfrac{1-\sqrt{2}}{2}$ **78.** $\dfrac{\sqrt{3}+\sqrt{2}}{\sqrt{3}}$ **79.** $\dfrac{\sqrt{x}-1}{3}$

80. $\dfrac{4-\sqrt{2y}}{2}$ **81.** $\dfrac{\sqrt{x}-\sqrt{y}}{x}$ **82.** $\dfrac{2\sqrt{x}+\sqrt{y}}{\sqrt{xy}}$

CHAPTER SUMMARY

[4.1–4.2] The laws of exponents are determined by the definitions adopted for powers:

(I) $$a^m \cdot a^n = a^{m+n}$$

(II) $$\frac{a^m}{a^n} = a^{m-n} = \frac{1}{a^{n-m}} \qquad (a \neq 0)$$

(III) $$(a^m)^n = a^{mn}$$

(IV) $$(ab)^n = a^n b^n$$

(V) $$\left(\frac{a}{b}\right)^n = \frac{a^n}{b^n} \qquad (b \neq 0)$$

[4.3] A number is expressed in **scientific notation** when it is expressed as a product of a number between 1 and 10 and a power of 10.

[4.4] For each natural number n and real number a, the number $a^{1/n}$, when it exists, is called an ***n*th root of *a***. If n is a natural number and a is a real number,

1. $a^{1/n}$ exists and is positive if a is positive;
2. $a^{1/n}$ does not exist in the set of real numbers if a is negative and n is even;
3. $a^{1/n}$ exists and is positive or negative as a is positive or negative if n is odd;
4. $a^{m/n} = (a^{1/n})^m = (a^m)^{1/n}$;
5. $a^{m/n} = |a|^{m/n}$ if m and n are even.

[4.5] The nth root of a is also designated by a **radical expression** $\sqrt[n]{a}$, where n is the **index** and a is the **radicand**.

$$\sqrt[n]{a} = a^{1/n} \qquad (a^{1/n} \in R)$$

$$\sqrt[n]{a^n} = |a| \qquad (n \text{ an even natural number})$$

$$\sqrt[n]{a^n} = a \qquad (n \text{ an odd natural number})$$

[4.6] The laws for radicals listed below follow from the corresponding laws for exponents:

$$\sqrt[n]{ab} = \sqrt[n]{a}\sqrt[n]{b} \qquad (\sqrt[n]{a}, \sqrt[n]{b} \in R)$$

$$\sqrt[n]{\frac{a}{b}} = \frac{\sqrt[n]{a}}{\sqrt[n]{b}} \qquad (\sqrt[n]{a}, \sqrt[n]{b} \in R, b \neq 0)$$

$$\sqrt[cn]{a^{cm}} = \sqrt[n]{a^m} \qquad (\sqrt[cn]{a^{cm}}, \sqrt[n]{a^m} \in R)$$

A radical expression is in **simplest form** if:

1. the radicand contains no polynomial factor raised to a power equal to or greater than the index of the radical;
2. the radicand contains no fractions;

3. no radical expressions are contained in denominators of fractions;
4. the index of the radical and exponents on factors in the radicand have no common factors.

[4.7] Expressions containing radicals can be rewritten using properties of real numbers. In particular, the fundamental principle of fractions can be used to rationalize the denominator (or the numerator) of a fraction containing radicals.

The symbols introduced in this chapter are listed on the inside of the front cover.

REVIEW EXERCISES

[4.1–4.2] *Simplify. In Problems* 1–20, *assume that no denominator equals* 0.

1. a. $\dfrac{x^4y^3}{xy}$ **b.** $(3x^2y)^3$

2. a. $\dfrac{x^{-2} \cdot y^{-1}}{x}$ **b.** $\left(\dfrac{x^2y^{-2}}{x^{-2}y}\right)^{-2}$

3. a. $\dfrac{3^{-2} + 2^{-1}}{3^{-2}}$ **b.** $(x^{-3} - y^{-2})^{-1}$

4. a. $\dfrac{(2x+1)^{-1}}{(2x)^{-1} + 1^{-1}}$ **b.** $\dfrac{x^{-2} - y^{-2}}{(x-y)^{-2}}$

[4.3] *Write each expression in scientific notation.*

5. a. 4200 **b.** 307,000,000,000

6. a. 0.0421 **b.** 0.000000000023

Simplify.

7. a. $\dfrac{(2 \times 10^{-3})(3 \times 10^4)}{6 \times 10^{-1}}$ **b.** $\dfrac{0.024 \times 0.00045}{0.0072}$

[4.4] *Simplify. Assume that all variables denote positive numbers.*

8. a. $(-27)^{2/3}$ **b.** $4^{-1/2}$

9. a. $x^{3/2} \cdot x^{1/2}$ **b.** $\left(\dfrac{x^{2/3}y^{1/2}}{x^{1/3}}\right)^6$

Multiply.

10. a. $x^{2/3}(x^{1/3} - x)$

b. $y^{-1/4}(y^{5/4} - y^{1/4})$

Factor as indicated.

11. a. $x^{-1/2} + x^{5/2} = x^{-1/2}(\underline{?})$

b. $y^{3/4} - y^{-1/4} = y^{-1/4}(\underline{?})$

[4.5] *Express in radical notation.*

12. a. $(1 - x^2)^{2/3}$

b. $(1 - x^2)^{-2/3}$

Express in exponential notation.

13. a. $\sqrt[3]{x^2y}$

b. $\dfrac{1}{\sqrt[3]{(a + b)^2}}$

Find each root indicated $(x, y > 0)$.

14. a. $\sqrt{4y^2}$

b. $\sqrt[3]{-8x^3y^6}$

Assume the variable is an element of the set of real numbers and write each expression using absolute value notation.

15. a. $\sqrt{2x^2}$

b. $\sqrt{y^2 - 2y + 1}$

[4.6] *Simplify. Assume that all variables are positive real numbers.*

16. a. $\sqrt{180}$

b. $\sqrt[4]{32x^4y^5}$

17. a. $\dfrac{x}{\sqrt{xy}}$

b. $\sqrt[3]{\dfrac{1}{2}}$

18. a. $\dfrac{\sqrt{xy}\sqrt{6x^3y}}{\sqrt{2xy}}$

b. $\sqrt[3]{\dfrac{24y^4}{x^3}}$

19. a. $\sqrt[3]{\dfrac{x^7y^5}{x^2y}}$

b. $\dfrac{\sqrt[3]{x^5y^4}}{\sqrt[3]{x^2y}}$

20. a. $\dfrac{\sqrt[3]{x^4}\sqrt[3]{y^2}}{\sqrt[3]{xy}}$

b. $\dfrac{\sqrt[3]{16x}\sqrt[3]{8y^2}}{\sqrt[3]{2xy}}$

[4.7] *Simplify. Assume that all variables are positive real numbers.*

21. a. $4\sqrt{12} + 2\sqrt{75}$ b. $3\sqrt{2x} - \sqrt{32x} + 4\sqrt{50x}$

22. a. $9\sqrt{2xy^2} - 3y\sqrt{8x}$ b. $\sqrt{75x^3} - x\sqrt{3x}$

Write each expression without parentheses and then write all radicals in simple form.

23. a. $\sqrt{3}(\sqrt{6} - 2)$ b. $\sqrt{5}(\sqrt{10} - \sqrt{5})$

24. a. $(2 - \sqrt{3})(3 - 2\sqrt{3})$ b. $(\sqrt{x} + 2)(\sqrt{x} - 2)$

Rationalize each denominator.

25. a. $\dfrac{4}{2 - \sqrt{3}}$ b. $\dfrac{2}{\sqrt{5} + 2}$

26. a. $\dfrac{y}{\sqrt{y} - 3}$ b. $\dfrac{x - y}{\sqrt{x} + \sqrt{y}}$

5

FIRST-DEGREE EQUATIONS AND INEQUALITIES

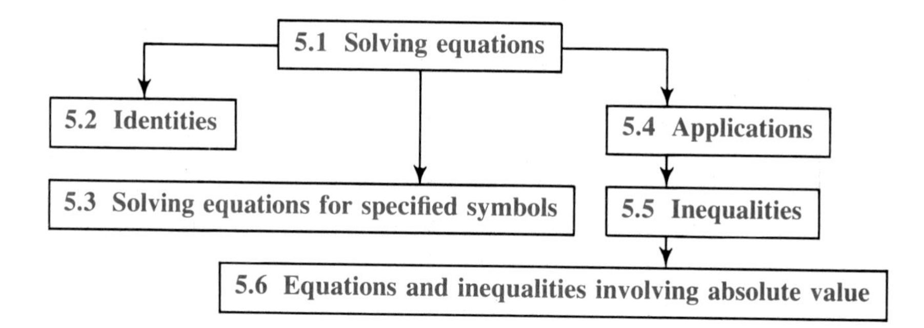

5.1

SOLVING EQUATIONS

In mathematics, sentences such as $3 + 2 = 5$ and $4 < 1 + 1$, which can be labeled true or false, are called **statements.** On the other hand, sentences containing variables, such as $x + 3 = 5$ and $x - 1 > 7$, which cannot be labeled true or false, are called **open sentences.** Statements or open sentences involving equality relationships are called **equations.** The expressions (or sometimes the values of these expressions) separated by the equality symbol (=) are called the **members** of the equation.

If we replace the variable in an equation with a numeral, and the resulting statement is true, the number represented by the numeral is called a **solution,** or **root,** of the equation and is said to *satisfy* the equation. For example, if we substitute 2 for x in $x + 3 = 5$, we obtain the true statement

$$2 + 3 = 5,$$

and 2 is a solution of the equation. The set of all numbers that satisfy an equation is called its **solution set.**

In discussing equations, a variable is frequently referred to as an **unknown,** although, again, its function is simply to represent an unspecified element of some set of numbers. The replacement set of any such variable will be the set of real numbers unless otherwise specified. In practical situations, however, it is always well to have a meaningful replacement set in mind. If we were investigating a situation involving the height of a man, we certainly would not consider negative numbers in the replacement set for any variable representing his height.

Equations that are satisfied by every element in the replacement set of the variable for which each member is defined are called **identities.** Equations that are not true for at least one element in the replacement set for which each member is defined are called **conditional equations**. For example, for x a real number,

$$x + 2 = x + 2$$

is an identity (the equation is satisfied by all real numbers), and

$$x + 2 = 5$$

is a conditional equation (the equation is not true, for example, for $x = 10$).

One major type of equation is the polynomial equation in one variable. This is an equation in which one member is 0 and the other member is a polynomial in simplest form. The equations

$$x^2 + 2x + 1 = 0, \tag{1}$$

$$2y - 3 = 0, \tag{2}$$

$$-z^4 + z^3 + 1 = 0 \tag{3}$$

are polynomial equations in x, y, and z, respectively; (1) is second-degree in x, (2) is first-degree in y, and (3) is fourth-degree in z. In this chapter we shall largely be concerned with first-degree polynomial equations in one variable.

The process of finding solution sets of conditional equations involves generating sets of equations that have identical solution sets. Such equations are called **equivalent equations**. Thus,

$$2x + 1 = x + 4,$$

$$2x = x + 3,$$

$$x = 3$$

are equivalent equations, because $\{3\}$ is the solution set of each. Since equations whose solution sets contain at least one element assert that, for some or all values of the variable, the left-hand and right-hand members are names for the same number, and since for each value of the variable a polynomial represents a number, the equality axioms for the real numbers imply the following properties:

▸ *1. The addition of the same expression representing a real number to each member of an equation produces an equivalent equation.*

▸ *2. The multiplication of each member of an equation by the same expression representing a nonzero real number produces an equivalent equation.*

Stated in symbols, these assertions are:

▸ *If $P(x)$, $Q(x)$, and $R(x)$ are expressions, then for all values of x for which these expressions represent real numbers, the equation*

$$P(x) = Q(x)$$

is equivalent to

$$P(x) + R(x) = Q(x) + R(x), \tag{4}$$

$$P(x) \cdot R(x) = Q(x) \cdot R(x), \quad \textit{for } R(x) \neq 0. \tag{5}$$

The proofs of these properties are omitted. Their application permits us to transform an equation whose solution set may not be obvious through a series of equivalent equations until we reach an equation that has an obvious solution. For example, consider the equation

$$2x + 1 = x + 4.$$

We can add $-x - 1$ to each member to obtain the equivalent equation

$$2x + 1 + (-x - 1) = x + 4 + (-x - 1),$$

from which

$$x = 3,$$

where the solution set $\{3\}$ is obvious.

Any application of property (4) or (5) is called an **elementary transformation**. An elementary transformation always results in an equivalent equation. Care must be exercised in the application of the second property, for we have specifically excluded *multiplication by* 0. For example, the equation $x = 3$ with solution set $\{3\}$ is not equivalent to $0 \cdot x = 0 \cdot 3$, with solution set $\{x \mid x \in R\}$. Or, as another example, to solve the equation

$$\frac{x}{x-3} = \frac{3}{x-3} + 2, \tag{6}$$

we might first multiply each member by $(x - 3)$ to attempt to produce an equivalent equation that is free of fractions. We have

$$(x-3)\frac{x}{x-3} = (x-3)\frac{3}{x-3} + (x-3)2$$

or

$$x = 3 + 2x - 6, \tag{7}$$

from which

$$x = 3,$$

and 3 *appears* to be a solution of (6). But, upon substituting 3 for x in (6), we have

$$\frac{3}{0} = \frac{3}{0} + 2,$$

where neither member is defined. In obtaining equation (7), each member of equation (6) was multiplied by $(x - 3)$, but if x is 3, then $(x - 3)$ is 0, and our second property is not applicable. Equation (7) is not equivalent to equation (6) for $x = 3$, and equation (6) has no solution.

We can always ascertain whether or not what we think is a solution of an equation is such in reality by substituting the suggested solution in the original equation and verifying that the resulting statement is true. If each of the equations in a sequence is obtained by means of an elementary transformation, the sole purpose for such a check is to detect arithmetic errors. We shall dispense with checking solution sets in the examples that follow except in cases where we apply what may be a nonelementary transformation—that is, where we multiply or divide by an expression containing a variable.

EXERCISE 5.1

A

a. *Verify that the given equation is satisfied by the given number.*
b. *Show that it is a conditional equation by substituting a different number.*

Example $3y + 6 = 4y - 4$, by 10

Solution

a. Substitute 10 for y and simplify each member.
Does $3(10) + 6 = 4(10) - 4$? Does $30 + 6 = 40 - 4$?
Yes.

b. Substitute a different number, say 1.
Does $3(1) + 6 = 4(1) - 4$? Does $3 + 6 = 4 - 4$?
No.

Hence, the equation is a conditional equation.

1. $x - 3 = 7$, by 10

2. $2x - 6 = 2$, by 4

3. $2x + 6 = 3x - 5$, by 11

4. $5x - 1 = 2x + 2$, by 1

5. $3a + 2 = 8 + a$, by 3

6. $3x - 5 = 2x + 7$, by 12

7. $3 = 6x + 3(x - 2)$, by 1

8. $6 = 2x + 6(2x + 1)$, by 0

9. $\dfrac{2x + 7}{5} + 3 = 6$, by 4

10. $\dfrac{3(2x + 5)}{9} - x = 0$, by 5

Solve, that is, find the solution set.

Examples **a.** $4 - (x - 1)(x + 2) = 8 - x^2$

b. $\frac{x}{3} + 4 = x - 2$

Solutions **a.** Apply the distributive property.

$$4 - (x^2 + x - 2) = 8 - x^2$$

$$4 - x^2 - x + 2 = 8 - x^2$$

$$-x^2 - x + 6 = 8 - x^2$$

Add $x^2 - 6$ to each member.

$$-x = 2$$

Multiply each member by -1.

$$x = -2$$

The solution set is $\{-2\}$.

b. Multiply each member by 3.

$$3\left(\frac{x}{3} + 4\right) = 3(x - 2)$$

$$x + 12 = 3x - 6$$

Add $-x + 6$ to each member.

$$18 = 2x$$

Multiply each member by $\frac{1}{2}$.

$$9 = x$$

The solution set is $\{9\}$.

[*Note:* Of course, the solution set can be specified at any time the solution becomes evident by inspection.]

11. $x + 5 = 7$

12. $x - 2 = 8$

13. $3 = x - 2$

14. $7 = x + 4$

15. $2x = 7 + x$

16. $x - 6 = 2x$

17. $3x + 1 = 5x - 2$

18. $4x - 3 = 2x + 1$

19. $2(x - 3) = 12$

20. $3(x + 4) = 6$

21. $x - (8 - x) = 4$

22. $3x - (3 + x) = 0$

23. $2[x - (2x + 1)] = 6$

24. $4[2x + (3x - 1)] = 5$

25. $-2[x + 3(x - 1)] = 4 + x$

26. $-5[2x - 2(x + 1)] = 6 - x$

27. $(x - 1)(x + 3) = x^2 + 4$

28. $(2x - 3)(x + 2) = x + 4 + 2x^2$

29. $(x + 3)^2 = x^2 + 4x$

30. $(2x - 3)^2 = 4x^2 - 8$

31. $x - (x + 2)(x - 3) = 4 - x^2$

32. $3x - (x - 3)(x + 1) = 6x - x^2$

33. $\frac{3x}{4} = 6$

34. $\frac{2x}{3} = 8$

35. $\frac{3}{5}x = -12$

36. $\frac{2}{7}x = -8$

37. $7 + \frac{5x}{3} = x - 2$

38. $x + 4 = \frac{2}{5}x - 3$

39. $1 + \frac{x}{9} = \frac{4}{3}$

40. $4 + \frac{x}{5} = \frac{5}{3}$

41. $\frac{x}{5} - \frac{x}{2} = 9$

42. $\frac{x}{4} = 2 - \frac{x}{3}$

43. $\frac{2x - 1}{5} - \frac{x + 1}{2} = 0$

44. $\frac{2x}{3} - \frac{2x + 5}{6} = \frac{1}{2}$

Example $\frac{2}{3} = 6 - \frac{x+10}{x-3}$

Solution Multiply each member by the LCD $3(x-3)$.

$$3(x-3)\frac{2}{3} = 3(x-3)6 - 3(x-3)\frac{x+10}{x-3}$$
$$2(x-3) = 18(x-3) - 3(x+10)$$

Apply the distributive property.

$$2x - 6 = 18x - 54 - 3x - 30$$

Combine like terms and add $-15x + 6$ to each member.

$$-13x = -78$$

Multiply each member by $-\frac{1}{13}$.

$$x = 6$$

The solution set is $\{6\}$.

Check A check is required because we multiplied each member of the equation by an expression containing the variable.

$$\frac{2}{3} = 6 - \frac{(6)+10}{(6)-3} = 6 - \frac{16}{3} = \frac{2}{3}$$

45. $\frac{x}{x-2} = \frac{2}{x-2} + 7$

46. $\frac{2}{x-9} = \frac{9}{x+12}$

47. $\frac{2}{y+1} + \frac{1}{3y+3} = \frac{1}{6}$

48. $\frac{5}{x-3} = \frac{x+2}{x-3} + 3$

49. $\frac{4}{2x-3} + \frac{4x}{4x^2-9} = \frac{1}{2x+3}$

50. $\frac{y}{y+2} - \frac{3}{y-2} = \frac{y^2+8}{y^2-4}$

B

51. Find a value for k in $3x - 1 = k$ so that the equation is equivalent to $2x + 5 = 1$.

52. For what value of k will the equation $2x - 3 = \frac{4+x}{k}$ have as its solution set $\{-1\}$?

53. For what value of k will the equation $\frac{2y+k}{y-5} + \frac{3}{4} = 8$ have as its solution set $\{9\}$?

54. For what value of k will the equation $\frac{y-3}{3y+k} + 1 = \frac{12}{7}$ have as its solution set $\{-7\}$?

5.2

IDENTITIES

We have observed that an equation such as

$$x + 2 = x + 2,$$

which is satisfied by *every element* in the replacement set of the variable (for which both members are defined), is called an identity. We can determine whether a given equation is an identity by changing one or both members of the equation to obtain a sequence of equivalent equations until we arrive at either an obvious identity—that is, an equation where both members are the same—or an equation that is obviously conditional. For example,

$$x^2 = (x - 1)(x + 1) + 1$$

may be written equivalently as

$$x^2 = (x^2 - 1) + 1,$$

from which

$$x^2 = x^2,$$

and the identity is established. If we attempt to write all the variables in one member by adding $-x^2$ to each member, we obtain

$$0 = 0,$$

which is true for any value of x.

As another example, consider the equation

$$(x - 2)^2 = x^2 - 3x + 2.$$

Upon performing the multiplication indicated in the left-hand member, we have

$$x^2 - 4x + 4 = x^2 - 3x + 2,$$

from which

$$x = 2.$$

Since this is clearly a conditional equation, the original equation is not an identity.

Now consider the equation

$$\frac{x^2 - 5x + 4}{x - 4} = x - 1, \tag{1}$$

in which a variable appears in the denominator of a fraction. Factoring the left-hand member yields

$$\frac{(x - 1)(x - 4)}{x - 4} = x - 1. \tag{1a}$$

Now applying the fundamental principle of fractions, with the restriction that $x - 4 \neq 0$, we obtain

$$(2) \qquad x - 1 = x - 1.$$

Note that the left-hand members in (1) and (1a) are not defined for $x = 4$,

$$\frac{(4)^2 - 5(4) + 4}{4 - 4} = \frac{0}{0},$$

although both members of (2) are defined for $x = 4$. Nevertheless, the orginal equation (1) is an identity, since it is satisfied by all numbers for which both members are defined.

We can also show that a given equation is not an identity by citing a single counterexample—that is, a single number that does not satisfy the equation. Thus, we can be sure that the equation

$$\frac{x - 2}{3} = x^2 - 1$$

is not an identity, since it is not true for $x = 0$; that is,

$$\frac{0 - 2}{3} \neq 0^2 - 1.$$

EXERCISE 5.2

A

Verify that each equation is an identity.

Example $\dfrac{3x - x}{2} + 3 = x + 3$

Solution Simplify the left-hand member.

$$\frac{2x}{2} + 3 = x + 3$$

$$x + 3 = x + 3$$

The left-hand member is identical to the right-hand member.

1. $\dfrac{x - 4x}{3} + 3x = 2x$

2. $\dfrac{y}{3} - \dfrac{2}{5} = \dfrac{5y - 6}{15}$

3. $(x + 2)(x - 2) - x^2 = -4$

4. $\dfrac{x(x-1)}{3} + x = \dfrac{x^2 + 2x}{3}$

5. $\dfrac{4x - 6x}{2} + 1 = 1 - x$

6. $(x + 2)(x + 3) - 6 = x(x + 5)$

7. $x(3 + x) - (x^2 + x) = 2x$

8. $x(x + 3) = (x + 1)^2 + x - 1$

9. $4x - \dfrac{5}{2}x - x = \dfrac{x}{2}$

10. $\dfrac{x^2}{4} + x + 1 = \dfrac{(x + 2)^2}{4}$

11. $\dfrac{y - 3}{3} = 2y - \dfrac{5y + 3}{3}$

12. $x(x + 1) = \dfrac{(2x + 1)^2 - 1}{4}$

13. $(x - 2)^2 - 3 = x^2 - 4x + 1$

14. $\dfrac{5y}{4} = \dfrac{2 + 3y}{4} - \dfrac{1 - y}{2}$

15. $14 = 3(x + 4) + \dfrac{4 - 6x}{2}$

16. $\dfrac{2z - 3}{2} + 1 = z - \dfrac{1}{2}$

Verify that each equation is an identity and specify any restrictions on the variables.

Example $\dfrac{x^2 - 2x - 3}{x + 1} = x - 3$

Solution Factor the left-hand member and reduce the fraction.

$$\frac{(x-3)(x+1)}{(x+1)} = x - 3$$

$$x - 3 = x - 3$$

The original equation is an identity. However, note that the left-hand member in the original equation is not defined for $x = -1$.

17. $y + 2 = \dfrac{(y + 3)(y - 1) + 3}{y}$

18. $z + 3 = \dfrac{(z + 1)(z + 2) - 2}{z}$

19. $\dfrac{x + 3}{x} + \dfrac{2}{x} = \dfrac{5}{x} + 1$

20. $\dfrac{(x + 2)^2 - 4(x + 1)}{x} = x$

21. $\dfrac{x^2 - 5x + 4}{x - 4} = x - 1$

22. $\dfrac{x^3 + 6x^2 + 5x}{x^2 + x} = x + 5$

B

Determine which of the equations are identities and which are conditional equations. Find the solution set for each conditional equation.

23. $(x + 2)^2 + (x - 1)^2 = (x + 1)^2 + x^2 + 4$

24. $x(x - 2) - (x - 2)(x + 2) = (x + 2)^2 - x(x + 6)$

25. $\dfrac{x}{x-2} - 7 = \dfrac{2}{x-2}$

26. $\dfrac{6}{x+1} - \dfrac{1}{2} = \dfrac{-1}{x+1}$

27. $\dfrac{1}{x-2} = \dfrac{2}{x-2} + 3$

28. $\dfrac{x^2 - x - 6}{x+2} = \dfrac{x^2 - 8x + 15}{x-5}$

29. $\dfrac{2}{3}(y-4) + \dfrac{2}{5}(y+3) = y - 1$

30. $\dfrac{1}{2x-3} + \dfrac{x}{4x^2 - 9} = \dfrac{1}{8x+12}$

5.3

SOLVING EQUATIONS FOR SPECIFIED SYMBOLS

An equation containing more than one variable, or containing symbols representing constants such as a, b, and c, can be solved for one of the symbols in terms of the remaining symbols by using the methods developed in Section 5.1. In general, we apply elementary transformations until the desired symbol stands alone as one member (ordinarily the left-hand member) of an equation.

EXERCISE 5.3

A

Solve for x or y. (Leave the results in the form of an equation equivalent to the given equation.)

Example $5by - 2a = 2ay$

Solution Add $2a - 2ay$ to each member.

$$5by - 2ay = 2a$$

Factor the left-hand member.

$$y(5b - 2a) = 2a$$

Multiply each member by $\dfrac{1}{5b - 2a}$.

$$y = \frac{2a}{5b - 2a}$$

1. $ax = b + c$

2. $b - c = ay$

3. $3ay - b = c$

4. $4by + c = b$

5. $ax = b - 2ax$

6. $4by = a + 2by$

7. $\frac{a}{b}y + c = 0$

8. $b = \frac{1}{c}y + a$

9. $\frac{x + a}{b} = c$

10. $b = \frac{y - a}{c}$

11. $\frac{c - 2y}{b} = a$

12. $\frac{b - 3x}{a} = c$

13. $ax = a - x$

14. $y = b + by$

15. $a(a - x) = b(b - x)$

16. $4y - 3(y - b) = 8b$

17. $(x - 4)(b + 2) = b$

18. $(y - 2)(a + 3) = 2a$

19. $\frac{2}{x} + \frac{2}{a} = 6$

20. $\frac{2}{y} - \frac{3}{b} = 4$

21. $\frac{a}{y} - \frac{1}{2} = \frac{a}{2y}$

22. $\frac{c}{3} + \frac{1}{x} = \frac{c}{6x}$

23. $\frac{1}{a} + \frac{1}{b} = \frac{1}{x}$

24. $\frac{1}{a} - \frac{1}{b} = \frac{1}{x}$

Solve.

Example $v = k + gt$, for g

Solution Add $-k$ to each member.

$$v - k = gt$$

Multiply each member by $\frac{1}{t}$.

$$\frac{v - k}{t} = g$$

$$g = \frac{v - k}{t}$$

25. $v = k + gt$, for k

26. $E = mc^2$, for m

27. $f = ma$, for m

28. $I = prt$, for p

29. $pv = K$, for v

30. $E = IR$, for R

31. $s = \frac{1}{2}at^2$, for a

32. $p = 2l + 2w$, for l

33. $v = k + gt$, for t

34. $y = \frac{k}{z}$, for z

35. $V = lwh$, for h

36. $W = I^2R$, for R

37. $180 = A + B + C$, for B

38. $S = \frac{a}{1 - r}$, for r

39. $A = \frac{h}{2}(b + c)$, for c

40. $S = 2r(r + h)$, for h

41. $S = 3\pi d + 5\pi D$, for d

42. $A = 2\pi rh + 2\pi r^2$, for h

43. $l = a + (n - 1)d$, for n

44. $\frac{1}{r} = \frac{1}{s} + \frac{2}{t}$, for r

B

45. An equation of the form

$$\frac{a}{b} = \frac{c}{d} \qquad (a, b, c, d \neq 0)$$

is called a **proportion**. The terms b and c are called **means** and the terms a and d are called **extremes**. Show that, in any proportion, the product of the means is equal to the product of the extremes.

46. If $\frac{a}{b} = \frac{c}{d}$, show that $\frac{b}{a} = \frac{d}{c}$. [*Hint:* Use the results of Problem 45.]

47. If $\frac{a}{b} = \frac{c}{d}$, show that $\frac{a}{c} = \frac{b}{d}$.

48. If $\frac{a}{b} = \frac{c}{d}$, show that $\frac{d}{b} = \frac{c}{a}$.

49. If $\frac{a}{b} = \frac{c}{d}$, show that $\frac{a+b}{b} = \frac{c+d}{d}$. $\left[\textit{Hint:} \text{ Consider } \frac{a}{b} + 1 = \frac{c}{d} + 1.\right]$

50. If $\frac{a}{b} = \frac{c}{d}$, show that $\frac{a-b}{b} = \frac{c-d}{d}$.

51. Show that for any $a, k \in R$, and $k \neq 1$, the equation $\frac{x}{x-a} = \frac{a}{x-a} + k$ has no solution.

5.4

APPLICATIONS

Word problems state relationships between numbers. Problems may be explicitly concerned with numbers, or they may be concerned with numerical measures of physical quantities. In either event, we seek a number or numbers for which the stated relationships hold. To express the quantitative ideas symbolically in the form of an equation is the most difficult part of solving word problems. Although, unfortunately there is no single means available to do this, the following suggestions are frequently helpful:

1. Represent the unknown quantities using phrases and symbols. Since at this time we are using one variable only, all relevant quantities should be represented in terms of this variable.
2. Where applicable, draw a sketch and label all known quantities in terms of symbols.
3. Find a quantity that can be represented in two different ways and write these representations as an equation (mathematical model). The equation may derive from

a. the problem itself, which may state a relationship explicitly; for example, "What number added to 4 gives 7?" produces the equation $x + 4 = 7$;

b. formulas or relationships that are part of your general mathematical background; for example, $A = \pi r^2$, $d = rt$, etc.

Summarizing information in tabular form is sometimes helpful in writing the equation.

4. Solve the resulting equation.
5. Check the results against the original problem. It is not sufficient to check the result in the equation, because the equation itself may be in error.

At this stage of your mathematical development, *as much or more attention should be given to the setting up of equations as is given to their solution.* Although some of the problems in this book can be solved without recourse to algebra, the practice obtained in setting up the equations involved will be helpful when more difficult problems are encountered. Therefore, we suggest that you *read carefully each step of the solutions of the illustrative examples in Exercise* 5.4.

EXERCISE 5.4

A

a. *Set up an equation for each problem.* ***b.*** *Solve.*

Example The first stage of a rocket burns 28 seconds longer than the second stage. If the total burning time for both stages is 152 seconds, how long does each stage burn?

Solution **a.** Express the *two* quantities asked for in *two* simple phrases and represent these quantities using symbols.

burning time of the first stage: t

burning time of the second stage: $t - 28$

Write an equation relating the unknown quantities.

$$t + (t - 28) = 152$$

b. Solve the equation.

$$t + t - 28 = 152$$
$$2t - 28 = 152$$
$$2t = 180$$
$$t = 90$$
$$t - 28 = 62$$

Therefore, the first stage burns 90 seconds and the second stage burns 62 seconds.

Check Does $90 + 62 = 152$? Yes.

1. The sum of two integers is 110. If one of the integers is 24 greater than the other, find the integers.
2. If $\frac{3}{5}$ of a number is 8 less than the number, what is the number?
3. In a student body election, 584 students voted for one or the other of two candidates for president. If the winner received 122 more votes than the loser, how many votes were cast for each candidate?
4. One side of a long-playing record contains 5 songs, with 15 second intervals between songs. If all the songs take an equal amount of time to play, and if the entire side takes 26 minutes, how many minutes does each song last?

Consecutive Integer Problems *Solving consecutive integer problems ordinarily involves using the fact that successive integers differ by 1; that is, successive integers can be represented by $x, x + 1, x + 2, \ldots$, etc. Of course, successive even integers, or successive odd integers are thus represented by $x, x + 2, x + 4, \ldots$, etc.*

Example The sum of three consecutive integers is 21 greater than 2 times the smallest of the three integers. Find the integers.

Solution **a.** Express the *three* quantities asked for in *three* simple phrases and then represent these quantities using symbols.

an integer: x

next consecutive integer: $x + 1$

next consecutive integer: $x + 2$

Write an equation representing the conditions given in the problem.

$$x + (x + 1) + (x + 2) = 2x + 21$$

b. Solve the equation.

$$3x + 3 = 2x + 21$$

$$x = 18$$

$$x + 1 = 19$$

$$x + 2 = 20$$

Therefore, the integers are 18, 19, and 20.

Check Does $18 + 19 + 20 = 2 \cdot 18 + 21$? Yes.

5. Find three consecutive integers whose sum is 78.
6. Find three consecutive integers such that the sum of the first and third is 146.
7. Find two consecutive even integers such that 7 times the lesser is equal to 6 times the greater.
8. Find three consecutive odd integers such that 2 times the sum of the first two is 1 less than 3 times the third.

Interest Problems *Simple interest problems involve the fact that the interest earned during a single year is equal to the amount invested times the annual rate of interest. Thus,* $I = A \times r$. *Tables are sometimes helpful in summarizing data for this type of problem.*

Example A man has an annual income of $12,000 from two investments. He has $10,000 more invested at 8% than he has invested at 6%. How much does he have invested at each rate?

Solution **a.** Express the *two* quantities asked for in *two* simple phrases and represent these quantities using symbols:

amount invested at 6%: A

amount invested at 8%: $A + 10{,}000$

Set up a table.

Investment	Rate	Amount	Interest
6%	0.06	A	$0.06A$
8%	0.08	$A + 10{,}000$	$0.08(A + 10{,}000)$

Write an equation relating the interest from each investment and the total interest received.

$$\begin{bmatrix}\text{interest from}\\ 6\%\text{ investment}\end{bmatrix} + \begin{bmatrix}\text{interest from}\\ 8\%\text{ investment}\end{bmatrix} = [\text{total interest}]$$

$$0.06A + 0.08(A + 10{,}000) = 12{,}000$$

b. Solve for A. First multiply each member by 100.

$$6A + 8(A + 10{,}000) = 1{,}200{,}000$$
$$6A + 8A + 80{,}000 = 1{,}200{,}000$$
$$14A = 1{,}120{,}000$$
$$A = 80{,}000$$
$$A + 10{,}000 = 90{,}000$$

Therefore, $80,000 is invested at 6% and $90,000 is invested at 8%.

Check Does $0.06(80{,}000) + 0.08(90{,}000) = 12{,}000$? Yes.

9. A man has invested $8000, part in a bank at 4% and part in a savings and loan association at 5%. If his annual return is $350, how much has he invested at each rate?

10. A sum of $2400 is split between an investment in a mutual fund paying 5% and one in corporate bonds paying 7%. If the return on the 5% investment exceeds that on the 7% investment by $12 per year, how much is invested at each rate?

11. If $3000 is invested in life insurance at 2%, how much additional money must be invested in stocks paying 5% to make the earnings on the total investment 4%?

12. If $6000 is invested in bonds at 6%, how much additional money must be invested in stocks at 3% to earn a return of 5% on the total investment?

Age Problems *Problems involving age are similar to integer problems, because, ordinarily, ages are considered in whole numbers of years. In solving age problems, it is often necessary to express ages at various times in terms of present age. A table is usually helpful in such cases.*

Example A man is 9 times as old as his son. In 9 years, he will be only 3 times as old as his son. How old is each now?

Solution **a.** Express the *two* quantities asked for in *two* simple phrases and represent these quantities using symbols.

son's age now: x

father's age now: $9x$

Set up a table.

Age now	Age 9 years from now
x $9x$	$x + 9$ $9x + 9$

Write an equation relating the ages in 9 years.

$$9x + 9 = 3(x + 9)$$

b. Solve the equation.

$$9x + 9 = 3x + 27$$
$$6x = 18$$
$$x = 3$$
$$9x = 27$$

Therefore, the son is now 3 years old and the father is now 27 years old.

Check Does $27 = 9 \cdot 3$? Yes.
Does $27 + 9 = 3(3 + 9)$? Yes.

13. In 3 years, a man will be 5 times as old as his son. If the man is now 9 times as old as his son, what are their present ages?

14. A boy is 3 times as old as his sister. In 4 years, he will be 2 times as old as his sister. How old is each now?

15. The sum of the ages of a father and his son is 60 years. In 4 years, the father will be 3 times as old as his son. How old is each now?

16. The sum of the ages of Joe and Pete is 45 years. In 5 years, Joe will be 4 times as old as Pete. How old is each now?

Using Formulas *In many practical problems it is necessary to use formulas to find one number when we know other numbers. In such cases, we substitute the known numbers and solve the resulting equation to determine the unknown number.*

Example The cutting speed S of a circular saw blade is given by

$$S = \frac{C\omega D}{12d},$$

where C is the circumference of the saw blade, ω is the angular velocity in revolutions per minute of the driving motor, D is the diameter of the driving pulley, and d is the diameter of the driven pulley. What angular velocity must the driving motor attain to achieve a cutting speed $S = 7000$ feet per minute if $C = 36$ inches, $D = 4$ inches, and $d = 3$ inches?

Solution In the given equation, replace S with 7000, C with 36, D with 4, and d with 3. Solve the resulting equation for ω.

$$(7000) = \frac{(36)\omega(4)}{12(3)}$$

$$7000 = \frac{\overset{4}{\cancel{144}}\omega}{\underset{1}{\cancel{36}}}$$

$$7000 = 4\omega$$

$$1750 = \omega$$

The motor must turn at 1750 revolutions per minute.

17. The relation between Fahrenheit and Celsius temperature is given by $F = \frac{9}{5}C + 32$. What is the Celsius temperature if the Fahrenheit temperature is 104°?

18. The monthly benefit B paid on a certain retirement plan is determined by

$$B = \frac{2}{5}w\left(1 + \frac{2n}{100}\right),$$

where w is the average monthly wage of the worker and n is the number of years worked. How many years must a person work to receive a monthly benefit of \$512 if his average monthly wage is \$800?

19. A projectile launched upward from the ground at a velocity of v feet per second is located at a height s above the ground at time t, as given by

$$s = vt - 16t^2.$$

With what velocity must an object be launched to obtain a height of 200 feet in 8 seconds?

20. The net resistance R_n of a parallel electrical circuit is given by

$$R_n = \frac{R_1 R_2}{R_1 + R_2},$$

where R_1 and R_2 are the individual resistances in parallel. If one of the individual resistances is 40 ohms, what must the other be if the net resistance is 30 ohms?

Geometry Problems *Solving problems involving geometric figures often requires the use of the formulas listed in the table Formulas from Geometry at the end of this book.*

Example The vertex angle of an isosceles triangle measures 30° more than the sum of the measures of its other two angles. What is the measure of each angle in the triangle?

Solution **a.** A sketch is helpful. Use the fact that in an isosceles triangle, the two angles opposite the two sides of equal length have equal measure. Express the *two* quantities asked for in *two* simple phrases and represent these quantities using symbols.

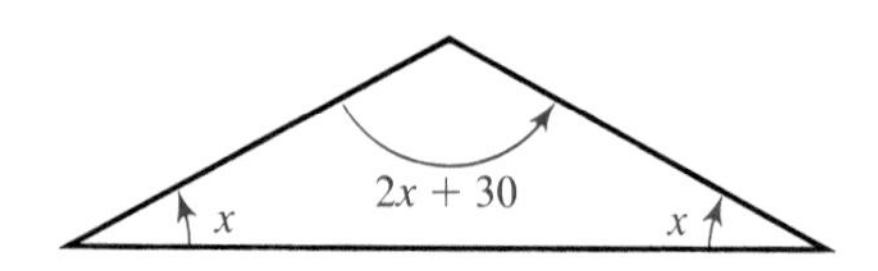

measure of each base angle: x

measure of the vertex angle: $2x + 30$

Write an equation using the fact that the sum of the measures of the angles in any triangle is 180°.

$$(x) + (x) + (2x + 30) = 180$$

b. Solve the equation.

$$4x + 30 = 180$$
$$4x = 150$$
$$x = 37\frac{1}{2}$$
$$2x + 30 = 105$$

Therefore, the angles in the triangle measure $37\frac{1}{2}°$, $37\frac{1}{2}°$, and 105°.

Check Does $37\frac{1}{2} + 37\frac{1}{2} + 105 = 180$? Yes.

21. One angle of a triangle measures 2 times another angle, and the third angle measures 12° more than the sum of the measures of the other two. Find the measure of each angle.

22. The measure of the smallest angle of a triangle is 25° less than the measure of another angle, and 50° less than the measure of the third angle. Find the measure of each angle.

23. Each of the equal sides of an isosceles triangle is 3 centimeters longer than 2 times the length of the base. Find the length of each side of the triangle if its perimeter is 86 centimeters.

24. When the length of each side of a square is increased by 5 centimeters, the area is increased by 85 square centimeters. Find the length of a side of the original square.

Lever Problems *A rigid bar rotating about a fixed point is called a* ***lever****. The fixed point is called the* ***fulcrum*** *of the lever. If a force F_1 is applied to one side of a lever at a distance d_1 from the fulcrum, and a force F_2 is applied on the other side of the lever at a distance d_2 from the fulcrum (see figure), then the lever will be in equilibrium if and only if*

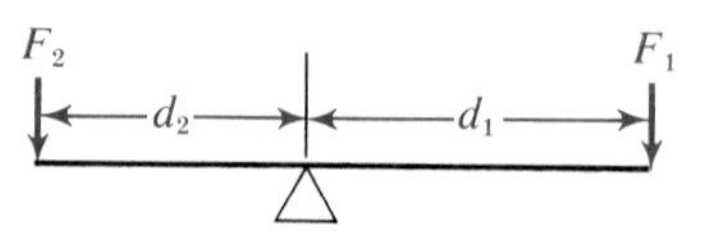

$$F_1 d_1 = F_2 d_2.$$

This relationship is called the ***law of the lever*** *and has many practical applications.*

Example What force must be applied to the end of a 10 centimeter beam to balance a weight of 6 grams hung on the other end if the beam rests on a fulcrum located $1\frac{1}{2}$ centimeters from the weight?

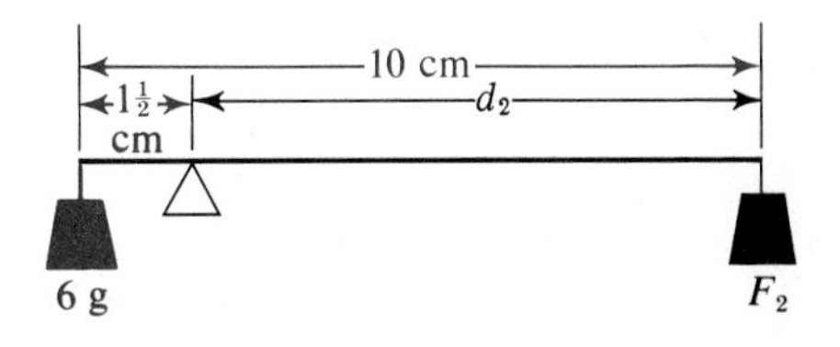

Solution **a.** The force to be applied is F_2. In the law of the lever, substitute

$$F_1 = 6, \qquad d_1 = \frac{3}{2}, \qquad d_2 = 10 - \frac{3}{2} = \frac{17}{2},$$

and write the resulting equation.

$$6 \cdot \frac{3}{2} = \frac{17}{2} \cdot F_2$$

b. Solve the equation.

$$9 = \frac{17}{2} \cdot F_2$$

$$F_2 = \frac{18}{17}$$

Therefore, the force to be applied is $1\frac{1}{17}$ grams.

Check Does $6 \cdot \frac{3}{2} = \frac{17}{2} \cdot \frac{18}{17}$? Yes.

25. A 24 gram weight is placed 2 centimeters further from the fulcrum on one side of a balanced lever than a 32 gram weight on the other side. How far is each weight from the fulcrum?

26. A 48 gram weight and a 36 gram weight are placed on opposite ends of a beam, each weight at a distance of 8 centimeters from the fulcrum. Where should a 24 gram weight be placed to balance the beam? [*Hint:* $F_1 d_1 = F_2 d_2 + F_3 d_3$.]

27. Where should the fulcrum of a 6 foot crowbar be placed so that a 200 pound man can just balance a 2200 pound weight?

28. Where should the fulcrum of a 9 foot crowbar be placed so that a 180 pound man can just balance a 900 pound weight?

Coin Problems *The basic idea of problems involving coins (or bills) is that the value of a number of coins (or bills) of the same denomination is equal to the product of the value of a single coin (or bill) and the total number of coins (or bills).*

$$\begin{bmatrix} \textit{value of} \\ n \\ \textit{coins} \end{bmatrix} = \begin{bmatrix} \textit{value of} \\ 1 \\ \textit{coin} \end{bmatrix} \times \begin{bmatrix} \textit{number} \\ \textit{of} \\ \textit{coins} \end{bmatrix}$$

It is helpful in most coin problems to think in terms of cents rather than dollars. Tables can also be helpful in solving coin problems.

Example A collection of coins consisting of dimes and quarters has a value of \$11.60. How many dimes and quarters are in the collection if there are 32 more dimes than quarters?

Solution **a.** Express the *two* quantities asked for in *two* simple phrases and represent the quantities using symbols.

$$\text{number of quarters: } x$$
$$\text{number of dimes: } x + 32$$

Set up a table.

Denomination	Value of 1 coin	Number of coins	Value of coins
dimes	10	$x + 32$	$10(x + 32)$
quarters	25	x	$25x$

Write an equation relating the value of the quarters and the value of the dimes to the value of the entire collection.

$$\begin{bmatrix}\text{value of}\\\text{quarters}\\\text{in cents}\end{bmatrix} + \begin{bmatrix}\text{value of}\\\text{dimes}\\\text{in cents}\end{bmatrix} = \begin{bmatrix}\text{value of}\\\text{collection}\\\text{in cents}\end{bmatrix}$$
$$25x \quad + 10(x + 32) = \quad 1160$$

b. Solve the equation.

$$25x + 10x + 320 = 1160$$
$$35x = 840$$
$$x = 24$$
$$x + 32 = 56$$

Therefore, there are 24 quarters and 56 dimes in the collection.

Check Do 24 quarters and 56 dimes have a value of \$11.60? Yes.

29. A man who has 12 more quarters than dimes has a total of \$12.45 in quarters and dimes. How many of each kind of coin does he have?

30. A savings bank containing \$1.75 in dimes, quarters, and nickels contains 3 more dimes than quarters, and 3 times as many nickels as dimes. How many coins of each kind are in the bank?

31. On an airplane flight for which first-class fare is \$80 and tourist fare is \$64 there were 42 passengers. If receipts for the flight totaled \$2880, how many first-class and how many tourist passengers were on the flight?

32. A vendor bought a supply of ice cream bars at 3 for 20¢. He ate one and sold the remainder at 10¢ each. If he made \$2 profit, how many bars did he buy?

B

Mixture Problems *The key to solving mixture problems is to recognize that the amount of a given substance in a mixture is obtained by multiplying the amount of the mixture by the percent (or rate) of the given substance in the mixture.*

$$\begin{bmatrix} \textit{amount} \\ \textit{of} \\ \textit{substance} \end{bmatrix} = \begin{bmatrix} \textit{percent (rate)} \\ \textit{of substance} \\ \textit{in mixture} \end{bmatrix} \times \begin{bmatrix} \textit{amount} \\ \textit{of} \\ \textit{mixture} \end{bmatrix}$$

Tables are especially helpful in mixture problems.

Example How many liters of a 10% solution of acid should be added to 20 liters of a 60% solution of acid to obtain a 50% solution?

Solution **a.** Express the quantity asked for in a simple phrase and represent the quantity using a symbol.

number of liters of 10% solution: n

Set up a table.

Mixture	Part of acid in mixture	Amount of mixture	Amount of acid
10%	0.10	n	$0.10n$
60%	0.60	20	$0.60(20)$
50%	0.50	$n + 20$	$0.50(n + 20)$

Write an equation relating the amount of *pure* acid before and after combining the solutions.

$$\begin{bmatrix} \text{pure acid in} \\ \text{10\% solution} \end{bmatrix} + \begin{bmatrix} \text{pure acid in} \\ \text{60\% solution} \end{bmatrix} = \begin{bmatrix} \text{pure acid in} \\ \text{50\% solution} \end{bmatrix}$$

$$0.10n \quad + \quad 0.60(20) \quad = \quad 0.50(n + 20)$$

b. Solve the equation. First multiply each member by 100.

$$10n + 60(20) = 50(n + 20)$$
$$10n + 1200 = 50n + 1000$$
$$-40n = -200$$
$$n = 5$$

Therefore, 5 liters of the 10% solution are needed.

Check Does 5 liters of a 10% solution when added to 20 liters of a 60% solution form a 50% solution?

Does $0.10(5) + 0.60(20) = 0.50(25)$?
Does $0.5 + 12.0 = 12.5$? Yes.

33. How many liters of a 30% salt solution must be added to 40 liters of a 12% salt solution to obtain a 20% solution?

34. How many grams of an alloy containing 45% silver must be melted with an alloy containing 60% silver to obtain 40 grams of an alloy containing 48% silver?

35. How much pure alcohol should be added to 12 liters of a 45% solution to obtain a 60% solution?

36. An automobile radiator contains 32 quarts of a 40% antifreeze solution. How many quarts of this solution should be drained from the radiator and replaced with water if the resulting solution is to be 20% antifreeze?

B

Uniform-Motion Problems *Solving uniform-motion problems requires applying the fact that the distance traveled at a uniform rate is equal to the product of the rate and the time traveled. This relationship may be expressed by any one of the three equations*

$$d = rt, \qquad r = \frac{d}{t}, \qquad t = \frac{d}{r}.$$

It is frequently helpful to use a table to solve these problems.

Example An express train travels 150 miles in the same time that a freight train travels 100 miles. If the express goes 20 miles per hour faster than the freight, find the rate of each.

Solution **a.** Express the *two* quantities asked for in *two* simple phrases and represent these quantities using symbols.

rate for the freight train: r

rate for the express train: $r + 20$

Set up a table.

	d	r	$t = \frac{d}{r}$
freight	100	r	$\frac{100}{r}$
express	150	$r + 20$	$\frac{150}{r + 20}$

The fact that the times are equal is the significant equality in the problem.

$$[t \text{ freight}] = [t \text{ express}]$$

$$\frac{100}{r} = \frac{150}{r + 20}$$

b. Solve the equation.

$$r(r+20)\left(\frac{100}{r}\right) = r(r+20)\left(\frac{150}{r+20}\right)$$

$$(r+20)100 = (r)150$$

$$100r + 2000 = 150r$$

$$-50r = -2000$$

$$r = 40$$

$$r + 20 = 60$$

Thus, the freight train's rate is 40 miles per hour and the express train's rate is 60 miles per hour.

Check The time of the freight ($^{100}/_{40}$) equals the time of the express train ($^{150}/_{60}$).

37. An airplane travels 1260 miles in the same time that an automobile travels 420 miles. If the rate of the airplane is 120 miles per hour greater than the rate of the automobile, find the rate of each.

38. Two planes leave an airport at the same time and travel in opposite directions. If one plane averages 440 miles per hour over the ground and the other 560 miles per hour, in how long will they be 2500 miles apart?

39. A ship traveling at 20 knots is 5 nautical miles out from a harbor when another ship leaves the harbor at 30 knots sailing on the same course. How long does it take the second ship to catch up to the first?

40. A boat sails due west from a harbor at 36 knots. An hour later, another boat leaves the harbor on the same course at 45 knots. How far out at sea will the second boat overtake the first?

Work Problems *Some problems involve the accomplishment of a task in some fixed time when a steady rate of work is assumed. For example, if it takes* 4 *hours for a pipe to empty a tank, then in* 1 *hour the pipe empties* $\frac{1}{4}$ *of the tank; or if it takes a man* 3 *days to paint a house, then in* 1 *day he paints* $\frac{1}{3}$ *of the house. Problems involving tasks of this kind can be represented by equations in which the right-hand member is* 1, *which denotes completion of the total task.*

Example One pipe can empty a tank in 6 hours and another can empty the same tank in 9 hours. How long will it take both pipes to empty the tank?

Solution **a.** Express the quantity asked for in a simple phrase and represent the quantity using a symbol.

time for both pipes to empty the tank: t

Write an equation expressing the conditions of the problem.

$$\begin{bmatrix}\text{portion of tank}\\ \text{emptied by pipe 1}\end{bmatrix} + \begin{bmatrix}\text{portion of tank}\\ \text{emptied by pipe 2}\end{bmatrix} = \begin{bmatrix}\text{portion of tank}\\ \text{emptied by both}\end{bmatrix}$$

$$\left(\frac{1}{6}\right)t \quad + \quad \left(\frac{1}{9}\right)t \quad = \quad 1$$

(solution continued)

b. Solve the equation.

$$(18)\left(\frac{1}{6}\right)t + (18)\left(\frac{1}{9}\right)t = (18)1$$

$$3t + 2t = 18$$

$$5t = 18$$

$$t = \frac{18}{5}$$

The tank will be emptied in $3\frac{3}{5}$ hours when both pipes are open.

Check Does $(\frac{1}{6})(\frac{18}{5}) + (\frac{1}{9})(\frac{18}{5}) = 1$?
Does $\frac{3}{5} + \frac{2}{5} = 1$? Yes.

41. It takes one pipe 30 hours to fill a tank, while a second pipe can fill the same tank in 45 hours. How long will it take both pipes running together to fill the tank?

42. One pipe can fill a tank in 4 hours and another can empty it in 6 hours. If both pipes are open, how long will it take to fill the empty tank?

43. A new billing machine can process a firm's monthly billings in 10 hours. By using an older machine together with the new machine, the billings can be completed in 6 hours. How long would it take the older machine to do the job alone?

44. A tractor plows $\frac{5}{9}$ of a field in 10 hours. By adding another tractor the job is finished in another 3 hours. How long would it take the second tractor to do the job alone?

5.5

INEQUALITIES

Open sentences of such forms as

(1) $$x + 3 \geq 10$$

and

(2) $$\frac{-2y - 3}{3} < 5$$

are called **inequalities**. For appropriate values of the variable, one member of an inequality represents a real number that is less than ($<$), less than or equal to ($\leq$), greater than or equal to ($\geq$), or greater than ($>$) the real number represented by the other member.

Any element of the replacement set of the variable for which an inequality is true is called a **solution**, and the set of all solutions of an inequality is called the **solution set** of the inequality. Inequalities that are true for every element in the replacement set of the variable, such as

$$x^2 + 1 > 0, \qquad x \in R,$$

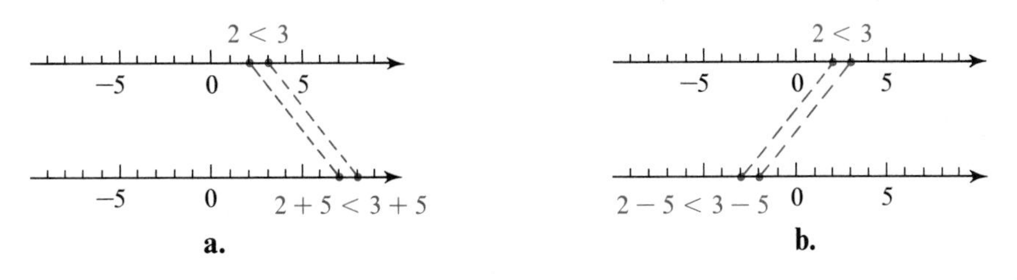

Figure 5.1

are called **absolute inequalities**. Inequalities that are not true for every element of the replacement set are called **conditional inequalities**—for example, inequalities (1) and (2).

As in the case with equations, we shall solve a given inequality by generating a series of equivalent inequalities (inequalities having the same solution set) until we arrive at one whose solution set is obvious. To do this we shall need some fundamental properties of inequalities. Notice that

$$2 < 3,$$

and

$$2 + 5 < 3 + 5,$$

and

$$2 - 5 < 3 - 5.$$

Parts a and b of Figure 5.1 demonstrate that the addition of 5 or -5 to each member of $2 < 3$ simply shifts the graphs of the members the same number of units to the right or left on the number line, with the order of the members left unchanged. This will be the case for the addition of any real number to each member of an inequality. Since for any real value of the variable for which an expression is defined the expression represents a real number, we generalize this idea and assert that:

► *1. The addition of the same expression representing a real number to each member of an inequality produces an equivalent inequality in the same sense.*

Next, if we multiply each member of

$$2 < 3$$

by 2, we have

$$4 < 6,$$

where the products form an inequality in the same sense. If, however, we multiply each member of

$$2 < 3$$

by -2, we have

$$-4 > -6,$$

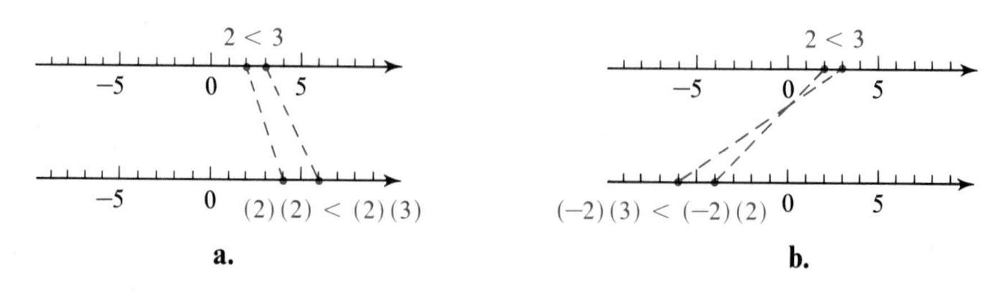

Figure 5.2

where the inequality is in the opposite sense. Figure 5.2 illustrates this. Multiplying each member of $2 < 3$ by 2 simply moves the graph of each member out twice as far in a positive direction (part a). Multiplying by -2 also doubles the absolute value of each member, but the products are negative and the sense of the inequality is reversed as shown in part b. In general, we assert that:

▸ 2. *If each member of an inequality is multiplied by the same expression representing a positive number, the result is an equivalent inequality in the same sense.*

▸ 3. *If each member of an inequality is multiplied by the same expression representing a negative number, the result is an equivalent inequality in the opposite sense.*

Statements 1, 2, and 3 above can be expressed in symbols as follows:

▸ *If $P(x)$, $Q(x)$, and $R(x)$ are expressions, then for all values of x for which these expressions represent real numbers*

$$P(x) < Q(x)$$

is equivalent to

$$P(x) + R(x) < Q(x) + R(x), \tag{3}$$

$$P(x) \cdot R(x) < Q(x) \cdot R(x), \quad \textit{for } R(x) > 0, \tag{4}$$

$$P(x) \cdot R(x) > Q(x) \cdot R(x), \quad \textit{for } R(x) < 0. \tag{5}$$

These relationships are also true with $<$ replaced by $\leq$ and $>$ replaced by $\geq$. Similar relationships are true with $<$ replaced by $>$ or $\geq$ and $>$ replaced by $<$ or $\leq$, respectively. The proofs of these properties are omitted.

Note that none of these assertions permits multiplying by 0 and that variables in expressions used as multipliers are restricted from values for which the expression vanishes or is not defined. The result of applying any of these properties is called an **elementary transformation**.

These properties can be applied to solve first-degree inequalities in the same way the equality properties are applied to solve first-degree equations. For example, let us find the solution set of

$$\frac{x-3}{4} < \frac{2}{3}.$$

By assertion 2, we can multiply each member by 12 to obtain

$$3(x-3) < 8,$$

or

$$3x - 9 < 8.$$

By assertion 1, we can add 9 to each member, giving us

$$3x < 17;$$

and finally, by assertion 2, we can multiply each member by $\frac{1}{3}$ to obtain

$$x < \frac{17}{3}.$$

The solution set is then $S = \{x \mid x < {}^{17}\!/_3\}$. Recall that this symbolism is read, "S is the set of all x such that x is less than ${}^{17}\!/_3$." This solution set can be graphed as shown in Figure 5.3, where the red line covers the points whose coordinates are in the solution set.

Figure 5.3

Figure 5.4

Sometimes two conditions are placed on a variable. For example, we may want to consider $\{x \mid -6 < 3x \text{ and } 3x \leq 15\}$, that is,

$$\{x \mid -6 < 3x\} \cap \{x \mid 3x \leq 15\}.$$

In this case, the two conditions may be expressed by

$$-6 < 3x \leq 15.$$

Using this form, each expression can be multiplied by $\frac{1}{3}$ to obtain

$$-2 < x \leq 5.$$

The solution set,

$$S = \{x \mid -2 < x \leq 5\},$$

is illustrated in Figure 5.4.

EXERCISE 5.5

A

Solve and graph each solution set.

Example $\frac{x+4}{3} \le 6 + x$

Solution Multiply each member by 3.

$$x + 4 \le 18 + 3x$$

Add $-4 - 3x$ to each member.

$$-2x \le 14$$

Multiply each member by $-\frac{1}{2}$ and reverse the sense of the inequality.

$$x \ge -7$$

The solution set is $\{x \mid x \ge -7\}$.

−10 −5 0 5 10

1. $3x < 6$
2. $x + 7 > 8$
3. $x - 5 \le 7$
4. $2x - 3 < 4$
5. $3x - 2 > 1 + 2x$
6. $2x + 3 \le x - 1$
7. $\frac{2x-6}{3} > 0$
8. $\frac{2x-3}{2} \le 5$
9. $\frac{5x-7x}{3} > 4$
10. $\frac{x-3x}{5} \le 6$
11. $\frac{2x-5x}{2} \le 7$
12. $\frac{x-6x}{2} < -20$
13. $\frac{x}{2} + 1 < \frac{x}{3} - x$
14. $\frac{1}{2}(x+2) \ge \frac{2x}{3}$
15. $2(x+2) \le \frac{3}{4}x - 1$
16. $\frac{2}{3}(x-1) + \frac{3}{4}(x+1) < 0$
17. $\frac{3}{4}(2x-1) - \frac{1}{2}(4x+3) \ge 0$
18. $\frac{3}{5}(3x+2) - \frac{2}{3}(2x-1) \le 2$

Example $4 < x + 4 \le 6$

Solution Add -4 to each member.

$$0 < x \le 2$$

The solution set is $\{x \mid 0 < x \le 2\}$.

0 5

19. $4 < x - 2 < 8$ **20.** $0 \leq 2x \leq 12$ **21.** $-3 < 2x + 1 \leq 7$

22. $2 \leq 3x - 4 \leq 8$ **23.** $6 < 4 - x < 10$ **24.** $-3 < 3 - 2x < 9$

Graph each set.

25. $\{x|x < 2\} \cap \{x|x > -2\}$

26. $\{x|x \leq 5\} \cap \{x|x \geq 1\}$

27. $\{x|x + 1 \leq 3\} \cap \{x|-(x + 1) \leq 3\}$

28. $\{x|2x - 3 < 5\} \cap \{x|-(2x - 3) < 5\}$

29. $\{x|x < -3\} \cup \{x|x > 3\}$

30. $\{x|x \leq 2\} \cup \{x|x \geq 6\}$

31. $\{x|x + 2 > 4\} \cup \{x|-(x + 2) > 4\}$

32. $\{x|x - 3 \geq 5\} \cup \{x|-(x - 3) \geq 5\}$

33. $\left\{x \middle| \frac{x-3}{4} - 1 < \frac{x}{2}\right\} \cap \left\{x \middle| -2 - \frac{x}{4} \geq \frac{1+x}{3}\right\}$

34. $\{x|2 \leq 3x - 7 < 14\} \cup \{x|-4 < 5x + 6 \leq 21\}$

Sentences describing inequalities are set up in the same way as those describing equalities, except that symbols such as $\leq$ or $>$ replace the symbol $=$.

Example A student must have an average of 80–90% on five tests in a course to receive a B. His grades on the first four tests were 98%, 76%, 86%, and 92%. What grade on the fifth test would give him a B in the course?

Solution **a.** Express the quantity asked for in a simple phrase and then represent the quantity symbolically.

grade (in percent) on the fifth test: x

Write an inequality expressing the word sentence.

$$80 \leq \frac{98 + 76 + 86 + 92 + x}{5} < 90$$

b. Solve.

$$400 \leq 352 + x < 450$$
$$48 \leq x < 98$$

The solution set is $\{x|48 \leq x < 98\}$. Therefore, any grade equal to or greater than 48 and less than 98 would give the student a B in the course.

35. In the preceding example, what grade on the fifth test would give the student a B if his grades on the first four tests were 78%, 64%, 88%, and 76%?

36. The Fahrenheit and Celsius temperatures are related by $C = \frac{5}{9}(F - 32)$. Within what range must the temperature be in degrees Fahrenheit for the temperature to lie between -10°C and 20°C?

37. A man wishes to invest \$10,000, part at 5% and part at 7%. What is the least amount he can invest at 7% if he wishes an annual return of at least \$616?

38. A man can sail upstream in a river at an average rate of 4 miles per hour, and downstream at an average rate of 6 miles per hour. What is the distance he can sail upstream if he starts at 8:00 AM and must be back no later than 6:00 PM?

5.6

EQUATIONS AND INEQUALITIES INVOLVING ABSOLUTE VALUE

In Section 1.4, we defined the absolute value of a real number by

$$|x| = \begin{cases} x, & \text{if } x \geq 0 \\ -x, & \text{if } x < 0 \end{cases}$$

and interpreted it in terms of distance on a number line. For example, $|-5| = 5$ by definition, but 5 also denotes the distance the graph of -5 is located from the origin. More generally, we have

$$|x - a| = \begin{cases} x - a, & \text{if } x - a \geq 0, \text{ or equivalently, if } x \geq a \\ -(x - a), & \text{if } x - a < 0, \text{ or equivalently, if } x < a \end{cases}$$

and $|x - a|$ can be interpreted on a line graph as denoting the distance the graph of x is located from the graph of a, as shown in Figure 5.5.

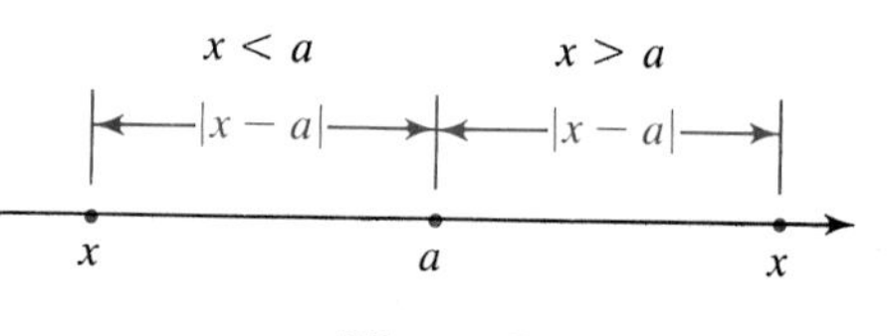

Figure 5.5

We can solve equations of the form

$$(1) \qquad |x - a| = b \qquad \text{or} \qquad |x + a| = |x - (-a)| = b$$

by appealing to the definition of absolute value. For example, the equation

$$|x - 3| = 5$$

implies that

$$x - 3 = 5,$$

or

$$-(x - 3) = 5.$$

From the equation $x - 3 = 5$ we have that $x = 8$, and from $-(x - 3) = 5$ we have that $x = -2$. Hence, the solution set we seek is

$$\{8\} \cup \{-2\} = \{8, -2\}.$$

In the equation $|x - 3| = 5$, we can obtain the solution set directly by comparing it with equation (1) and observing that the graph of x is 5 units from the graph of 3. Hence, the solutions are 8 and -2.

The foregoing example suggests that in general an equation of the form $|x - a| = b$ can be written equivalently as

$$x - a = b \quad \text{or} \quad -(x - a) = b. \tag{1a}$$

Any equation of the form $|x - a| = b$ is not a polynomial equation, because the variable appears within the absolute value symbols. Hence, such an equation cannot be assigned a degree. For example, although the equation

$$|x - 3| = 5$$

contains the first-degree polynomial $x - 3$, it has *two* solutions, 8 and -2. The equation is a contraction of two first-degree equations, and the presence of two solutions in no way contradicts our earlier discussion concerning the number of roots of a first-degree equation.

Inequalities involving absolute value notation are encountered quite frequently and require some additional discussion. Consider the inequality

$$|x + 1| < 3. \tag{2}$$

From the definition of absolute value, this inequality is equivalent to

$$x + 1 < 3, \qquad \text{for } x + 1 \geq 0 \qquad (x \geq -1),$$

and

$$-(x + 1) < 3, \qquad \text{for } x + 1 < 0 \qquad (x < -1).$$

The inequality $-(x + 1) < 3$ is equivalent to $x + 1 > -3$ or

$$-3 < x + 1.$$

Hence, the solution set of equation (2) is the intersection of the solution sets of $-3 < x + 1$ and $x + 1 < 3$, and therefore (2) can be written in the compact form

$$-3 < x + 1 < 3. \tag{3}$$

Any inequality of the form (2) should be written in the form (3) before undertaking elementary transformations to find the solution set. In general, the inequality $|x - a| < b$ $(b > 0)$ is equivalent to $-b < x - a < b$. In the present example, we then add -1 to each expression in (3), to obtain $-4 < x < 2$, from which the solution set is

$$S = \{x \mid -4 < x < 2\},$$

or

$$S = \{x \mid x > -4\} \cap \{x \mid x < 2\}.$$

The graph of this set is shown in Figure 5.6 on page 162.

Figure 5.6

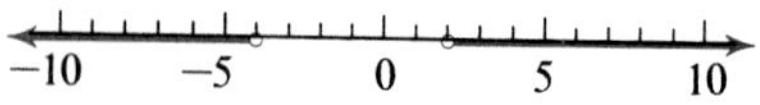

Figure 5.7

Now consider the inequality

$$|x + 1| > 3.$$

From the definition of absolute value, this inequality is equivalent to

$$x + 1 > 3, \qquad \text{for } x + 1 \geq 0 \qquad (x \geq -1),$$

or

$$-(x + 1) > 3, \qquad \text{for } x + 1 < 0 \qquad (x < -1).$$

The inequality $x + 1 > 3$ is equivalent to $x > 2$, and the inequality $-(x + 1) > 3$ is equivalent to $x < -4$. Hence, the solution set is given by

$$S = \{x | x > 2 \quad \text{or} \quad x < -4\}.$$

The graph of S is shown in Figure 5.7. Observe that S can also be given by

$$S = \{x | x > 2\} \cup \{x | x < -4\}.$$

Note that "$x > 2$ or $x < -4$" represents two disjoint intervals and hence cannot be written in the form $a < b < c$. It is not correct to write $2 < x < -4$ because 2 is not less than -4.

The two preceding examples suggest that in general an equation of the form $|x - a| < b$ can be written equivalently as

(4) $$-b < x - a < b$$

and $|x - a| > b$ can be written equivalently as

(5) $$x - a > b \qquad \text{or} \qquad -(x - a) > b.$$

EXERCISE 5.6

A

Solve.

Example $|x + 5| = 8$

Solution Write the equation as two first-degree equations.

$$x + 5 = 8 \qquad \text{or} \qquad -(x + 5) = 8$$

Solve each equation.

$$x = 3 \qquad \text{or} \qquad -x - 5 = 8$$
$$-x = 13$$
$$x = -13$$

The solution set is $\{3, -13\}$.

1. $|x| = 5$
2. $|x| = 7$
3. $|x - 4| = 9$
4. $|x - 3| = 7$
5. $|2x + 1| = 13$
6. $|3x - 1| = 5$
7. $|4 - 3x| = 1$
8. $|6 - 5x| = 4$
9. $|4x + 3| = 0$
10. $|3x - 7| = 0$
11. $\left|x - \frac{3}{4}\right| = \frac{1}{4}$
12. $\left|x + \frac{2}{3}\right| = \frac{1}{3}$
13. $\left|1 + \frac{3}{2}x\right| = \frac{1}{4}$
14. $\left|1 - \frac{1}{2}x\right| = \frac{3}{4}$
15. $\left|\frac{1}{3} - 4x\right| = \frac{5}{6}$
16. $\left|\frac{2}{3} + 2x\right| = \frac{7}{12}$
17. $\left|\frac{2}{3}x - \frac{1}{2}\right| = 0$
18. $\left|\frac{3}{4}x + \frac{3}{8}\right| = 0$

Solve and graph each solution set.

Example $|1 - 2x| \leq 7$

Solution Using inequality (4), rewrite without the absolute value symbol.

$$-7 \leq 1 - 2x \leq 7$$

Add -1 to each expression.

$$-8 \leq -2x \leq 6$$

Multiply each expression by $-\frac{1}{2}$ and change the inequality symbols.

$$4 \geq x \geq -3$$

The solution set is

$$\{x \mid -3 \leq x \leq 4\};$$

that is,

$$\{x \mid -3 \leq x\} \cap \{x \mid x \leq 4\}.$$

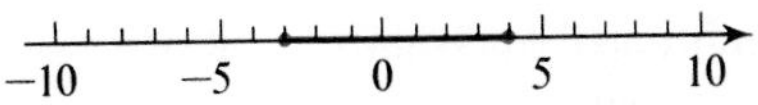

19. $|x| < 2$
20. $|x| < 5$
21. $|x + 3| \leq 4$
22. $|x + 1| \leq 8$
23. $|2x - 5| < 3$
24. $|2x + 4| < 6$
25. $|4 - x| \leq 8$
26. $|5 - 2x| \leq 15$

Example $|3x - 6| > 9$

Solution Since the inequality is of the form $|x - a| > b$, the intervals involved are disjoint, and we must consider two inequalities separately. Using inequalities (5), on page 162, rewrite without the absolute value symbol and solve.

$$3x - 6 > 9 \quad \text{or} \quad -(3x - 6) > 9$$
$$3x > 15 \qquad\qquad 3x - 6 < -9$$
$$x > 5 \qquad\qquad 3x < -3$$
$$x < -1$$

The solution set is

$$\{x|x < -1 \quad \text{or} \quad x > 5\} \quad \text{or} \quad \{x|x < -1\} \cup \{x|x > 5\}.$$

−10 −5 0 5 10

27. $|x| > 3$
28. $|x| \geq 5$
29. $|x - 2| > 5$
30. $|x + 5| > 2$
31. $|3 - 2x| \geq 7$
32. $|4 - 3x| > 10$

State the values of the variables for which each equation is true.

33. $|x - 3| = x - 3$
34. $|x + 2| = x + 2$
35. $|y + 6| = -(y + 6)$
36. $|y - 5| = -(y - 5)$
37. $|2x - 3| = 2x - 3$
38. $|3y + 4| = -(3y + 4)$

CHAPTER SUMMARY

[5.1] A replacement for the variable in an equation or inequality that results in a true statement is called a **solution** (or **root**) of the equation or inequality; the set of all solutions is called the **solution set**.

A **conditional equation** is not true for at least one member in the replacement set of the variable.

Equations or inequalities having identical solution sets are called **equivalent equations** or **equivalent inequalities**, respectively. To solve a given conditional equation, we apply **elementary transformations** to generate a sequence of equivalent equations until we arrive at one that can be solved by inspection.

The addition of the same expression representing a real number to each member of an equation or the multiplication of each member of an equation by the same expression representing a nonzero real number produces an equivalent equation. This notion is expressed symbolically as follows:

If $P(x)$, $Q(x)$, and $R(x)$ are expressions, then for all values of x for which these expressions represent real numbers, the equation

$$P(x) = Q(x)$$

is equivalent to

$$P(x) + R(x) = Q(x) + R(x),$$
$$P(x) \cdot R(x) = Q(x) \cdot R(x), \quad \text{for } R(x) \neq 0.$$

[5.2] An identity is an equation that yields a true statement for every member of the replacement set of the variable for which each member is defined. We can verify that an equation is an identity by changing one or both members to obtain a sequence of equivalent equations until we arrive at an equation in which both members are the same.

[5.3] An equation containing more than one variable can be rewritten in other forms by generating equivalent equations.

[5.4] Equations can be used as mathematical models to solve practical problems.

[5.5] The addition of the same expression representing a real number to each member of an inequality produces an equivalent inequality in the *same* sense. If each member of an inequality is multiplied by the same expression representing a positive number, the result is an equivalent inequality in the *same* sense. If each member of an inequality is multiplied by the same expression representing a negative number, the result is an equivalent inequality in the *opposite* sense. This notion is expressed symbolically as follows:

If $P(x)$, $Q(x)$, and $R(x)$ are expressions, then for all values of x for which these expressions represent real numbers,

$$P(x) < Q(x)$$

is equivalent to

$$P(x) + R(x) < Q(x) + R(x),$$
$$P(x) \cdot R(x) < Q(x) \cdot R(x), \quad \text{for } R(x) > 0,$$
$$P(x) \cdot R(x) > Q(x) \cdot R(x), \quad \text{for } R(x) < 0.$$

[5.6] The equation $|x - a| > b$ for $b > 0$ is equivalent to the pair of equations $x - a < -b$ or $x - a > b$; the equation $|x - a| < b$ for $b > 0$ is equivalent to $-b < x - a < b$.

REVIEW EXERCISES

[5.1] *Solve.*

1. a. $2[x - (2x + 1)] = 6$ **b.** $2 + \frac{x}{3} = \frac{5}{6}$

2. a. $\frac{x}{x + 1} + \frac{4}{5} = 6$ **b.** $1 - \frac{x - 2}{x - 3} = \frac{3}{x - 1}$

[5.2] *Verify that each equation is an identity.*

3. a. $[(a - b)^2 + 2ab] = a^2 + b^2$ **b.** $\frac{x^2 + 7x + 12}{x + 3} - 4 = x$

[5.3] **4.** Solve $\frac{x - y}{3} = \frac{x + y}{2}$ for y in terms of x.

5. Solve $\frac{x - y}{3} = \frac{x + y}{2}$ for x in terms of y.

6. Solve $s = \frac{1}{2}at^2 + 2t$ for a.

7. Solve $l = a + nd - d$ for d.

[5.4] *Solve.*

8. The sum of two integers is 29. If one of the integers is 7 less than the other, what are the integers?

9. How many liters of a disinfectant that is 74% alcohol must be mixed with 5 liters of a disinfectant that is 90% alcohol to obtain a disinfectant that contains 84% alcohol?

10. A man invested $7200, part at 5% and part at $5\frac{1}{2}$%. How much did he invest at each rate if the yearly return from both investments amounted to $387?

[5.5] *Solve and graph each solution set.*

11. a. $\frac{x - 3}{4} \leq 6$ **b.** $2(x - 1) \leq \frac{2}{3}x$

12. a. $1 < 2x - 3 \leq 7$ **b.** $-5 < 4 - 3x < 1$

13. Graph $\{x \mid 2x + 3 < 7\} \cap \{x \mid 4x + 1 > 3\}$.

14. A baseball team wins 40 of its first 50 games. How many games must it win of the remaining 40 it must play in order to have an average between 60% and 70% wins in its entire schedule?

[5.6] *Solve.*

15. a. $|2x + 1| = 7$ b. $|3 - 4x| = 8$

16. a. $\left|x + \frac{1}{2}\right| \geq \frac{5}{2}$ Graph the solution set. b. $|x - 2| < 5$ Graph the solution set.

6

SECOND-DEGREE EQUATIONS AND INEQUALITIES

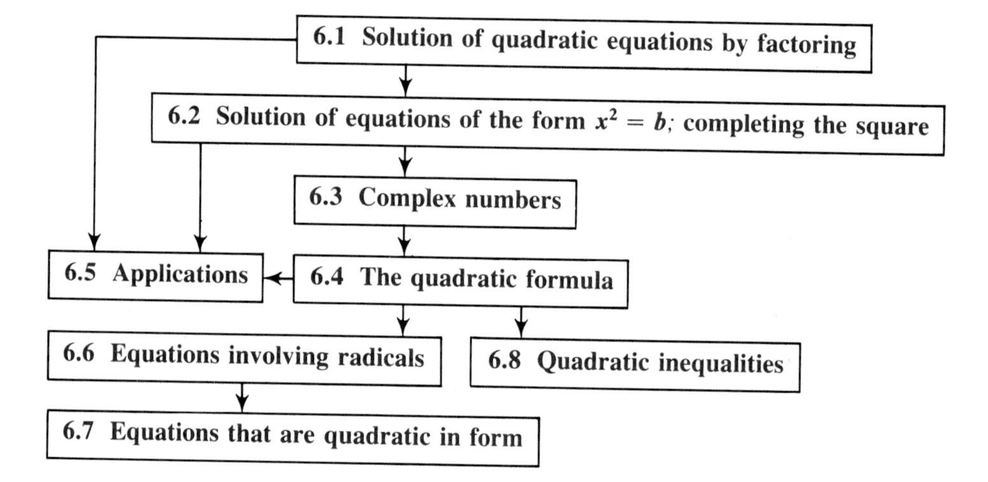

6.1

SOLUTION OF QUADRATIC EQUATIONS BY FACTORING

An open sentence in one variable that in simplest form is a second-degree equation is called a **quadratic equation** in that variable. We shall designate as standard form for such equations

$$\boldsymbol{ax^2 + bx + c = 0},$$

where a, b, and c are constants representing real numbers and $a \neq 0$. If $b = 0$ or $c = 0$, then the equations are of the form

$$ax^2 + c = 0 \qquad \text{or} \qquad ax^2 + bx = 0,$$

and are called **incomplete quadratic equations.**

In Chapter 5 we solved linear equations by performing certain elementary transformations. These transformations are equally applicable to equations of higher degree and, in particular, to quadratic equations. For example,

$$2x = 6 - 8x^2$$

is equivalent to

$$8x^2 + 2x - 6 = 0,$$

which is equivalent to

$$4x^2 + x - 3 = 0.$$

If the left-hand member of a quadratic equation in standard form is factorable, we may solve the equation by making use of the principle:

▶ *The product of two factors equals* 0 *if and only if one or both of the factors equals* 0.

That is,

$$ab = 0 \quad \textbf{if and only if} \quad a = 0 \quad \textbf{or} \quad b = 0.$$

For example, $(x - 1)(x + 2) = 0$ if and only if

$$x - 1 = 0 \qquad (x = 1)$$

or

$$x + 2 = 0 \qquad (x = -2).$$

The word "or" in the above statement is used in an inclusive sense to mean either one *or* the other *or* both. The truth of this assertion is a consequence of the zero factor property. The "if" part stems directly from this property. That is, if either a or b is 0, then ab must be 0, because $0 \cdot b = 0$ for every real number b. To establish the "only if" part, we assume that either $a = 0$ or $a \neq 0$. If $a = 0$, the conclusion of the theorem is true, and if $a \neq 0$, then we can multiply each member of $ab = 0$ by $1/a$ to obtain

$$\left(\frac{1}{a}\right)ab = \left(\frac{1}{a}\right)0 \qquad \text{or} \qquad b = 0,$$

and the conclusion is also true. Since these are the only possibilities, the assertion is established.

For an application of the theorem, consider the equation

$$x^2 + 2x - 15 = 0, \tag{1}$$

from which

$$(x + 5)(x - 3) = 0, \tag{2}$$

which will be true if and only if

$$x + 5 = 0 \qquad \text{or} \qquad x - 3 = 0.$$

We observe that if $x = -5$, then $x + 5$ equals 0, and if $x = 3$, then $x - 3$ equals 0, and either -5 or 3, when substituted for x in (2) or (1), will make the left-hand member 0. The solution set is $\{-5, 3\}$.

In general, the solution set of a quadratic equation can be expected to contain two elements. However, if the left-hand member of a quadratic equation in standard form is the square of a binomial, we find that the solution set contains but one member. Consider the equation

$$x^2 - 2x + 1 = 0.$$

Factoring the left-hand member, we have

$$(x - 1)(x - 1) = 0,$$

and the solution set is $\{1\}$, which contains only one element. For reasons of convenience and consistency in more advanced work, a solution of this sort is said to be of **multiplicity two**.

Notice that the solution set of the quadratic equation

$$(x - r_1)(x - r_2) = 0 \tag{3}$$

is $\{r_1, r_2\}$. Therefore, if r_1 and r_2 are given as solutions of a quadratic equation, the equation can be written directly as (3). By completing the indicated multiplication, the equation can be transformed to standard form. For example, if 2 and -3 are the solutions of a quadratic equation, then

$$[x - (2)][x - (-3)] = 0,$$

or

$$(x - 2)(x + 3) = 0,$$

from which

$$x^2 + x - 6 = 0.$$

To find a quadratic equation whose solutions are $\frac{1}{4}$ and $\frac{3}{2}$, we write

$$\left(x - \frac{1}{4}\right)\left(x - \frac{3}{2}\right) = 0,$$

from which

$$x^2 - \frac{7}{4}x + \frac{3}{8} = 0.$$

This equation can be transformed to one with integral coefficients by multiplying each member by 8 to obtain the equivalent equation,

$$8x^2 - 14x + 3 = 0.$$

EXERCISE 6.1

A

Solve for x. Determine your answer by inspection or follow the procedure in the example.

Example $x(x + 3) = 0$

Solution Set each factor equal to 0 and solve the equations.

$$x = 0 \qquad x + 3 = 0$$
$$x = -3$$

The solution set is $\{0, -3\}$.

1. $(x + 2)(x - 5) = 0$ **2.** $(x + 3)(x - 4) = 0$ **3.** $(2x + 5)(x - 2) = 0$
4. $(x + 1)(3x - 1) = 0$ **5.** $x(2x + 1) = 0$ **6.** $x(3x - 7) = 0$
7. $4(x - 6)(2x + 3) = 0$ **8.** $5(2x - 7)(x + 1) = 0$ **9.** $3(x - 2)(2x + 1) = 0$
10. $7(x + 5)(3x - 1) = 0$ **11.** $4(2x - 5)(3x + 2) = 0$ **12.** $2(3x - 4)(2x + 5) = 0$

Example $x^2 + x = 30$

Solution Write in standard form (the right-hand member is 0).

$$x^2 + x - 30 = 0$$

Factor the left-hand member.

$$(x + 6)(x - 5) = 0$$

Set each factor equal to 0 and solve the equations.

$$x + 6 = 0 \qquad x - 5 = 0$$

$$x = -6 \qquad x = 5$$

The solution set is $\{-6, 5\}$.

13. $x^2 - 3x = 0$ **14.** $x^2 + 5x = 0$ **15.** $2x^2 = 6x$
16. $3x^2 = 3x$ **17.** $x^2 - 9 = 0$ **18.** $x^2 - 4 = 0$
19. $2x^2 - 18 = 0$ **20.** $3x^2 - 3 = 0$ **21.** $9x^2 = 4$
22. $4x^2 = 1$ **23.** $\frac{4}{9}x^2 - 1 = 0$ **24.** $\frac{9x^2}{25} - 1 = 0$
25. $x^2 - 5x + 4 = 0$ **26.** $x^2 + 5x + 6 = 0$ **27.** $x^2 - 5x - 14 = 0$
28. $x^2 - x - 42 = 0$ **29.** $3x^2 - 6x = -3$ **30.** $12x^2 = 8x + 15$

Example $3x(x + 1) = 2x + 2$

Solution Write in standard form.

$$3x^2 + x - 2 = 0$$

Factor the left-hand member.

$$(3x - 2)(x + 1) = 0$$

Set each factor equal to 0 and solve the equations.

$$3x - 2 = 0 \qquad x + 1 = 0$$

$$x = \frac{2}{3} \qquad x = -1$$

The solution set is $\{2/3, -1\}$.

31. $x(2x - 3) = -1$ **32.** $2x(x - 2) = x + 3$ **33.** $(x - 2)(x + 1) = 4$
34. $x(3x + 2) = (x + 2)^2$ **35.** $(x - 1)^2 = 2x^2 + 3x - 5$ **36.** $x(x + 1) = 4 - (x + 2)^2$

Example $x^2 - \frac{17}{3}x = 2$

Solution Write in standard form.

$$x^2 - \frac{17}{3}x - 2 = 0$$

Multiply each member by 3 and simplify.

$$3(x^2) - 3\left(\frac{17}{3}x\right) - 3(2) = 3(0)$$

$$3x^2 - 17x - 6 = 0$$

Factor the left-hand member.

$$(3x + 1)(x - 6) = 0$$

Set each factor equal to 0 and solve the equations.

$$3x + 1 = 0 \qquad x - 6 = 0$$

$$x = -\frac{1}{3} \qquad x = 6$$

The solution set is $\{-\frac{1}{3}, 6\}$.

37. $\frac{2x^2}{3} + \frac{x}{3} - 2 = 0$

38. $x - 1 = \frac{x^2}{4}$

39. $\frac{x^2}{6} + \frac{x}{3} = \frac{1}{2}$

40. $\frac{x}{4} - \frac{3}{4} = \frac{1}{x}$

41. $3 = \frac{10}{x^2} - \frac{7}{x}$

42. $\frac{4}{3x} + \frac{3}{3x + 1} + 2 = 0$

43. $\frac{2}{x - 3} - \frac{6}{x - 8} = -1$

44. $\frac{x}{x - 1} - \frac{x}{x + 1} = \frac{4}{3}$

Given the solutions of a quadratic equation, r_1 and r_2, write the equation in standard form with integral coefficients.

Example $\frac{3}{4}$ and -2

Solution Write in the form $(x - r_1)(x - r_2) = 0$.

$$\left(x - \frac{3}{4}\right)[x - (-2)] = 0$$

Write in standard form.

$$\left(x - \frac{3}{4}\right)(x + 2) = 0$$

$$x^2 + \frac{5}{4}x - \frac{3}{2} = 0$$

$$4x^2 + 5x - 6 = 0$$

45. -2 and 1 **46.** -4 and 3 **47.** 0 and -5 **48.** 0 and 5

49. 4 and $-\frac{3}{4}$ **50.** $-\frac{2}{3}$ and 3 **51.** $\frac{1}{2}$ and $\frac{3}{5}$ **52.** $\frac{2}{3}$ and $\frac{1}{5}$

B

Solve each equation for x in terms of a and b.

53. $x^2 - 4b^2 = 0$

54. $x^2 - (a + b)^2 = 0$

55. $x^2 - 3ax - 4a^2 = 0$

56. $x^2 + 4bx - 12b^2 = 0$

57. $x^2 - (a + b)x + ab = 0$

58. $x^2 + (2a - b)x - 2ab = 0$

59. $x^2 + \left(a + \frac{b}{2}\right)x + \frac{ab}{2} = 0$

60. $x^2 - \frac{1}{2}(a + b)x + \frac{ab}{4} = 0$

6.2

SOLUTION OF EQUATIONS OF THE FORM $x^2 = b$; COMPLETING THE SQUARE

Quadratic equations of the form

$$x^2 = b \qquad (b \geq 0)$$

may be solved by a method often termed the **extraction of roots**. If the equation has a solution, then from the definition of a square root, x must be a square root of b. Since each positive real number b has two square roots, we have two solutions if $b > 0$. These are given by

$$\boldsymbol{x = \sqrt{b}, \qquad x = -\sqrt{b},}$$

and the solution set is $\{\sqrt{b}, -\sqrt{b}\}$. If $b = 0$, we have one number, 0, satisfying the equation, and the solution set of $x^2 = 0$ is $\{0\}$. The same conclusion can be reached by noting that, if we factor

$$x^2 - b = 0 \qquad (b \geq 0),$$

we have

$$(x - \sqrt{b})(x + \sqrt{b}) = 0,$$

which, when solved by the methods of Section 6.1, also leads to the solution set $\{\sqrt{b}, -\sqrt{b}\}$. Equations of the form

$$(x - a)^2 = b \qquad (b \geq 0)$$

can be solved by the same method. For example,

$$(x - 2)^2 = 16$$

implies that

$$x - 2 = 4 \quad \text{or} \quad x - 2 = -4,$$

from which we have

$$x = 6 \quad \text{or} \quad x = -2,$$

and the solution set is $\{6, -2\}$.

The method of extraction of roots can be used to find the solution set of any quadratic equation. Let us first consider a specific example,

$$x^2 - 4x - 12 = 0,$$

which can be written

$$x^2 - 4x \qquad = 12.$$

If the square of one-half of the coefficient of the first-degree term,

$$\left[\frac{1}{2}(-4)\right]^2,$$

is added to each member, we obtain

$$x^2 - 4x + 4 = 12 + 4,$$

in which the left-hand member is the square of $(x - 2)$. Therefore, the equation can be written

$$(x - 2)^2 = 16$$

and the solution set obtained as above.

Now consider the general quadratic equation in standard form,

$$ax^2 + bx + c = 0,$$

for the special case where $a = 1$; that is,

(1) $$x^2 + bx + c = 0.$$

If we can factor the left-hand member of (1), we can solve the equation by factoring; if not, we can write it in the form

$$(x - p)^2 = q \qquad (q \geq 0),$$

which we can solve by the extraction of roots. We begin the latter process by adding $-c$ to each member of (1), which yields

(2) $$x^2 + bx \qquad = -c.$$

If we then add the square of one-half of the coefficient of x, $(b/2)^2$, to each member of (2), the result is

(3) $$x^2 + bx + \left(\frac{b}{2}\right)^2 = -c + \left(\frac{b}{2}\right)^2,$$

where the left-hand member is equivalent to $(x + b/2)^2$, and we have

$$(4) \qquad \left(x + \frac{b}{2}\right)^2 = -c + \frac{b^2}{4}.$$

Since we have performed only elementary transformations, (4) is equivalent to (2) and we can solve (4) as in the example at the beginning of this section.

The technique used to obtain equations (3) and (4) is called **completing the square**. We can determine the term necessary to complete the square in (2) by dividing the coefficient b of the linear term by 2 and squaring the result. The expression obtained,

$$x^2 + bx + \left(\frac{b}{2}\right)^2,$$

can then be written in the form $(x + b/2)^2$.

We began with the special case,

$$(5) \qquad x^2 + bx + c = 0,$$

rather than the general form,

$$ax^2 + bx + c = 0,$$

because the term necessary to complete the square is more obvious when $a = 1$. However, a quadratic equation in standard form can always be written in the form (5) by multiplying each member of

$$ax^2 + bx + c = 0$$

by $1/a$ $(a \neq 0)$ and obtaining

$$x^2 + \frac{b}{a}x + \frac{c}{a} = 0.$$

In this section we have considered quadratic equations with real solutions only. An equation of the form $x^2 = b$ does not have real solutions if $b < 0$, because the square of any real number is positive. In Section 6.3 we will consider a new set of numbers which contains solutions of all quadratic equations with real coefficients.

EXERCISE 6.2

A

Solve for x by the extraction of roots.

Example $7x^2 - 63 = 0$

Solution Obtain an equivalent equation with x^2 as the only term in the left-hand member.

$$x^2 = 9$$

Set x equal to each square root of 9.

$$x = 3 \qquad x = -3$$

The solution set is $\{3, -3\}$.

1. $x^2 = 100$ **2.** $x^2 = 16$ **3.** $9x^2 = 25$ **4.** $4x^2 = 9$

5. $2x^2 = 14$ **6.** $3x^2 = 15$ **7.** $4x^2 - 24 = 0$ **8.** $3x^2 - 9 = 0$

9. $\frac{2x^2}{3} = 4$ **10.** $\frac{3x^2}{5} = 3$ **11.** $\frac{4x^2}{3} = 27$ **12.** $\frac{9x^2}{2} = 50$

Example $(x + 3)^2 = 7$

Solution Set $x + 3$ equal to each square root of 7.

$$x + 3 = \sqrt{7} \qquad x + 3 = -\sqrt{7}$$

$$x = -3 + \sqrt{7} \qquad x = -3 - \sqrt{7}$$

The solution set is $\{-3 + \sqrt{7}, -3 - \sqrt{7}\}$.

13. $(x - 2)^2 = 9$ **14.** $(x + 3)^2 = 4$ **15.** $(2x - 1)^2 = 16$

16. $(3x + 1)^2 = 25$ **17.** $(x + 2)^2 = 3$ **18.** $(x - 5)^2 = 7$

19. $(x - 2)^2 = 12$ **20.** $(x + 3)^2 = 18$ **21.** $(x - 7)^2 = 4$

22. $(x + 3)^2 = 36$ **23.** $(2x - 5)^2 = 9$ **24.** $(3x + 4)^2 = 16$

a. *What term must be added to each expression to make a perfect square?*
b. *Write the resulting expression as the square of a binomial.*

Example $x^2 - 9x$

Solution **a.** Square one-half of the coefficient of the first-degree term -9.

$$\left(-\frac{9}{2}\right)^2 = \frac{81}{4}$$

b. Rewrite $x^2 - 9x + \frac{81}{4}$ as the square of an expression: $\left(x - \frac{9}{2}\right)^2$.

25. $x^2 + 2x$ **26.** $x^2 - 4x$ **27.** $x^2 - 6x$ **28.** $x^2 + 10x$

29. $x^2 + 3x$ **30.** $x^2 - 5x$ **31.** $x^2 - 7x$ **32.** $x^2 + 11x$

33. $x^2 - x$ **34.** $x^2 + 15x$ **35.** $x^2 + \frac{1}{2}x$ **36.** $x^2 - \frac{3}{4}x$

Solve by completing the square.

Example $2x^2 + x - 1 = 0$

Solution Rewrite the equation with the constant term as the right-hand member and the coefficient of x^2 equal to 1.

$$x^2 + \frac{1}{2}x \qquad = \frac{1}{2}$$

Add the square of one-half of the coefficient of the first-degree term to each member.

$$x^2 + \frac{1}{2}x + \frac{1}{16} = \frac{1}{2} + \frac{1}{16}$$

Rewrite the left-hand member as the square of an expression.

$$\left(x + \frac{1}{4}\right)^2 = \frac{9}{16}$$

Set $x + \frac{1}{4}$ equal to each square root of $\frac{9}{16}$.

$$x + \frac{1}{4} = \frac{3}{4} \qquad x + \frac{1}{4} = -\frac{3}{4}$$

$$x = \frac{1}{2} \qquad x = -1$$

The solution set is $\{1/2, -1\}$.

37. $x^2 + 4x - 12 = 0$ **38.** $x^2 - x - 6 = 0$ **39.** $x^2 - 2x + 1 = 0$

40. $x^2 + 4x + 4 = 0$ **41.** $x^2 + 9x + 20 = 0$ **42.** $x^2 - x - 20 = 0$

43. $x^2 - 2x - 1 = 0$ **44.** $x^2 + 3x - 1 = 0$ **45.** $2x^2 = 4 - 3x$

46. $2x^2 = 6 - 5x$ **47.** $2x^2 + 4x = 3$ **48.** $3x^2 + x = 4$

B

Solve for x in terms of a, b, and c.

49. $x^2 - a = 0$ **50.** $x^2 - 2a = 0$ **51.** $\frac{ax^2}{b} = c$ **52.** $\frac{bx^2}{c} - a = 0$

53. $(x - a)^2 = 16$ **54.** $(x + a)^2 = 36$ **55.** $(ax + b)^2 = 9$ **56.** $(ax - b)^2 = 25$

57. Solve $a^2 + b^2 = c^2$ for b in terms of a and c.

58. Solve $S = h - r^2$ for r in terms of S and h.

59. Solve $A = P(1 + r)^2$ for r in terms of A and P.

60. Solve $s = \frac{1}{2}gt^2 + c$ for t in terms of s, g, and c.

61. Solve $ax^2 + bx + c = 0$ for x in terms of a, b, and c by using the method of completing the square.

6.3

COMPLEX NUMBERS

In Section 6.2 we noted that some quadratic equations do not have solutions in the set of real numbers. For example, if $b > 0$, then $x^2 = -b$ has no real number solution, because there is no real number whose square is negative. For this reason, the expression $\sqrt{-b}$, for $b \in R$, $b > 0$, is undefined in the set of real numbers. In this section, we wish to consider a set of numbers that contains members whose squares are negative real numbers and that also contains a subset of members that can be identified with the set of real numbers. We shall see that this new set of numbers, called the set C of **complex numbers**, provides solutions for all quadratic equations with real coefficients in one variable.

Let us first assume a set of numbers among whose members are square roots of negative real numbers. We define $\sqrt{-b}$ and $-\sqrt{-b}$, where b is a positive real number, to be numbers whose squares are equal to $-b$. That is, for $b > 0$,

$$\sqrt{-b}\sqrt{-b} = -b \quad \textbf{and} \quad (-\sqrt{-b})(-\sqrt{-b}) = -b.$$

In particular,

$$\sqrt{-1}\sqrt{-1} = -1 \quad \textbf{and} \quad (-\sqrt{-1})(-\sqrt{-1}) = -1.$$

It is customary to use the symbol i for $\sqrt{-1}$. With this convention and assuming that $-i = -1 \cdot i$,

$$i = \sqrt{-1}, \qquad -i = -\sqrt{-1}, \qquad i^2 = -1, \qquad (-i)^2 = -1.$$

Thus, in this new set of numbers, -1 has two square roots, i and $-i$. Next, assuming that $-\sqrt{-b} = -1 \cdot \sqrt{-b}$ and that

$$(i\sqrt{b})(i\sqrt{b}) = i^2b = -1 \cdot b = -b,$$

it follows from the definition of $\sqrt{-b}$ that

$$\sqrt{-b} = \sqrt{-1}\sqrt{b} = i\sqrt{b} \quad \textbf{and} \quad -\sqrt{-b} = -1 \cdot \sqrt{-1}\sqrt{b} = -i\sqrt{b}.$$

Hence, a square root of any negative real number can be represented as the product of a real number and the number $\sqrt{-1}$, or i. For example,

$$\sqrt{-4} = \sqrt{-1}\sqrt{4} = i\sqrt{4} = 2i$$

and

$$-\sqrt{-3} = -\sqrt{-1}\sqrt{3} = -i\sqrt{3} = -\sqrt{3}\,i.$$

The numbers represented by the symbols $\sqrt{-b}$ and $-\sqrt{-b}$, where $b \in R$ and $b > 0$, are called **pure imaginary numbers.**

Now, consider all possible expressions of the form $a + bi$, where $a, b \in R$ and $i = \sqrt{-1}$, which are the sums of all real numbers and all pure imaginary numbers. Such an expression names a complex number, that is, a number in the set C. If $b = 0$, then $a + bi = a$, and it is evident that the set R of real numbers is a subset of the set C of complex numbers. If $b \neq 0$, then $a + bi$ is called an **imaginary number** (see Figure 6.1). For example, the numbers -7, $3 + 2i$, and $4i$ are all complex numbers. However, -7 is also a real number, and $3 + 2i$ and $4i$ are also imaginary numbers. Furthermore, $4i$ is a pure imaginary number.

Complex numbers: $C = \{a + bi \mid a, b \in R\}$

$(b = 0)$
Real numbers: $a + bi = a$

$(b \neq 0)$
Imaginary numbers: $a + bi$

$(a = 0)$
Pure imaginary numbers: $a + bi = bi$

Figure 6.1

Although the following definitions for sums and products of complex numbers may appear arbitrary and unusual (particularly in the case of products), we shall see that they relate to the corresponding operations in R in a direct and useful way.

If $a_1 + b_1 i$ and $a_2 + b_2 i$ are complex numbers, where the variables with subscripts denote specific real numbers, then

$$\mathbf{(a_1 + b_1 i) + (a_2 + b_2 i) = (a_1 + a_2) + (b_1 + b_2)i}$$
$$\mathbf{(a_1 + b_1 i) \cdot (a_2 + b_2 i) = (a_1 a_2 - b_1 b_2) + (a_1 b_2 + a_2 b_1)i.}$$

With the operations of addition and multiplication thus defined, it can be shown that the axioms on page 17 are valid properties for complex numbers.

Note that the definition for a product conforms to the result of multiplying complex numbers as though they were polynomials. For example, from the definition,

$$\begin{aligned}(2-i)(1+3i) &= [2\cdot 1-(-1)(3)]+[2\cdot 3+(-1)(1)]i\\ &= [2+3]+[6-1]i = 5+5i.\end{aligned}$$

However, using conventional polynomial multiplication, we have

$$\begin{aligned}(2-i)(1+3i) &= 2(1+3i)-i(1+3i)\\ &= 2+6i-i-3i^2.\end{aligned}$$

Since $i^2 = -1$, we find that

$$\begin{aligned}2+6i-i-3i^2 &= 2+6i-i-3(-1)\\ &= 2+6i-i+3,\end{aligned}$$

or

$$(2-i)(1+3i) = 5+5i.$$

Ordinarily we compute such products by multiplying rather than from the definition.

The symbol $\sqrt{-b}$, $b > 0$, should be used with care, since certain relationships involving the square root symbol that are valid for real numbers are not valid when the symbol does not represent a real number. For instance,

$$\sqrt{-2}\sqrt{-3} = (i\sqrt{2})(i\sqrt{3}) = i^2\sqrt{6} = -\sqrt{6} \neq \sqrt{(-2)(-3)}.$$

To avoid difficulty with this point, you should rewrite all expressions of the form $\sqrt{-b}$, $b > 0$, in the form $i\sqrt{b}$ or $\sqrt{b}\,i$ before engaging in any computations. For example,

$$\begin{aligned}(2+\sqrt{-3})(2-\sqrt{-3}) &= (2+i\sqrt{3})(2-i\sqrt{3})\\ &= 4-3i^2\\ &= 4-3(-1) = 7.\end{aligned}$$

The quotient of two complex numbers can be found by using the following property, which is analogous to the *fundamental principle of fractions* in the set of real numbers. First recall from page 122 that, for $b > 0$, the *conjugate* of $a+\sqrt{b}$ is $a-\sqrt{b}$. Similarly, the conjugate of $a+\sqrt{-b}$ is $a-\sqrt{-b}$. For example, the conjugate of $2+3i$ is $2-3i$ and the conjugate of $-3-i$ is $-3+i$.

The quotient

$$\frac{a+bi}{c+di}$$

of two complex numbers can be written in standard form by multiplying the numerator and the denominator by $c-di$, the conjugate of the denominator.

Thus, for example,

$$\frac{4+i}{2+3i} = \frac{(4+i)(2-3i)}{(2+3i)(2-3i)} = \frac{8-10i-3i^2}{4-9i^2}$$

$$= \frac{8-10i+3}{4+9} = \frac{11}{13} - \frac{10}{13}i.$$

EXERCISE 6.3

A

Write each expression in the form $a + bi$ or $a + ib$.

Examples **a.** $3\sqrt{-18}$ **b.** $2 - 3\sqrt{-16}$

Solutions **a.** $3\sqrt{-18} = 3\sqrt{-1 \cdot 9 \cdot 2}$
$= 3\sqrt{-1}\sqrt{9}\sqrt{2}$
$= 3i(3)\sqrt{2}$
$= 9i\sqrt{2}$

b. $2 - 3\sqrt{-16} = 2 - 3\sqrt{-1 \cdot 16}$
$= 2 - 3\sqrt{-1}\sqrt{16}$
$= 2 - 3i(4)$
$= 2 - 12i$

1. $\sqrt{-4}$ **2.** $\sqrt{-9}$ **3.** $\sqrt{-32}$

4. $\sqrt{-50}$ **5.** $3\sqrt{-8}$ **6.** $4\sqrt{-18}$

7. $4 + 2\sqrt{-1}$ **8.** $5 - 3\sqrt{-1}$ **9.** $3\sqrt{-50} + 2$

10. $5\sqrt{-12} - 1$ **11.** $\sqrt{4} + \sqrt{-4}$ **12.** $\sqrt{20} - \sqrt{-20}$

Examples **a.** $(2 + 3i) - (4 - i)$ **b.** $\dfrac{3}{3+i}$

Solutions **a.** $2 + 3i + (-4 + i)$
$= [2 + (-4)] + (3 + 1)i$
$= -2 + 4i$

b. $\dfrac{3}{3+i} = \dfrac{3(3-i)}{(3+i)(3-i)}$
$= \dfrac{9-3i}{9-i^2} = \dfrac{9-3i}{9+1}$
$= \dfrac{9}{10} - \dfrac{3}{10}i$

13. $(2 + 4i) + (3 + i)$ **14.** $(2 - i) + (3 - 2i)$

15. $(4 - i) - (6 - 2i)$ **16.** $(2 + i) - (4 - 2i)$

17. $3 - (4 + 2i)$

18. $(2 - 6i) - 3$

19. $(2 - i)(3 + 2i)$

20. $(1 - 3i)(4 - 5i)$

21. $(1 - 3i)^2$

22. $(2 + i)^2$

23. $\dfrac{5}{2i}$

24. $\dfrac{3 - i}{4i}$

25. $\dfrac{2}{1 - i}$

26. $\dfrac{2 + i}{1 + 3i}$

Write the given product or quotient in the form a + bi or a + ib.

Examples **a.** $(1 + \sqrt{-2})(3 - \sqrt{-2})$ **b.** $\dfrac{1}{3 - \sqrt{-4}}$

Solutions **a.**

$$\begin{aligned}(1 + \sqrt{-2})(3 - \sqrt{-2}) &= (1 + i\sqrt{2})(3 - i\sqrt{2}) \\ &= 3 - i\sqrt{2} + 3i\sqrt{2} - i^2 \cdot 2 \\ &= 3 + 2i\sqrt{2} - i^2 \cdot 2 \\ &= 5 + 2i\sqrt{2}\end{aligned}$$

b.

$$\begin{aligned}\frac{1}{3 - \sqrt{-4}} &= \frac{1(3 + 2i)}{(3 - 2i)(3 + 2i)} \\ &= \frac{3 + 2i}{9 - 4i^2} \\ &= \frac{3}{13} + \frac{2}{13}i\end{aligned}$$

27. $\sqrt{-4}(1 - \sqrt{-4})$

28. $\sqrt{-9}(3 + \sqrt{-16})$

29. $(2 + \sqrt{-9})(3 - \sqrt{-9})$

30. $(4 - \sqrt{-2})(3 + \sqrt{-2})$

31. $(2 + \sqrt{-5})^2$

32. $(3 - \sqrt{-2})^2$

33. $\dfrac{3}{\sqrt{-4}}$

34. $\dfrac{-1}{\sqrt{-25}}$

35. $\dfrac{3}{2 - \sqrt{-9}}$

36. $\dfrac{2}{3 + \sqrt{-5}}$

37. $\dfrac{2 - \sqrt{-1}}{2 + \sqrt{-1}}$

38. $\dfrac{1 + \sqrt{-2}}{3 - \sqrt{-3}}$

39. Simplify. [*Hint:* $i^2 = -1$ and $i^4 = 1$.]

a. i^6 **b.** i^{12} **c.** i^{15}

40. Express with a positive exponent and simplify.

a. i^{-1} **b.** i^{-2} **c.** i^{-3}

B

41. Evaluate $x^2 + 2x + 3$ for $x = 1 + i$.

42. Evaluate $2y^2 - y + 2$ for $y = 2 - i$.

43. For what values of x will $\sqrt{x - 5}$ be real? Imaginary?

44. For what values of x will $\sqrt{x + 3}$ be real? Imaginary?

6.4

THE QUADRATIC FORMULA

In Section 6.1 we considered quadratic equations which we were able to solve by factoring over the integers. Let us now develop a formula which can be used to solve all quadratic equations, and, in particular, quadratic equations having solutions in the set of complex numbers.

We can complete the square on the general quadratic equation

$$ax^2 + bx + c = 0 \qquad (a \neq 0)$$

as follows:

$$x^2 + \frac{b}{a}x + \frac{c}{a} = 0$$

$$x^2 + \frac{b}{a}x + \left(\frac{b}{2a}\right)^2 = -\frac{c}{a} + \left(\frac{b}{2a}\right)^2$$

$$\left(x + \frac{b}{2a}\right)^2 = \frac{b^2}{4a^2} - \frac{c}{a}$$

$$\left(x + \frac{b}{2a}\right)^2 = \frac{b^2 - 4ac}{4a^2}$$

$$x + \frac{b}{2a} = \pm\sqrt{\frac{b^2 - 4ac}{4a^2}}$$

$$x = -\frac{b}{2a} \pm \frac{\sqrt{b^2 - 4ac}}{2a}$$

$$x = \frac{-b \pm \sqrt{b^2 - 4ac}}{2a}.$$

The result is a formula for the roots of a quadratic equation expressed in terms of the coefficients. The symbol $\pm$ is used to condense the two equations

$$\boldsymbol{x = \frac{-b + \sqrt{b^2 - 4ac}}{2a}} \quad \text{and} \quad \boldsymbol{x = \frac{-b - \sqrt{b^2 - 4ac}}{2a}}$$

into a single equation. We need only substitute the coefficients a, b, and c of a given quadratic equation in the formula to find the solution set for the equation.

In the quadratic formula, the number represented by $b^2 - 4ac$ is called the **discriminant** of the equation. If a, b, and c are real numbers, the discriminant affects the solution set in the following ways:

1. If $b^2 - 4ac = 0$, there is one real solution.
2. If $b^2 - 4ac > 0$, there are two unequal real solutions.
3. If $b^2 - 4ac < 0$, there are two imaginary solutions.

We note from the form

$$x = \frac{-b}{2a} \pm \frac{\sqrt{b^2 - 4ac}}{2a}$$

that, if the solutions of a quadratic equation are irrational or imaginary, they are conjugates;

$$-\frac{b}{2a} + \frac{\sqrt{b^2 - 4ac}}{2a} \quad \text{and} \quad -\frac{b}{2a} - \frac{\sqrt{b^2 - 4ac}}{2a}.$$

EXERCISE 6.4

A

In Problems 1–20 solve for x, y, or z using the quadratic formula.

Example $\frac{x^2}{4} + x = \frac{5}{4}$

Solution Write in standard form.

$$x^2 + 4x = 5$$
$$x^2 + 4x - 5 = 0$$

Substitute 1 for a, 4 for b, and -5 for c in the quadratic formula.

$$x = \frac{-(4) \pm \sqrt{(4)^2 - 4(1)(-5)}}{2(1)}$$
$$= \frac{-4 \pm \sqrt{16 + 20}}{2}$$
$$= \frac{-4 \pm 6}{2}$$

The solution set is $\{-5, 1\}$.

1. $x^2 - 5x + 4 = 0$ **2.** $x^2 - 4x + 4 = 0$ **3.** $y^2 + 3y = 4$

4. $y^2 - 5y = 6$ **5.** $z^2 = 3z - 1$ **6.** $2z^2 = 7z - 6$

7. $0 = 3x^2 - 5x + 1$ **8.** $0 = 2x^2 - x + 1$ **9.** $5z + 6 = 6z^2$

10. $13z + 5 = 6z^2$ **11.** $x^2 - 5x = 0$ **12.** $y^2 + 3y = 0$

13. $3y^2 - 6 = 0$ **14.** $4y^2 + 8 = 0$ **15.** $x^2 + x = \frac{15}{4}$

16. $\dfrac{y^2}{2} - \dfrac{y}{2} = -1$ **17.** $\dfrac{x^2}{4} = 3 - \dfrac{x}{4}$ **18.** $\dfrac{x^2 - 3}{2} + \dfrac{x}{4} = 1$

19. $\dfrac{x^2 - 1}{2} = x + \dfrac{1}{2}$ **20.** $\dfrac{x^2 - 2}{3} + x = \dfrac{2}{3}$

In Problems 21–26 find only the discriminant and determine whether the solution(s) are: **a.** *one real;* **b.** *real and unequal;* **c.** *imaginary and unequal.*

Example $x^2 - x - 3 = 0$

Solution Substitute 1 for a, -1 for b, and -3 for c in the discriminant.

$$b^2 - 4ac = (-1)^2 - 4(1)(-3) = 1 + 12 = 13$$

Because $13 > 0$, the solutions are real and unequal.

21. $x^2 - 7x + 12 = 0$ **22.** $y^2 - 2y - 3 = 0$ **23.** $5x^2 + 2x - 1 = 0$

24. $2y^2 + 3y + 7 = 0$ **25.** $x^2 - 2x + 1 = 0$ **26.** $2z^2 - z = 12$

B

Solve for x in terms of the other variables or constants.

Example $x^2 - xy + y = 2$

Solution Write the equation in standard form.

$$x^2 - yx + (y - 2) = 0$$

Substitute 1 for a, $-y$ for b, and $y - 2$ for c in the quadratic formula.

$$x = \frac{-(-y) \pm \sqrt{(-y)^2 - 4(1)(y - 2)}}{2(1)}$$

$$x = \frac{y \pm \sqrt{y^2 - 4y + 8}}{2}$$

27. $x^2 - kx - 2k^2 = 0$ **28.** $2x^2 - kx + 3 = 0$ **29.** $ax^2 - x + c = 0$

30. $x^2 + 2x + c + 3 = 0$ **31.** $x^2 + 2x - y = 0$ **32.** $2x^2 - 3x + 2y = 0$

33. $2x^2 - x + y = 2$ **34.** $3x^2 + 2x - 3y = 4$ **35.** $x^2 - 2xy - 3y = 2$

36. $3x^2 + xy + 5y = 1$ **37.** $3x^2 + xy + y^2 = 2$ **38.** $x^2 - 3xy + y^2 = 3$

39. In Problem 37 solve for y in terms of x.

40. In Problem 38 solve for y in terms of x.

Example Determine k so that the solutions of $x^2 - 3x + k - 1 = 0$ are imaginary.

Solution The solutions are imaginary if the discriminant $b^2 - 4ac < 0$. Substituting -3 for b, 1 for a, and $k - 1$ for c in $b^2 - 4ac$, we have

$$(-3)^2 - 4(1)(k-1) < 0$$
$$9 - 4k + 4 < 0$$
$$-4k < -13$$
$$k > \frac{13}{4}.$$

41. Determine $k \neq 0$ so that $kx^2 + 4x + 1 = 0$ has one solution.

42. Determine k so that $x^2 - kx + 9 = 0$ has one solution.

43. Determine k so that the solutions of $x^2 + 2x + k + 3 = 0$ are real numbers.

44. Determine k so that the solutions of $x^2 - x + k = 2$ are not real numbers.

45. If r_1 and r_2 are solutions of a quadratic equation, show that

$$r_1 + r_2 = -\frac{b}{a} \quad \text{and} \quad r_1 \cdot r_2 = \frac{c}{a}.$$

Hint: Let

$$r_1 = \frac{-b + \sqrt{b^2 - 4ac}}{2a} \quad \text{and} \quad r_2 = \frac{-b - \sqrt{b^2 - 4ac}}{2a}.$$

46. Solve for y.

$$a\left(\frac{-b + \sqrt{b^2 - 4ac}}{2a}\right)^2 + b\left(\frac{-b + \sqrt{b^2 - 4ac}}{2a}\right) + y = 0$$

6.5 APPLICATIONS

Some of the procedures suggested in Chapter 5 (page 142) for writing equations for word problems should be reviewed at this time. In some cases, the mathematical model we obtain for a physical situation is a quadratic equation and consequently has two solutions. It may be that one but not both of the solutions of the equation fits the physical situation. For example, if we were asked to find two consecutive *natural numbers* whose product is 72, we would write the equation

$$x(x + 1) = 72$$

as our model. Solving this equation, we have

$$x^2 + x - 72 = 0$$
$$(x + 9)(x - 8) = 0,$$

where the solution set is $\{8, -9\}$. Since -9 is not a natural number, we must reject it as a possible answer to the original question; however, the solution 8 leads to the consecutive natural numbers 8 and 9. As additional examples, we observe that we would not accept -6 feet as the height of a man or $27/4$ for the number of people in a room.

A quadratic equation used as a model for a physical situation may have two, one, or no meaningful solutions—meaningful, that is, in a physical sense. Answers to word problems should always be checked in the original problem.

EXERCISE 6.5

A

Solve.

1. Find two numbers that differ by 3 and have a product of 40.
2. Find two consecutive positive integers whose product is 56.
3. A rectangular lawn is 2 meters longer than it is wide. If its area is 63 square meters, what are the dimensions of the lawn?
4. The length of a rectangular steel plate is 2 centimeters greater than 2 times its width. If the area of the plate is 40 square centimeters, find its dimensions.
5. The sum of a number and its reciprocal is $17/4$. What is the number?
6. The difference of a number and 2 times its reciprocal is $17/3$. What is the number?
7. A ball thrown vertically upward reaches a height h in feet, given by the equation $h = 64t - 16t^2$, where t is the time in seconds after the throw. How long will it take the ball to reach a height of 48 feet on its way up?
8. Using the information in Problem 7, determine how long after the ball was thrown it will return to the ground.
9. The distance s a body falls in a vacuum is given by $s = v_0 t + (1/2)gt^2$, where s is in feet, t is in seconds, v_0 is the initial velocity in feet per second, and g is the constant of acceleration due to gravity (approximately 32 feet per second per second). How long will it take a body to fall 150 feet if v_0 is 20 feet per second?
10. Using the information in Problem 9, determine how long it will take a body to fall 150 feet if the body starts from rest ($v_0 = 0$).
11. The ball in Problem 7 lands on top of a building 60 feet high on its way down. How long had it been in the air when it struck the roof?
12. How long will it take an object to fall 280 feet in a vacuum if it is given an initial velocity of 24 feet per second? (See Problem 9.)
13. The numerator of a fraction is 1 less than the denominator. The sum of the fraction and 2 times its reciprocal is $41/12$. Find the numerator and the denominator.
14. The numerator of a fraction is 2 less than the denominator. The sum of the fraction and 3 times its reciprocal is $28/5$. Find the numerator and the denominator.

15. The length of a rectangle is 1 meter less than 4 times the width. Its area is 33 square meters. Find the dimensions.

16. The area of a triangle is 70 square centimeters. Find the lengths of the base and the altitude if the base is 13 centimeters longer than the altitude.

17. The sum of the squares of three consecutive negative integers is 434. Find the integers.

18. Find two consecutive positive integers whose cubes differ by 919.

19. The number of diagonals, D, of a polygon of n sides is given by

$$D = \frac{n}{2}(n - 3).$$

How many sides does a polygon with 90 diagonals have?

20. The formula

$$s = \frac{n}{2}(n + 1)$$

gives the sum of the first n natural numbers 1, 2, 3, ... How many consecutive natural numbers starting with 1 will give a sum of 406?

B

21. A rectangle measuring 25 by 50 centimeters has its area increased by 318 square centimeters by a border of uniform width along both shorter sides and one longer side. Find the width of the border.

22. A tray is formed from a rectangular piece of metal whose length is 2 centimeters greater than its width by cutting a square with sides 2 centimeters in length from each corner, and then bending up the sides. Find the dimensions of the tray if the volume is 160 cubic centimeters.

23. A man sailed a boat across a lake and back in $2\frac{1}{2}$ hours. If his rate returning was 2 miles per hour less than his rate going, and if the distance each way was 6 miles, find his rate each way. [*Hint:* distance (d) = rate $(r) \times$ time (t).]

24. A man rode a bicycle for 6 miles and then walked an additional 4 miles. The total time for his trip was 6 hours. If his rate walking was 2 miles per hour less than his rate on the bicycle, what was each rate?

25. A riverboat that travels at 18 miles per hour in still water can go 30 miles up a river in 1 hour less time than it can go 63 miles down the river. What is the speed of the current in the river?

26. If a boat travels 20 miles per hour in still water and takes 3 hours longer to go 90 miles up a river than it does to go 75 miles down the river, what is the speed of the current in the river?

27. A commuter takes a train 10 miles to his job in a city. The train returns him home at a rate 10 miles per hour greater than the rate of the train that takes him to work. If he spends a total 50 minutes a day commuting, what is the rate of each train?

28. On a 50 mile trip, a man traveled 10 miles in heavy traffic and then 40 miles in less congested traffic. If his average rate in heavy traffic was 20 miles per hour less than his average rate in light traffic, what was each rate if the trip took 1 hour and 30 minutes?

6.6

EQUATIONS INVOLVING RADICALS

To solve equations containing radicals, we shall make use of the fact that:

▸ *If each member of an equation is raised to the same natural number power, the solution set of the original equation is a subset of the solution set of the resulting equation.*

For example, if $x = 3$, then the solution set of $x^2 = 9$ contains 3 as an element. In general, for $n \in N$, the solution set of

(1) $$[P(x)]^n = [Q(x)]^n$$

contains all the solutions of

(2) $$P(x) = Q(x).$$

The application of this assumption does not, however, result in an equivalent equation. Equation (1) actually may have additional solutions that are not solutions of (2). With respect to (2), these are called **extraneous solutions**. Thus, the solution set of the equation $x^2 = 9$, obtained from $x = 3$ by squaring each member, contains -3 as an extraneous solution, since -3 does not satisfy the original equation. Because the result of applying the foregoing assumption is not an equivalent equation, each solution obtained through its use *must* be checked in the original equation to verify its validity.

EXERCISE 6.6

A

Solve and check. If there is no solution, so state.

Example $\sqrt{x+2} + 4 = x$

Solution Obtain $\sqrt{x+2}$ as the only term in one member.

$$\sqrt{x+2} = x - 4$$

Square each member.

$$(\sqrt{x+2})^2 = (x-4)^2$$

$$x + 2 = x^2 - 8x + 16$$

Solve the quadratic equation.

$$x^2 - 9x + 14 = 0$$
$$(x - 7)(x - 2) = 0$$
$$x = 7 \qquad x = 2$$

Check Does $\sqrt{7 + 2} + 4 = 7$? Does $7 = 7$? Yes. The solution set is $\{7\}$.

Does $\sqrt{2 + 2} + 4 = 2$? Does $6 = 2$? No. 2 is not a solution.

1. $\sqrt{x} - 5 = 3$
2. $\sqrt{x} - 4 = 1$
3. $\sqrt{y + 6} = 2$
4. $\sqrt{y - 3} = 5$
5. $3z + 4 = \sqrt{3z + 10}$
6. $2z - 3 = \sqrt{7z - 3}$
7. $\sqrt{x - 3\sqrt{x}} = 2$
8. $\sqrt{x\sqrt{x - 5}} = 6$
9. $\sqrt[3]{x} = -3$
10. $\sqrt[3]{x} = -4$
11. $\sqrt[4]{x} - 1 = 2$
12. $\sqrt[4]{x - 1} = 3$
13. $(2x - 1)^{1/2} = 5$
14. $(3x + 1)^{1/3} = -1$

Example $\sqrt{y - 5} - \sqrt{y} = 1$

Solution Write an equivalent equation with $\sqrt{y - 5}$ as the left-hand member.

$$\sqrt{y - 5} = 1 + \sqrt{y}$$

Square each member.

$$(\sqrt{y - 5})^2 = (1 + \sqrt{y})^2$$
$$y - 5 = 1 + 2\sqrt{y} + y$$

Write an equivalent equation with $\sqrt{y}$ as the right-hand member.

$$-3 = \sqrt{y}$$

[*Note:* At this point it is obvious that the equation has no solution, since $\sqrt{y}$ cannot be negative. We continue the solution process, however, for illustrative purposes.]

Square each member.

$$9 = y$$

Check Does $\sqrt{4} - \sqrt{9} = 1$? Does $2 - 3 = 1$? No.
9 is not a solution of the original equation. Hence, the solution set is $\varnothing$.

15. $\sqrt{y + 4} = \sqrt{y + 20} - 2$
16. $4\sqrt{y} + \sqrt{1 + 16y} = 5$
17. $\sqrt{x} + \sqrt{2} = \sqrt{x + 2}$
18. $\sqrt{4x + 17} = 4 - \sqrt{x + 1}$
19. $(5 + x)^{1/2} + x^{1/2} = 5$
20. $(y + 7)^{1/2} + (y + 4)^{1/2} = 3$
21. $(y^2 - 3y + 5)^{1/2} - (y + 2)^{1/2} = 0$
22. $(z - 3)^{1/2} + (z + 5)^{1/2} = 4$

Solve. Leave the results in the form of an equation. Assume that no variable takes a value for which any denominator is 0.

23. $r = \sqrt{\dfrac{A}{\pi}}$, for A

24. $t = \sqrt{\dfrac{2v}{g}}$, for g

25. $x\sqrt{xy} = 1$, for y

26. $P = \pi\sqrt{\dfrac{l}{g}}$, for g

27. $r = \sqrt{t^2 - s^2}$, for t

28. $q - 1 = 2\sqrt{\dfrac{r^2 - 1}{3}}$, for r

B

29. In the first example in this exercise set, the second equation in our sequence is $\sqrt{x + 2} = x - 4$. How can we tell by inspection that this equation will not be satisfied by any real number less than 4?

30. In Problem 23, specify the restrictions on each variable in the set of real numbers.

31. In Problem 25, specify the restrictions on each variable in the set of real numbers.

32. In Problem 27, specify the restrictions on each variable in the set of real numbers.

6.7

EQUATIONS THAT ARE QUADRATIC IN FORM

Some equations that are not quadratic equations are nevertheless quadratic in form—that is, they are of the form

(1) $$au^2 + bu + c = 0,$$

where u represents some expression in terms of another variable. For example,

(2)
$$x^4 - 10x^2 + 9 = 0 \quad \text{is quadratic in } x^2,$$
$$(x^2 - 1)^2 + 3(x^2 - 1) + 2 = 0 \quad \text{is quadratic in } (x^2 - 1),$$
$$y + 2\sqrt{y} - 8 = 0 \quad \text{is quadratic in } \sqrt{y},$$
$$\left(\frac{z + 1}{z}\right)^2 - 5\left(\frac{z + 1}{z}\right) + 6 = 0 \quad \text{is quadratic in } \left(\frac{z + 1}{z}\right).$$

We can solve an equation of the form (1) by first finding the equations

$$u = r_1 \quad \text{and} \quad u = r_2$$

and then solving these equations for the desired roots. For example, if we let $x^2 = u$, then (2) above reduces to

$$u^2 - 10u + 9 = 0.$$

Solving this equation, we obtain

$$(u - 9)(u - 1) = 0$$
$$u = 9 \qquad u = 1.$$

Since $u = x^2$, we have

$$x^2 = 9 \qquad x^2 = 1,$$

from which we obtain the solution set $\{3, -3, 1, -1\}$.

As an alternative method, we could factor the left-hand member of (2) directly as

$$(x^2 - 9)(x^2 - 1) = 0,$$

and then set each factor equal to 0. Thus,

$$x^2 - 9 = 0 \qquad \text{or} \qquad x^2 - 1 = 0$$
$$x = \pm 3 \qquad \text{or} \qquad x = \pm 1,$$

and we obtain the solution set $\{3, -3, 1, -1\}$.

EXERCISE 6.7

A

Solve for x, y, or z.

Example $y - 2\sqrt{y} - 8 = 0$

Solution Let $\sqrt{y} = u$; therefore, $y = u^2$. Substitute for y and $\sqrt{y}$.

$$u^2 - 2u - 8 = 0$$

Solve for u.

$$(u - 4)(u + 2) = 0$$
$$u = 4 \qquad u = -2$$

Replace u with $\sqrt{y}$ and solve for y. Since $\sqrt{y}$ cannot be negative, only 4 need be considered.

$$\sqrt{y} = 4$$
$$y = 16$$

Check Does $16 - 2\sqrt{16} - 8 = 0$? Does $0 = 0$? Yes.
Hence, $\{16\}$ is the solution set.

1. $x^4 - 5x^2 + 4 = 0$
2. $x^4 - 13x^2 + 36 = 0$
3. $2x^4 + 17x^2 - 9 = 0$
4. $z^4 - 2z^2 - 24 = 0$
5. $x - 2\sqrt{x} - 15 = 0$
6. $y + 3\sqrt{y} - 10 = 0$
7. $y^2 + 7 - \sqrt{y^2 + 7} - 12 = 0$
8. $y^2 - 5 - 5\sqrt{y^2 - 5} + 6 = 0$
9. $y^{2/3} - 2y^{1/3} - 8 = 0$
10. $z^{2/3} - 2z^{1/3} = 35$
11. $x^{3/2} - 9x^{3/4} + 8 = 0$
12. $x^{3/4} - 2x^{3/8} + 1 = 0$
13. $x - 9x^{1/2} + 18 = 0$
14. $z + z^{1/2} = 72$
15. $2x - 9x^{1/2} = -4$
16. $8x^{1/2} + 7x^{1/4} = 1$

B

17. $y^{-2} - y^{-1} - 12 = 0$
18. $z^{-2} + 9z^{-1} - 10 = 0$
19. $(x - 1)^{1/2} - 2(x - 1)^{1/4} - 15 = 0$
20. $(x - 2)^{1/2} - 11(x - 2)^{1/4} + 18 = 0$

Solve Problems 21–24 in two ways:
a. By the method of Section 6.6.
b. By the method of this section.

21. $y - 7\sqrt{y} + 12 = 0$
22. $x + \sqrt{x} - 6 = 0$
23. $x + 18 = 11\sqrt{x}$
24. $y + 2\sqrt{y} = 15$

6.8

QUADRATIC INEQUALITIES

As in the case with first-degree inequalities, we can generate equivalent second-degree inequalities by applying Properties 1–3 and equations (3)–(5) of Section 5.5. Additional procedures are necessary, however, to obtain the solution sets of such inequalities. For example, consider the inequality

$$x^2 + 4x < 5.$$

To determine values of x for which this condition holds, we might first rewrite the sentence equivalently as

$$x^2 + 4x - 5 < 0,$$

and then as

$$(x + 5)(x - 1) < 0.$$

In this case it is clear that only those values of x for which the factors $x + 5$ and $x - 1$ are opposite in sign will be in the solution set. These can be determined analytically by noting that $(x+5)(x-1)<0$ implies either

(1) $\quad x+5<0 \quad$ and $\quad x-1>0$

or

(2) $\quad x+5>0 \quad$ and $\quad x-1<0.$

Each of these two cases can be considered separately.

The inequalities for (1),

$$x+5<0 \quad \text{and} \quad x-1>0,$$

imply

$$x<-5 \quad \textit{and} \quad x>1,$$

a condition which is not satisfied by any values of x. Therefore, the solution set of (1) is $\varnothing$. The inequalities for (2),

$$x+5>0 \quad \text{and} \quad x-1<0,$$

imply

$$x>-5 \quad \textit{and} \quad x<1,$$

which leads to the solution set

$$\{x \mid -5<x<1\}.$$

Thus, the complete solution set is

$$S=\{x \mid -5<x<1\} \cup \varnothing=\{x \mid -5<x<1\}.$$

Notice in the foregoing solution process that the numbers -5 and 1 separate the set of real numbers into the three intervals,

$$\{x \mid x<-5\}, \quad \{x \mid -5<x<1\}, \quad \{x \mid x>1\},$$

shown in Figure 6.2. Each of these intervals either is or is not a part of the solution set of the inequality $x^2+4x<5$. To determine which, we need only

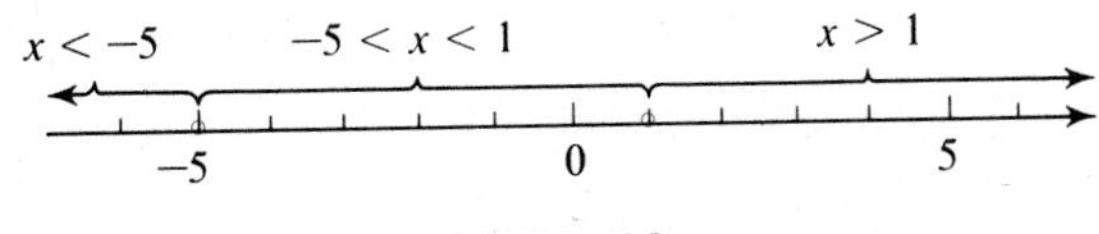

Figure 6.2

substitute an arbitrarily selected number from each interval and test it in the inequality. Let us use -6, 0, and 2.

$(-6)^2+4(-6)\overset{?}{<}5$	$0^2+4(0)\overset{?}{<}5$	$2^2+4(2)\overset{?}{<}5$
$36-24\overset{?}{<}5$	$0+0\overset{?}{<}5$	$4+8\overset{?}{<}5$
$12\overset{?}{<}5$	$0\overset{?}{<}5$	$12\overset{?}{<}5$
no	yes	no

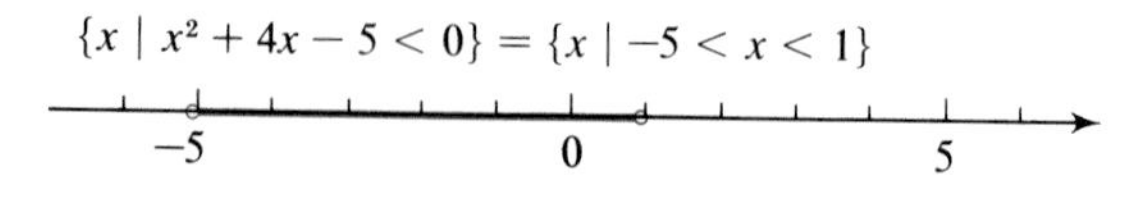

Figure 6.3

Clearly, the only interval involved that is in the solution set is $\{x \mid -5 < x < 1\}$, and hence this interval is the solution set. The graph is shown in Figure 6.3. If the inequality in this example were $x^2 + 4x - 5 \leq 0$, the end points on the graph would be shown as closed dots.

Numbers such as -5 and 1 in the preceding example, for which either $Q(x) = 0$ or $Q(x)$ is undefined in an inequality of the form $Q(x) < 0$ or $Q(x) > 0$, are called **critical numbers.** For example, critical numbers for

$$2x^2 - x - 1 > 0$$

are $-\frac{1}{2}$ and 1, because

$$2x^2 - x - 1 = (2x + 1)(x - 1) = 0$$

for these values; and critical numbers for

$$\frac{1}{x^2 - 4} < 0$$

are 2 and -2, because

$$\frac{1}{x^2 - 4} = \frac{1}{(x - 2)(x + 2)}$$

is not defined for either number.

Let us use the notion of critical numbers to solve another inequality,

$$x^2 - 3x - 4 \geq 0. \tag{3}$$

Factoring the left-hand member yields

$$(x + 1)(x - 4) \geq 0,$$

and we observe that -1 and 4 are critical numbers because $(x + 1)(x - 4) = 0$ for these values. Hence, we wish to check the intervals shown on the number line in Figure 6.4. We substitute selected arbitrary values in each interval, say -2, 0, and 5 for the variable in (3).

$$(-2)^2 - 3(-2) - 4 \overset{?}{\geq} 0 \qquad (0)^2 - 3(0) - 4 \overset{?}{\geq} 0 \qquad (5)^2 - 3(5) - 4 \overset{?}{\geq} 0$$

yes no yes

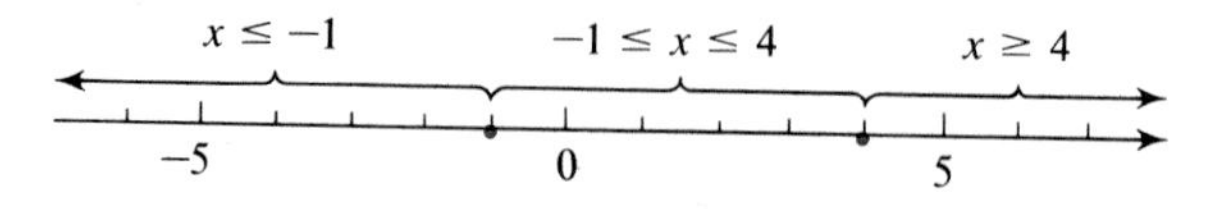

Figure 6.4

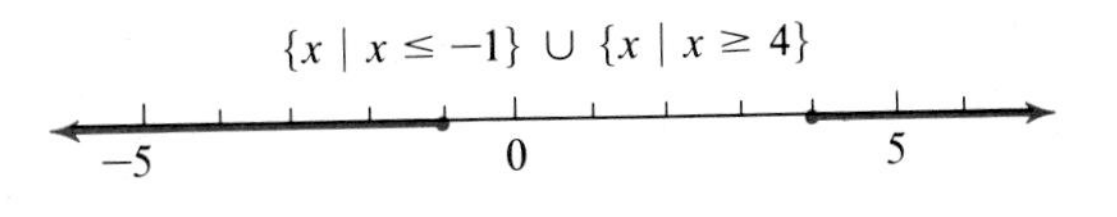

Figure 6.5

Hence, the solution set is

$$\{x \mid x \leq -1\} \cup \{x \mid x \geq 4\}.$$

The graph is shown in Figure 6.5.

Inequalities involving fractions have to be approached with care if any fraction contains a variable in the denominator. If each member of such an inequality is multiplied by an expression containing the variable, we have to be careful either to distinguish between those values of the variable for which the expression denotes a positive and a negative number, or to make certain that the expression by which we multiply is always positive. However, we can use the notion of critical numbers to avoid these complications. For example, consider the inequality

$$\frac{x}{x-2} \geq 5. \tag{4}$$

We first write (4) equivalently as

$$\frac{x}{x-2} - 5 \geq 0,$$

from which

$$\frac{x - 5(x-2)}{x-2} \geq 0,$$

$$\frac{-4x + 10}{x-2} \geq 0. \tag{4a}$$

In this case the critical numbers are $5/2$ and 2, because

$$-4x + 10 = 0 \qquad \text{and hence} \qquad \frac{-4x+10}{x-2} = 0$$

for $x = 5/2$, and

$$\frac{-4x+10}{x-2}$$

is undefined for $x = 2$. Thus, we want to check the intervals shown on the number line in Figure 6.6. Substituting arbitrary values for the variable in (4) or (4a) in

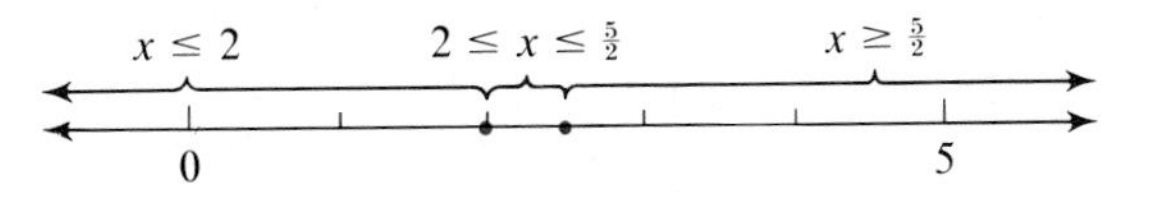

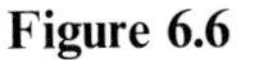
Figure 6.6

each of the three intervals, say 0, $2\frac{1}{4}$, or $\frac{9}{4}$, and 3, we can identify the solution set

$$\left\{x \mid 2 < x \leq \frac{5}{2}\right\},$$

as we did in the previous examples. The graph is shown in Figure 6.7. Note that the left-hand end point is an open dot (2 is not a member of the solution set), because the left-hand member of (4) is undefined for $x = 2$.

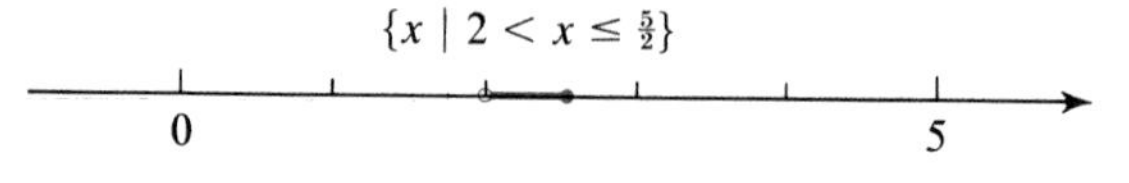

Figure 6.7

EXERCISE 6.8

A

Solve and represent each solution set on a line graph.

1. $(x + 1)(x - 2) > 0$
2. $(x + 2)(x + 5) < 0$
3. $x(x - 2) \leq 0$
4. $x(x + 3) \geq 0$
5. $x^2 - 3x - 4 > 0$
6. $x^2 - 5x - 6 \geq 0$
7. $x^2 < 5$
8. $4x^2 + 1 < 0$
9. $x^2 > -5$
10. $x^2 + 1 > 0$
11. $\dfrac{2}{x} \leq 4$
12. $\dfrac{3}{x - 6} > 8$
13. $\dfrac{x}{x + 2} > 4$
14. $\dfrac{x + 2}{x - 2} \geq 6$

B

15. $\dfrac{4}{x + 1} < \dfrac{3}{x}$
16. $\dfrac{3}{x - 1} \geq \dfrac{1}{x}$
17. $\dfrac{2}{x - 2} \geq \dfrac{4}{x}$
18. $\dfrac{3}{4x + 1} > \dfrac{2}{x - 5}$
19. $5 < x^2 + 1 < 10$
20. $-2 < y^2 - 3 < 13$

CHAPTER SUMMARY

[6.1] The **standard form** for a quadratic equation in one variable is

$$ax^2 + bx + c = 0,$$

where a, b, and c are constants with $a \neq 0$.

To solve an equation by factoring, we use the fact that the product of two factors is 0 if and only if one or both of the factors equals 0. That is,

$$ab = 0 \quad \text{if and only if} \quad a = 0 \quad \text{or} \quad b = 0.$$

[6.2] To solve an equation by the **extraction of roots** method, we use the fact that:

$$\text{If } x^2 = b, \quad \text{then} \quad x = \sqrt{b} \quad \text{or} \quad x = -\sqrt{b}.$$

To solve an equation by completing the square, we add $(b/2)^2$ to complete the square in an expression of the form $x^2 + bx$.

[6.3] A number of the form $a + bi$, where $a, b \in R$, is called a **complex number**. It is a real number if $b = 0$, it is an **imaginary number** if $b \neq 0$, and it is a **pure imaginary number** if $a = 0$ and $b \neq 0$.

The sum and the product of two complex numbers are defined so that they conform with the sum and the product of two binomials, respectively.

[6.4] The solutions of $ax^2 + bx + c = 0$ $(a \neq 0)$ are given by

$$x = \frac{-b + \sqrt{b^2 - 4ac}}{2a} \quad \text{and} \quad x = \frac{-b - \sqrt{b^2 - 4ac}}{2a}.$$

The number represented by $b^2 - 4ac$ is the **discriminant** of the quadratic equation $ax^2 + bx + c = 0$.

[6.5] Quadratic equations are sometimes useful as mathematical models in physical situations. In such cases, all the solutions of an equation may not be meaningful.

[6.6] To solve equations containing radicals, we use the fact that if each member of an equation is raised to the same power, the solution set of the original equation is a subset of the solution set of the resulting equation. That is:

For $n \in N$, the solution set of $[P(x)]^n = [Q(x)]^n$ contains all the solutions of $P(x) = Q(x)$.

[6.7] An equation of the form

$$au^2 + bu + c = 0,$$

where u represents some expression in another variable, is said to be **quadratic in form**.

[6.8] Quadratic inequalities can be solved by identifying critical numbers.

The symbols introduced in this chapter are listed on the inside of the front cover.

REVIEW EXERCISES

[6.1] *Solve for x.*

1. a. $x^2 - 2x = 0$ **b.** $x^2 - 5x + 6 = 0$

2. a. $(x - 6)(x + 4) = -9$ **b.** $x(x + 1) = 6$

3. a. $x - 1 = \frac{1}{4}x^2$ **b.** $x + \frac{3}{x} = 4$

Given the solutions of a quadratic equation, write the equation in standard form with integral coefficients.

4. a. -2 and 5 **b.** $\frac{1}{3}$ and $-\frac{1}{4}$

[6.2] *Solve for x by the extraction of roots.*

5. a. $2x^2 = 50$ **b.** $3x^2 - 7 = 0$

6. a. $(x + 3)^2 = 25$ **b.** $(x - 4)^2 = 15$

Solve by completing the square.

7. a. $x^2 - 4x - 6 = 0$ **b.** $2x^2 + 3x - 3 = 0$

[6.3] *Write each expression in the form a + bi or a + ib.*

8. a. $4 + 2\sqrt{-9}$ **b.** $5 - 3\sqrt{-12}$

9. a. $(2 - 3i) + (4 + 2i)$ **b.** $(6 + 2i) - (1 - i)$

10. a. $(5 + i)(2 + 3i)$ **b.** $i(3 - 4i)$

11. a. $\frac{4}{3i}$ **b.** $\frac{1 - 3i}{2 - i}$

[6.4] *Solve for x using the quadratic formula.*

12. a. $\frac{1}{2}x^2 + 1 = \frac{3}{2}x$ **b.** $x^2 - 3x + 7 = 0$

13. a. $x^2 - 3x + 1 = 0$ **b.** $2x^2 + x - 3 = 0$

[6.5] **14.** The base of a triangle is 1 inch longer than 2 times the length of its altitude and its area is 18 square inches. Find the length of the base and the altitude.

[6.6] *Solve.*

15. a. $x - 3\sqrt{x} + 2 = 0$ **b.** $\sqrt{x+1} + \sqrt{x+8} = 7$

[6.7] *Solve.*

16. a. $x^4 - 3x^2 - 4 = 0$ **b.** $x - x^{1/2} = 12$

[6.8] **17.** Solve $x^2 - 9x > 0$ and graph the solution set on a line graph.
18. Solve $x^2 + 6x + 8 < x + 2$ and graph the solution set on a line graph.

7

FUNCTIONS, RELATIONS, AND THEIR GRAPHS: PART I

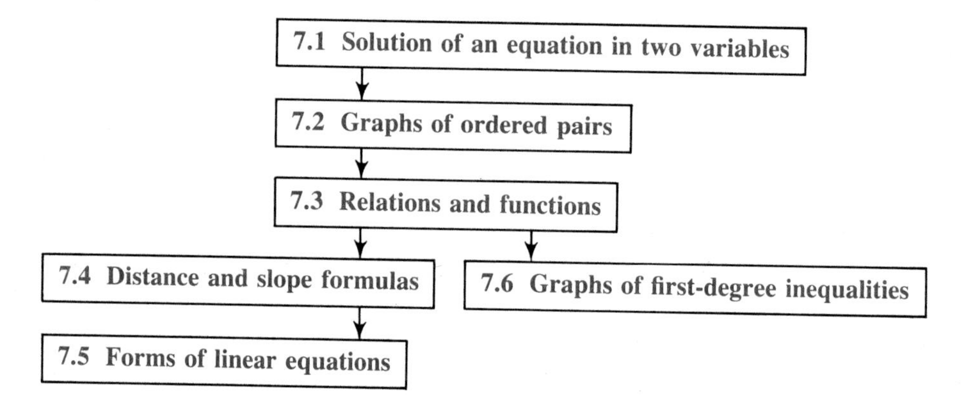

7.1

SOLUTION OF AN EQUATION IN TWO VARIABLES

An equation or inequality in two variables, such as

$$y = 2x + 3 \qquad \text{or} \qquad y < 2x + 3,$$

is said to be *satisfied* if the variables are replaced with a pair of numbers—one from the replacement set of x and one from the replacement set of y—which make the resulting statement true. The pair of numbers, usually written in the form (x, y), is a **solution** of the equation or inequality. The pair (x, y) is called an **ordered pair**, because it is understood that the numbers are considered in a particular order, x first and y second. These numbers are called the **first** and **second components** of the ordered pair, respectively. Although any letters may be used as variables, in this book we shall usually use x and y.

In our work we shall be interested primarily in the set of all ordered pairs which can be formed with real number components; that is, the infinite set of ordered pairs

$$\{(x, y) \mid x \in R \quad \text{and} \quad y \in R\}.$$

This set is called the **Cartesian product** of R and R, and is denoted by $\boldsymbol{R} \times \boldsymbol{R}$ or $\boldsymbol{R}^2$.

To find ordered pairs in $R \times R$ that are solutions of a given equation, we can assign a real number value to one of the variables and then determine the related value, if any, of the other. Thus, for

$$y - x = 1,$$

we can obtain solutions (ordered pairs) by assigning to x any real number as a value and then determining the corresponding value of y. For example, substituting 2, 3, and 4 for x, we have

$$y - (2) = 1, \quad \text{from which} \quad y = 3;$$
$$y - (3) = 1, \quad \text{from which} \quad y = 4;$$
$$y - (4) = 1, \quad \text{from which} \quad y = 5.$$

Thus, $(2, 3)$, $(3, 4)$, and $(4, 5)$ are solutions of $y - x = 1$ in $R \times R$.

In finding ordered pairs that satisfy a given equation or inequality, it is sometimes helpful if one variable is first expressed in terms of the other. The assertions on page 132 that are used to generate equivalent equations in one variable apply also in transforming equations in two or more variables. For example, before assigning values to x in

(1) $$y - 3x = 4,$$

we can transform the equation to the equivalent equation

(2) $$y = 3x + 4.$$

Both equations have the same solution set. In (1) the variables x and y are said to be **implicitly** related, while in (2), y is said to be expressed **explicitly** in terms of x.

EXERCISE 7.1

A

Find each missing component so that each ordered pair is a solution of the equation.

Examples $y - 2x = 4$

a. $(0, \underline{?})$ **b.** $(\underline{?}, 0)$ **c.** $(3, \underline{?})$

Solutions

a. $y - 2x = 4$
$y - 2(0) = 4$
$y = 4$
$(0, 4)$

b. $y - 2x = 4$
$(0) - 2x = 4$
$x = -2$
$(-2, 0)$

c. $y - 2x = 4$
$y - 2(3) = 4$
$y = 10$
$(3, 10)$

1. $y = x + 7$	**a.** $(0, \underline{?})$	**b.** $(2, \underline{?})$	**c.** $(-2, \underline{?})$
2. $y = 6 - 2x$	**a.** $(0, \underline{?})$	**b.** $(\underline{?}, 0)$	**c.** $(-1, \underline{?})$
3. $3x - 4y = 6$	**a.** $(0, \underline{?})$	**b.** $(\underline{?}, 0)$	**c.** $(-5, \underline{?})$
4. $x + 2y = 5$	**a.** $(0, \underline{?})$	**b.** $(5, \underline{?})$	**c.** $(-3, \underline{?})$

List the elements in each set.

Example $\{(x, y) \mid y = 3x + 1, \quad x \in \{1, 2, 3\}\}$

Solution For $x = 1$, $\quad y = 3(1) + 1 = 4$.

For $x = 2$, $\quad y = 3(2) + 1 = 7$.

For $x = 3$, $\quad y = 3(3) + 1 = 10$.

Thus, $\{(x, y) \mid y = 3x + 1, \quad x \in \{1, 2, 3\}\} = \{(1, 4), (2, 7), (3, 10)\}$.

5. $\{(x, y) \mid y = x - 4, \quad x \in \{-3, 0, 3\}\}$

6. $\{(x, y) \mid y = 2x + 6, \quad x \in \{-2, 0, 2\}\}$

7. $\{(x, y) \mid y = 4 + 2x, \quad x \in \{1, 2, 3\}\}$

8. $\{(x, y) \mid y = 6 - 3x, \quad x \in \{5, 10, 15\}\}$

9. $\left\{(x, y) \mid x + y = 6, \quad x \in \left\{\frac{1}{2}, \frac{3}{2}, \frac{5}{2}\right\}\right\}$

10. $\left\{(x, y) \mid 2x + 3y = 12, \quad x \in \left\{-\frac{1}{4}, 0, \frac{1}{4}\right\}\right\}$

11. Find k if $(1, 2)$ is a solution of $x + ky = 8$.

12. Find k if $(-6, 1)$ is a solution of $kx - 2y = 6$.

Transform the given equation into one in which y is expressed explicitly in terms of x.

Examples **a.** $3y - xy = 4$ **b.** $x^2 + 4y^2 = 5$

Solutions

a.
$$3y - xy = 4$$
$$y(3 - x) = 4$$
$$y = \frac{4}{3 - x}$$

b.
$$x^2 + 4y^2 = 5$$
$$4y^2 = 5 - x^2$$
$$y^2 = \frac{1}{4}(5 - x^2)$$
$$y = \pm\frac{1}{2}\sqrt{5 - x^2}$$

13. $xy - x = 2$

14. $3x - xy = 6$

15. $xy - y = 4$

16. $x^2y - xy + 5 = 0$

17. $x^2 - 4y = 7$

18. $x^2y - xy + 3 = 5y$

19. $4 = \dfrac{x}{y^2 - 2} \quad (y \neq -\sqrt{2}, \sqrt{2})$

20. $9 = \dfrac{2x^3}{1 - y^2} \quad (y \neq -1, 1)$

21. $3x^2 - 4y = 4$

22. $x^2y - 3x^2 + 2y = 0$

B

Find the solutions of each equation for each specified component.

23. $y = x^2 - 1$, for $x = 3$

24. $y = x^2 - 5x + 7$, for $x = -1$

25. $y = x^2 - 1$, for $y = 3$

26. $y = x^2 - 5x + 7$, for $y = 1$

27. $y = |x| - 3$, for $x = -4$

28. $y = |x| + 5$, for $x = -2$

29. $y = |x - 1| + |x|$, for $x = -1$

30. $y = |2x + 1| - |x|$, for $x = -3$

31. $y = 2^x$, for $x = 3$

32. $y = 2^x$, for $x = -3$

33. $y = 2^{-x}$, for $x = 3$

34. $y = 2^{-x}$, for $x = -3$

7.2

GRAPHS OF ORDERED PAIRS

The graph of $R \times R$, an infinite set of points, is the entire geometric plane. We shall assume that the ordered pairs (x, y) in $R \times R$ can be placed in a one-to-one correspondence with the geometric points in the plane by any arbitrary scaling of the axes of the familiar **Cartesian** (or **rectangular**) **coordinate system** shown in Figure 7.1. That is, for each point in the plane, there corresponds a unique ordered pair and vice versa. This correspondence is established by using the first component, x, of the ordered pair to denote the directed (perpendicular) distance of the point from the vertical axis (the **abscissa** of the point); x is positive if to the right of the vertical axis and negative if to the left. The second component, y, of the ordered pair is used to denote the directed (perpendicular) distance of the point from the horizontal axis (the **ordinate** of the point); y is positive if above the horizontal axis and negative if below. Each axis is labeled with the variable it represents. Most commonly, the horizontal axis is called the ***x* axis** and the vertical axis is called the ***y* axis**. The components of the ordered pair that corresponds to a given point are called the **coordinates** of the point, and the point is called the **graph** of the ordered pair. Each of the four regions into which the axes divide the plane is called a **quadrant**, and they are referred to by Roman numerals, as illustrated in Figure 7.1.

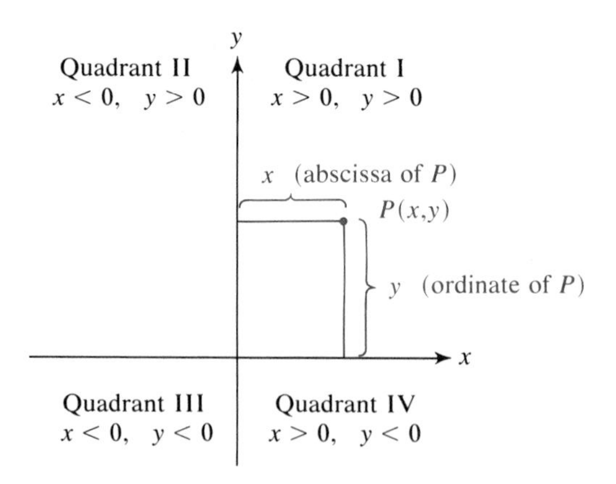

Figure 7.1

Equations in two variables have solutions that are ordered pairs of real numbers. Hence, these solutions can be displayed on a Cartesian coordinate system. For example,

$$\{(x, y) \mid y = x + 2, \quad x \in \{1, 2\}\}$$

contains two ordered pairs that are solutions of the equation $y = x + 2$. Substituting 1 and 2, respectively, for x, we have

$$y = (1) + 2 = 3,$$
$$y = (2) + 2 = 4,$$

from which we obtain the two ordered pairs $(1, 3)$ and $(2, 4)$, whose graphs (two points) are shown in Figure 7.2, part a. The graph of

$$\{(x, y) \mid y = x + 2, \qquad x \in \{-2, -1, 0, 1, 2\}\}$$

(five points) is shown in Figure 7.2, part b. In general, we are interested in the graphs of equations in two variables over the set of all real numbers x for which y is a real number. The graph of $y = x + 2$ for $x \in R$ (an infinite set of points) is shown in Figure 7.2, part c. It can be shown that the coordinates of each point on the line in part c satisfy $y = x + 2$ and, conversely, that every solution of the equation corresponds to a point on the line. The line is referred to as the graph of the solution set of the equation, or as the graph of the equation in $R \times R$. Although we will not prove it here, any first-degree equation in two variables—that is, any equation that can be written equivalently in the form

$$ax + by + c = 0 \qquad (a \text{ and } b \text{ not both } 0),$$

where a, b, and c are real numbers—has a graph that is a straight line. For this reason, such equations are often called **linear equations**. Since any two distinct points determine a straight line, it is evident that we need find only two solutions of such an equation to determine its graph.

In practice, the two solutions that are easiest to find are those with first and second components, respectively, equal to 0—that is, the solutions $(x_1, 0)$ and

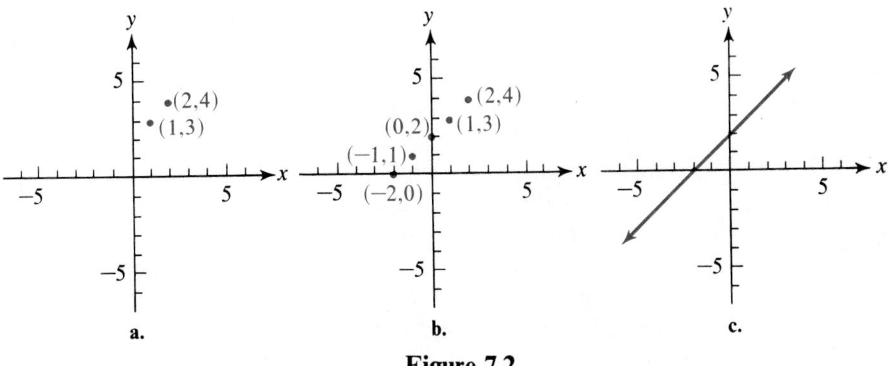

Figure 7.2

$(0, y_1)$. Since these two points are the points where the graph intersects the x and y axes, they are easy to locate. The numbers x_1 and y_1 are called the ***x* and *y* intercepts** of the graph. For example, consider the equation

$$3x + 4y = 12.$$

If $y = 0$,

$$3x = 12$$
$$x = 4,$$

and the x intercept is 4. If $x = 0$,

$$4y = 12$$
$$y = 3,$$

and the y intercept is 3. Thus, the graph of the equation appears as shown in Figure 7.3.

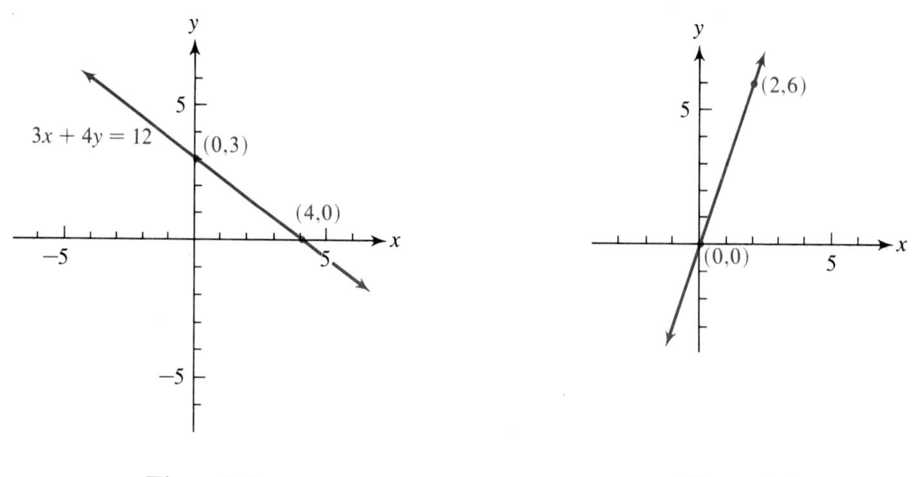

Figure 7.3 **Figure 7.4**

If the graph intersects the axes at or near the origin, the intercepts either do not represent two separate points, or the points are too close together to be of much use in drawing the graph. It is then necessary to plot at least one other point at a distance far enough removed from the origin to establish the line with accuracy. For example, consider the equation $y = 3x$. If $x = 0$, then $y = 0$, and both intercepts of its graph are at the point $(0, 0)$. Assigning any other replacement for x, say 2, we obtain a second solution $(2, 6)$. We first graph the ordered pairs $(0, 0)$ and $(2, 6)$ and then complete the graph, as shown in Figure 7.4.

There are two special cases of linear equations worth noting. First, an equation such as

$$y = 4$$

can be considered an equation in two variables in $R \times R$,

$$0x + y = 4.$$

For each x, this equation assigns $y = 4$. That is, any ordered pair of the form $(x, 4)$ is a solution of the equation. For example,

$$(-1, 4), \qquad (2, 4), \qquad (4, 4),$$

are all solutions of the equation. If we draw a straight line through the graphs of these points, we obtain the graph shown in Figure 7.5.

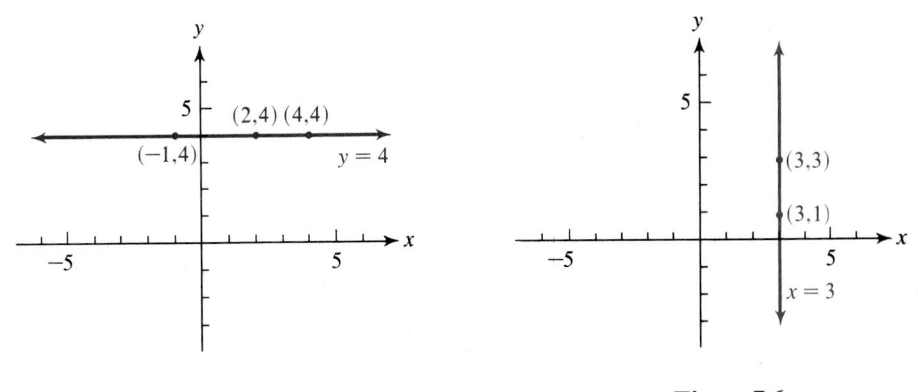

Figure 7.5

Figure 7.6

The other special case of a linear equation is of the type

$$x = 3,$$

which, in $R \times R$, may be looked upon as an equation in two variables,

$$x + 0y = 3.$$

Here, only one value is permissible for x, namely 3, while any value may be assigned to y. That is, any ordered pair of the form $(3, y)$ is a solution of this equation. If we choose two solutions, say $(3, 1)$ and $(3, 3)$, and draw a straight line through the graphs of these points, we have the graph of the equation shown in Figure 7.6.

If k represents a constant (real number), then, in general, the graph of $y = k$ is a horizontal line and the graph of $x = k$ is a vertical line.

EXERCISE 7.2

A

Graph each finite set of ordered pairs.

Example $\{(-2, 1), (0, 4), (2, -3)\}$

Solution

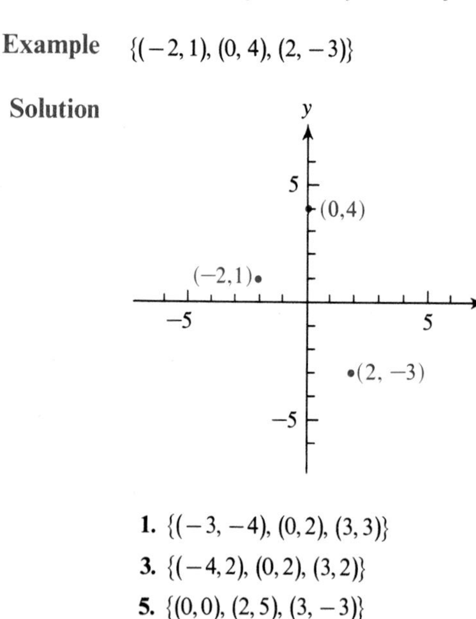

1. $\{(-3, -4), (0, 2), (3, 3)\}$
2. $\{(-2, 2), (2, 4), (2, -4)\}$
3. $\{(-4, 2), (0, 2), (3, 2)\}$
4. $\{(2, -3), (2, 2), (2, 5)\}$
5. $\{(0, 0), (2, 5), (3, -3)\}$
6. $\{(0, 0), (3, -2), (5, 1)\}$
7. $\{(0, -3), (3, 0), (5, 2)\}$
8. $\{(-3, 0), (0, 2), (2, 2)\}$

Graph each set in $R \times R$.

Example $\{(x, y) \mid x \geq 0\}$

Solution Shade the region where the value of x is nonnegative.

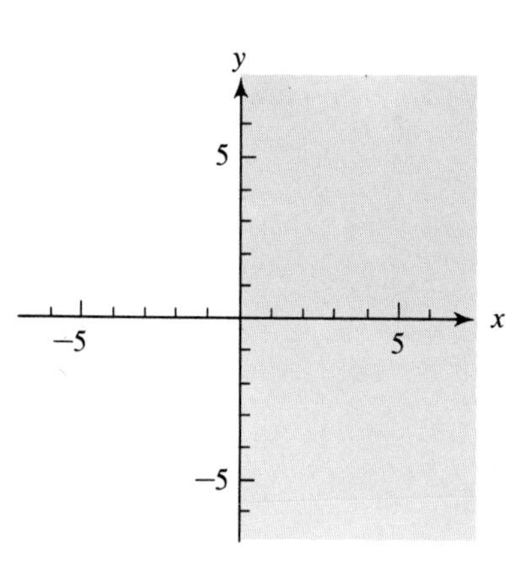

9. $\{(x, y) \mid x \geq 0 \text{ and } y \geq 0\}$

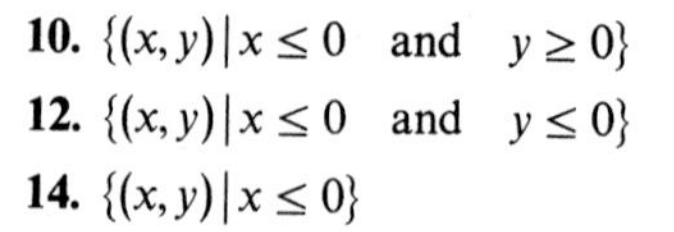

10. $\{(x, y) \mid x \leq 0 \text{ and } y \geq 0\}$
11. $\{(x, y) \mid x \geq 0 \text{ and } y \leq 0\}$
12. $\{(x, y) \mid x \leq 0 \text{ and } y \leq 0\}$
13. $\{(x, y) \mid y \leq 0\}$
14. $\{(x, y) \mid x \leq 0\}$

Graph each finite set.

Example $\{(x, y) \mid y = 2x + 1, \quad x \in \{-2, 0, 2\}\}$

Solution Substitute -2, 0, and 2 for x in $y = 2x + 1$.

$$y = 2(-2) + 1 = -3$$
$$y = 2(0) + 1 \quad = 1$$
$$y = 2(2) + 1 \quad = 5$$

Graph the ordered pairs

$$(-2, -3), \qquad (0, 1), \qquad (2, 5).$$

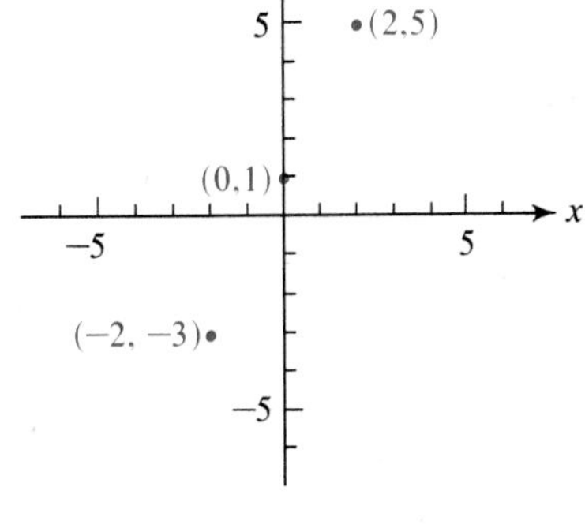

15. $\{(x, y) \mid y = x + 2, \quad x \in \{-4, -2, 0\}\}$

16. $\{(x, y) \mid y = x - 3, \quad x \in \{0, 2, 4\}\}$

17. $\{(x, y) \mid y = 2x - 1, \quad x \in \{-2, -1, 0, 1, 2\}\}$

18. $\{(x, y) \mid y = 3x - 2, \quad x \in \{-2, -1, 0, 1, 2\}\}$

19. $\{(x, y) \mid y = 4 - x, \quad x \in \{-2, -1, 0, 1, 2\}\}$

20. $\{(x, y) \mid y = 3 - 2x, \quad x \in \{-2, -1, 0, 1, 2\}\}$

Graph each equation in $R \times R$.

Example $3x - 4y = 24$

Solution Determine the intercepts.

If $x = 0$, then $y = -6$.
If $y = 0$, then $x = 8$.

Hence, -6 is the y intercept and 8 is the x intercept.

Graph these intercepts and draw a line through the points.

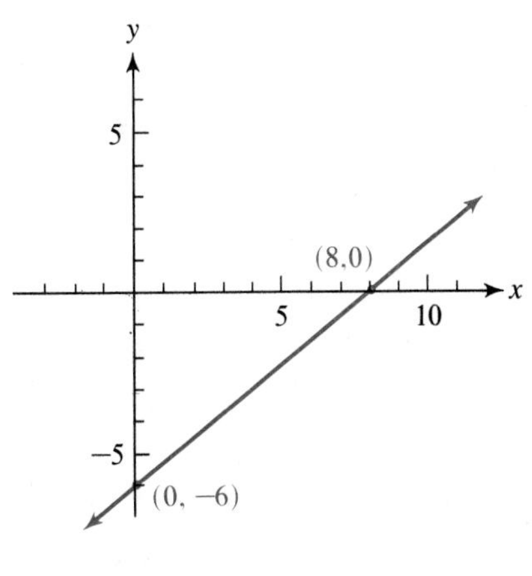

21. $y = x - 5$

22. $y = x + 3$

23. $y = 3x + 6$

24. $y = 4x - 8$

25. $x + 2y = 8$

26. $3x - y = 6$

27. $3x - 4y = 12$

28. $2x + 6y = 6$

29. $2x - y = 0$

30. $x + 3y = 0$

31. $y = -3$

32. $x = -2$

33. $x = 4$

34. $y = 5$

35. $y = -3x$

36. $x = 2y$

37. $4x - y = 0$

38. $4x + y = 0$

Example $y = x + 2 \quad (x \geq 0)$

Solution First sketch the graph of $y = x + 2$ for $x \in R$, as shown in graph a. Then delete the part of the graph for which $x < 0$. The remaining graph of $y = x + 2$ for $x \geq 0$ is shown in graph b.

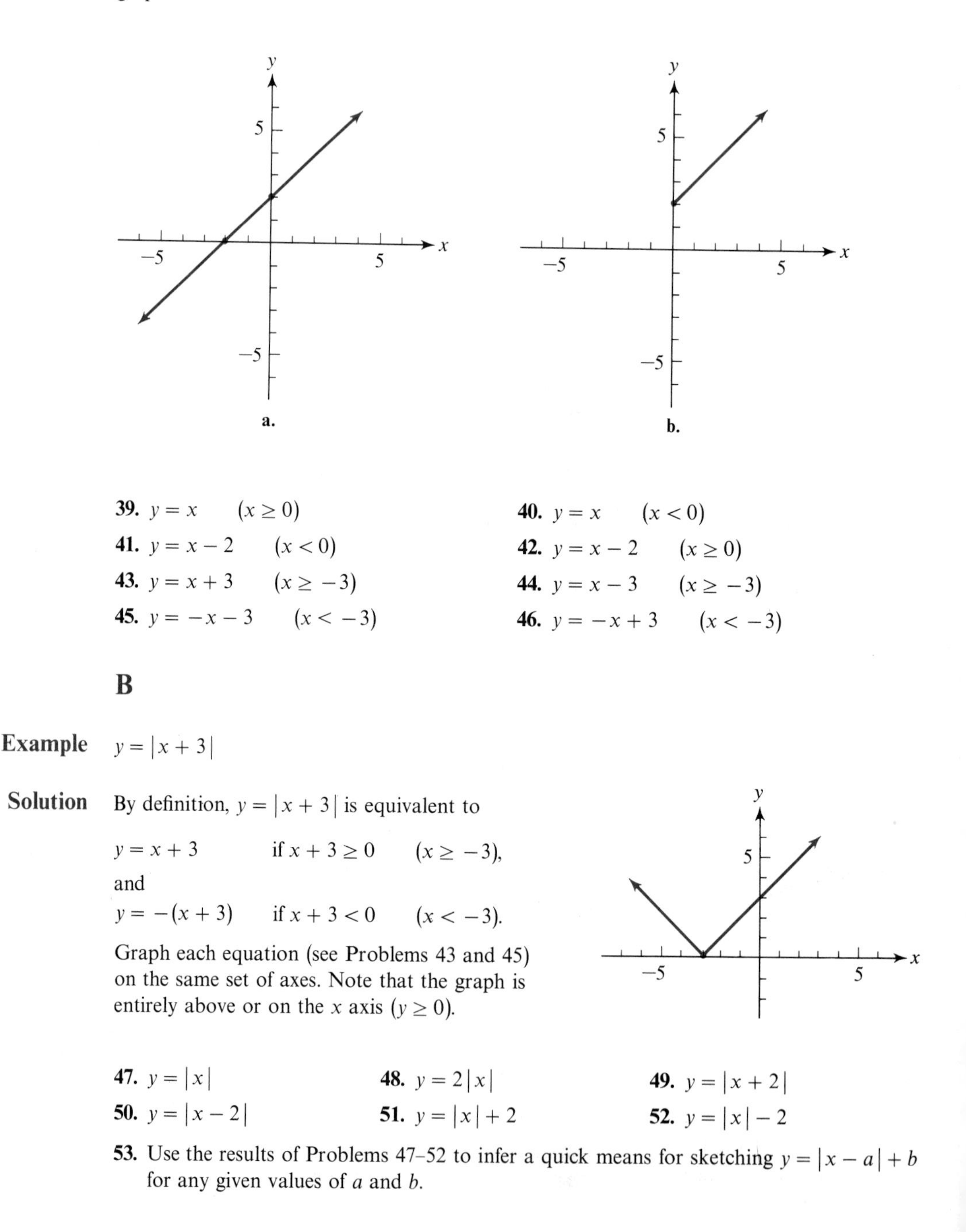

39. $y = x \quad (x \geq 0)$

40. $y = x \quad (x < 0)$

41. $y = x - 2 \quad (x < 0)$

42. $y = x - 2 \quad (x \geq 0)$

43. $y = x + 3 \quad (x \geq -3)$

44. $y = x - 3 \quad (x \geq -3)$

45. $y = -x - 3 \quad (x < -3)$

46. $y = -x + 3 \quad (x < -3)$

B

Example $y = |x + 3|$

Solution By definition, $y = |x + 3|$ is equivalent to

$y = x + 3 \qquad \text{if } x + 3 \geq 0 \qquad (x \geq -3),$

and

$y = -(x + 3) \qquad \text{if } x + 3 < 0 \qquad (x < -3).$

Graph each equation (see Problems 43 and 45) on the same set of axes. Note that the graph is entirely above or on the x axis ($y \geq 0$).

47. $y = |x|$

48. $y = 2|x|$

49. $y = |x + 2|$

50. $y = |x - 2|$

51. $y = |x| + 2$

52. $y = |x| - 2$

53. Use the results of Problems 47–52 to infer a quick means for sketching $y = |x - a| + b$ for any given values of a and b.

7.3

RELATIONS AND FUNCTIONS

In Sections 7.1 and 7.2 we considered sets of ordered pairs of real numbers. The notion of a set of ordered pairs is particularly useful in mathematics and is given a special name—**relation**.

▶ *Any set of ordered pairs is a relation.*

The set of all first components of the ordered pairs in a relation is the **domain** of the relation, and the set of all second components of the ordered pairs is the **range** of the relation. For example,

$$\{(1,3),(2,4),(3,5)\}$$

(first components: elements in the domain; second components: elements in the range)

is a relation with domain $\{1,2,3\}$ and range $\{3,4,5\}$. Each element in the range of a relation is called a **value** of the relation. The term "relation" stems directly from the fact that a set of ordered pairs displays a precise relationship between the elements of two sets, the domain and the range.

A special kind of relation with which we shall be concerned is called a **function**.

▶ *A function is a relation in which no two ordered pairs have the same first components.*

For example, $\{(2,3),(3,5),(4,5)\}$ is a function because no two ordered pairs have the same first component. $\{(3,4),(3,5),(4,7)\}$ is a relation, but is not a function because two ordered pairs have the same first component.

In Section 7.2 a line was referred to as the graph of the solution set of the equation, or simply the graph of the equation. The line is also referred to as the graph of the function *defined* by the equation *in* or *over* $R \times R$. Hereafter, we shall be interested mainly in relations and functions in $R \times R$. *If the domain of a relation is not specified, we shall assume that the relation has as domain the subset of* R *containing all real numbers for which real numbers exist in the range.*

Functions are usually designated by means of a single symbol, P, R, f, g, or some other letter. The symbol for the function can be used in conjunction with the variable representing an element in the domain to represent the associated element in the range. Thus, $f(x)$ (read "f of x" or "the value of f at x") is the element in the range of f associated with the element x in the domain. This is precisely the same use that we made of $P(x)$ in Section 2.1 when we discussed expressions, and, in particular, polynomial expressions. For example, suppose f is the function defined by the equation

$$y = x - 3;$$

then we can just as well write

$$f(x) = x - 3,$$

where $f(x)$ plays the same role as y. In this case, several values of the function, say for

$$x = 6, \qquad x = -2, \qquad x = -5$$

are given by

$$f(6) = 3, \qquad f(-2) = -5, \qquad f(-5) = -8.$$

Function notation can be used to denote the ordinate associated with a given abscissa of a graph. For example, the graph of

$$f(x) = x - 3$$

is shown in Figure 7.7, and specific ordinates are shown as line segments labeled with their lengths in function notation. We mention this here because it is often useful to think of the ordinate of a point in terms of the directed length of the perpendicular line segment from the x axis to the point.

The ordinate concept also provides a means of mentally checking to see whether a particular graph does or does not represent a function. If we imagine a line parallel to the y axis to pass across the graph from left to right, we can examine whether or not it intersects the graph at more than one point at each

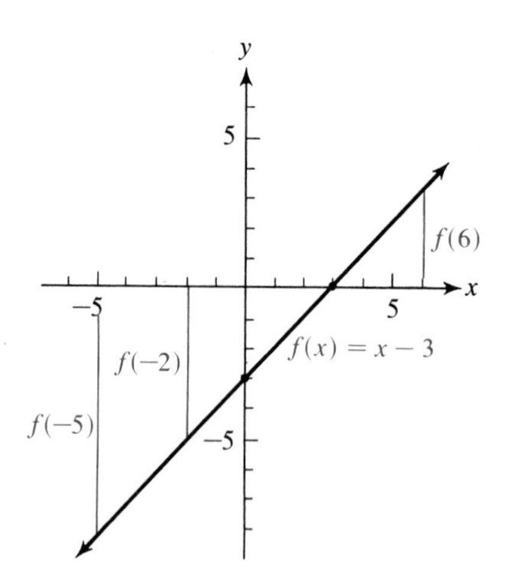

Figure 7.7

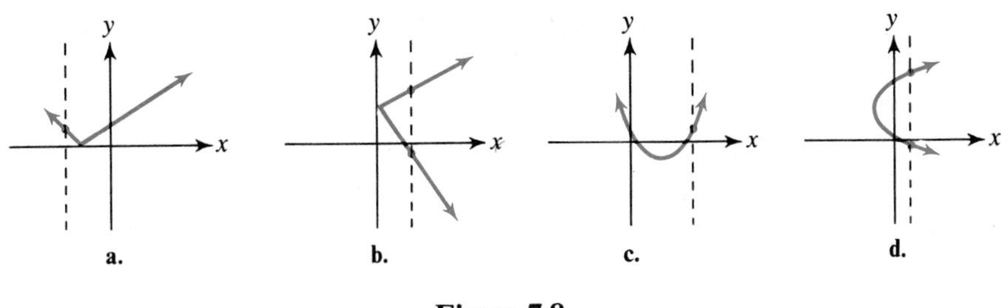

Figure 7.8

position on the x axis; that is, whether or not there are two or more different ordinates for any given abscissa. If there are, the graph is not the graph of a function.

Parts a and c of Figure 7.8 are graphs of functions, and parts b and d are graphs of relations that are not functions.

EXERCISE 7.3

A

a. Specify the domain and the range of each relation.
b. State whether or not each relation is a function.

Example $\{(2,3), (3,5), (3,6), (4,6), (5,6)\}$

Solution **a.** Domain, set of first components: $\{2,3,4,5\}$
Range, set of second components: $\{3,5,6\}$

b. Relation is not a function. Two ordered pairs, $(3,5)$ and $(3,6)$, have the same first component.

1. $\{(-2,3), (-1,4), (0,5), (1,6)\}$

2. $\{(-1,-1), (0,0), (1,1), (2,2)\}$

3. $\{(5,1), (6,1), (7,1), (8,1)\}$

4. $\{(4,1), (4,2), (4,3), (4,4)\}$

5. $\{(2,3), (2,4), (3,3), (3,4)\}$

6. $\{(-5,2), (2,-5), (3,6), (6,3)\}$

7. $\{(0,0), (2,1), (2,2), (3,3)\}$

8. $\{(2,-1), (3,-1), (3,1), (4,2)\}$

Graph the function defined by each equation over $R \times R$.

9. $y = 3x + 4$

10. $y = 2x - 5$

11. $x + 3y = 7$

12. $x - 2y = 5$

13. $y = 6$

14. $y = -4$

Find the value for each expression.

Example $f(3) - f(1)$, if $f(x) = 5x + 2$

Solution $f(3) = 5(3) + 2 = 15 + 2 = 17$
$f(1) = 5(1) + 2 = 5 + 2 = 7$

Hence, $f(3) - f(1) = 17 - 7 = 10$.

15. $f(4)$, if $f(x) = x^2 - 2x + 1$

16. $g(3)$, if $g(x) = 2x^2 + 3x - 1$

17. $g(5) - g(2)$, if $g(x) = x + 3$

18. $f(2) - f(0)$, if $f(x) = 3x - 1$

19. $f(3) - f(-3)$, if $f(x) = x^2 - x + 1$

20. $f(0) - f(-2)$, if $f(x) = x^2 + 3x - 2$

In Problems 21–26, find: ***a.*** $f(x + h)$; ***b.*** $f(x + h) - f(x)$; ***c.*** $\dfrac{f(x + h) - f(x)}{h}$.

Example $f(x) = x^2 + 1$

Solution **a.** $f(x + h) = (x + h)^2 + 1 = x^2 + 2xh + h^2 + 1$

b. $f(x + h) - f(x) = (x^2 + 2xh + h^2 + 1) - (x^2 + 1) = 2xh + h^2$

c. $\dfrac{f(x + h) - f(x)}{h} = \dfrac{2xh + h^2}{h} = 2x + h$

21. $f(x) = 3x - 4$

22. $f(x) = 2x + 5$

23. $f(x) = x^2 - 3x + 5$

24. $f(x) = x^2 + 2x$

25. $f(x) = x^3 + 2x - 1$

26. $f(x) = x^3 + 3x^2 - 1$

B

Example Graph $f(x) = x - 1$. Represent $f(5)$ and $f(3)$ by drawing line segments from $(5, 0)$ to $(5, f(5))$ and from $(3, 0)$ to $(3, f(3))$.

Solution $f(5)$ is the ordinate at $x = 5$
$f(3)$ is the ordinate at $x = 3$

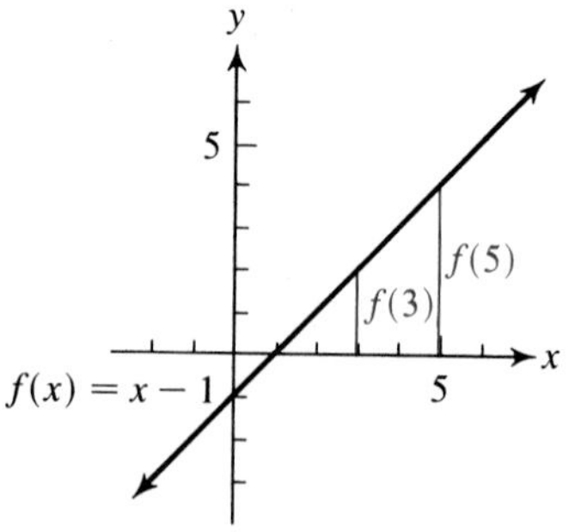

27. Graph $f(x) = 2x + 1$. Represent $f(3)$ and $f(-2)$ by drawing line segments from $(3, 0)$ to $(3, f(3))$ and from $(-2, 0)$ to $(-2, f(-2))$.

28. Graph $f(x) = 2x - 5$. Represent $f(4)$ and $f(0)$ by drawing line segments from $(4, 0)$ to $(4, f(4))$ and from $(0, 0)$ to $(0, f(0))$.

29. Graph $f(x) = 4 + x$. Represent $f(5)$ and $f(3)$ by drawing line segments from $(5,0)$ to $(5, f(5))$ and from $(3,0)$ to $(3, f(3))$.

30. Graph $f(x) = 3x - 6$. Represent $f(6)$ and $f(-3)$ by drawing segments from $(6,0)$ to $(6, f(6))$ and from $(-3,0)$ to $(-3, f(-3))$.

31. Using function notation, represent the length of line segment CB in the figure in terms of x and $x + h$.

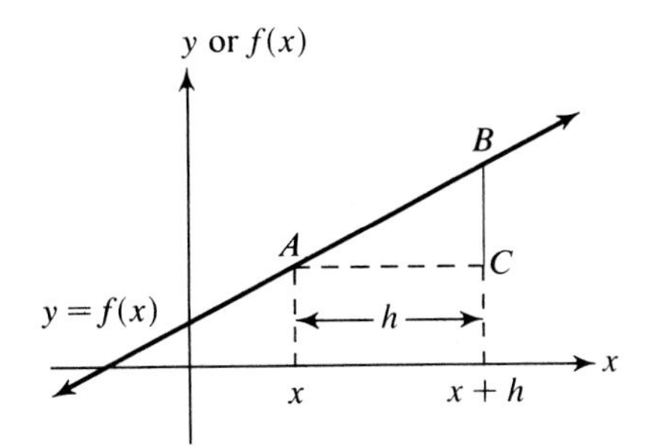

32. Using the results of Problem 31, represent the ratio of the length of line segment BC to the length of line segment AC in terms of x and $x + h$.

7.4

DISTANCE AND SLOPE FORMULAS

Any two distinct points in a plane can be looked upon as the end points of a line segment. We shall discuss two fundamental properties of a line segment—its **length** and its **inclination** with respect to the x axis. We first observe that any two distinct points either lie on the same vertical line or one is to the right of the other; the point on the right has a greater x coordinate than the point on the left. If we construct a line parallel to the y axis through P_2, and a line parallel to the x axis through P_1, the lines will meet at a point P_3, as shown in either part a or b of Figure 7.9. The x coordinate of P_3 is evidently the same as the x coordinate of P_2, while the y coordinate of P_3 is the same as that of P_1; hence, the coordinates of P_3 are (x_2, y_1). By inspection, we observe that the distance between P_2 and P_3 is $(y_2 - y_1)$ and the distance between P_1 and P_3 is $(x_2 - x_1)$.

In general, since $y_2 - y_1$ is positive or negative as $y_2 > y_1$ or $y_2 < y_1$, respectively, and $x_2 - x_1$ is positive or negative as $x_2 > x_1$ or $x_2 < x_1$, respectively, we may designate the distances represented by $x_2 - x_1$ and $y_2 - y_1$ as positive or negative.

The Pythagorean theorem can be used to find the length of the line segment from P_1 to P_2. This theorem asserts that the square of the length of the

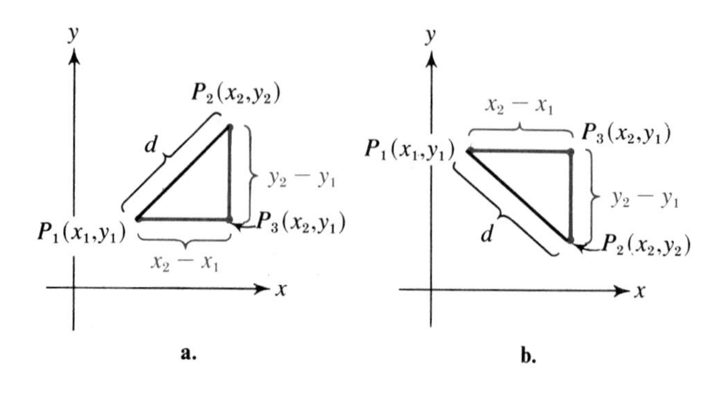

Figure 7.9

hypotenuse of any right triangle is equal to the sum of the squares of the lengths of the legs. Thus,

$$d^2 = (x_2 - x_1)^2 + (y_2 - y_1)^2,$$

and if we consider only the positive square root of the right-hand member, we have what is called the **distance formula,**

$$\text{(1)} \qquad \boldsymbol{d = \sqrt{(x_2 - x_1)^2 + (y_2 - y_1)^2}.}$$

If the points P_1 and P_2 lie on the same horizontal line ($y_2 = y_1$) and if we are concerned only with distance and not direction, then

$$d = \sqrt{(x_2 - x_1)^2 + 0^2} = |x_2 - x_1|,$$

and, if they lie on the same vertical line ($x_2 = x_1$), then

$$d = \sqrt{0^2 + (y_2 - y_1)^2} = |y_2 - y_1|.$$

The second useful property of the line segment joining two points is its inclination. The inclination of a line segment can be measured by comparing the *rise* (difference of y coordinates) of the segment with a given *run* (difference of x coordinates) as shown in Figure 7.10. The ratio of *rise* to *run* is called the **slope** of the line segment and is designated by the letter m. Thus,

$$m = \frac{\text{rise}}{\text{run}}.$$

Since the rise is $y_2 - y_1$ and the run is $x_2 - x_1$, the slope of the line segment joining P_1 and P_2 is given by

$$\text{(2)} \qquad \boldsymbol{m = \frac{y_2 - y_1}{x_2 - x_1} \qquad (x_2 \neq x_1).}$$

Taking P_2 to the right of P_1, the run, $x_2 - x_1$, is positive and the slope is positive or negative as the rise, $y_2 - y_1$, is positive or negative—that is, according to whether the line slopes upward or downward from P_1 to P_2. A positive

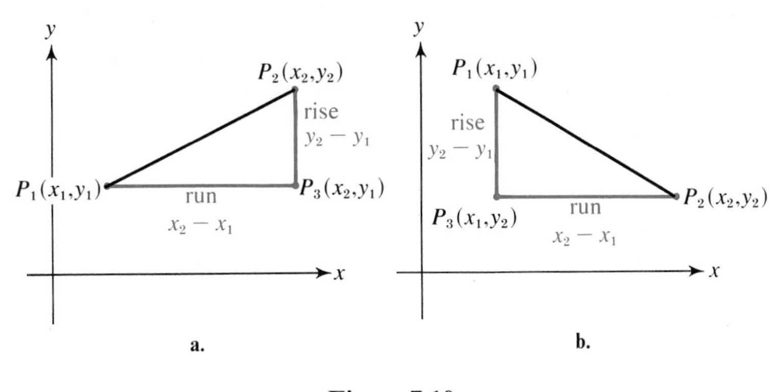

Figure 7.10

slope indicates that a line is rising to the right; a negative slope indicates that it is falling to the right. Since

$$\frac{y_2 - y_1}{x_2 - x_1} = \frac{-(y_1 - y_2)}{-(x_1 - x_2)} = \frac{y_1 - y_2}{x_1 - x_2},$$

the restriction that P_2 be to the right of P_1 is not necessary, and the order in which the points are considered is immaterial.

If a segment is parallel to the x axis, as shown in part a of Figure 7.11, then $y_2 - y_1 = 0$ and it will have a slope of 0. If a segment is parallel to the y axis, as shown in part b of Figure 7.11, $x_2 - x_1 = 0$ and its slope is not defined.

Two line segments with slopes m_1 and m_2 are parallel if and only if $m_1 = m_2$ is true, with both m_1 and m_2 defined. It can be shown that they are perpendicular if neither m_1 nor m_2 is 0 and $m_1 = -1/m_2$, or, equivalently, $m_1 m_2 = -1$. For example, in Figure 7.12, line segment AB with slope

$$m_1 = \frac{5-3}{5-2} = \frac{2}{3}$$

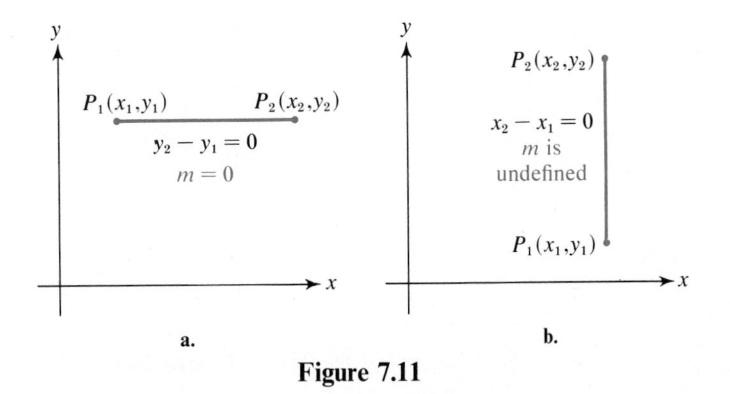

Figure 7.11

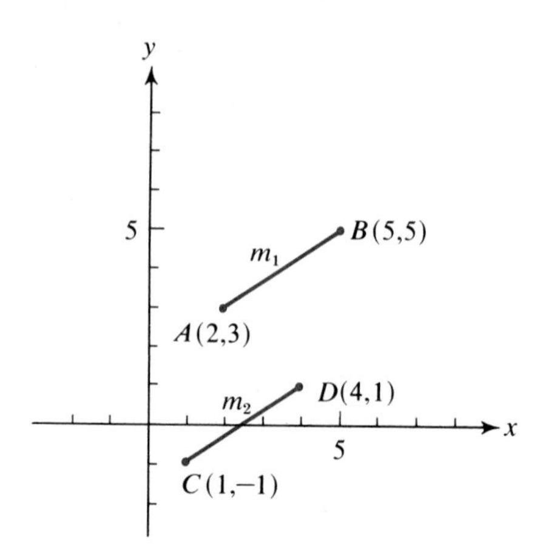

Figure 7.12

is parallel to line segment CD with slope

$$m_2 = \frac{1 - (-1)}{4 - 1} = \frac{2}{3}.$$

In Figure 7.13, line segment DE with slope

$$m_2 = \frac{1 - 7}{4 - 0} = \frac{-6}{4} = \frac{-3}{2}$$

is perpendicular to line segment AB with slope

$$m_1 = \frac{5 - 3}{5 - 2} = \frac{2}{3},$$

because

$$m_1 m_2 = \frac{2}{3}\left(\frac{-3}{2}\right) = -1.$$

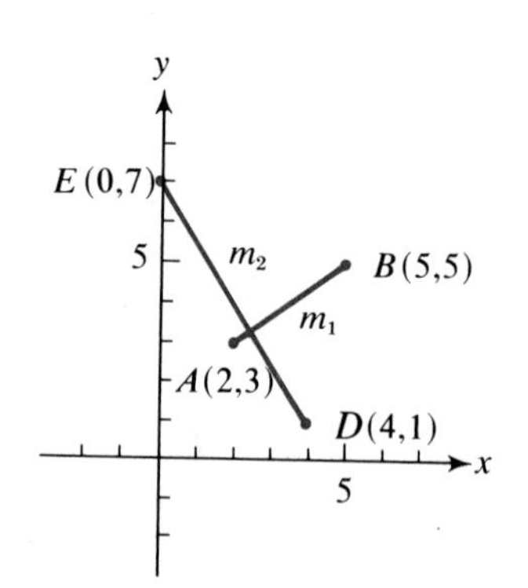

Figure 7.13

EXERCISE 7.4

A

Find the distance between each of the given pairs of points, and find the slope of the line segment joining them. Sketch each line segment on the coordinate plane.

Example $(3, -5), (2, 4)$

Solution Consider $(3, -5)$ as P_1 and $(2, 4)$ as P_2.

$$d = \sqrt{(x_2 - x_1)^2 + (y_2 - y_1)^2} \qquad m = \frac{y_2 - y_1}{x_2 - x_1}$$

$$= \sqrt{[2 - 3]^2 + [4 - (-5)]^2} \qquad = \frac{4 - (-5)}{2 - 3}$$

$$= \sqrt{1 + 81} = \sqrt{82} \qquad = \frac{9}{-1} = -9$$

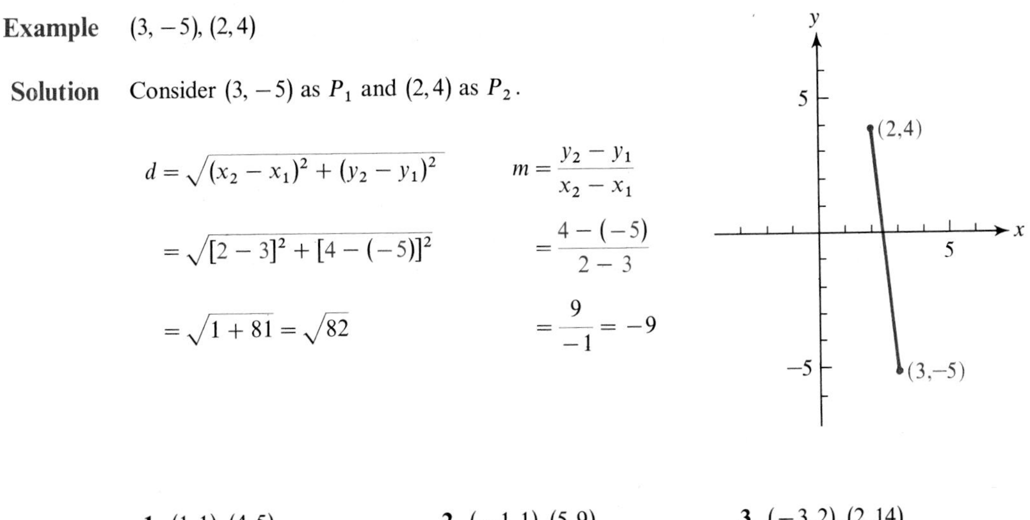

1. $(1, 1), (4, 5)$
2. $(-1, 1), (5, 9)$
3. $(-3, 2), (2, 14)$
4. $(-4, -3), (1, 9)$
5. $(2, 1), (4, 0)$
6. $(-3, 2), (0, 0)$
7. $(5, -4), (-1, 1)$
8. $(2, -3), (-2, -1)$
9. $(3, 5), (-2, 5)$
10. $(2, 0), (-2, 0)$
11. $(0, 5), (0, -5)$
12. $(-2, -5), (-2, 3)$

Find the perimeter of the triangle whose vertices are given. Sketch each triangle on the coordinate plane.

13. $(0, 6), (9, -6), (-3, 0)$
14. $(10, 1), (3, 1), (5, 9)$
15. $(5, 6), (11, -2), (-10, -2)$
16. $(-1, 5), (8, -7), (4, 1)$
17. Show that the two line segments whose end points are $(5,4)$, $(3,0)$ and $(-1,8)$, $(-4, 2)$ are parallel.
18. Show that the two line segments whose end points are $(-4, 2)$, $(2, -2)$ and $(3, 0)$, $(-3, 4)$ are parallel.
19. Show that the two line segments whose end points are $(0, -7)$, $(8, -5)$ and $(5, 7)$, $(8, -5)$ are perpendicular.
20. Show that the two line segments whose end points are $(8, 0)$, $(6, 6)$ and $(-3, 3)$, $(6, 6)$ are perpendicular.

B

21. Show that the triangle described in Problem 13 is a right triangle. [*Hint:* Use the converse of the Pythagorean theorem—that is, if $c^2 = a^2 + b^2$, the triangle is a right triangle; or show that two sides are perpendicular.]

22. Show that the triangle with vertices at $(0,0)$, $(6,0)$, and $(3,3)$ is a right isosceles triangle—that is, a right triangle with two sides that have the same length.

23. Show that the points $(2,4)$, $(3,8)$, $(5,1)$, and $(4,-3)$ are the vertices of a parallelogram. [*Hint:* A four-sided figure is a parallelogram if the opposite sides are parallel.]

24. Show that the points $(-5,4)$, $(7,-11)$, $(12,25)$, and $(0,40)$ are the vertices of a parallelogram.

25. Given the points $P_1(4,-1)$, $P_2(2,7)$, and $P_3(-3,4)$, find the value of k in the ordered pair $P_4(5,k)$ that makes P_1P_2 parallel to P_3P_4.

26. Using the points in Problem 25, find the value of k that makes P_1P_2 perpendicular to P_3P_4.

7.5

FORMS OF LINEAR EQUATIONS

Let us designate

$$(1) \qquad \mathbf{ax + by + c = 0} \qquad (\mathbf{b \neq 0})$$

as **standard form** for a linear equation, and then consider two alternative forms that display useful aspects.

Assuming that the slope of the line segment joining any two points on a line does not depend upon the points, consider a line on the plane with given slope m and passing through a given point (x_1, y_1), as shown in Figure 7.14. If we choose any other point on the line and assign to it the coordinates (x, y), the slope of the line is given by

$$(2) \qquad \frac{y - y_1}{x - x_1} = m \qquad (x \neq x_1),$$

from which

$$(3) \qquad y - y_1 = m(x - x_1) \qquad (x \neq x_1).$$

Notice that (2) and (3) are equivalent for all values of x except x_1, in which case the left-hand member of (2) is not defined. The ordered pair (x_1, y_1) corresponds to a point on the line, and the coordinates of (x_1, y_1) also satisfy (3). Hence, the graph of (3) contains all points in the graph of (2) and, in addition, the point (x_1, y_1). Since x and y are the coordinates of *any* point on the line,

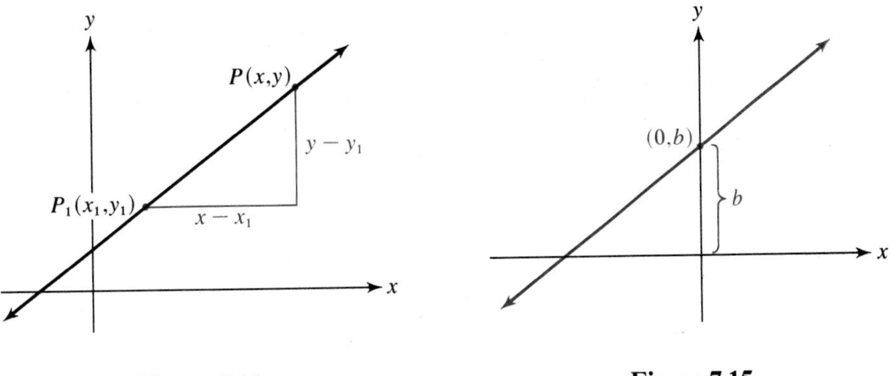

Figure 7.14

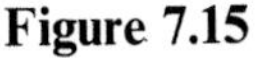
Figure 7.15

without the restriction $x \neq x_1$, (3) is the equation of the line passing through (x_1, y_1) with slope m. The equation

$$(4) \qquad \boldsymbol{y - y_1 = m(x - x_1)}$$

is called the **point-slope** form for a linear equation.

Now consider the equation of the line passing through a given point on the y axis with coordinates $(0, b)$ and slope m, as shown in Figure 7.15. Substituting $(0, b)$ in the point-slope form of a linear equation,

$$y - y_1 = m(x - x_1),$$

we obtain

$$y - b = m(x - 0),$$

from which

$$(5) \qquad \boldsymbol{y = mx + b.}$$

Equation (5) is called the **slope-intercept form** for a linear equation.

Any equation in standard form, such as

$$4x + 3y - 6 = 0,$$

can be written in the slope-intercept form by solving explicitly for y in terms of x; thus,

$$3y = -4x + 6,$$

from which

$$y = -\frac{4}{3}x + 2.$$

The slope of the line, $-4/3$, and the y intercept, 2, can now be read directly from the last form of the equation by comparing it with (5).

EXERCISE 7.5

A

Sketch the line that goes through each of the given points and has the given slope. Find the equation and write it in standard form.

Example $(-3, -2), \quad m = -\frac{2}{3}$

Solution Substitute given values in the point-slope form of the linear equation.

$$y - y_1 = m(x - x_1)$$

$$y - (-2) = -\frac{2}{3}[x - (-3)]$$

$$3(y + 2) = -2(x + 3)$$

$$3y + 6 = -2x - 6$$

$$2x + 3y + 12 = 0$$

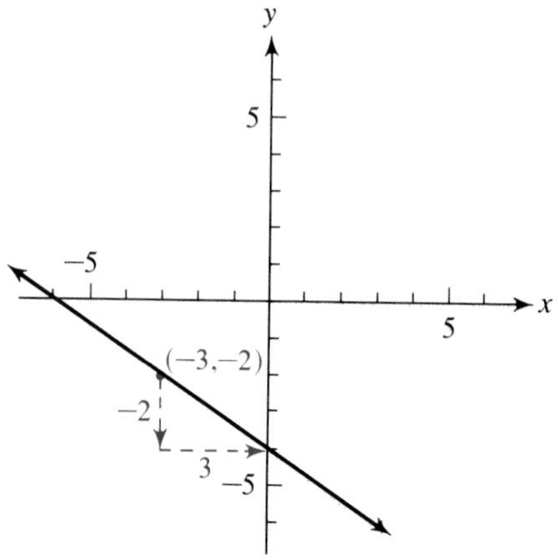

1. $(2, 1), \quad m = 4$
2. $(-2, 3), \quad m = 5$
3. $(5, 5), \quad m = -1$
4. $(-3, -2), \quad m = \frac{1}{2}$
5. $(0, 0), \quad m = 3$
6. $(-1, 0), \quad m = 1$
7. $(0, -1), \quad m = -\frac{1}{2}$
8. $(2, -1), \quad m = \frac{3}{4}$
9. $(-2, -2), \quad m = -\frac{3}{4}$
10. $(2, -3), \quad m = 0$
11. $(-4, 2), \quad m = 0$
12. $(-1, -2)$, parallel to x axis

Write each equation in slope-intercept form; specify the slope of the line and the y intercept.

Example $2x - 5y = 5$

Solution Solve explicitly for y.

$$-5y = 5 - 2x$$

$$5y = 2x - 5$$

$$y = \frac{2}{5}x + (-1)$$

Compare with the general slope-intercept form $y = mx + b$.

$$\text{slope: } \frac{2}{5}; \qquad y \text{ intercept: } -1$$

13. $x + y = 3$

14. $2x + y = -1$

15. $3x + 2y = 1$

16. $3x - y = 7$

17. $x - 3y = 2$

18. $2x - 3y = 0$

19. $8x - 3y = 0$

20. $-x = 2y - 5$

21. $y + 2 = 0$

22. $y - 3 = 0$

23. $x + 5 = 0$

24. $x - 4 = 0$

B

25. Write the equation of the line that is parallel to the graph of $x - 2y = 5$ and passes through the origin. Draw the graphs of both equations.

26. Write the equation of the line that passes through (0, 5) parallel to $2y - 3x = 5$. Draw the graphs of both equations.

27. Write the equation of the line that is perpendicular to the graph of $x - 2y = 5$ and passes through the origin. Draw the graphs of both equations.

28. Write the equation of the line that is perpendicular to the graph of $2y - 3x = 5$ and passes through (0, 5). Draw the graphs of both equations.

29. Show that

$$y - y_1 = \left(\frac{y_2 - y_1}{x_2 - x_1}\right)(x - x_1)$$

is the equation of the line joining the points (x_1, y_1) and (x_2, y_2). [*Hint:* Use the point-slope form and the slope formula from Section 7.4.]

30. Using the form

$$y - y_1 = \left(\frac{y_2 - y_1}{x_2 - x_1}\right)(x - x_1),$$

find an equation in standard form for the line through the given points.

a. (2, 1) and (−1, 3)

b. (3, 0) and (4, 2)

c. (−2, 1) and (3, −2)

d. (−1, −1) and (1, 1)

31. Find the slope of the line with equation $ax + by = 0$ in terms of a and b.

32. Find the slope of the line with equation $ax + by + c = 0$ in terms of a, b, and c.

33. Find the slope of the line that is perpendicular to the line with equation $ax + by = 0$.

34. Find the slope of the line that is perpendicular to the line with equation $ax + by + c = 0$.

7.6

GRAPHS OF FIRST-DEGREE INEQUALITIES

An open sentence of the form

$$\boldsymbol{ax + by + c \leq 0},$$

where a, b, and c are real numbers, is an inequality of the first degree. Such inequalities can be graphed on the plane, but the graph will be a region of the plane rather than a straight line. For example, consider the inequality

$$2x + y - 3 < 0. \tag{1}$$

Rewritten in the form

$$y < -2x + 3, \tag{2}$$

we have that y is less than $-2x + 3$ for each x. The graph of the equation

$$y = -2x + 3 \tag{3}$$

is simply a straight line, as illustrated in Figure 7.16. Therefore, to graph (2), we need only observe that any point below this line has x and y coordinates that satisfy (2), and consequently the solution set of (2) corresponds to the entire region below the line. The region is indicated on the graph by shading. The broken line in Figure 7.17 indicates that the points on the line do not correspond to elements in the solution set of the inequality. If the original inequality were

$$2x + y - 3 \leq 0,$$

the line would be a part of the graph of the solution set and would be shown as a solid line.

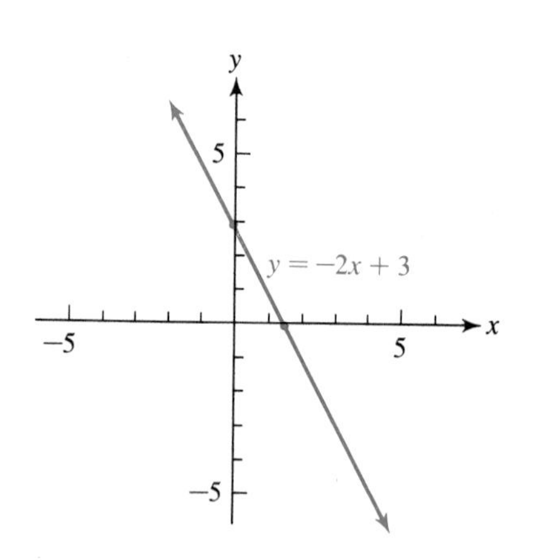

Figure 7.16

In general, if $b \neq 0$,

$$ax + by + c < 0 \qquad \text{or} \qquad ax + by + c > 0$$

will have as a solution set all ordered pairs associated with the points in a **half-plane** either above or below the line with equation

$$ax + by + c = 0,$$

depending upon the inequality symbols involved. We can determine which of the half-planes should be shaded by substituting the coordinates of any point not on the line and noting whether or not they satisfy the inequality. If they do, then the half-plane containing the point is shaded; if they do not, then the other half-plane is shaded. A good point to use in this process is the origin, with coordinates (0,0) if the origin does not lie in the line.

Inequalities do not define functions according to our definition in Section 7.3, because each element in the domain is not associated with a unique element in the range. For example, in the inequality

$$y < -2x + 3,$$

if $x = -3$,

$$y < -2(-3) + 3$$
$$y < 9,$$

which is certainly not a unique value for y. The infinite number of values for $y < 9$ is indicated by the heavy red vertical line in Figure 7.18 on page 228.

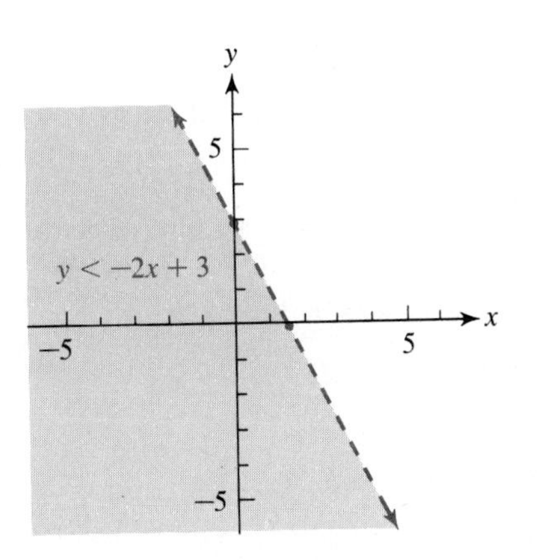

Figure 7.17

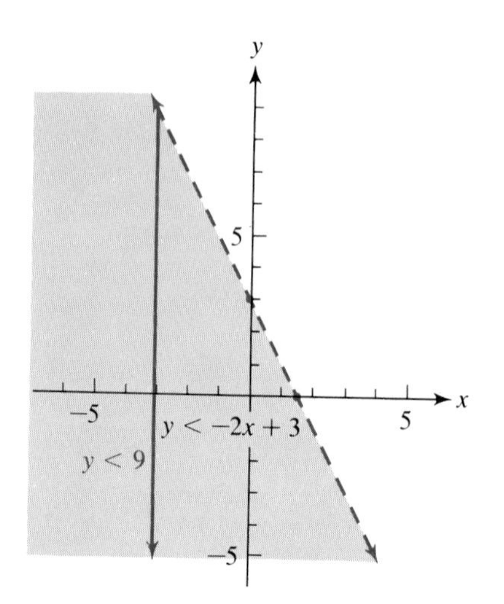

Figure 7.18

EXERCISE 7.6

A

Graph the solution set of the inequality.

Example $2x + y \geq 4$

Solution Graph the equality.

$$2x + y = 4$$

Substitute 0 for x and y in the inequality.

$$2(0) + (0) \geq 4$$

Since this is not a true statement, shade the half-plane not containing the origin.

The line is included in the graph.

1. $y < x$ **2.** $y > x$ **3.** $y \leq x + 2$ **4.** $y \geq x - 2$

5. $x + y < 5$ 6. $x - y < 3$ 7. $2x + y < 2$ 8. $x - 2y < 5$
9. $x \leq 2y + 4$ 10. $2x \leq y + 1$ 11. $0 \geq x - y$ 12. $0 \geq x + 3y$

Graph each set of ordered pairs.

Example $\{(x, y) \mid x > 2\}$

Solution Graph the equality $x = 2$.

Shade the region to the right of the line representing $x = 2$.

The line is excluded from the graph.

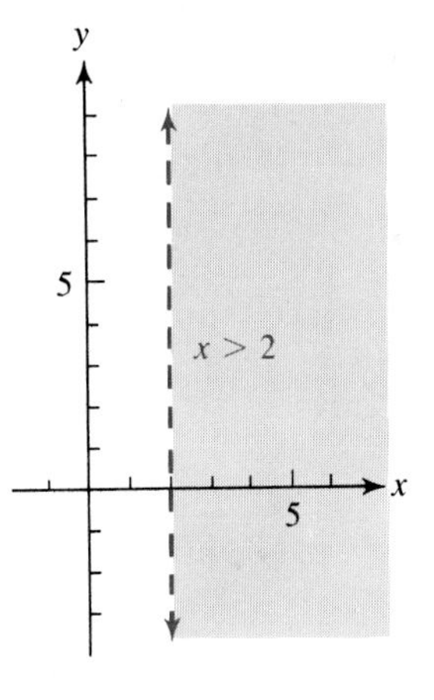

13. $\{(x, y) \mid x > 0\}$ 14. $\{(x, y) \mid y < 0\}$ 15. $\{(x, y) \mid y \geq 3\}$
16. $\{(x, y) \mid x < -2\}$ 17. $\{(x, y) \mid -1 < x < 5\}$ 18. $\{(x, y) \mid 0 \leq y \leq 1\}$
19. $\{(x, y) \mid |x| < 3\}$ 20. $\{(x, y) \mid |y| > 1\}$

Graph each of the set intersections using double shading.

Example $\{(x, y) \mid y > x\} \cap \{(x, y) \mid y > 2\}$

Solution

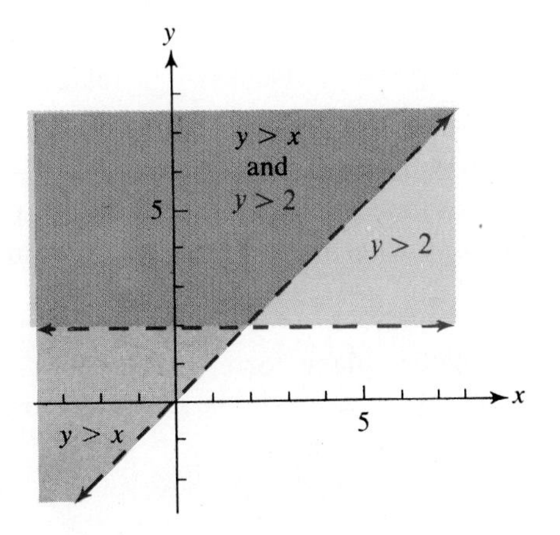

21. $\{(x, y) \mid x \geq 4\} \cap \{(x, y) \mid y \geq 2\}$

22. $\{(x, y) \mid x \leq 2\} \cap \{(x, y) \mid y \leq 2\}$

23. $\{(x, y) \mid x + y \leq 6\} \cap \{(x, y) \mid x + y \geq 4\}$

24. $\{(x, y) \mid y \geq x\} \cap \{(x, y) \mid y \leq -x\}$

25. $\{(x, y) \mid |x| \leq 2\} \cap \{(x, y) \mid |y| \leq 2\}$

26. $\{(x, y) \mid |x| > 3\} \cap \{(x, y) \mid |y| > 3\}$

B

Graph each relation.

27. $\{(x, y) \mid y \geq |x|\}$

28. $\{(x, y) \mid y < |x|\}$

29. $\{(x, y) \mid y < |x| - 2\}$

30. $\{(x, y) \mid y \geq |x| + 5\}$

CHAPTER SUMMARY

[7.1] The solution of an equation or inequality in two variables is an **ordered pair** of numbers. The replacement set for possible solutions is the infinite set of ordered pairs $\{(x, y) \mid x \in R \text{ and } y \in R\}$, denoted by $\boldsymbol{R \times R}$ or $\boldsymbol{R^2}$.

[7.2] We can use a **Cartesian** (or **rectangular**) **coordinate system** to graph ordered pairs of numbers on a plane. The components of a given ordered pair are called the **coordinates** of its graph.

The graph of a first-degree equation in two variables is a straight line. The x coordinate of a point of intersection of a graph with the x axis is an $\boldsymbol{x}$ **intercept** of the graph and the y coordinate of a point of intersection of a graph with the y axis is a $\boldsymbol{y}$ **intercept** of the graph; these are obtained by setting $y = 0$ and $x = 0$, respectively, in the equation.

[7.3] A **relation** is a set of ordered pairs. A **function** is a relation in which no two ordered pairs have the same first components.

The **domain** of a relation (or function) is the set of all first components in the ordered pairs in the relation (or function), and the **range** of a relation (or function) is the set of all second components in the ordered pairs in the relation (or function).

[7.4] For any two points in a geometric plane corresponding to (x_1, y_1) and (x_2, y_2), the distance d between the points is given by the **distance formula**,

$$d = \sqrt{(x_2 - x_1)^2 + (y_2 - y_1)^2},$$

and the slope m of the line containing the points is given by

$$m = \frac{y_2 - y_1}{x_2 - x_1} \qquad (x_2 \neq x_1).$$

The slopes of parallel lines are equal, and the nonzero slopes of perpendicular lines are the negative reciprocals of each other.

[7.5] The **standard form** of an equation of a line is

$$ax + by + c = 0.$$

The **point-slope form** of an equation of a line is

$$y - y_1 = m(x - x_1),$$

where the line has slope m and contains the point (x_1, y_1). The **slope-intercept form** for an equation of a line is

$$y = mx + b,$$

where the line has slope m and y intercept b.

[7.6] The graph of a first-degree (linear) inequality in two variables is a **half-plane**.

The symbols introduced in this chapter are listed on the inside of the front cover.

REVIEW EXERCISES

[7.1] **1.** Find the missing component in each solution of $2x - 6y = 12$.

a. $(0, \underline{?})$ **b.** $(\underline{?}, 0)$ **c.** $(3, \underline{?})$

2. List the ordered pairs in $\{(x, y) \mid y = 2x - 3, \quad x \in \{2, 4, 6\}\}$.

3. Find k if $(3, 2)$ is a solution of the equation $2x - ky = 6$.

4. In the equation $xy = 2x^2y + 3$, express y explicitly in terms of x.

[7.2] *Graph.*

5. a. $\{(2, 4), (4, 5), (6, 4)\}$ **b.** $\{(x, y) \mid y = 2x - 5, \quad x \in \{-2, -1, 0, 1, 2\}\}$

6. a. $3x - y = 6$ in $R \times R$ **b.** $3x - y = 0$ in $R \times R$

[7.3] 7. Specify the domain and range of each relation. Is the relation a function?

a. $\{(3,5), (4,6), (4,6)\}$ b. $\{(2,3), (3,4), (4,5), (5,5)\}$

8. Specify the domain and range of each relation.

a. $\{(x, y) | y = 2x, \quad x \in \{2, 4\}\}$ b. $\{(x, y) | y = 3x + 2, \quad x \in \{-2, 0, 2\}\}$

9. Given that $f(x) = x - 4$, find each of the following:

a. $f(6)$ b. $f(x+h) - f(x)$

10. Given that $f(x) = 2x^2 - 3x + 1$, find each of the following:

a. $f(3)$ b. $\dfrac{f(x+h) - f(x)}{h}$

[7.4] 11. a. Find the distance between the points $(3, -5)$ and $(6, 8)$.

b. Find the slope of the line segment joining the points in part a.

12. a. Find the distance between the points $(4, 2)$ and $(7, -4)$.

b. Find the slope of the line segment joining the points in part a.

13. Find the perimeter of each triangle with the given vertices.

a. $(5, 2), (-1, 3), (4, 7)$ b. $(-2, -4), (6, 3), (4, -1)$

14. a. Given points $P_1(2,2)$, $P_2(-2,-2)$, $P_3(0,4)$, and $P_4(-3,1)$, show that the line segments P_1P_2 and P_3P_4 are parallel.

b. Show that the line segments P_1P_2 and P_1P_3 are perpendicular.

[7.5] 15. a. Find the equation in standard form of the line through $(3,5)$ with slope 2.

b. Write the equation in slope-intercept form.

16. a. Write $2x + 3y = 6$ in slope-intercept form.

b. Specify the slope and the y intercept of its graph.

17. a. Write the equation in standard form of the line through $(-1,1)$ that is parallel to the graph of $2x = 5 - 3y$.

b. Write the equation of the line through $(-1,1)$ that is perpendicular to the graph of the equation.

18. a. Write the equation in standard form of the line through $(4, -2)$ that is parallel to the graph of $4x - 6y = 5$.

b. Write the equation of the line through $(-1, 1)$ that is perpendicular to the graph of the equation.

[7.6] *Graph.*

19. **a.** $\{(x, y) \mid y > x + 2\}$

b. $\{(x, y) \mid y \leq 4x + 4\}$

20. **a.** $\{(x, y) \mid y \leq 4\} \cap \{(x, y) \mid x \geq 1\}$

b. $\{(x, y) \mid x + y \leq 4\} \cap \{(x, y) \mid y \geq 0\}$

8

FUNCTIONS, RELATIONS, AND THEIR GRAPHS: PART II

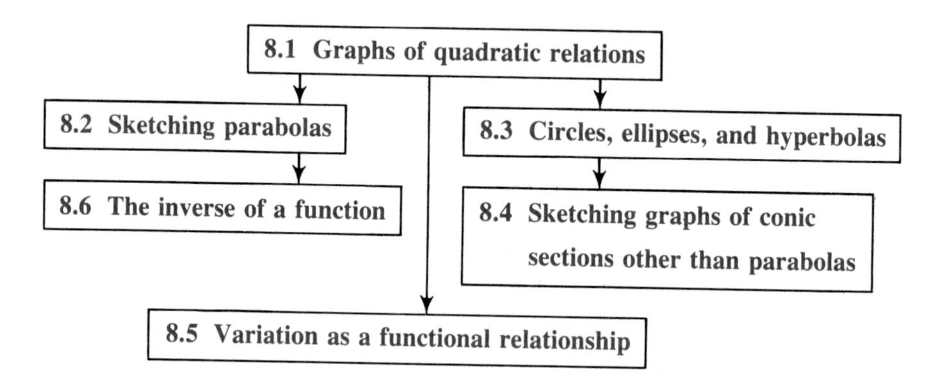

8.1

GRAPHS OF QUADRATIC RELATIONS

Consider the quadratic equation in two variables

$$(1) \qquad y = x^2 - 4.$$

As with linear equations in two variables, solutions of this equation are ordered pairs. We need replacements for both x and y to obtain a statement we can judge to be true or false. Such ordered pairs can be found by arbitrarily assigning values to x and computing related values for y. For instance, assigning the value -3 to x in (1), we have

$$\begin{aligned} y &= (-3)^2 - 4 \\ &= 5, \end{aligned}$$

and $(-3, 5)$ is a solution. Similarly, we find that

$$(-2, 0), \quad (-1, -3), \quad (0, -4), \quad (1, -3), \quad (2, 0), \quad (3, 5)$$

are also solutions of (1). Plotting the corresponding points on the plane, we have the graph shown in Figure 8.1. Clearly, these points do not lie on a straight line, and we might reasonably inquire whether the graph of the solution set of (1),

$$S = \{(x, y) \mid y = x^2 - 4\},$$

forms any kind of meaningful pattern on the plane. By plotting additional solutions of (1)—solutions with x components between those already found—we may be able to obtain a clearer picture. Accordingly, we find the solutions

$$\left(\frac{-5}{2}, \frac{9}{4}\right), \quad \left(\frac{-3}{2}, \frac{-7}{4}\right), \quad \left(\frac{-1}{2}, \frac{-15}{4}\right), \quad \left(\frac{1}{2}, \frac{-15}{4}\right), \quad \left(\frac{3}{2}, \frac{-7}{4}\right), \quad \left(\frac{5}{2}, \frac{9}{4}\right),$$

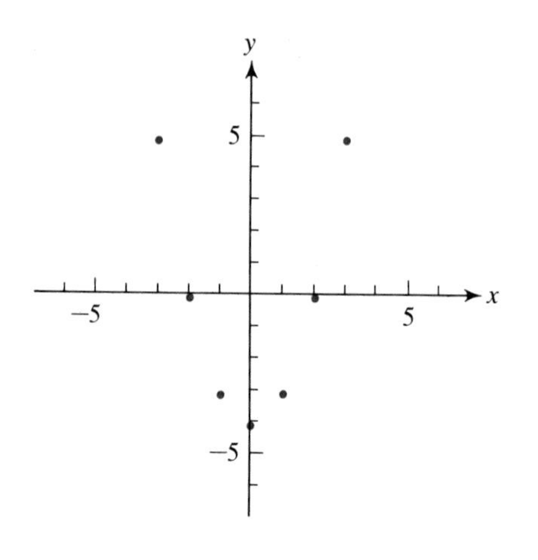

Figure 8.1

and by plotting these points in addition to those found earlier, we have the graph shown in Figure 8.2. It now appears reasonable to connect these points in sequence by a smooth curve, as in Figure 8.3, and to assume that the curve is a good approximation to the graph of (1). This curve is an example of a **parabola**. Note that, regardless of the number of individual points we plot in graphing (1), we have no absolute assurance that connecting these points with a smooth curve will produce the correct graph, although the more points we plot and locate, the more reasonable it should seem that connecting them will do this. Proving that the graph is indeed a parabola requires the use of concepts from the calculus and is not attempted here.

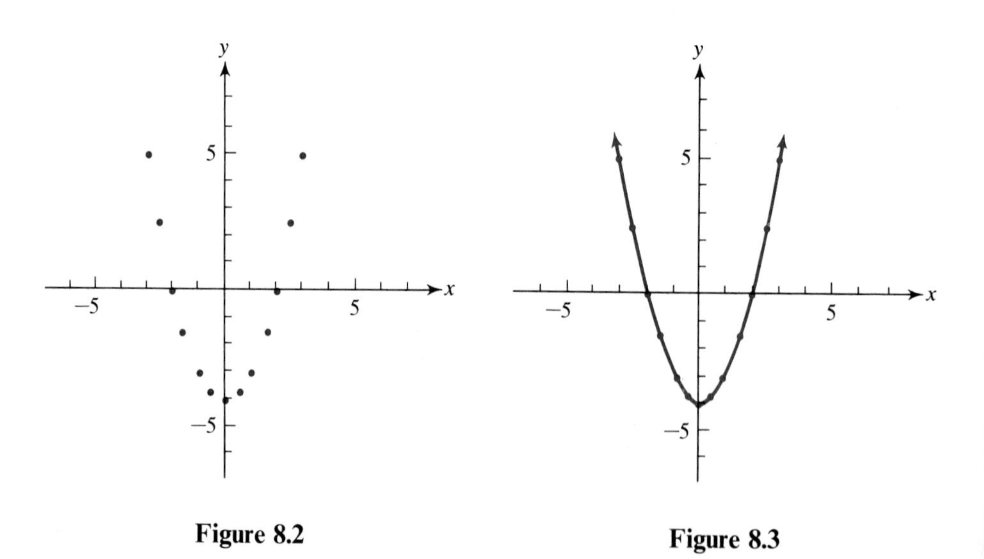

Figure 8.2

Figure 8.3

More generally, the graph of any quadratic equation of the form

$$y = ax^2 + bx + c, \tag{2}$$

where a, b, and c are real and $a \neq 0$, is a parabola. Since with each x an equation of the form (2) associates only one y, such an equation defines a function whose domain is the set of real numbers and whose range is some subset of the real numbers. For example, by inspecting the graph in Figure 8.3, we observe that the range of the function defined by (1) is

$$\{y \mid y \geq -4\}.$$

The graph of a quadratic equation of the form

$$x = ay^2 + by + c \qquad (a \neq 0), \tag{3}$$

is also a parabola. In this case, in order to graph the equation, it is easier to assign arbitrary values to y to obtain associated values of x. For example, if y is assigned a value 2 in

$$x = y^2 - 4y + 3, \tag{4}$$

we have

$$\begin{aligned} x &= (2)^2 - 4(2) + 3 \\ &= -1, \end{aligned}$$

and $(-1, 2)$ is in the solution set of (4). The graphs of $(-1, 2)$ and other ordered pairs in the solution set obtained in a similar way are shown in the graph of (4) in Figure 8.4. Note that for all but one value of x, (4) associates two values of y. Any vertical line with equation $x = k$, where $k > -1$, will intersect the graph of (4) in two places. Hence, this equation and, more generally, equations of the form (3), do not define functions.

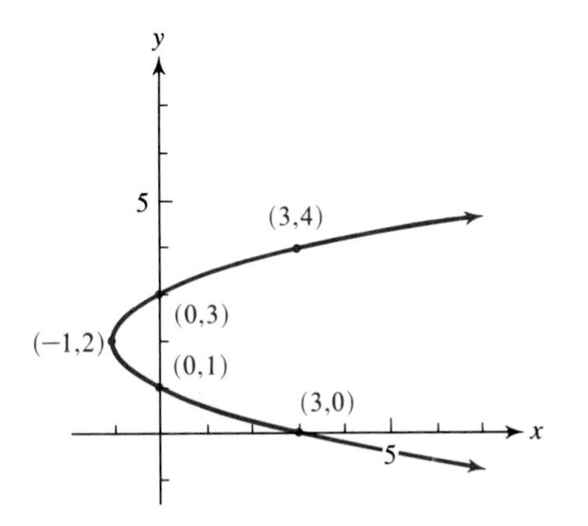

Figure 8.4

Observe that the graph of equation (1), the parabola in Figure 8.3, opens upward and that the graph of (4), the parabola in Figure 8.4, opens to the right. As you graph the equations in Exercise 8.1 you will be able to observe that the graphs of all equations of the form (2) open upward if $a > 0$ and downward if $a < 0$, and that graphs of all equations of the form (3) open to the right if $a > 0$ and to the left if $a < 0$.

As in the case of the graph of a linear equation in Section 7.2, the ordinate of a point on the graph of a quadratic equation is the length of the line segment drawn from the point on the curve perpendicular to the x axis. Graphing the equation

$$f(x) = -x^2 + 16$$

by first obtaining the ordered pairs $(-5, -9)$, $(-4, 0)$, $(-3, 7)$, $(-2, 12)$, $(-1, 15)$, $(0, 16)$, $(1, 15)$, $(2, 12)$, $(3, 7)$, $(4, 0)$, and $(5, -9)$, we have the parabola shown in Figure 8.5. Five representative ordinates are shown.

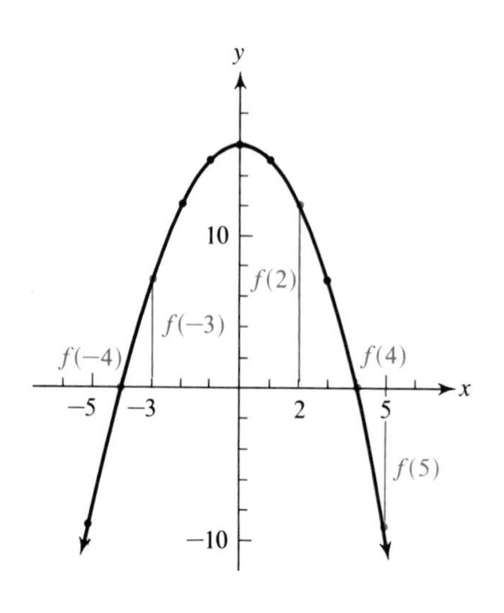

Figure 8.5

EXERCISE 8.1

A

a. *Find the solutions of the equation using integral values for x where* $-4 \leq x \leq 4$.
b. *Use the solutions to graph the equation.*

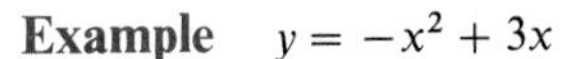

Example $y = -x^2 + 3x$

Solution **a.** $(-4, -28)$
$(-3, -18)$
$(-2, -10)$
$(-1, -4)$
$(0, 0)$
$(1, 2)$
$(2, 2)$
$(3, 0)$
$(4, -4)$

b.

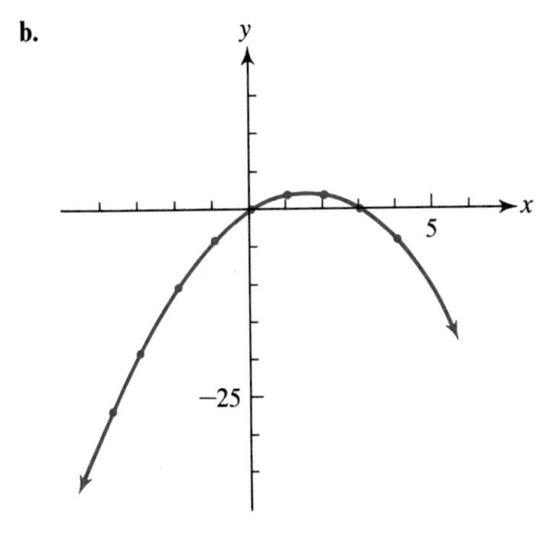

1. $y = x^2 + 1$ **2.** $y = x^2 + 4$ **3.** $f(x) = x^2 - 3$

4. $f(x) = x^2 - 5$ **5.** $y = -x^2 + 4$ **6.** $y = -x^2 + 5$

7. $y = 3x^2 + x$ **8.** $y = 5x^2 - x$ **9.** $y = x^2 + 2x + 1$

10. $y = x^2 - 2x + 1$ **11.** $f(x) = -2x^2 + x - 3$ **12.** $f(x) = -x^2 + 2x - 1$

13. $f(x) = -x^2 + x + 1$ **14.** $f(x) = -3x^2 + x - 2$

a. *Find the solutions of the equation using integral values for* y *where* $-4 \leq y \leq 4$.
b. *Use the solutions to graph the equation.*

15. $x = y^2$ **16.** $x = y^2 - 4$ **17.** $x = -4y^2 + 4$

18. $x = -2y^2 - 6$ **19.** $x = y^2 + 3y + 2$ **20.** $x = -y^2 + y + 2$

B

21. Graph the equation $f(x) = x^2 + 1$. Represent $f(0)$ and $f(4)$ by drawing line segments from $(0, 0)$ to $(0, f(0))$ and from $(4, 0)$ to $(4, f(4))$.

22. Graph the equation $f(x) = x^2 - 1$. Represent $f(-3)$ and $f(2)$ by drawing line segments from $(-3, 0)$ to $(-3, f(-3))$ and from $(2, 0)$ to $(2, f(2))$.

23. Graph $x = y^2$ for $y > 0$.

24. Graph $x = y^2 - 4$ for $y > 0$.

25. Graph $\{(x, y) | x = -y^2, \quad y > 0\} \cup \{(x, y) | x = y^2, \quad y \geq 0\}$.

26. Graph $\{(x, y) | x = -y^2 + 4, \quad x \leq 0\} \cup \{(x, y) | x = y^2 - 4, \quad x \geq 0\}$.

27. Graph the equation $y^2 = 4$ in $R \times R$. [*Hint:* Consider the equation $0x + y^2 = 4$.]

28. Does the equation $y^2 = 4$ define a function? Why or why not?

29. Graph each of the following equations on a separate coordinate system:

a. $y = \sqrt{16 - x^2}$ **b.** $y = -\sqrt{16 - x^2}$ **c.** $y = \pm\sqrt{16 - x^2}$

Which of these equations does not define a function? Why not?

30. If for each x, $f(x) = f(-x)$, the graph of f is said to be **symmetric** with respect to the y axis. Which of the following have graphs that are symmetric with respect to the y axis?

a. $f(x) = x$ **b.** $f(x) = x^2$ **c.** $f(x) = \sqrt{x^2 + 1}$ **d.** $f(x) = |x|$

8.2

SKETCHING PARABOLAS

When graphing a quadratic equation in two variables it is desirable to select first components for the ordered pairs that ensure that the more significant parts of the graph are displayed. For a parabola, these parts include the intercepts, if they exist, and the maximum or minimum (highest or lowest) point on the curve.

In locating the graphs of the x and y intercepts of the function defined by

$$f(x) = ax^2 + bx + c,$$

we identify the values of $f(x)$ for which $x = 0$ and the values of x for which $f(x) = 0$. For example, the y intercept of

$$y = -2x^2 - 5x + 3 \tag{1}$$

can be found by inspection to be 3 by assigning the value 0 to x.

The x intercepts of (1) are the solutions of

$$-2x^2 - 5x + 3 = 0,$$

which can be easily found by first writing the equation equivalently as

$$-1(x + 3)(2x - 1) = 0;$$

then, by inspection, the solutions, and therefore the x intercepts of the graph of (1), are -3 and $\frac{1}{2}$.

For any function f, values of x for which $f(x) = 0$ are called **zeros** of the function. Thus, we have three different names for a single idea:

1. The elements of the solution set of the equation

$$ax^2 + bx + c = 0.$$

2. The zeros of the function defined by

$$y = ax^2 + bx + c.$$

3. The x intercepts of the graph of

$$y = ax^2 + bx + c.$$

Recall from Section 6.4 that a quadratic equation in one variable may have no real solution, one real solution, or two real solutions. If the equation has no real solution, we find that the graph of the related quadratic equation in two variables does not touch the x axis; if there is one solution, the graph is tangent to the x axis; if there are two real solutions, the graph crosses the x axis at two distinct points, as shown in Figure 8.6.

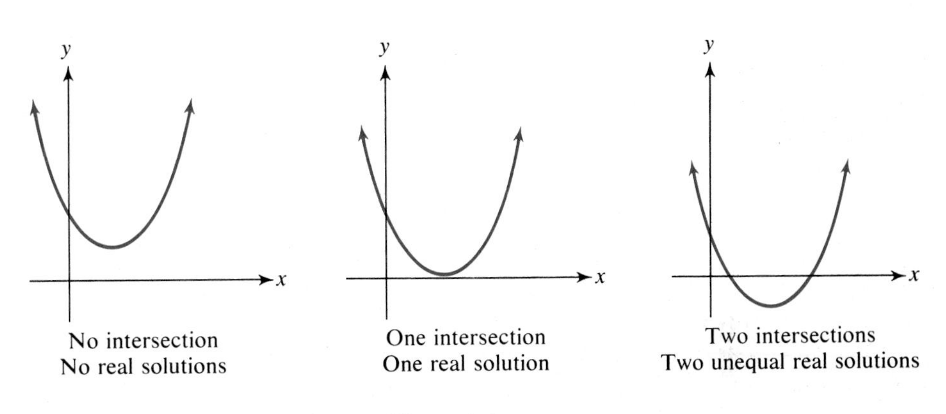

Figure 8.6

We can obtain the coordinates of the lowest (or the highest) point on a parabola by the procedure of completing the square which was discussed in Section 6.2. For example, (1) can be rewritten as

(2) $$y = -2\left(x^2 + \frac{5}{2}x \qquad \right) + 3$$

and then as

$$y = -2\left(x^2 + \frac{5}{2}x + \frac{25}{16}\right) + 3 + 2\left(\frac{25}{16}\right),$$

where $2(^{25}/_{16})$ was both subtracted from and added to the right-hand member. Since

$$x^2 + \frac{5}{2}x + \frac{25}{16} = \left(x + \frac{5}{4}\right)^2$$

and

$$3 + 2\left(\frac{25}{16}\right) = \frac{49}{8},$$

(2) is equivalent to

$$y = -2\left(x + \frac{5}{4}\right)^2 + \frac{49}{8}.$$

For $x = -^5/_4$, y has a maximum value $^{49}/_8$. Therefore, the high point corresponds to $(-^5/_4, {}^{49}/_8)$. The graphs of the intercepts and high point are shown in part a of Figure 8.7. These points are sufficient to complete the graph shown in part b.

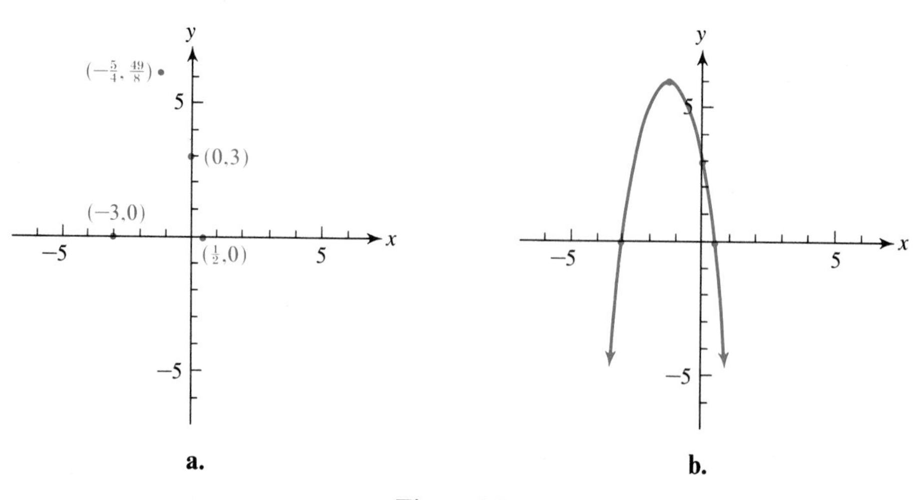

Figure 8.7

To sketch parabolas quickly, recall from Section 8.1 that for $a < 0$, the graph of $y = ax^2 + bx + c$ has the form shown in Figure 8.7. If $a > 0$, the graph has the form of any of the graphs shown in Figure 8.6.

Quadratic inequalities of the form

$$y < ax^2 + bx + c$$

or

$$y > ax^2 + bx + c,$$

which define quadratic relations, can be graphed in the same manner in which we graphed linear inequalities in two variables in Section 7.6. We first graph the equation having the same members and then shade an appropriate region as required. For instance, to graph

(3) $$y < x^2 + 2,$$

we first graph

(4) $$y = x^2 + 2,$$

and then shade the region below the curve. As we did in Section 7.6, we can determine which part of the plane should be shaded by substituting the coordinates of the origin $(0, 0)$ in the inequality and noting whether or not the result is true. Substituting $(0, 0)$ in (3), we obtain $0 < 0 + 2$, which is true, so the part of the plane including the origin is shaded. Since the graph of (4) is not part of the graph of (3), a broken curve is used in Figure 8.8.

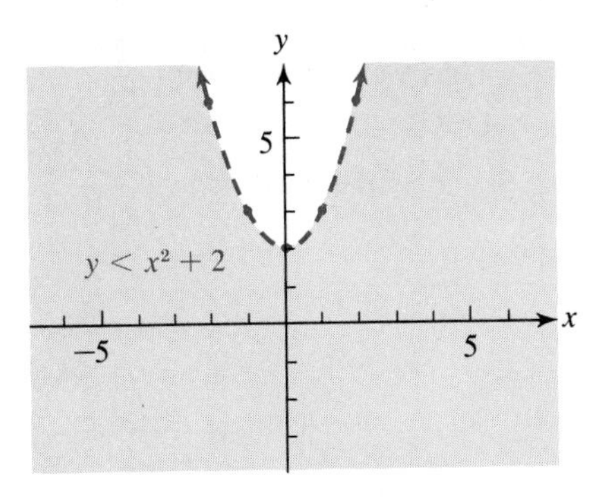

Figure 8.8

EXERCISE 8.2

A

Use analytic methods to find the x intercepts (if they exist), the y intercept, and the coordinates of the maximum or minimum point for the graph of each equation. Use this information to sketch each graph.

Example $y = x^2 - 7x + 6$

Solution The solution set of

$$x^2 - 7x + 6 = 0$$
$$(x - 1)(x - 6) = 0,$$

is $\{1, 6\}$; hence, the x intercepts are 1 and 6.

Substituting 0 for x in

$$y = x^2 - 7x + 6,$$

we have $y = 6$; thus, the y intercept is 6.

Completing the square in the right-hand member by adding and subtracting $^{49}/_4$, we obtain

$$y = \left(x^2 - 7x + \frac{49}{4}\right) - \frac{49}{4} + 6$$

$$y = \left(x - \frac{7}{2}\right)^2 - \frac{25}{4}.$$

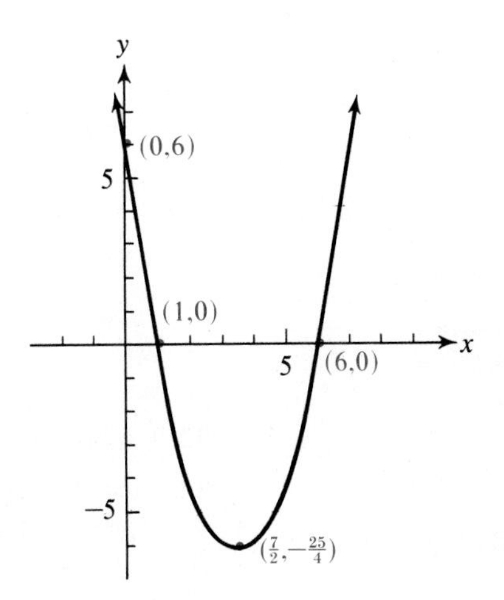

(*solution continued*)

For $x = \frac{7}{2}$, y has a minimum value $-\frac{25}{4}$.

Therefore, the minimum point is at $\left(\frac{7}{2}, -\frac{25}{4}\right)$.

1. $y = x^2 - 5x + 4$
2. $y = x^2 + x - 6$
3. $f(x) = x^2 - 4x$
4. $f(x) = x^2 + 6x$
5. $g(x) = x^2 + 4x - 5$
6. $g(x) = x^2 + 6x + 8$
7. $y = 2x^2 + 3x - 9$
8. $y = 3x^2 - 5x - 2$
9. $f(x) = -x^2 + 7x - 6$
10. $f(x) = -x^2 - 4x + 5$
11. $y = -2x^2 - 7x - 3$
12. $y = -3x^2 + 7x - 2$

Example Find two numbers whose sum is 18 and whose product is as large as possible.

Solution Let x represent the first number; and $18 - x$ represent the second number. Then their product, P, is

$$P = x(18 - x)$$
$$P = -x^2 + 18x$$
$$P = -1(x^2 - 18x + 81) + 81$$
$$P = -1(x - 9)^2 + 81,$$

and $x = 9$ yields the maximum value for P. Since $18 - x = 9$, the numbers are 9 and 9.

Check

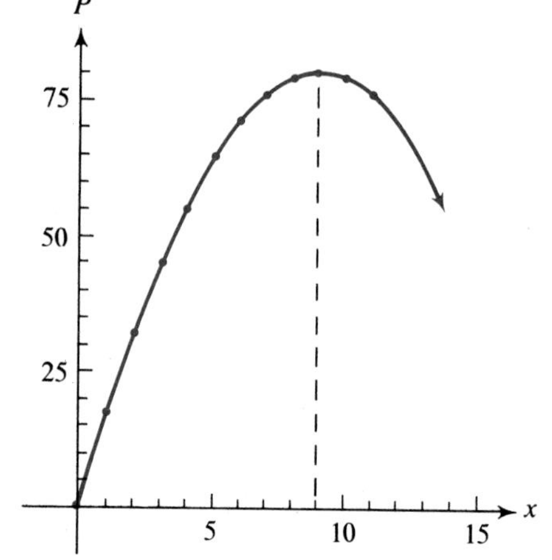

13. Find two numbers whose sum is 12 and whose product is a maximum.
14. Find the maximum area of a rectangle whose perimeter is 100 inches. [*Hint:* Let x represent the length, then $50 - x$ represents the width and $A = x(50 - x)$.]

Graph.

Example $\{(x, y) | y = x^2 - 1\} \cap \{(x, y) | y = 1 - x^2\}$

Solution The graph of the intersection of the given sets consists of the two points shown in red.

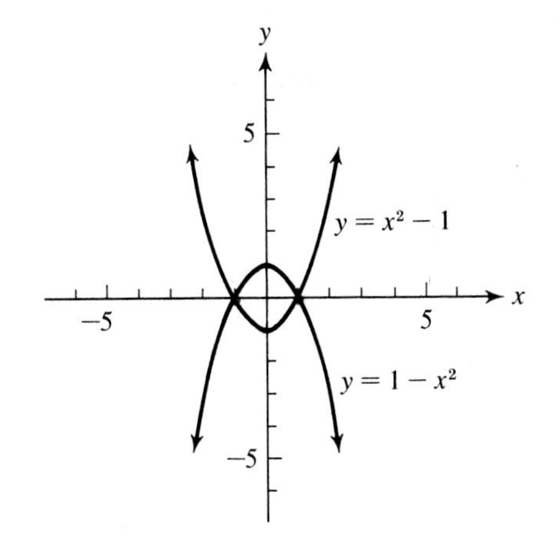

15. $\{(x, y) | y = -x^2\} \cap \{(x, y) | y = x^2 - 4\}$
16. $\{(x, y) | y = x^2\} \cap \{(x, y) | y = -x^2 + 4\}$
17. $\{(x, y) | y = x^2 - 2x + 1\} \cup \{(x, y) | y = -x^2 + 2x - 1\}$
18. $\{(x, y) | y = x^2 + 1\} \cup \{(x, y) | y = -x^2 - 1\}$

Graph.

Example $y \geq x^2 + 2x$

Solution Graph $y = x^2 + 2x$.

We can find the region to shade by selecting any ordered pair whose graph is not on the curve and determining whether or not the ordered pair is a solution of the inequality. Arbitrarily selecting $(0, 6)$ and substituting in $y \geq x^2 + 2x$, we obtain

$$(6) \geq (0)^2 + 2(0),$$

which is a true statement. Hence, we shade the region above the graph of $y = x^2 + 2x$ because it includes the graph of $(0, 6)$.

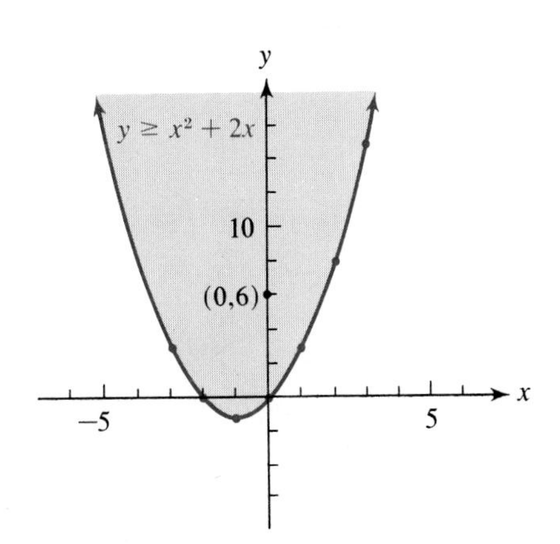

19. $y \geq x^2 - 6x + 8$
20. $y \leq x^2 - 6x + 8$
21. $y < x^2 + x - 6$
22. $y > x^2 + x - 6$
23. $y \leq x^2 + 3x + 2$
24. $y \geq x^2 + 3x + 2$

B

25. $\{(x, y) | y \leq 1 - x^2\} \cap \{(x, y) | y \geq -1\}$
26. $\{(x, y) | y \leq 1 - x^2\} \cup \{(x, y) | y \geq -1\}$
27. $\{(x, y) | y \leq 4 - x^2\} \cup \{(x, y) | y \geq x^2 - 4\}$
28. $\{(x, y) | y \leq 4 - x^2\} \cap \{(x, y) | y \geq x^2 - 4\}$

29. Sketch the family of four curves $y = kx^2$ ($k = 1, 2, 3, 4$) on a single set of axes.

30. Sketch the family of four curves $y = kx^2$ ($k = -1, -2, -3, -4$) on a single set of axes.

31. Sketch the family of four curves $y = x^2 + k$ ($k = -2, 0, 2, 4$) on a single set of axes.

32. Sketch the family of four curves $y = x^2 + kx$ ($k = -2, 0, 2, 4$) on a single set of axes.

33. The equation $d = 32t - 8t^2$ relates the distance d (feet) above the ground obtained in time t (seconds), by an object thrown vertically upward. Sketch the graph of the equation for $0 \leq t \leq 4$ and estimate the time it will take the object to reach its greatest height.

34. In Problem 33, how long is the object in the air?

35. Sketch the graph of the quadratic function $\{(x, f(x))\}$ such that $f(x) > 0$ for all x, where $1 < x < 3$, $f(1) = f(3) = 0$, and $f(0) = -3$.

36. Sketch the graph of the quadratic function $\{(x, g(x))\}$ such that $g(x) < 0$ for all x, where $-1 < x < 5$, $g(-1) = g(5) = 0$, and $g(3) = -3$.

8.3

CIRCLES, ELLIPSES, AND HYPERBOLAS

In addition to $y = ax^2 + bx + c$ and $x = ay^2 + by + c$, there are three other types of quadratic equations in two variables whose graphs are of particular interest. We shall discuss each of them separately. First, consider the equation

$$x^2 + y^2 = 25. \tag{1}$$

Solving this equation explicitly for y, we have

$$y = \pm\sqrt{25 - x^2}. \tag{2}$$

Assigning values to x, we find the following ordered pairs in the solution set:

$$\begin{array}{llllll} (-5, 0), & (-4, 3), & (-3, 4), & (0, 5), & (3, 4), & (4, 3), \\ (5, 0), & (-4, -3), & (-3, -4), & (0, -5), & (3, -4), & (4, -3). \end{array}$$

Plotting these points on the plane, we have the graph shown in part a of Figure 8.9. Connecting these points with a smooth curve, we have the graph shown in part b of Figure 8.9. This graph is a circle with radius 5 and center at the origin.

Since the number 25 in the right-hand member of (1) clearly determines the length of the radius of the circle, we can generalize and observe that any equation of the form

$$x^2 + y^2 = r^2$$

graphs into a circle with radius r and center at the origin. (See Problem 29, Exercise 8.3 for a more general approach.)

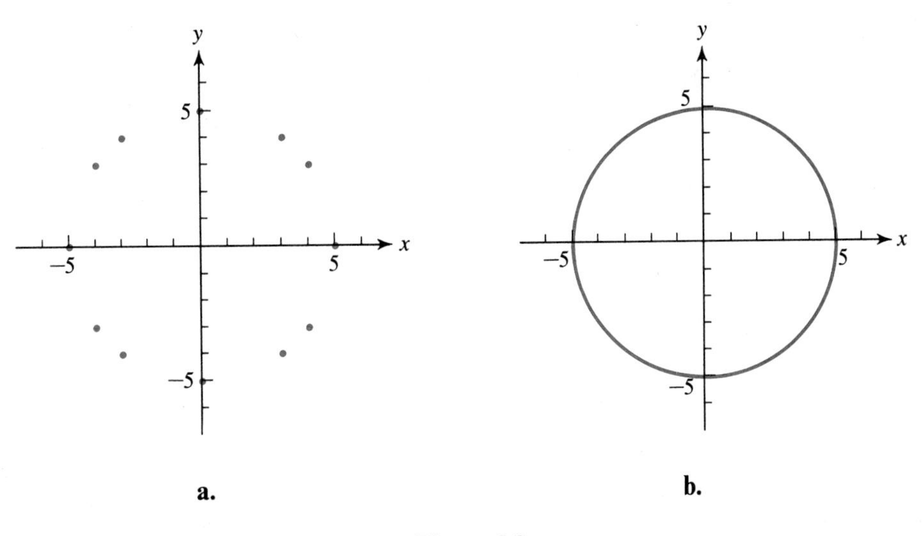

Figure 8.9

Note that in the preceding example it is not necessary to assign any values to x such that $|x| > 5$, because y is imaginary for these values. Since, except for -5 and 5, each permissible value for x is associated with two values for y—one positive and one negative—equations (1) and (2) do not define functions. We could represent relationship (2) by two equations,

(3a) $$y = f(x) = \sqrt{25 - x^2}$$

and

(3b) $$y = g(x) = -\sqrt{25 - x^2},$$

whose graphs would appear as in Figure 8.10 on page 248. These equations do define functions. In both cases, the domain of the function in $R \times R$ is

$$\{x \mid |x| \le 5\},$$

while the ranges differ. By inspecting the graph, we observe that the ranges of the functions defined by (3a) and (3b) are, respectively,

$$\{y \mid 0 \le y \le 5\} \quad \text{and} \quad \{y \mid -5 \le y \le 0\}.$$

The second quadratic equation in two variables that is of special interest is typified by

(4) $$4x^2 + 9y^2 = 36.$$

We obtain solutions of (4) by first solving explicitly for y,

(5) $$y = \pm\frac{2}{3}\sqrt{9 - x^2},$$

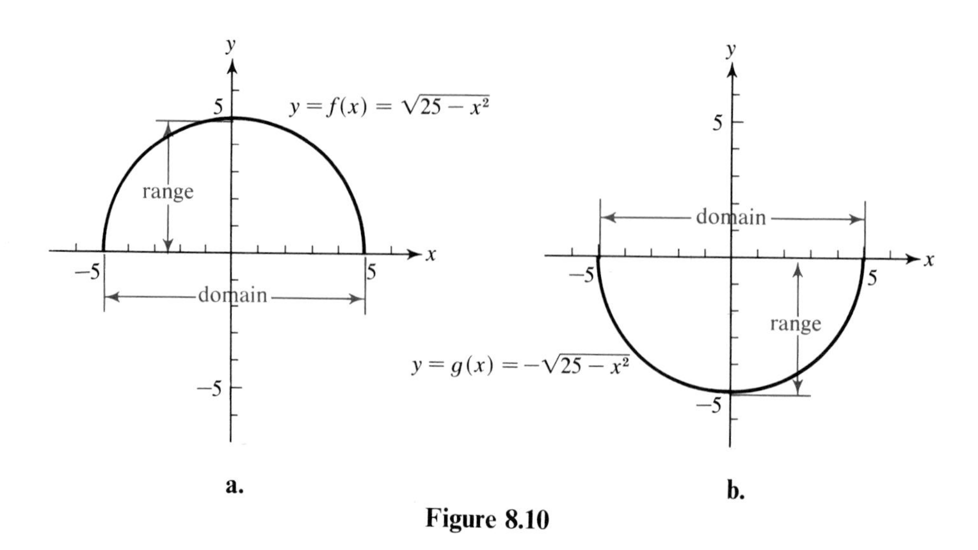

Figure 8.10

and assigning to x some values $-3 \le x \le 3$. We obtain, for example,

$$(-3, 0),\quad \left(-2, \frac{2}{3}\sqrt{5}\right),\quad \left(-1, \frac{4}{3}\sqrt{2}\right),\quad (0, 2),\quad \left(1, \frac{4}{3}\sqrt{2}\right),\quad \left(2, \frac{2}{3}\sqrt{5}\right),$$

$$(3, 0),\quad \left(-2, -\frac{2}{3}\sqrt{5}\right),\ \left(-1, -\frac{4}{3}\sqrt{2}\right),\ (0, -2),\ \left(1, -\frac{4}{3}\sqrt{2}\right),\ \left(2, -\frac{2}{3}\sqrt{5}\right).$$

Locating the corresponding points on the plane (decimal approximations for irrational numbers can be obtained from the table on page 390) and connecting them with a smooth curve, we have the graph shown in Figure 8.11. This curve is called an **ellipse**.

The third type of quadratic equation is exemplified by

$$x^2 - y^2 = 9.$$

Solving this equation for y, we obtain

$$y = \pm\sqrt{x^2 - 9}.$$

We note that the domain in $R \times R$ is $\{x \mid x \ge 3 \text{ or } x \le -3\}$, because members of the interval $-3 < x < 3$ yield imaginary values for y. We obtain, as a part of the solution set, the ordered pairs

$$(-5, 4),\quad (-4, \sqrt{7}),\quad (-3, 0),\quad (4, \sqrt{7}),\quad (5, 4),$$
$$(-5, -4),\quad (-4, -\sqrt{7}),\quad (3, 0),\quad (4, -\sqrt{7}),\quad (5, -4).$$

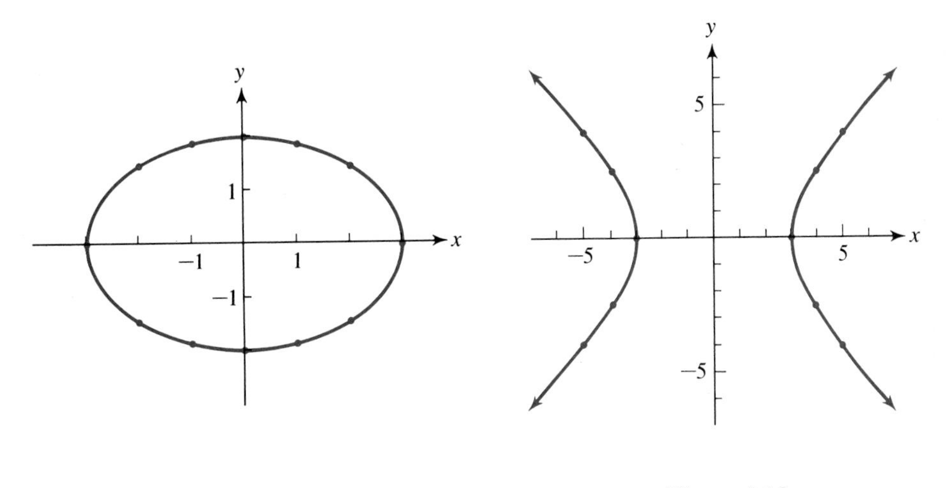

Figure 8.11 **Figure 8.12**

If we plot the corresponding points and connect them with a smooth curve, we obtain Figure 8.12. This curve is called a **hyperbola**.

The graphs of the equations dealt with in this section, together with the parabola of Sections 8.1 and 8.2, are called **conic sections**, or **conics**, because such curves are intersections of a plane and a cone, as shown in Figure 8.13.

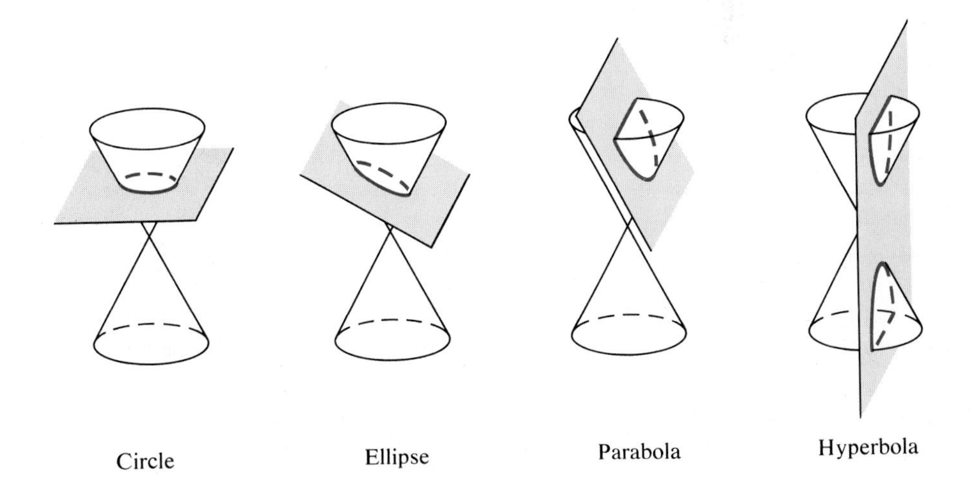

Figure 8.13

EXERCISE 8.3

A

a. Rewrite each equation with y as the left-hand member.
b. State the domain of the relations defined by the equation in $R \times R$.

Example $4x^2 + y^2 = 36$

Solution $y^2 = 36 - 4x^2$

$y^2 = 4(9 - x^2)$

a. $y = \pm 2\sqrt{9 - x^2}$ **b.** domain: $\{x \mid -3 \le x \le 3\}$

1. $x^2 + y^2 = 4$ **2.** $x^2 + y^2 = 9$ **3.** $9x^2 + y^2 = 36$ **4.** $4x^2 + y^2 = 4$
5. $x^2 + 4y^2 = 16$ **6.** $x^2 + 9y^2 = 4$ **7.** $x^2 - y^2 = 1$ **8.** $4x^2 - y^2 = 1$
9. $y^2 - x^2 = 9$ **10.** $4y^2 - 9x^2 = 36$ **11.** $2x^2 + 3y^2 = 24$ **12.** $4x^2 + 3y^2 = 12$

Graph the equation of the problem cited.

Example $4x^2 + y^2 = 36$

Solution $y^2 = 36 - 4x^2$

$y = \pm 2\sqrt{9 - x^2}$

Some solutions are

$(3, 0)$, $(-3, 0)$,
$(2, \pm 2\sqrt{5})$, $(-2, \pm 2\sqrt{5})$,
$(1, \pm 4\sqrt{2})$, $(-1, \pm 4\sqrt{2})$,
$(0, \pm 6)$.

13. Problem 1 **14.** Problem 2 **15.** Problem 3 **16.** Problem 4
17. Problem 5 **18.** Problem 6 **19.** Problem 7 **20.** Problem 8
21. Problem 9 **22.** Problem 10 **23.** Problem 11 **24.** Problem 12

B

25. Graph $xy = 4$ and $xy = 12$ on the same set of axes. (These curves are called **rectangular hyperbolas.**)

26. Graph $xy = -4$ and $xy = -12$ on the same set of axes.

27. Graph $4x^2 - y^2 = 0$. Generalize from the result and discuss the graph of any equation of the form $ax^2 - by^2 = c$, where $a, b > 0$ and $c = 0$.

28. Graph $4x^2 + y^2 = 0$. Generalize from the result and discuss the graph of any equation of the form $ax^2 + by^2 = c$, where $a, b > 0$ and $c = 0$.

29. Use the distance formula to show that the graph of $\{(x, y) | x^2 + y^2 = r^2\}$ is the set of all points located a distance r from the origin.

30. Use the distance formula to show that the graph of the relation

$$\{(x, y) | (x - h)^2 + (y - k)^2 = r^2\}$$

is the set of all points located a distance r from the graph of (h, k).

8.4

SKETCHING GRAPHS OF CONIC SECTIONS OTHER THAN PARABOLAS

In Section 8.2, we stressed the use of certain properties to help sketch parabolas. These properties include the x and y intercepts and the coordinates of the highest or lowest points on the graph. We can also use intercepts, together with the form of the equation, to help sketch other conic sections. To do this, we first observe that the ideas developed in Section 8.3 can be summarized and extended as given below.

A quadratic equation of the form

$$ax^2 + by^2 = c \qquad \text{(both } a \text{ and } b \text{ not equal to 0)}$$

has a graph in $R \times R$ with center at the origin that is one of the following:

1. A circle if $a = b$ and a, b, and c have like signs. For example,

$$x^2 + y^2 = 9$$

or

$$4x^2 + 4y^2 = 12.$$

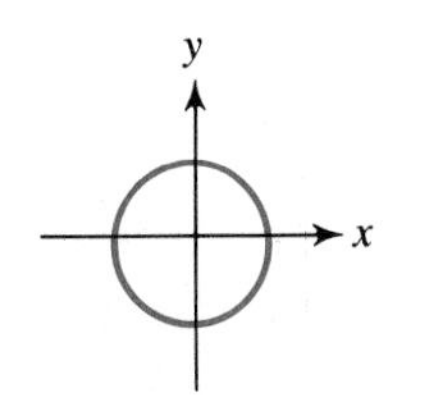

2. An ellipse if $a \neq b$ and a, b, and c have like signs. For example,

$$4x^2 + 9y^2 = 36$$

or

$$x^2 + 8y^2 = 12.$$

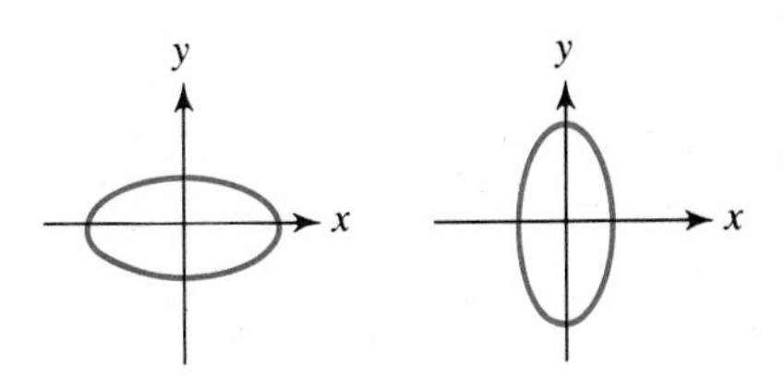

3. A hyperbola if a and b are opposite in sign and $c \neq 0$. For example,

$$2x^2 - 6y^2 = 9$$

or

$$4y^2 - x^2 = 2.$$

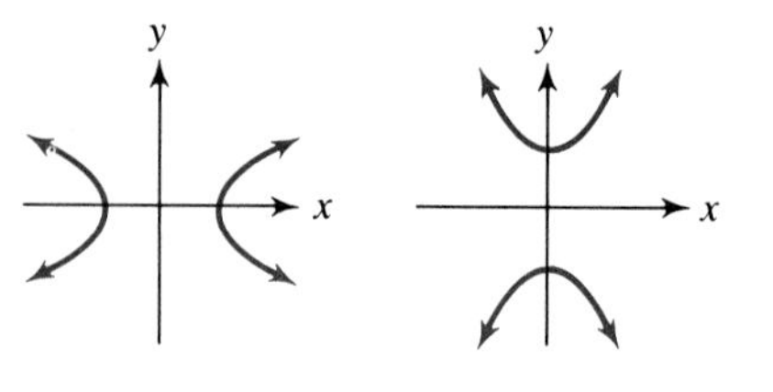

4. Two distinct lines through the origin if a and b are opposite in sign and $c = 0$ (see Problem 27, Exercise 8.3). For example,

$$4x^2 - y^2 = 0$$

or

$$2y^2 - 7x^2 = 0.$$

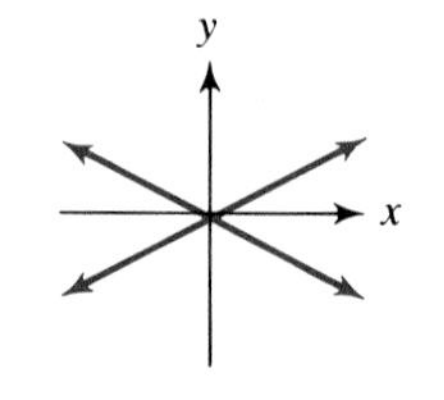

5. A point if a and b have the same sign and $c = 0$ (see Problem 28, Exercise 8.3). For example,

$$x^2 + 4y^2 = 0$$

or

$$-2x^2 - 5y^2 = 0.$$

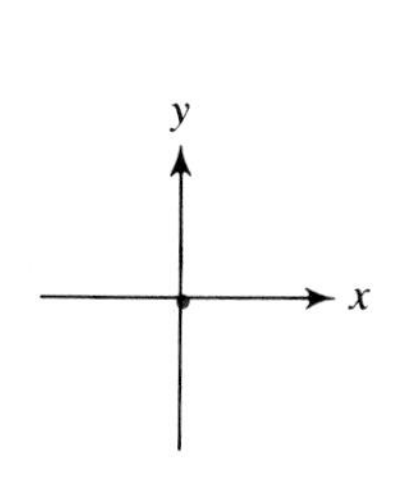

6. If a and b are both greater than 0 and $c < 0$ or if a and b are both less than or equal to 0 and $c > 0$, there is no graph in $R \times R$. For example,

$$x^2 + 3y^2 = -4$$

or

$$-2x^2 - 3y^2 = 6$$

Although cases 4 and 5 were not mentioned in the discussion of conic sections, they are frequently referred to as **degenerate conic sections.**

When you have recognized the general form of the curve, the graph of a few points should suffice to sketch the graph. The intercepts, for instance, are always easy to locate. Consider the equation

$$x^2 + 4y^2 = 8. \tag{1}$$

By comparing this equation with case 2 above, we note immediately that its graph is an ellipse. If $y = 0$, then $x = \pm\sqrt{8}$, and if $x = 0$, then $y = \pm\sqrt{2}$. We can then sketch the graph of (1) as in Figure 8.14.

As another example, consider the equation

$$x^2 - y^2 = 3. \tag{2}$$

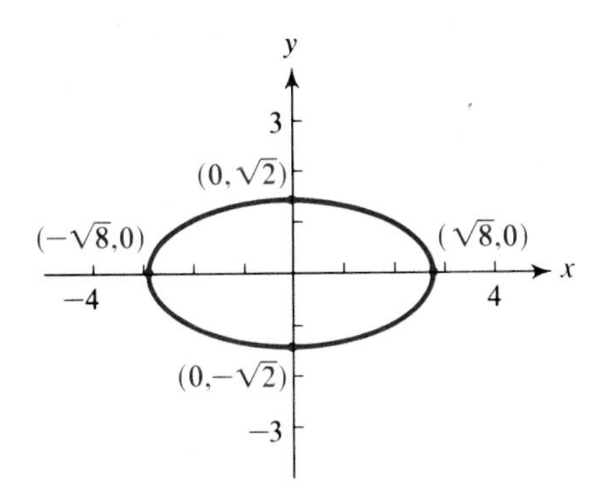

Figure 8.14

By comparing this equation with case 3 above, we see that its graph is a hyperbola. If $y = 0$, then $x = \pm\sqrt{3}$, and if $x = 0$, then y is imaginary (the graph will not cross the y axis). By assigning a few other arbitrary values to one of the variables—say, (4,) and (−4,) to x—we can find the additional ordered pairs

$$(4, \sqrt{13}), \quad (4, -\sqrt{13}), \quad (-4, \sqrt{13}), \quad (-4, -\sqrt{13}),$$

which satisfy (2). The graph can then be sketched as shown in Figure 8.15. The dashed lines shown in the figure are called **asymptotes** of the graph, and they comprise the graph of $x^2 - y^2 = 0$. While we shall not discuss the notion in detail here, it is true that, in general, the graph of $ax^2 - by^2 = c$ will "approach" the two straight lines in the graph of $ax^2 - by^2 = 0$ for each $a, b, c > 0$. (See Problems 27 and 28, Exercise 8.4.) Furthermore, any equation of the form

$$ax^2 - by^2 = c \qquad (a, b, c > 0)$$

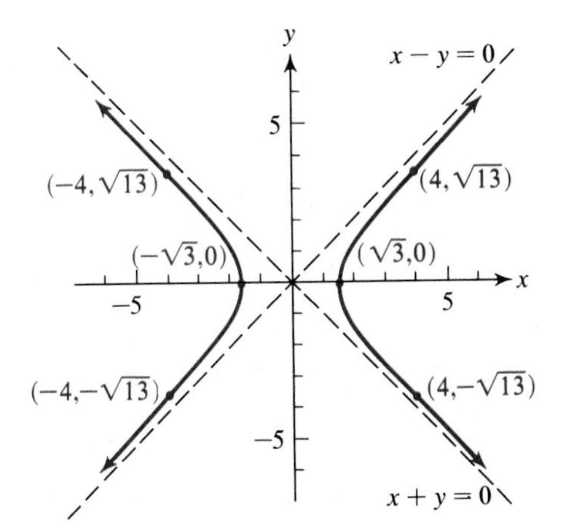

Figure 8.15

graphs into a hyperbola with center at the origin and x intercepts $\pm\sqrt{c/a}$. Any equation of the form

$$by^2 - ax^2 = c \qquad (a, b, c > 0)$$

graphs into a hyperbola with center at the origin and y intercepts $\pm\sqrt{c/b}$.

EXERCISE 8.4

A

Name and sketch the graph of each equation.

Example $4x^2 = 36 - 9y^2$

Solution Rewrite in standard form.

$$4x^2 + 9y^2 = 36$$

By inspection, the graph is an ellipse.

The x intercepts are 3 and -3;
the y intercepts are 2 and -2.

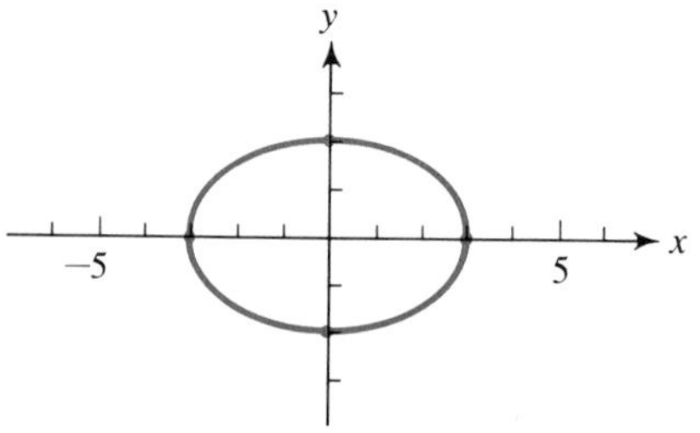

1. $x^2 + y^2 = 49$
2. $x^2 + y^2 = 64$
3. $4x^2 + 25y^2 = 100$
4. $x^2 + 2y^2 = 8$
5. $x^2 - y^2 = 9$
6. $x^2 - 2y^2 = 8$
7. $4x^2 - 4y^2 = 0$
8. $x^2 - 9y^2 = 0$
9. $x^2 + 4y^2 = 0$
10. $3x^2 + 2y^2 = 0$
11. $4x^2 = 1 - 4y^2$
12. $9y^2 = 2 - 9x^2$
13. $3x^2 = 12 - 4y^2$
14. $4x^2 = 12 - 3y^2$
15. $y^2 = 16 + 4x^2$
16. $x^2 = 25 + 5y^2$
17. $y^2 + 4 = -2x^2$
18. $4x^2 + 6 = -y^2$

19. Graph the set of points satisfying *both* $x^2 + y^2 = 25$ *and* $y = x^2$.

20. Graph the set of points satisfying *either* $x^2 + y^2 = 16$ *or* $y = x^2 - 4$.

21. Graph $\{(x, y) | y = x + 2\} \cap \{(x, y) | 4x^2 + y^2 = 36\}$.

22. Graph $\{(x, y) | y = x + 2\} \cup \{(x, y) | 4x^2 + y^2 = 36\}$.

B

23. Graph $\{(x, y) | x^2 + 4y^2 \le 9\}$.

24. Graph $\{(x, y) | x^2 - y^2 \ge 1\}$.

25. Graph $\{(x, y) | 4x^2 + 9y^2 \ge 36\}$.

26. Graph $\{(x, y) | 9x^2 - 4y^2 \le 36\}$.

27. Show that the ordinates to the graph of $ax^2 - by^2 = c$ for any value of x are given by

$$y = \pm\sqrt{\frac{a}{b}}\,x\left(\sqrt{1 - \frac{c}{ax^2}}\right).$$

28. Explain why, for large values of x, the value of the expression

$$\sqrt{1 - \frac{c}{ax^2}}$$

approaches 1, and hence, explain why

$$y = \pm\sqrt{\frac{a}{b}}\,x$$

are equations for the asymptotes to the graph of $ax^2 - by^2 = c$.

8.5

VARIATION AS A FUNCTIONAL RELATIONSHIP

There are two types of widely used functional relationships to which custom has assigned special names. First, any function defined by the equation

(1) $\quad y = kx \quad$ (**k a positive constant**)

is an example of **direct variation**. The variable y is said to **vary directly** as the variable x. Another example of direct variation is

(1a) $\quad y = kx^2 \quad$ (k a positive constant),

where we say that y varies directly as the square of x and, in general,

(1b) $\quad y = kx^n \quad$ (k a positive constant and $n > 0$)

asserts that y varies directly as the nth power of x. We find examples of such variation in the relationships existing between the length of the radius of a circle and the circumference and area. Thus,

(2) $$C = 2\pi r$$

asserts that the circumference of a circle varies directly as the length of the radius, while

(3) $$A = \pi r^2$$

expresses the fact that the area of a circle varies directly as the square of the length of the radius. Since for each r, (2) and (3) associate only one value of C or A, both of these equations define functions—(2) a linear function and (3) a quadratic function.

The second important type of function is defined by the equation

(4) $\quad xy = k \quad$ (k a positive constant),

where x and y are said to **vary inversely**. When (4) is written in the form

(5) $$y = \frac{k}{x},$$

y is said to vary inversely as x. Similarly, if

(5a)
$$y = \frac{k}{x^2},$$

y is said to vary inversely as the square of x, etc. As an example of inverse variation, consider the set of rectangles with area 24 square units. Since the area of a rectangle is given by $lw = A$, we have

$$lw = 24,$$

and the length and width of the rectangle can be seen to vary inversely. Equation (5) or (5a) associates only one y with each x ($x \neq 0$). Hence, an inverse variation defines a function with domain

$$\{x \mid x \neq 0\}.$$

The names *direct* and *inverse* as applied to variation arise from the facts that in direct variation an assignment of increasing absolute values to x results in increasing absolute values of y, whereas in inverse variation an assignment of increasing absolute values to x results in decreasing absolute values of y.

The constant involved in equations describing direct or inverse variation is called the **constant of variation**. If we know that one variable varies directly or inversely as another, and if we have one set of associated values for the variables, we can find the constant of variation involved. For example, suppose we know that y varies directly as x^2, and that $y = 4$ when $x = 7$. The fact that y varies directly as x^2 tells us that

(6)
$$y = kx^2.$$

The fact that $y = 4$ when $x = 7$ tells us that (7, 4) is a solution of (6), and hence,

$$4 = k(7)^2 = k(49),$$

from which

$$k = \frac{4}{49}.$$

Substituting $4/49$ for k in (6), we obtain

$$y = \frac{4}{49}x^2,$$

the equation that specifically expresses the direct variation.

In the event that one variable varies as the product of two or more other variables, we refer to the relationship as **joint variation**. Thus, if y varies jointly as u, v, and w, we have

(7)
$$\boldsymbol{y = kuvw}.$$

Also, direct and inverse variation may take place concurrently. That is, y may vary directly as x and inversely as z, giving rise to the equation

$$y = k\frac{x}{z}.$$

It should be pointed out that the way in which the word "variation" is used here is technical, and when the ideas of direct, inverse, or joint variation are encountered, we should always think of equations of the form (1), (5), or (7). For instance, the equations

$$y = 2x + 1, \qquad y = \frac{1}{x} - 2, \qquad y = xz + 2$$

do not describe examples of variation within our meaning of the word.

An alternative expression—**proportional to**—is used to describe the relationships discussed in this section. To say that "y is directly proportional to x" or "y is inversely proportional to x" is another way of describing direct and inverse variation. The use of the word "proportion" arises from the fact that any two solutions (a, b) and (c, d) of an equation expressing a direct variation satisfy an equation of the form

$$\frac{a}{b} = \frac{c}{d},$$

which is commonly called a **proportion**. For example, consider the situation in which the volume of a gas varies directly with the absolute temperature and inversely with the pressure; this can be represented by the relationship

$$V = \frac{kT}{P}. \tag{8}$$

For any set of values T_1, P_1, and V_1,

$$k = \frac{V_1 P_1}{T_1}, \tag{8a}$$

and for any other set of values T_2, P_2, and V_2,

$$k = \frac{V_2 P_2}{T_2}. \tag{8b}$$

Equating the right-hand members of (8a) and (8b),

$$\frac{V_1 P_1}{T_1} = \frac{V_2 P_2}{T_2},$$

from which the value of any variable can be determined if the values of the other variables are known.

EXERCISE 8.5

A

Write an equation expressing the relationship between the variables, using k as the constant of variation.

Example At a constant temperature, the volume (V) of a gas varies inversely as the pressure (P).

Solution $V = \frac{k}{P}$

1. The distance (d) traveled by a car moving at a constant rate varies directly as the time (t).
2. The tension (T) on a spring varies directly as the distance (s) it is stretched.
3. The current (I) in an electrical circuit with constant voltage varies inversely as the resistance (R) of the circuit.
4. The time (t) required by a car to travel a fixed distance varies inversely as the rate (r) at which it travels.
5. The volume (V) of a rectangular box of fixed depth varies jointly as its length (l) and width (w).
6. The power (P) in an electric circuit varies jointly as the resistance (R) and the square of the current (I).

Find the constant of variation for each of the stated conditions.

Example V varies inversely as P, and $V = 100$ when $P = 30$

Solution Write an equation expressing the relationship between the variables.

$$V = \frac{k}{P}$$

Substitute the known values for the variables and solve for k.

$$100 = \frac{k}{30}$$

$$k = 3000$$

7. y varies directly as x, and $y = 6$ when $x = 2$
8. y varies directly as x, and $y = 2$ when $x = 5$
9. u varies inversely as the square of v, and $u = 2$ when $v = 10$

10. r varies inversely as the cube of t, and $r = 8$ when $t = 10$

11. z varies jointly as x and y, and $z = 8$ when $x = 2$ and $y = 2$

12. p varies jointly as q and r, and $p = 5$ when $q = 2$ and $r = 7$

13. z varies directly as the square of x and inversely as the cube of y, and $z = 4$ when $x = 3$ and $y = 2$

14. z varies directly as the cube of x and inversely as the square of y, and $z = 2$ when $x = 2$ and $y = 4$

15. z varies inversely as the sum of x^2 and y, and $z = 12$ when $x = 4$ and $y = 6$

16. z varies directly as the sum of x and y and inversely as their product, and $z = 8$ when $x = 3$ and $y = 4$

Solve.

Example If V varies directly as T and inversely as P, and $V = 40$ when $T = 300$ and $P = 30$, find V when $T = 324$ and $P = 24$.

Solution Write an equation expressing the relationship between the variables.

$$V = \frac{kT}{P} \tag{1}$$

Substitute the initially known values for V, T, and P. Solve for k.

$$40 = \frac{k300}{30}$$

$$4 = k$$

Rewrite (1) with k replaced by 4.

$$V = \frac{4T}{P}$$

Substitute the second set of values for T and P and solve for V.

$$V = \frac{4(324)}{24} = 54$$

17. If y varies directly as x^2 and $y = 9$ when $x = 3$, find y when $x = 4$.

18. If r varies directly as s and inversely as t and $r = 12$ when $s = 8$ and $t = 2$, find r when $s = 3$ and $t = 6$.

19. The distance a particle falls in a certain medium is directly proportional to the square of the length of time it falls. If the particle falls 16 feet in 2 seconds, how far will it fall in 10 seconds?

20. In Problem 19, how far will the body fall in 20 seconds?

21. The pressure exerted by a liquid at a given point varies directly as the depth of the point beneath the surface of the liquid. If a certain liquid exerts a pressure of 40 pounds per square foot at a depth of 10 feet, what would be the pressure at 40 feet?

22. The volume (V) of a gas varies directly as its temperature (T) and inversely as its pressure (P). A gas occupies 20 cubic feet at a temperature of 300°K (Kelvin) and a pressure of 30 pounds per square inch. What will the volume be if the temperature is raised to 360°K and the pressure is decreased to 20 pounds per square inch?

23. The maximum safe uniformly distributed load (L) for a horizontal beam varies jointly as its breadth (b) and square of the depth (d), and inversely as the length (l). An 8 foot beam with $b = 2$ and $d = 4$ will safely support a uniformly distributed load of up to 750 pounds. How many uniformly distributed pounds will an 8 foot beam support if $b = 2$ and $d = 6$?

24. The resistance (R) of a wire varies directly as the length (l) and inversely as the square of its diameter (d); 50 feet of wire of diameter 0.012 inch has a resistance of 10 ohms. What is the resistance of 50 feet of the same type of wire if the diameter is increased to 0.015 inch?

Represent the relationship of the problem cited as a proportion by eliminating the constant of variation, and then solve for the required variable.

Example If V varies directly as T and inversely as P, and $V = 40$ when $T = 300$ and $P = 30$, find V when $T = 324$ and $P = 24$.

Solution Write an equation expressing the relationship between the variables.

$$V = \frac{kT}{P}$$

Solve for k.

$$k = \frac{VP}{T}$$

Write a proportion relating the variables for two different sets of conditions.

$$\frac{V_1 P_1}{T_1} = \frac{V_2 P_2}{T_2}$$

Substitute the known values of the variables and solve for V_2.

$$\frac{(40)(30)}{300} = \frac{V_2(24)}{324},$$

$$V_2 = \frac{(324)(40)(30)}{300(24)} = 54$$

25. Problem 17 **26.** Problem 18 **27.** Problem 19 **28.** Problem 20

29. Problem 21 **30.** Problem 22 **31.** Problem 23 **32.** Problem 24

B

33. From the formula for the circumference of a circle, $C = \pi D$, show that the ratio of the circumference of two circles equals the ratio of their respective diameters.

34. From the formula for the area of a circle, $A = \pi r^2$, show that the ratio of the areas of two circles equals the ratio of the squares of their respective radii.

35. The intensity of light on a surface varies inversely as the square of its distance from a light source. What is the effect on the intensity of light on an object if the distance between the light source and the object is doubled?

36. The frequency of vibration of a guitar string varies directly as the square root of the tension and inversely as the length of the string. What is the effect on the frequency if the tension is increased fourfold and the length of the string is doubled?

37. Graph on the same set of axes the linear functions defined by $y = kx$, $x \geq 0$, when the constant $k = 1, 2, 3$. What effect does a variation in k have on the graph of $y = kx$?

38. Graph on the same set of axes the function defined by $xy = k$, for $k = 1, 2$, and $x > 0$. What effect does a variation in k have on the graph of $xy = k$?

39. Graph on the same set of axes the equations $y = kx$, $y = kx^2$, and $y = kx^3$, where $k = 2$ and $x \geq 0$. What effect does increasing the degree of the equation $y = kx^n$ have on the graph of the equation?

8.6

THE INVERSE OF A FUNCTION

Let us now consider an extension of the relation concept. If the components of each ordered pair in a given relation are interchanged, the resulting relation and the given relation are called **inverses** of each other, and each is said to be the inverse of the other. Thus,

$$\{(1,2), (3,4), (5,6)\} \qquad \text{and} \qquad \{(2,1), (4,3), (6,5)\}$$

are inverse relations.

The inverse of a relation F is denoted by F^{-1} (read "F inverse" or "the inverse of F"). It is evident from the definition of inverse relations that the domain and range of F^{-1} are the range and domain, respectively, of F. If $y = F(x)$ or $y < F(x)$ defines a relation F, then $x = F(y)$ or $x < F(y)$ defines the inverse of F. For example, the inverse of the relation defined by

$$y = 4x - 3 \tag{1}$$

is defined by

$$x = 4y - 3, \tag{2}$$

or, when y is expressed in terms of x, by

$$y = \frac{1}{4}(x + 3). \tag{2a}$$

Equations (2) and (2a) are equivalent. In general, the equation $x = F(y)$ is equivalent to $y = F^{-1}(x)$.

The graphs of inverse relations are related in an interesting and useful way. To see this, we first observe in Figure 8.16 that the graphs of the ordered pairs (a, b) and (b, a) are always located symmetrically with respect to the graph of $y = x$. Therefore, because for every ordered pair (a, b) in F, the ordered pair (b, a) is in F^{-1}, the graphs of $y = F^{-1}(x)$ and $y = F(x)$ are reflections of each other about the graph of $y = x$.

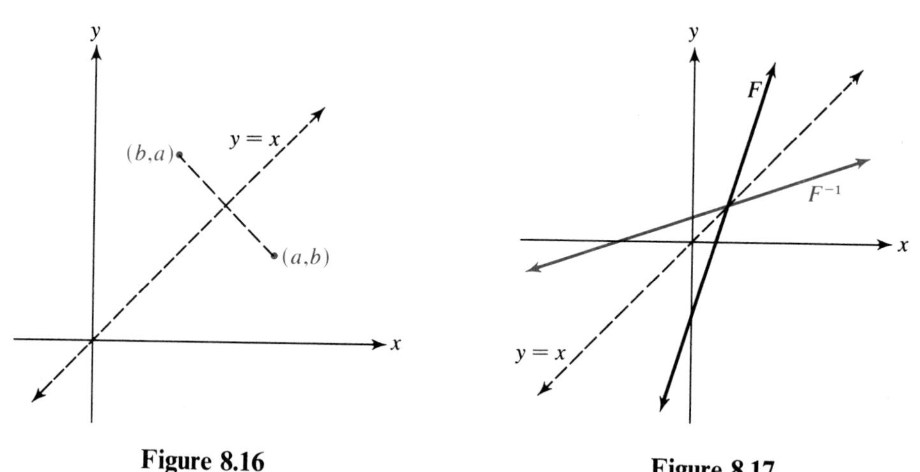

Figure 8.16

Figure 8.17

Figure 8.17 shows the graphs of

$$F = \{(x, y) \mid y = 4x - 3\}$$

and its inverse,

$$F^{-1} = \{(x, y) \mid x = 4y - 3\} = \left\{(x, y) \mid y = \frac{1}{4}(x + 3)\right\},$$

together with the graph of $y = x$.

Every relation has an inverse which may or may not be a function. For example, the inverse of the function

$$F = \{(1, 2), (2, 5), (3, 5)\}$$

is

$$F^{-1} = \{(2, 1), (5, 2), (5, 3)\},$$

which is not a function because two ordered pairs have the same first component. As another example, Figure 8.18 shows the graph of the function

$$F = \{(x, y) \mid y = x^2\},$$

together with the graph of its inverse,

$$F^{-1} = \{(x, y) \mid x = y^2\} = \{(x, y) \mid y = \pm\sqrt{x}\}.$$

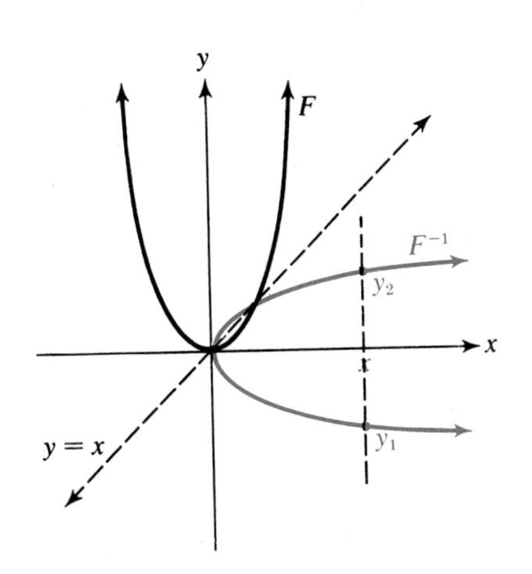

Figure 8.18

Since, for all but one value in its domain ($x = 0$), F^{-1} associates two different y's with each x, F^{-1} is not a function.

In order for a function F to have an inverse that is a function, not only must each element in the domain of F be associated with just one element in its range, but, also, each element in its range must be associated with just one element in its domain. Such a function F is called a **one-to-one function**. Then if F and F^{-1} are both functions, F associates the number a with the *unique* number b, and F^{-1} associates the number b with the *unique* number a. Therefore, it must be true that, for every x in the domain of F,

$$F^{-1}[F(x)] = x$$

(read "F inverse of F of x is equal to x") and, for every x in the domain of F^{-1},

$$F[F^{-1}(x)] = x$$

(read "F of F inverse of x is equal to x"). For example, if $F = \{(2, 5)\}$, then $F(2) = 5$, $F^{-1} = \{(5, 2)\}$, and $F^{-1}(5) = 2$. Note that

$$F^{-1}[F(2)] = F^{-1}(5) = 2$$

and

$$F[F^{-1}(5)] = F(2) = 5.$$

As another example, consider (1) and note that if F is the linear function defined by

$$F(x) = 4x - 3,$$

then F is a one-to-one function. Now F^{-1} is defined by

$$F^{-1}(x) = \frac{1}{4}(x + 3),$$

and we see that

$$F^{-1}[F(x)] = \frac{1}{4}[(4x - 3) + 3] = x$$

and

$$F[F^{-1}(x)] = 4\left[\frac{1}{4}(x + 3)\right] - 3 = x.$$

Whether or not a function is a one-to-one function can be readily determined from its graph. Recall (page 215) that any vertical line will intersect the graph of a function in at most one point. If the function F is one-to-one, any horizontal line will also intersect its graph in at most one point, as shown in part a of Figure 8.19. If it is not one-to-one, a horizontal line will intersect the graph at more than one point, as shown in part b of Figure 8.19.

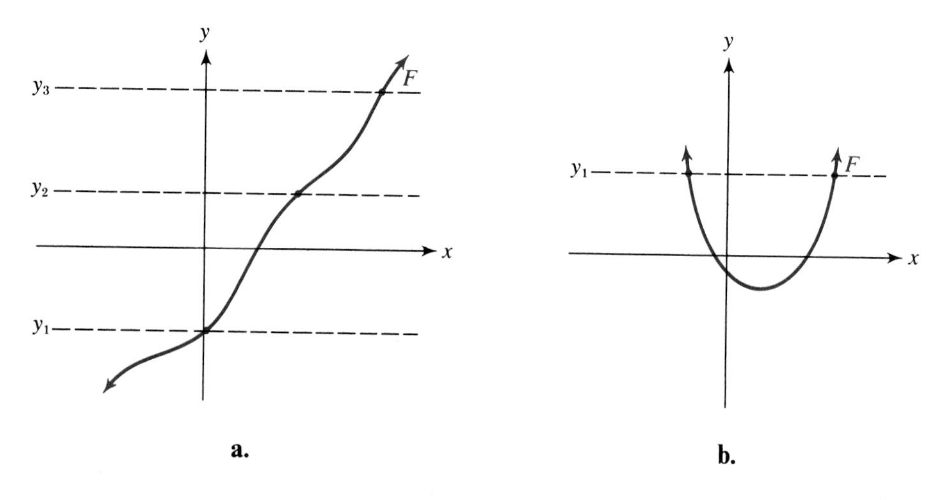

Figure 8.19

EXERCISE 8.6

A

Find the inverse of each function and state whether the inverse is also a function.

Examples a. $F = \{(2, 1), (4, 6)\}$

b. $Q = \{(-2, 1), (3, 4), (5, 4)\}$

Solutions a. Interchanging the components in each ordered pair, we obtain

$$F^{-1} = \{(1, 2), (6, 4)\},$$

which is a function because the ordered pairs do not contain the same first component.

b. Interchanging the components in each ordered pair, we obtain

$$Q^{-1} = \{(1, -2), (4, 3), (4, 5)\},$$

which is not a function because two ordered pairs contain the same first component, 4.

1. $F = \{(-2, -2), (2, 2)\}$

2. $F = \{(-5, 1), (5, 2)\}$

3. $Q = \{(1, 3), (2, 3), (3, 4)\}$

4. $Q = \{(2, 2), (3, 3), (4, 3)\}$

5. $G = \{(1, 1), (2, 2), (3, 3)\}$

6. $G = \{(-2, 0), (0, 0), (4, -2)\}$

In Problems 7–18, each equation defines a relation, F, in $R \times R$.
a. Write the equation defining F^{-1}.
b. Sketch the graphs of F and F^{-1} on the same set of axes.
c. State whether F^{-1} is a function.

Example $y = x^2 - 9$

Solution a. Interchange variables x and y. The equation defining F^{-1} is

$$x = y^2 - 9.$$

c. F^{-1} is not a function because for each $x > -9$, there are two values of y.

b.

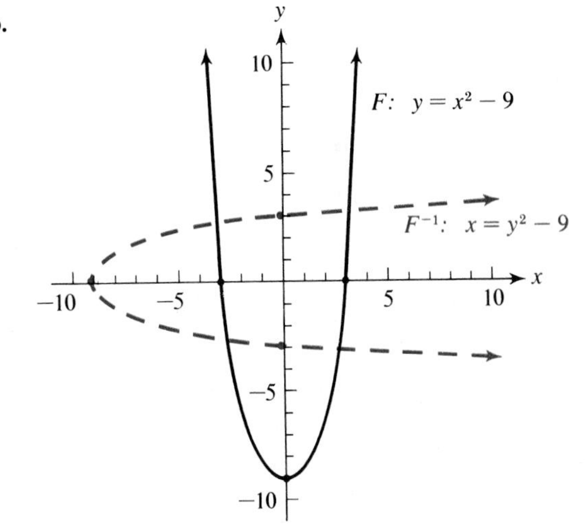

7. $2x + 4y = 7$

8. $3x - 2y = 5$

9. $y = x^2 - 4x$

10. $y = x^2 - 4$

11. $x^2 + 4y^2 = 36$

12. $x^2 + y^2 = 4$

13. $x^2 - y^2 = 3$

14. $9x^2 + y^2 = 36$

15. $y = \sqrt{4 + x^2}$

16. $y = -\sqrt{x^2 - 4}$

17. $y = |x|$

18. $y \leq |x| + 1$

B

In Problems 19–24, each equation defines a one-to-one function F in $R \times R$. Find the equation defining F^{-1} and show that $F[F^{-1}(x)] = F^{-1}[F(x)] = x$. [Hint: Solve explicitly for y and let $y = F(x)$.]

19. $y = x$
20. $y = -x$
21. $2x + y = 4$
22. $x - 2y = 4$
23. $3x - 4y = 12$
24. $3x + 4y = 12$

25. Using the graphs of the equations you obtained for Problems 7–18, state which equations define one-to-one functions.

26. Explain why every nonconstant linear function has a function for an inverse.

27. Consider the function F with domain $\{x \mid x \leq 1\}$ defined by $y = x^2 - 2x + 1$ in $R \times R$.

a. What is the range of F?

b. Find the equation defining F^{-1} and state its domain.

c. Is F^{-1} a function?

28. The equation $y = \sqrt{4 - x^2}$ defines a function F with domain $\{x \mid |x| \leq 2\}$.

a. What is the range of F?

b. Find the equation defining F^{-1} and state its domain.

c. Is F^{-1} a function?

CHAPTER SUMMARY

[8.1] The graph of $y = ax^2 + bx + c$, where a, b, and c are constants with $a \neq 0$, is a **parabola**. The parabola opens upward if $a > 0$ and downward if $a < 0$. The graph of $x = ay^2 + by + c$ is a parabola that opens to the right if $a > 0$ and to the left if $a < 0$.

[8.2] A quick sketch of a parabola can be made by first identifying its form by inspecting the coefficient of the second-degree term, and then finding the intercepts and the high or low point by algebraic methods.

[8.3] The graph of $ax^2 + by^2 = c$ is an **ellipse**, a **circle**, a **hyperbola**, or, exceptionally, a pair of lines, a point, or $\varnothing$. These graphs, together with parabolas, are called **conic sections**.

[8.4] A quick sketch of the graph of an equation of the form $ax^2 + by^2 = c$ can be made by first identifying its form by inspecting the coefficients, and then using such aids as intercepts and asymptotes to draw the curve.

[8.5] The equation $y = kx$ (k a positive constant) defines a function called a **direct variation.** The equation $xy = k$ (k a positive constant) defines a function called an **inverse variation.** In each case, the constant k is called the **constant of variation.**

[8.6] The **inverse** F^{-1} of the relation F can be obtained by interchanging the components of each ordered pair in F. If the relation F is a one-to-one function, then the inverse F^{-1} is also a function. The graphs of F^{-1} and F are reflections of each other about the graph of $y = x$.

The symbols introduced in this chapter are listed on the inside of the front cover.

REVIEW EXERCISES

[8.1–8.2]

1. a. Use analytic methods to find the x and y intercepts of the graph of $y = x^2 - 6x + 5$.

b. Find the coordinates of the minimum point of the graph.

2. Graph the equation in Problem 1.

3. Graph $y = -x^2 + 6x - 8$.

4. Graph $y \geq x^2 + 1$.

[8.3–8.4]

5. State the domain of the relation defined by $9x^2 - y^2 = 9$.

6. Name the graph of each equation.

a. $x^2 - 3y^2 = 8$ **b.** $x^2 = 4 - y^2$

c. $x^2 - y + 4 = 0$ **d.** $2y^2 = 4 - x^2$

Graph each relation.

7. $\{(x, y) \mid x^2 + y^2 = 36\}$ **8.** $\{(x, y) \mid 4x^2 + y^2 = 36\}$

9. $\{(x, y) \mid 9x^2 - y^2 = 0\}$ **10.** $\{(x, y) \mid 4x^2 - y^2 = 16\}$

[8.5]

11. If y varies inversely as t^2, and $y = 16$ when $t = 3$, find y when $t = 4$.

12. The weight of a body above the surface of the earth varies inversely as the square of its distance from the center of the earth. If we assume the radius of the earth to be 4000 miles, how much would a man weigh 500 miles above the earth's surface if he weighed 200 pounds on the surface?

[8.6]

13. Find the inverse of each function and state whether the inverse is also a function.

a. $F = \{(3, 7), (4, 8), (8, 4)\}$ **b.** $G = \{(2, 6), (3, 8), (5, 8)\}$

14. Find the inverse of the function defined by $y = x^2 + 9x$ and state whether the inverse is a function.

15. Graph the function defined by $y = 2x + 6$ and its inverse on the same set of axes.

16. Graph the function defined by $y = x^2 + 1$ and its inverse on the same set of axes.

9

EXPONENTIAL AND LOGARITHMIC FUNCTIONS

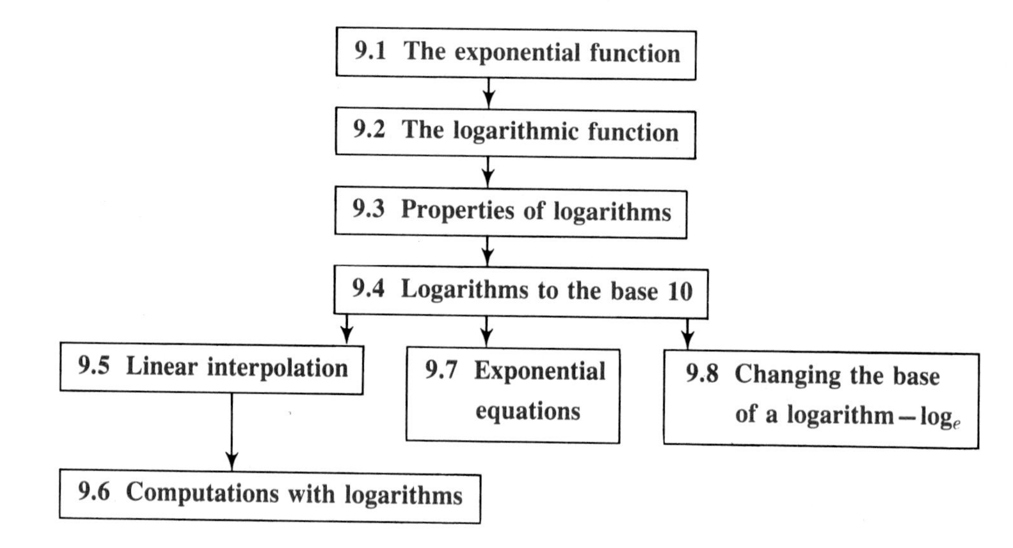

9.1

THE EXPONENTIAL FUNCTION

In Chapter 2, powers a^x were defined for any real number a, and x a natural number. In Chapter 4, the definition was extended to include x negative or 0, and then x a rational number m/n where the base a was restricted to positive values to ensure that $a^{m/n}$ is a real number for n an even integer.

We now inquire whether we can interpret powers with irrational exponents, such as

$$b^{\pi}, \qquad b^{\sqrt{2}}, \qquad b^{-\sqrt{3}}$$

to be real numbers. In Section 4.5 we observed that irrational numbers can be approximated by rational numbers to as great a degree of accuracy as desired. That is, $\sqrt{2} \approx 1.4$ or $\sqrt{2} \approx 1.414$, etc. Also, although we do not prove it here, if x and y are rational numbers such that $x > y$,

$$\text{if} \qquad b > 1, \qquad \text{then} \qquad b^x > b^y,$$

and

$$\text{if} \qquad 0 < b < 1, \qquad \text{then} \qquad b^x < b^y.$$

Now, because 2^x is defined for rational x, we can write the sequence of inequalities

$$2^1 < 2^{\sqrt{2}} < 2^2$$
$$2^{1.4} < 2^{\sqrt{2}} < 2^{1.5}$$
$$2^{1.41} < 2^{\sqrt{2}} < 2^{1.42}$$
$$2^{1.414} < 2^{\sqrt{2}} < 2^{1.415},$$

and so on, where $2^{\sqrt{2}}$ is a number lying between the number on the left and that on the right. It is clear that this process can be continued indefinitely, and that the difference between the number on the left and that on the right can be made as small as we please. This being the case, we assume that there is just one number, $2^{\sqrt{2}}$, that will satisfy each inequality if this process is carried on indefinitely. Since we can produce the same type of argument for any irrational exponent x, we shall assume that b^x $(b > 0)$ is defined for all real values of x.

Since for each real x there is one and only one number b^x, the equation

$$f(x) = b^x \qquad (b > 0) \tag{1}$$

defines a function. Because $1^x = 1$ for all real values of x, (1) defines a constant function if $b = 1$. If $b \neq 1$, we say that (1) defines an **exponential function.**

Exponential functions can perhaps be visualized more clearly by considering their graphs. We illustrate two typical examples in which $0 < b < 1$ and $b > 1$, respectively. Assigning values to x in the equations

$$f(x) = \left(\frac{1}{2}\right)^x \qquad \text{and} \qquad f(x) = 2^x,$$

we find some ordered pairs in each function and sketch the graphs in Figure 9.1. Notice that a line parallel to the y axis will not intersect the graphs at more than one point, and hence each equation defines a function. Notice also that the graph of the function determined by $f(x) = (\frac{1}{2})^x$ goes *down* to the right, and the graph of the function determined by $f(x) = 2^x$ goes *up* to the right. For this reason, we say that the former function is a **decreasing function** and the latter is an **increasing function**. In each case, the domain is the set of real numbers, the range is the set of positive real numbers, and both functions are one-to-one.

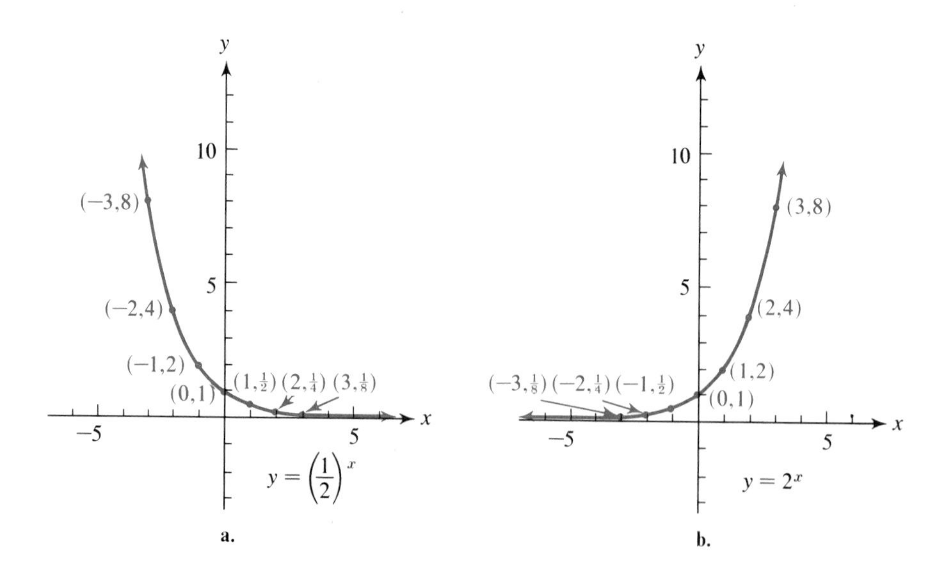

Figure 9.1

Furthermore, in respect to defining powers with irrational exponents, the fact that the definition leads to one-to-one functions suggests that the laws of exponents I through V on pages 97 and 98 are also valid for such powers.

EXERCISE 9.1

A

Find the second component of each of the ordered pairs that makes the pair a solution of the equation.

Example $y = 2^x$; $(-3, \ )$, $(0, \ )$, $(3, \ )$

Solution For $x = -3$, $y = 2^{(-3)} = \dfrac{1}{2^3} = \dfrac{1}{8}$.

For $x = 0$, $y = 2^0 = 1$.

For $x = 3$, $y = 2^3 = 8$.

The ordered pairs are $(-3, \frac{1}{8})$, $(0, 1)$, and $(3, 8)$.

1. $y = 3^x$; $(0, \ )$, $(1, \ )$, $(2, \ )$

2. $y = 4^x$; $\left(-\frac{1}{2}, \ \right)$, $(0, \ )$, $\left(\frac{1}{2}, \ \right)$

3. $y = 2^x$; $(-4, \ )$, $(0, \ )$, $(4, \ )$

4. $y = 5^x$; $(-2, \ )$, $(0, \ )$, $(2, \ )$

5. $y = \left(\frac{1}{2}\right)^x$; $(-4, \ )$, $(0, \ )$, $(4, \ )$

6. $y = \left(\frac{1}{3}\right)^x$; $(-3, \ )$, $(0, \ )$, $(3, \ )$

7. $y = 10^x$; $(-2, \ )$, $(-1, \ )$, $(0, \ )$

8. $y = 10^x$; $(0, \ )$, $(1, \ )$, $(2, \ )$

Graph the function defined by each exponential equation. Use selected integral values $-5 < x < 5$.

Example $y = 3^x$

Solution For convenience, arbitrarily select integral values of x, say,

$(-2, \ )$, $(-1, \ )$, $(0, \ )$, $(1, \ )$, $(2, \ )$, $(3, \ )$.

Determine the y components of each ordered pair.

$\left(-2, \frac{1}{9}\right)$, $\left(-1, \frac{1}{3}\right)$, $(0, 1)$, $(1, 3)$, $(2, 9)$, $(3, 27)$

Plot the points and connect them with a smooth curve.

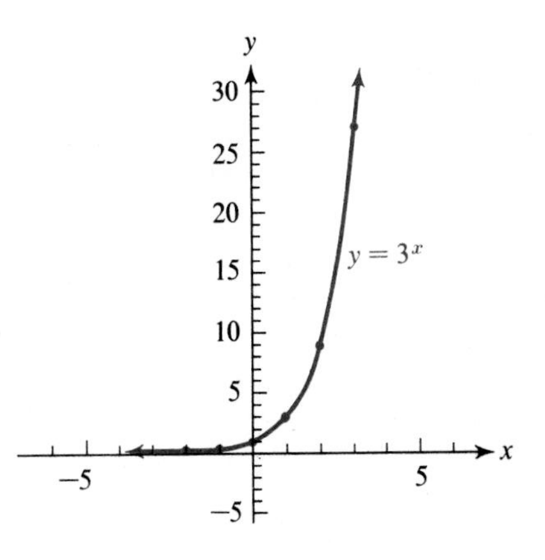

9. $y = 4^x$ **10.** $y = 5^x$ **11.** $y = 10^x$ **12.** $y = 2^{-x}$

13. $y = 3^{-x}$ **14.** $y = 2^{2x}$ **15.** $y = 3^{2x}$ **16.** $y = \left(\frac{1}{3}\right)^x$

17. $y = \left(\frac{1}{4}\right)^x$ **18.** $y = \left(\frac{1}{10}\right)^x$ **19.** $y = \left(\frac{1}{2}\right)^{-x}$ **20.** $y = \left(\frac{1}{3}\right)^{-x}$

B

Estimate the solution of each exponential equation by graphic methods.

21. $2^x = 5$ [*Hint:* Graph $y_1 = 2^x$ and $y_2 = 5$, and approximate the value of x at their point of intersection.]

22. $3^x = 4$

23. For what set of positive real numbers a will $y = a^x$ define an increasing function? A decreasing function?

24. For what set of positive real numbers a will $y = a^{-x}$ define an increasing function? A decreasing function? [*Hint:* $a^{-x} = (1/a)^x$.]

9.2

THE LOGARITHMIC FUNCTION

Recall from Section 8.6 that the inverse of a function can be obtained by interchanging the components in each ordered pair of the function and that this can be accomplished by interchanging the variables in the defining equation of the function. Recall also that the graph of the inverse of a function is symmetric to the graph of the function about the line with equation $y = x$.

In the exponential function

(1) $$\{(x, y) \mid y = b^x, \quad b > 0, \quad b \neq 1\},$$

for $b = \frac{1}{2}$ and $b = 2$ (see the graphs in Figure 9.1), there is only one y associated with each x as well as only one x associated with each y. Therefore, the inverse of function (1) is also a function, namely,

(2) $$\{(x, y) \mid x = b^y, \quad b > 0, \quad b \neq 1\}.$$

Since the domain and range of (1) are the same as the range and domain of (2), respectively, we have for the domain of (2), $\{x \mid x > 0\}$, while the range of (2) is R, the set of real numbers. The graphs of functions of the form (2) can be illustrated by the example

$$x = 10^y \qquad (x > 0).$$

We assign arbitrary values to x, say, 0.01, 0.1, 1, 10, and 100, and obtain the ordered pairs which can be plotted and connected with a smooth curve as in Figure 9.2. Alternatively, we can reflect the graph of $y = 10^x$ about the graph of $y = x$ and obtain the same result.

It is always useful to be able to express the variable y explicitly in terms of the variable x. To do this in equations such as (2), we use the notation

$$(3) \qquad \mathbf{y = \log_b x \qquad (x > 0, b > 0, b \neq 1),}$$

where $\log_b x$ is read "logarithm to the base b of x" or "logarithm of x to the base b." The functions defined by such equations are called **logarithmic functions**.

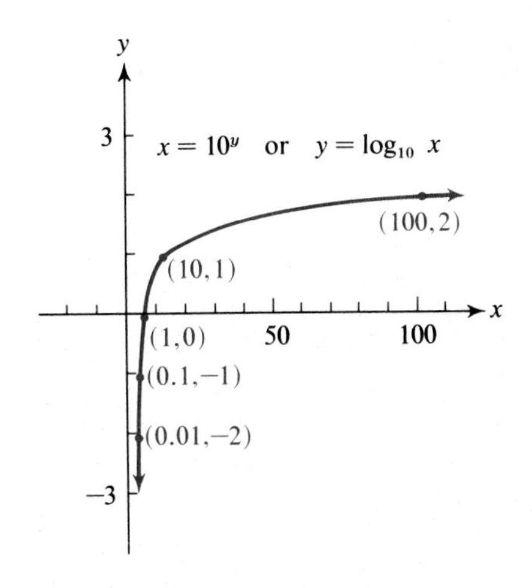

Figure 9.2

It should be recognized that

$$(4) \qquad \mathbf{x = b^y} \quad \text{and} \quad \mathbf{y = \log_b x}$$

are two different forms of an equation defining the same function, in the same way that $x - y = 4$ and $y = x - 4$ define the same function, and we may use whichever equation suits our purpose. Thus, exponential statements may be written in logarithmic form; for example,

$$5^2 = 25 \quad \text{is equivalent to} \quad \log_5 25 = 2,$$

$$8^{1/3} = 2 \quad \text{is equivalent to} \quad \log_8 2 = \frac{1}{3},$$

$$3^{-2} = \frac{1}{9} \quad \text{is equivalent to} \quad \log_3 \frac{1}{9} = -2,$$

and so on. Also, logarithmic statements may be written in exponential form; for example,

$$\log_{10} 100 = 2 \quad \text{is equivalent to} \quad 10^2 = 100,$$
$$\log_3 81 = 4 \quad \text{is equivalent to} \quad 3^4 = 81,$$
$$\log_2 \frac{1}{2} = -1 \quad \text{is equivalent to} \quad 2^{-1} = \frac{1}{2}.$$

Since $x = b^y$ and $y = \log_b x$ are equivalent, if we substitute $\log_b x$ for the exponent y in the left-hand equation in (4), we obtain the important identity

(5) $$b^{\log_b x} = x.$$

In other words, we can think of $\log_b x$ as an exponent on b. For example,

$$2^{\log_2 8} = 8.$$

EXERCISE 9.2

A

Express each equation in logarithmic notation.

Examples **a.** $3^2 = 9$ **b.** $16^{1/4} = 2$ **c.** $64^{-1/3} = \frac{1}{4}$

Solutions **a.** $\log_3 9 = 2$ **b.** $\log_{16} 2 = \frac{1}{4}$ **c.** $\log_{64} \frac{1}{4} = -\frac{1}{3}$

1. $4^2 = 16$ **2.** $5^3 = 125$ **3.** $3^3 = 27$ **4.** $8^2 = 64$

5. $\left(\frac{1}{2}\right)^2 = \frac{1}{4}$ **6.** $\left(\frac{1}{3}\right)^2 = \frac{1}{9}$ **7.** $8^{-1/3} = \frac{1}{2}$ **8.** $64^{-1/6} = \frac{1}{2}$

9. $10^2 = 100$ **10.** $10^0 = 1$ **11.** $10^{-1} = 0.1$ **12.** $10^{-2} = 0.01$

Express each equation in exponential notation.

Examples **a.** $\log_6 36 = 2$ **b.** $\log_{1/5} 125 = -3$ **c.** $\log_{10} 10{,}000 = 4$

Solutions **a.** $6^2 = 36$ **b.** $\left(\frac{1}{5}\right)^{-3} = 125$ **c.** $10^4 = 10{,}000$

13. $\log_2 64 = 6$ **14.** $\log_5 25 = 2$ **15.** $\log_3 9 = 2$

16. $\log_{16} 256 = 2$ **17.** $\log_{1/3} 9 = -2$ **18.** $\log_{1/2} 8 = -3$

19. $\log_{10} 1000 = 3$ **20.** $\log_{10} 1 = 0$ **21.** $\log_{10} 0.01 = -2$

Find the value of each logarithm.

Examples **a.** $\log_4 16$ **b.** $\log_3 81$

Solutions **a.** 4 raised to what power equals 16?
2

b. 3 raised to what power equals 81?
4

22. $\log_5 5$ **23.** $\log_7 49$ **24.** $\log_2 32$ **25.** $\log_4 64$

26. $\log_5 \sqrt{5}$ **27.** $\log_3 \sqrt{3}$ **28.** $\log_3 \frac{1}{3}$ **29.** $\log_5 \frac{1}{5}$

30. $\log_3 3$ **31.** $\log_2 2$ **32.** $\log_{10} 10$ **33.** $\log_{10} 100$

34. $\log_{10} 1$ **35.** $\log_{10} 0.1$ **36.** $\log_{10} 0.01$

Solve for the unknown value.

Examples **a.** $\log_2 x = 3$ **b.** $\log_b 2 = \frac{1}{2}$

Solutions **a.** Write in exponential form.

$$2^3 = x$$

Solve for the variable.

$$x = 8$$

b. Write in exponential form.

$$b^{1/2} = 2$$

Solve for the variable.

$$(b^{1/2})^2 = 2^2$$
$$b = 4$$

37. $\log_3 9 = y$ **38.** $\log_5 125 = y$ **39.** $\log_b 8 = 3$ **40.** $\log_b 625 = 4$

41. $\log_4 x = 3$ **42.** $\log_{1/2} x = -5$ **43.** $\log_2\left(\frac{1}{8}\right) = y$ **44.** $\log_5 5 = y$

45. $\log_b 10 = \frac{1}{2}$ **46.** $\log_b 0.1 = -1$ **47.** $\log_2 x = 2$ **48.** $\log_{10} x = -3$

B

Simplify each expression.

Example $\log_2(\log_3 3)$

Solution Since $\log_3 3 = 1$,

$$\log_2(\log_3 3) = \log_2 1$$
$$= 0.$$

49. $\log_2(\log_4 16)$
50. $\log_5(\log_5 5)$
51. $\log_{10}[\log_3(\log_5 125)]$
52. $\log_{10}[\log_2(\log_3 9)]$
53. $\log_2[\log_2(\log_2 16)]$
54. $\log_4[\log_2(\log_3 81)]$
55. $\log_b(\log_b b)$
56. $\log_b(\log_a a^b)$

57. What is $\log_2 1$? Show that $\log_b 1 = 0$ for $b > 0$.

58. Show that $\log_b b = 1$.

59. For what values of x is $\log_b(x - 9)$ defined?

60. For what values of x is $\log_b(x^2 - 4)$ defined?

9.3

PROPERTIES OF LOGARITHMS

Because a logarithm is an exponent by definition, the following three laws are valid for positive real numbers b $(b \neq 1)$, x_1, x_2, and all real numbers m:

(1) $$\log_b(x_1 x_2) = \log_b x_1 + \log_b x_2,$$

(2) $$\log_b \frac{x_2}{x_1} = \log_b x_2 - \log_b x_1,$$

(3) $$\log_b(x_1)^m = m \log_b x_1.$$

The validity of (1) is established as follows: Since

$$x_1 = b^{\log_b x_1} \quad \text{and} \quad x_2 = b^{\log_b x_2},$$

then

$$x_1 x_2 = b^{\log_b x_1} \cdot b^{\log_b x_2}$$
$$= b^{\log_b x_1 + \log_b x_2},$$

and, by the definition of a logarithm,

$$\log_b(x_1 x_2) = \log_b x_1 + \log_b x_2.$$

The validity of (2) and (3) can also be established (see Problems 49 and 50, Exercise 9.3).

EXERCISE 9.3

A

Express as the sum or difference of simpler logarithmic quantities. Assume that all variables denote positive real numbers.

Example $\log_b \sqrt{\frac{xy}{z}}$

Solution First express $\sqrt{\frac{xy}{z}}$ in exponential notation.

$$\log_b \sqrt{\frac{xy}{z}} = \log_b \left(\frac{xy}{z}\right)^{1/2}$$

By the third law of logarithms (3),

$$\log_b \left(\frac{xy}{z}\right)^{1/2} = \frac{1}{2} \log_b \left(\frac{xy}{z}\right)$$

Use laws (1) and (2).

$$\frac{1}{2} \log_b \left(\frac{xy}{z}\right) = \frac{1}{2} (\log_b x + \log_b y - \log_b z).$$

Therefore,

$$\log_b \sqrt{\frac{xy}{z}} = \frac{1}{2} (\log_b x + \log_b y - \log_b z).$$

1. $\log_b(2x)$ **2.** $\log_b(xy)$ **3.** $\log_b(3xy)$ **4.** $\log_b(4yz)$

5. $\log_b\left(\frac{x}{y}\right)$ **6.** $\log_b\left(\frac{y}{x}\right)$ **7.** $\log_b\left(\frac{xy}{z}\right)$ **8.** $\log_b\left(\frac{x}{yz}\right)$

9. $\log_b x^3$ **10.** $\log_b x^{1/3}$ **11.** $\log_b \sqrt{x}$ **12.** $\log_b \sqrt[5]{y}$

13. $\log_b \sqrt[3]{x^2}$ **14.** $\log_b \sqrt{x^3}$ **15.** $\log_b(x^2y^3)$ **16.** $\log_b(x^{1/3}z^2)$

17. $\log_b\left(\frac{x^{1/2}y}{z^2}\right)$ **18.** $\log_b \frac{xy^3}{z^{1/2}}$ **19.** $\log_{10} \sqrt[3]{\frac{xy^2}{z}}$ **20.** $\log_{10} \sqrt[5]{\frac{x^2y}{z^3}}$

21. $\log_{10}(\sqrt{x}\sqrt[3]{y^2})$ **22.** $\log_{10}(\sqrt[3]{x}\sqrt{y^3})$ **23.** $\log_{10} 2\pi\sqrt{\frac{l}{g}}$ **24.** $\log_{10}\sqrt{\frac{2L}{R^2}}$

25. $\log_{10}\sqrt{(s-a)(s-b)}$ **26.** $\log_{10}\sqrt{s^2(s-a)^3}$

Express as a single logarithm with a coefficient of 1.

Example $\frac{1}{2}(\log_b x - \log_b y)$

Solution By the second law of logarithms (2),

$$\frac{1}{2}(\log_b x - \log_b y) = \frac{1}{2} \log_b\left(\frac{x}{y}\right).$$

By the third law of logarithms (3),

$$\frac{1}{2} \log_b\left(\frac{x}{y}\right) = \log_b\left(\frac{x}{y}\right)^{1/2}.$$

Therefore,

$$\frac{1}{2}(\log_b x - \log_b y) = \log_b\left(\frac{x}{y}\right)^{1/2}.$$

27. $\log_b x + \log_b y$

28. $\log_b x - \log_b y$

29. $2\log_b x - 3\log_b y$

30. $\frac{1}{4}\log_b x + \frac{3}{4}\log_b y$

31. $3\log_b x + \log_b y - 2\log_b z$

32. $\frac{1}{3}(\log_b x + \log_b y - 2\log_b z)$

33. $\frac{1}{2}(\log_{10} y + \log_{10} x - 2\log_{10} z)$

34. $\frac{1}{2}(\log_{10} x - 3\log_{10} y - \log_{10} z)$

35. $-2\log_b x$

36. $-\log_b x$

Solve each logarithmic equation.

Example $\log_{10}(x+9) + \log_{10} x = 1$

Solution Use the first law of logarithms (1).

$$\log_{10}[(x+9)(x)] = 1$$

Write in exponential form.

$$x^2 + 9x = 10^1$$

Solve for x.

$$x^2 + 9x - 10 = 0$$
$$(x+10)(x-1) = 0$$
$$x = -10 \quad \text{or} \quad x = 1$$

Note that the left-hand member of the original equation is meaningful only if $x + 9 > 0$ and $x > 0$; -10 does not satisfy these conditions. Hence, the solution set is $\{1\}$.

37. $\log_{10} x + \log_{10} 2 = 3$

38. $\log_{10}(x-1) - \log_{10} 4 = 2$

39. $\log_{10} x + \log_{10}(x+21) = 2$

40. $\log_{10}(x+3) + \log_{10} x = 1$

41. $\log_{10}(x+2) + \log_{10}(x-1) = 1$

42. $\log_{10}(x+3) - \log_{10}(x-1) = 1$

B

Verify that each statement is true.

43. $\log_b 4 + \log_b 8 = \log_b 64 - \log_b 2$

44. $\log_b 24 - \log_b 2 = \log_b 3 + \log_b 4$

45. $2\log_b 6 - \log_b 9 = 2\log_b 2$

46. $4\log_b 3 - 2\log_b 3 = \log_b 9$

47. $\frac{1}{2}\log_b 12 - \frac{1}{2}\log_b 3 = \frac{1}{3}\log_b 8$

48. $\frac{1}{4}\log_b 8 + \frac{1}{4}\log_b 2 = \log_b 2$

49. Prove that $\log_b \frac{x_2}{x_1} = \log_b x_2 - \log_b x_1$. [*Hint:* See the proof of the first law of logarithms (1) on page 276.]

50. Prove that $\log_b (x_1)^m = m \log_b x_1$.

51. Show by an example that for all $x, y > 0$,

$$\log_{10}(x + y) \neq \log_{10} x + \log_{10} y.$$

52. Show by an example that for all $x, y > 0$,

$$\log_{10} \frac{x}{y} \neq \frac{\log_{10} x}{\log_{10} y}.$$

9.4

LOGARITHMS TO THE BASE 10

There are two logarithmic functions of special interest in mathematics; one is defined by

$$y = \log_{10} x, \tag{1}$$

and the other, by

$$y = \log_e x, \tag{2}$$

where e is an irrational number whose decimal approximation to eight digits is 2.7182818. Because these functions possess similar properties and because we are more familiar with the number 10 as the base of our system of numeration, we shall, for the present, confine our attention to (1).

Values for $\log_{10} x$ are called **logarithms to the base 10**, or **common logarithms**. From equation (5) on page 274,

$$10^{\log_{10} x} = x \qquad (x > 0);$$

that is, $\log_{10} x$ is the exponent that must be placed on 10 so that the resulting power is x. The problem we are concerned with in this section is that of finding, for each positive x, $\log_{10} x$. First, $\log_{10} x$ can easily be determined for all values of x that are integral powers of 10:

$\log_{10} 1000 = 3$ since $10^3 = 1000$
$\log_{10} 100 = 2$ since $10^2 = 100$
$\log_{10} 10 = 1$ since $10^1 = 10$
$\log_{10} 1 = 0$ since $10^0 = 1$
$\log_{10} 0.1 = -1$ since $10^{-1} = 0.1$
$\log_{10} 0.01 = -2$ since $10^{-2} = 0.01$
$\log_{10} 0.001 = -3$ since $10^{-3} = 0.001$

Notice that the logarithm of a power of 10 is simply the exponent on the base 10 when the power is written in scientific notation. For example,

$$\log_{10} 100 = \log_{10} 10^2 = 2$$
$$\log_{10} 0.01 = \log_{10} 10^{-2} = -2,$$

and so on.

A table of logarithms is used to find $\log_{10} x$ where $1 \leq x \leq 10$ (see inside the back cover). Consider the excerpt from this table shown here. Each number in

x	0	1	2	3	4	5	6	7	8	9
3.8	.5798	.5809	.5821	.5832	.5843	.5855	.5866	.5877	.5888	.5899
3.9	.5911	.5922	.5933	.5944	.5955	.5966	.5977	.5988	.5999	.6010
4.0	.6021	.6031	.6042	.6053	.6064	.6075	.6085	.6096	.6107	.6117
4.1	.6128	.6138	.6149	.6160	.6170	.6180	.6191	.6201	.6212	.6222
4.2	.6232	.6243	.6253	.6263	.6274	.6284	.6294	.6304	.6314	.6325
4.3	.6335	.6345	.6355	.6365	.6375	.6385	.6395	.6405	.6415	.6425
4.4	.6435	.6444	.6454	.6464	.6474	.6484	.6493	.6503	.6513	.6522
4.5	.6532	.6542	.6551	.6561	.6571	.6580	.6590	.6599	.6609	.6618
4.6	.6628	.6637	.6646	.6656	.6665	.6675	.6684	.6693	.6702	.6712

the column headed x represents the first two significant digits of the numeral for x, while each of the other column-head numbers represents the third significant digit of the numeral for x. The digits located at the intersection of a row and a column form the logarithm of x. For example, to find $\log_{10} 4.25$, we look at the intersection of the row containing 4.2 under x and the column containing 5. Thus,

$$\log_{10} 4.25 = 0.6284.$$

Similarly,

$$\log_{10} 4.02 = 0.6042$$
$$\log_{10} 4.49 = 0.6522,$$

and so on. The equals sign is used here in a very loose sense. More properly $\log_{10} 4.25 \approx 0.6284$, $\log_{10} 4.02 \approx 0.6042$, and $\log_{10} 4.49 \approx 0.6522$, because these numbers are irrational and cannot be precisely represented by a finite decimal numeral. However, we shall follow customary usage and write $=$ instead of $\approx$.

Now suppose we wish to find $\log_{10} x$ for values of x outside the range of the table—that is, for $0 < x < 1$ or $x > 10$ (see Figure 9.3). This can be done quite readily by first representing the number in scientific notation and applying the first law of logarithms. For example,

$$\begin{aligned}\log_{10} 42.5 = \log_{10}(4.25 \times 10^1) &= \log_{10} 4.25 + \log_{10} 10^1 \\ &= 0.6284 + 1 \\ &= 1.6284,\end{aligned}$$

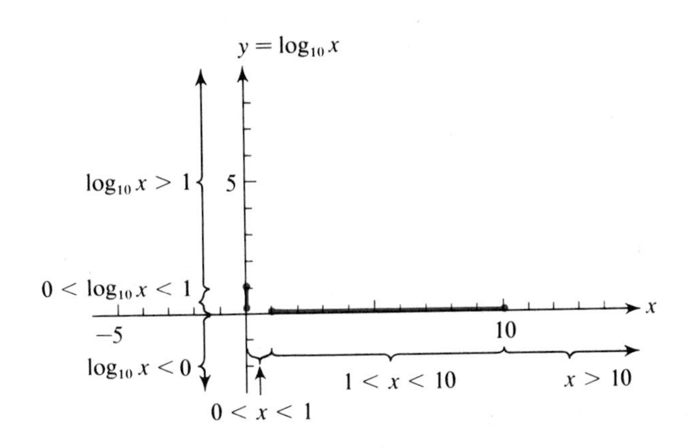

Figure 9.3

$$\begin{aligned}\log_{10} 425 = \log_{10}(4.25 \times 10^2) &= \log_{10} 4.25 + \log_{10} 10^2 \\ &= 0.6284 + 2 \\ &= 2.6284,\end{aligned}$$

$$\begin{aligned}\log_{10} 4250 = \log_{10}(4.25 \times 10^3) &= \log_{10} 4.25 + \log_{10} 10^3 \\ &= 0.6284 + 3 \\ &= 3.6284.\end{aligned}$$

Observe that the decimal portion of the logarithm is always 0.6284, and *the integral portion is the exponent on* 10 *when the number is written in scientific notation.*

This process can be reduced to a mechanical one by considering $\log_{10} x$ to consist of two parts, an integral part (called the **characteristic**) and a nonnegative decimal fraction part (called the **mantissa**). Thus, the table of values for $\log_{10} x$ for $1 < x < 10$ can be looked upon as a table of mantissas for $\log_{10} x$ for all $x > 0$.

First let us consider an example where $x > 10$. To find $\log_{10} 4370$, we write

$$\log_{10} 4370 = \log_{10}(4.37 \times 10^3).$$

Upon examining the table of logarithms, we find $\log_{10} 4.37 = 0.6405$, so that

$$\log_{10} 4370 = 3.6405,$$

where we have prefixed the characteristic 3, the exponent on the base 10.

Now consider an example of the form $\log_{10} x$ for $0 < x < 1$. To find $\log_{10} 0.00402$, we write

$$\log_{10} 0.00402 = \log_{10}(4.02 \times 10^{-3}).$$

We find from the table that $\log_{10} 4.02 = 0.6042$. Upon adding 0.6042 to the characteristic -3, we obtain

$$\log_{10} 0.00402 = -2.3958,$$

where the decimal portion of the logarithm is no longer 0.6042 as it is in the case of all numbers $x > 1$ for which the first three significant digits of x are 402. To circumvent this situation, and thus provide access to the table, it is customary to write the logarithm in a form in which the decimal part is positive. In the foregoing example, we write

$$\begin{aligned}\log_{10} 0.00402 &= 0.6042 - 3\\ &= 0.6042 + (7 - 10)\\ &= 7.6042 - 10,\end{aligned}$$

and the decimal part is positive. The logarithms

$$6.6042 - 9 \qquad \text{and} \qquad 12.6042 - 15,$$

for example, are equally valid representations, but $7.6042 - 10$ is customary in most cases.

It is possible to reverse the process described in this section and find x when $\log_{10} x$ is given. In this event, x is referred to as the **antilogarithm** (antilog_{10}) of $\log_{10} x$. For example, $\text{antilog}_{10}\, 1.6395$ can be obtained by locating the mantissa, 0.6395, in the body of the $\log_{10}$ table and observing that the associated antilog_{10} is 4.36. Thus,

$$\text{antilog}_{10}\, 1.6395 = \text{antilog}_{10}(0.6395 + 1) = 4.36 \times 10^1 = 43.6.$$

If we seek the common logarithm of a number that is not an entry in the table (for example, $\log_{10} 3712$) or if we seek x when $\log_{10} x$ is not an entry in the table, it is customary to use a procedure called **linear interpolation**, which is discussed in Section 9.5.

EXERCISE 9.4

A

Write the characteristic of the logarithm of each number.

Examples

a. $\log_{10} 348$

b. $\log_{10} 0.0057$

Solutions

a. Represent the number in scientific notation.

$$\log_{10}(3.48 \times 10^2)$$

The exponent on the base 10 is the characteristic.

$$2$$

b. Represent the number in scientific notation.

$$\log_{10}(5.7 \times 10^{-3})$$

The exponent on the base 10 is the characteristic.

$$-3 \quad \text{or} \quad 7 - 10$$

1. $\log_{10} 312$ **2.** $\log_{10} 8.12$ **3.** $\log_{10} 7912$
4. $\log_{10} 31$ **5.** $\log_{10} 0.02$ **6.** $\log_{10} 0.00851$
7. $\log_{10} 8.012$ **8.** $\log_{10} 752.31$ **9.** $\log_{10} 0.00031$
10. $\log_{10} 0.0004$ **11.** $\log_{10}(15 \times 10^3)$ **12.** $\log_{10}(820 \times 10^4)$

Find each logarithm.

Examples **a.** $\log_{10} 16.8$ **b.** $\log_{10} 0.043$

Solutions

a. Represent the number in scientific notation.

$$\log_{10}(1.68 \times 10^1)$$

Determine the mantissa from the table of logarithms.

$$0.2253$$

Add the characteristic as determined by the exponent on the base 10.

$$1.2253$$

b. Represent the number in scientific notation.

$$\log_{10}(4.3 \times 10^{-2})$$

Determine the mantissa from the table of logarithms.

$$0.6335$$

Add the characteristic as determined by the exponent on the base 10.

$$8.6335 - 10$$

13. $\log_{10} 6.73$ **14.** $\log_{10} 891$ **15.** $\log_{10} 83.7$
16. $\log_{10} 21.4$ **17.** $\log_{10} 317$ **18.** $\log_{10} 219$
19. $\log_{10} 0.813$ **20.** $\log_{10} 0.00214$ **21.** $\log_{10} 0.08$
22. $\log_{10} 0.000413$ **23.** $\log_{10}(2.48 \times 10^2)$ **24.** $\log_{10}(5.39 \times 10^{-3})$

Find each antilogarithm.

Example $\text{antilog}_{10} 2.7364$

Solution Locate the mantissa in the body of the table of mantissas and determine the associated antilog_{10} (a number between 1 and 10); write the characteristic as an exponent on the base 10.

$$\text{antilog}_{10} 2.7364 = \text{antilog}_{10}(0.7364 + 2) = 5.45 \times 10^2 = 545$$

25. $\text{antilog}_{10} 0.6128$ **26.** $\text{antilog}_{10} 0.2504$ **27.** $\text{antilog}_{10} 1.5647$
28. $\text{antilog}_{10} 3.9258$ **29.** $\text{antilog}_{10}(8.8075 - 10)$ **30.** $\text{antilog}_{10}(3.9722 - 5)$
31. $\text{antilog}_{10} 1.2041$ **32.** $\text{antilog}_{10} 2.6590$ **33.** $\text{antilog}_{10} 3.7388$
34. $\text{antilog}_{10} 2.0086$ **35.** $\text{antilog}_{10}(6.8561 - 10)$ **36.** $\text{antilog}_{10}(1.8156 - 4)$

Find each number by means of the table of logarithms to the base 10.

Example $10^{0.6263}$

Solution Since

$$\log_{10} 10^{0.6263} = 0.6263,$$

we have

$$10^{0.6263} = \text{antilog}_{10} 0.6263 = 4.23.$$

37. $10^{0.9590}$ **38.** $10^{0.8241}$ **39.** $10^{3.6990}$

40. $10^{2.3874}$ **41.** $10^{2.0531}$ **42.** $10^{1.7396}$

B

43. Between which two consecutive integers will $\log_5 33$ lie? $\log_4 33$? $\log_{30} 33$?

44. Between which two consecutive integers will $\log_2 \frac{1}{5}$ lie? $\log_3 \frac{1}{7}$?

Without using the tables of logarithms, find the value of each of the following.

45. $\text{antilog}_{10}(\log_{10} 49)$ **46.** $\log_{10}(\text{antilog}_{10} 2.5761)$

Values for $\log_{10} x$, *where* $0 < x < 1$, *are negative numbers. These negative numbers have been shown as the sum of a positive decimal part and a negative characteristic. Find the antilogarithm of each number in which both the characteristic and the decimal part are negative.*

Example -2.2182

Solution

$$\begin{aligned} -2.2182 &= -2 + (-0.2182) \\ &= -2 + (0.7818 - 1) \\ &= 0.7818 - 3 \\ &= 7.7818 - 10 \end{aligned}$$

Hence, using the tables we have

$$\begin{aligned} \text{antilog}_{10}(-2.2182) &= \text{antilog}_{10}(7.7818 - 10) \\ &= 6.05 \times 10^{-3} \end{aligned}$$

47. -0.5272 **48.** -1.2984 **49.** -2.6882 **50.** -3.0670

9.5

LINEAR INTERPOLATION

Let us recall from Section 7.3 that a function is a set of ordered pairs. A table of common logarithms represents just such a set. For each number x there is an associated number $\log_{10} x$, and we have a set of ordered pairs $(x, \log_{10} x)$ displayed in convenient tabular form. Only three digits for the number x and four for the number $\log_{10} x$ appear in the table. By means of a process called **linear interpolation**, however, the table can be used to find approximations to logarithms for numbers with four-digit numerals.

Let us examine geometrically the concepts involved. A portion of the graph of

$$y = \log_{10} x$$

is shown in Figure 9.4. The curvature is exaggerated to illustrate the principle involved. We propose to use the straight line joining the points P_1 and P_2 as an approximation to the curve passing through the points. If an enlarged graph of $y = \log_{10} x$ were available, the value of $\log_{10} 2.257$ could be found by using the ordinate (RT) to the curve for $x = 2.257$. Since there is no way to accomplish this with a table of values only, we shall use the ordinate (RS) to the straight line as an approximation to the ordinate of the curve.

This can be accomplished directly from the set of numbers available in the table of logarithms inside the back cover. Consider Figure 9.5, where line segments $P_2 P_3$ and $P_4 P_5$ are perpendicular to line segment $P_1 P_3$. From geometry, we know that $\Delta P_1 P_4 P_5$ is similar to $\Delta P_1 P_2 P_3$, where the corresponding lengths of the sides are proportional, and hence

$$\frac{x}{X} = \frac{y}{Y}. \tag{1}$$

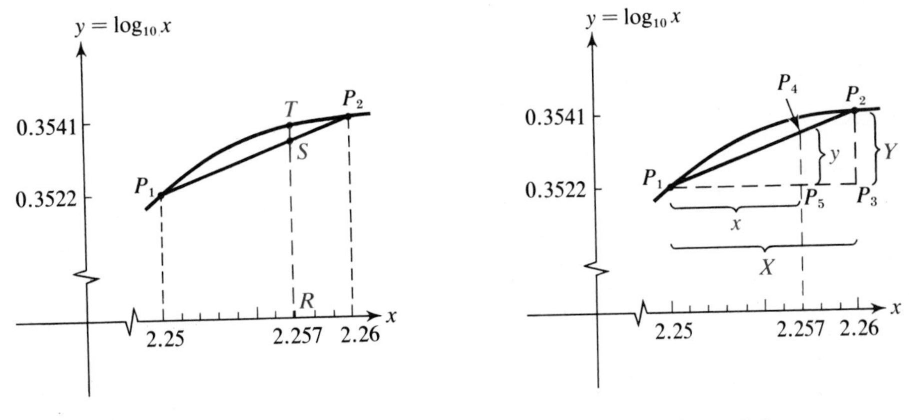

Figure 9.4

Figure 9.5

If we know any three of these numbers, the fourth can be determined. For the purpose of interpolation, we assume that all the members of the domain of the logarithmic function noted in the table now have four-digit numerals; that is, we consider 2.250 instead of 2.25 and 2.260 instead of 2.26. We note in Figure 9.5 that x (0.007) is the distance between 2.250 and 2.257, X (0.010) is the distance between 2.250 and 2.260, and Y (0.0019) is the distance between the logarithms 0.3522 and 0.3541. This data can be conveniently arranged as follows:

$$0.010\left\{\begin{array}{l}0.007\left\{\begin{array}{l}\log_{10} 2.250 = 0.3522\\ \log_{10} 2.257 = \quad ?\end{array}\right\}y\\ \log_{10} 2.260 = 0.3541\end{array}\right\}0.0019$$

To find the value of y, we use (1):

$$\frac{0.007}{0.010} = \frac{y}{0.0019}$$

$$y = \frac{7}{10}(0.0019) = 0.00133 \approx 0.0013.$$

We can now add 0.0013 to 0.3522 to obtain a good approximation of the required logarithm. That is,

$$\log_{10} 2.257 = 0.3522 + 0.0013 = 0.3535$$

The first example in Exercise 9.5 shows another convenient arrangement for the calculations involved in the example presented here. The antilogarithm of a number can be found by a similar procedure. The second example in Exercise 9.5 illustrates the process. However, with practice, it is possible to interpolate mentally in both procedures.

EXERCISE 9.5

Find each logarithm.

Example $\log_{10} 32.54$

Solution

x	$\log_{10} x$
$10\left\{\begin{array}{l}4\left\{\begin{array}{l}32.50\\ 32.54\end{array}\right.\\ 32.60\end{array}\right.$	$\left.\begin{array}{l}\left.\begin{array}{l}1.5119\\ ?\end{array}\right\}y\\ 1.5132\end{array}\right\}0.0013$

Note that for convenience we have written 4 and 10 for the differences for values of x instead of 0.04 and 0.10, respectively, since only their ratios are involved. We now set up a proportion and solve for y.

$$\frac{4}{10} = \frac{y}{0.0013}$$

$$y = \frac{4}{10}(0.0013) = 0.00052 \approx 0.0005$$

Adding this value of y to 1.5119, we have

$$\log_{10} 32.54 = 1.5119 + 0.0005 = 1.5124.$$

1. $\log_{10} 4.213$ **2.** $\log_{10} 8.184$ **3.** $\log_{10} 6.219$ **4.** $\log_{10} 10.31$

5. $\log_{10} 1522$ **6.** $\log_{10} 203.4$ **7.** $\log_{10} 37{,}110$ **8.** $\log_{10} 72.36$

9. $\log_{10} 0.5123$ **10.** $\log_{10} 0.09142$ **11.** $\log_{10} 0.008351$ **12.** $\log_{10} 0.03741$

Find each antilogarithm.

Example $\text{antilog}_{10} 2.8472$

Solution We first find an approximation to the antilogarithm of the mantissa 0.8472. This number is not an entry in the table, but we note that the entries closest to 0.8472 are 0.8470 and 0.8476. The necessary data from the table can be arranged for interpolation in the following way:

		x	$\text{antilog}_{10} x$		
6	2	0.8470	7.030	y	0.010
		0.8472	?		
		0.8476	7.040		

$$\frac{2}{6} = \frac{y}{0.010},$$

where we have written 2 and 6 for 0.0002 and 0.0006, respectively.

$$y = \frac{2}{6}(0.010) \approx 0.003.$$

Adding this value of y to 7.030, we have

$$\text{antilog}_{10} 0.8472 = 7.030 + 0.003 = 7.033.$$

However, since the characteristic of the original number 2.8472 is 2, we have

$$\text{antilog}_{10}(0.8472 + 2) = 7.033 \times 10^2 = 703.3.$$

13. $\text{antilog}_{10} 0.5085$
14. $\text{antilog}_{10} 0.8087$
15. $\text{antilog}_{10} 1.9512$
16. $\text{antilog}_{10} 2.2620$
17. $\text{antilog}_{10} 1.0220$
18. $\text{antilog}_{10} 3.0759$
19. $\text{antilog}_{10}(8.7055 - 10)$
20. $\text{antilog}_{10}(3.6112 - 5)$
21. $\text{antilog}_{10}(9.8742 - 10)$
22. $\text{antilog}_{10}(20.9979 - 22)$
23. $\text{antilog}_{10}(2.8748 - 3)$
24. $\text{antilog}_{10}(7.7397 - 10)$

9.6

COMPUTATIONS WITH LOGARITHMS

The use of the slide rule and the advent of electronic computing devices have almost removed the need to perform routine numerical computations with pencil and paper by logarithms. Nevertheless, we introduce the techniques involved in making such computations because the writing of the logarithmic equations sheds light on the properties of the logarithmic function and the usefulness of the laws of logarithms. We reproduce these laws here using the base 10: If x_1 and x_2 are positive real numbers, then

(1) $$\log_{10}(x_1 x_2) = \log_{10} x_1 + \log_{10} x_2,$$

(2) $$\log_{10} \frac{x_2}{x_1} = \log_{10} x_2 - \log_{10} x_1,$$

(3) $$\log_{10}(x_1)^m = m \log_{10} x_1.$$

We also state two assumptions that are helpful in making computations using logarithms.

L-1 If $x_1 = x_2$ $(x_1, x_2 > 0)$, then $\log_{10} x_1 = \log_{10} x_2$.

L-2 If $\log_{10} x_1 = \log_{10} x_2$, then $x_1 = x_2$.

These assumptions should seem plausible, because exponential and logarithmic functions are one-to-one functions.

Now, consider the product

$$(3.826)(0.00729).$$

If we set

$$N = (3.826)(0.00729),$$

then by assumption L-1,

$$\log_{10} N = \log_{10}[(3.826)(0.00729)].$$

Now, by the first law of logarithms (1),

$$\log_{10} N = \log_{10} 3.826 + \log_{10} 0.00729,$$

and by using the table, we obtain

$$\log_{10} 3.826 = 0.5828,$$
$$\log_{10} 0.00729 = 7.8627 - 10,$$

so that

$$\log_{10} N = (0.5828) + (7.8627 - 10)$$
$$= 8.4455 - 10.$$

The computation is completed by referring to the table for

$$N = \text{antilog}_{10}(8.4455 - 10) = 2.789 \times 10^{-2}$$
$$= 0.02789.$$

Hence,

$$N = (3.826)(0.00729) = 0.02789.$$

Actual computation shows the product to be 0.02789154. Some error should be expected because we are using approximations to irrational numbers when we use a table of logarithms.

Consider a more complicated example. Setting

$$N = \frac{(8.21)^{1/2}(2.17)^{2/3}}{(3.14)^3},$$

we have

$$\log_{10} N = \log_{10} \frac{(8.21)^{1/2}(2.17)^{2/3}}{(3.14)^3}$$
$$= \log_{10}(8.21)^{1/2} + \log_{10}(2.17)^{2/3} - \log_{10}(3.14)^3$$
$$= \frac{1}{2}\log_{10} 8.21 + \frac{2}{3}\log_{10} 2.17 - 3\log_{10} 3.14.$$

The table provides values for the logarithms involved here, and the remainder of the computation is routine. To avoid confusion in computations of this sort, a systematic approach of some kind is desirable (see the example in Exercise 9.6).

In computations with logarithms it is sometimes necessary to use alternative forms for characteristics in order to keep the decimal part of the logarithm positive so that we can continue to use the tables. For example, given

$$Q = \frac{2.43}{7.83},$$

we have

$$\log_{10} Q = \log_{10} \frac{2.43}{7.83} = \log_{10} 2.43 - \log_{10} 7.83 \tag{4}$$
$$= 0.3856 - 0.8938.$$

Now by using 10.3856 − 10 in place of 0.3856, we can avoid the negative mantissa that would result from simplifying (4). That is, we can obtain

$$(10.3856 - 10) - 0.8938 = 9.4918 - 10$$

as the logarithm of the quotient instead of

$$0.3856 - 0.8938 = -0.5082.$$

The latter is not convenient to use, because it is not an entry in the logarithm table.

Similarly, in computing

$$N = \sqrt[3]{0.043} = (0.043)^{1/3},$$

we have

$$\log_{10} N = \log_{10}(0.043)^{1/3} = \frac{1}{3}\log_{10} 0.043,$$

and from the logarithm table,

$$\log_{10} N = \frac{1}{3}(8.6335 - 10).$$

Now, if we use 7.6335 − 9, 4.6335 − 6, or 1.6335 − 3 in place of 8.6335 − 10, the negative portion of the characteristic is evenly divisible by 3. Thus, we have

$$\log_{10} N = \frac{1}{3}(7.6335 - 9) = 2.5445 - 3$$

instead of

$$\log_{10} N = \frac{1}{3}(8.6335 - 10) = 2.8778 - \frac{10}{3},$$

where the latter has a nonintegral characteristic.

EXERCISE 9.6

A

Compute by means of logarithms.

Example $\sqrt{\dfrac{(23.4)(0.681)}{4.13}}$

Solution Let $P = \sqrt{\dfrac{(23.4)(0.681)}{4.13}}$.

Then

$$\log_{10} P = \frac{1}{2}(\log_{10} 23.4 + \log_{10} 0.681 - \log_{10} 4.13).$$

Set-up:

$$\left.\begin{aligned} \log_{10} 23.4 &= 1.3692 \\ \log_{10} 0.681 &= 9.8331 - 10 \end{aligned}\right\} \text{add}$$

$$\left.\begin{aligned} \log_{10}(23.4)(0.681) &= 11.2023 - 10 \\ \log_{10} 4.13 &= 0.6160 \end{aligned}\right\} \text{subtract}$$

$$10.5863 - 10$$

$$\log_{10} \frac{(23.4)(0.681)}{4.13} = 10.5863 - 10 = 0.5863$$

$$\frac{1}{2} \log_{10} \frac{(23.4)(0.681)}{4.13} = \frac{1}{2}(0.5863) = 0.2932$$

Hence, $\log_{10} P = 0.2932$, from which

$$P = \text{antilog}_{10} 0.2932 = 1.964.$$

1. $(2.32)(1.73)$

2. $(82.3)(6.12)$

3. $\dfrac{3.15}{1.37}$

4. $\dfrac{1.38}{2.52}$

5. $\dfrac{0.0149}{32.3}$

6. $\dfrac{0.00214}{3.17}$

7. $(2.3)^5$

8. $(4.62)^3$

9. $\sqrt[3]{8.12}$

10. $\sqrt[5]{75}$

11. $(0.0128)^5$

12. $(0.0021)^6$

13. $\sqrt{0.0021}$

14. $\sqrt[5]{0.0471}$

15. $\sqrt[3]{0.0214}$

16. $\sqrt[4]{0.0018}$

17. $\dfrac{(8.12)(8.74)}{7.19}$

18. $\dfrac{(0.421)^2(84.3)}{\sqrt{21.7}}$

19. $\dfrac{(6.49)^2\sqrt[3]{8.21}}{17.9}$

20. $\dfrac{(2.61)^2(4.32)}{\sqrt{7.83}}$

21. $\dfrac{(0.3498)(27.16)}{6.814}$

22. $\dfrac{(4.813)^2(20.14)}{3.612}$

23. $\sqrt{\dfrac{(4.71)(0.00481)}{(0.0432)^2}}$

24. $\sqrt{\dfrac{(2.85)^3(0.97)}{(0.035)}}$

25. $\sqrt{25.1(25.1 - 18.7)(25.1 - 4.3)}$

26. $\sqrt{\dfrac{(4.17)^3(68.1 - 4.7)}{(68.1 - 52.9)}}$

27. $\dfrac{\sqrt{(23.4)^3(0.0064)}}{\sqrt[3]{69.1}}$

28. $\dfrac{\sqrt{38.7}\sqrt[3]{491}}{\sqrt[4]{9.21}}$

B

29. The period T of a simple pendulum is given by the formula $T = 2\pi\sqrt{L/g}$, where T is measured in seconds, L is the length of the pendulum in feet, $\pi \approx 3.14$, and $g \approx 32$ feet per second per second. Find the period of a pendulum 1 foot long.

30. The area A of a triangle in terms of the sides is given by the formula

$$A = \sqrt{s(s-a)(s-b)(s-c)},$$

where a, b, and c are the lengths of the sides of the triangle and s equals one-half the perimeter. Find the area of a triangle in which the lengths of the three sides are 2.314 inches, 4.217 inches, and 5.618 inches.

9.7

EXPONENTIAL EQUATIONS

An equation in which a variable occurs in an exponent is called an **exponential equation**. Solution sets of some such equations in one variable can be found by means of logarithms. Consider the equation

$$5^x = 7.$$

Because $5^x > 0$ for all x, we can apply assumption L-1 from Section 9.6, and write

$$\log_{10} 5^x = \log_{10} 7,$$

and, from the third law of logarithms,

$$x \log_{10} 5 = \log_{10} 7.$$

Multiplying each member by $1/\log_{10} 5$, we get

$$x = \frac{\log_{10} 7}{\log_{10} 5} = \frac{0.8451}{0.6990} = 1.209,$$

and the solution set is $\{1.209\}$. Note that when we seek a numerical approximation to this solution, the logarithms are *divided*, not subtracted. The quotient

$$\frac{\log_{10} 7}{\log_{10} 5} \neq \log_{10} \frac{7}{5},$$

from which we have

$$\frac{\log_{10} 7}{\log_{10} 5} \neq \log_{10} 7 - \log_{10} 5.$$

As another example of an exponential equation, consider

$$6^{3x-4} = 3.$$

We have from L-1,

$$\log_{10} 6^{3x-4} = \log_{10} 3,$$

and, from the third law of logarithms,

$$(3x-4)\log_{10} 6 = \log_{10} 3.$$

Multiplying each member by $1/\log_{10} 6$ gives

$$3x - 4 = \frac{\log_{10} 3}{\log_{10} 6},$$

from which

$$3x = \frac{\log_{10} 3}{\log_{10} 6} + 4$$

$$x = \frac{\log_{10} 3}{3 \log_{10} 6} + \frac{4}{3},$$

and the solution set is

$$\left\{\frac{\log_{10} 3}{3 \log_{10} 6} + \frac{4}{3}\right\}.$$

A decimal numeral approximation can be obtained by using the logarithm tables and then evaluating the expression. Thus,

$$\frac{\log_{10} 3}{3 \log_{10} 6} + \frac{4}{3} = \frac{0.4771}{3(0.7782)} + \frac{4}{3}$$

$$= 0.2044 + 1.3333$$

$$= 1.5377.$$

In Exercise 9.7, we show examples of solutions using both logarithmic notation (for students who do not have electronic calculators) and decimal notation (for students who have calculators).

EXERCISE 9.7

A

Solve. Show solutions in logarithmic form and in decimal form.

Example $3^{x-2} = 16$

Solution Use assumption L-1.

$$\log_{10} 3^{x-2} = \log_{10} 16$$

Use the third law of logarithms.

$$(x - 2)\log_{10} 3 = \log_{10} 16$$

(*solution continued*)

Multiply each member by $\frac{1}{\log_{10} 3}$.

$$x - 2 = \frac{\log_{10} 16}{\log_{10} 3}$$

$$x = \frac{\log_{10} 16}{\log_{10} 3} + 2$$

Or, in decimal form,

$$x = \frac{1.2041}{0.4771} + 2 = 4.5238$$

1. $2^x = 7$ **2.** $3^x = 4$ **3.** $3^{x+1} = 8$ **4.** $2^{x-1} = 9$

5. $7^{2x-1} = 3$ **6.** $3^{x+2} = 10$ **7.** $4^{x^2} = 15$ **8.** $3^{x^2} = 21$

9. $3^{-x} = 10$ **10.** $2.13^{-x} = 8.1$ **11.** $3^{1-x} = 15$ **12.** $4^{2-x} = 10$

Solve each exponential equation. Leave the results in the form of an equation equivalent to the given equation.

13. $y = x^n$ for n **14.** $y = Cx^{-n}$ for n

15. $y = e^{kt}$ for t **16.** $y = k(1 - e^{-t})$ for t

Solve Problems 17–22 using the following information: P dollars invested at an interest rate r compounded yearly yields an amount A after n years, given by $A = P(1 + r)^n$. If the interest is compounded t times yearly, the amount is given by

$$A = P\left(1 + \frac{r}{t}\right)^{tn}.$$

Example One dollar compounded annually for 12 years yields \$1.127. What is the rate of interest to the nearest $\frac{1}{2}\%$?

Solution Using the given equation, we obtain $(1 + r)^{12} = 1.127$. Equate $\log_{10}$ of each member and apply the third law of logarithms.

$$12 \log_{10}(1 + r) = \log_{10} 1.127 = 0.0519$$

Multiply each member by $\frac{1}{12}$.

$$\log_{10}(1 + r) = \frac{1}{12}(0.0519) = 0.0043$$

Determine $\text{antilog}_{10} 0.0043$ and solve for r.

$$\text{antilog}_{10} 0.0043 = 1 + r = 1.01$$

$$r = 0.01 \quad \text{or} \quad r = 1\%$$

17. One dollar compounded annually for 10 years yields \$1.48. What is the rate of interest to the nearest $\frac{1}{2}\%$?

18. How many years (nearest year) would it take for \$1 to yield \$2.19 if compounded annually at 4%?

19. What rate of interest (nearest $\frac{1}{2}\%$) is required so that \$100 would yield \$113 after 5 years if the money were compounded semiannually?

20. What rate of interest (nearest $\frac{1}{2}\%$) is required so that \$40 would yield \$50.90 after 3 years if the money were compounded quarterly?

21. Find the compounded amount of \$5000 invested at 4% for 10 years when compounded annually. When compounded semiannually.

22. Two investors, A and B, each invested \$10,000 at 4% for 20 years with a bank that computed interest quarterly. Investor A withdrew interest at the end of each 3 month period, but B allowed the investment to be compounded. How much more than A did B earn over the period of 20 years?

Solve Problems 23–26 using the following information: The chemist defines the pH *(hydrogen potential) of a solution by*

$$\begin{aligned} \text{pH} &= \log_{10}\frac{1}{[\text{H}^+]} \\ &= \log_{10}[\text{H}^+]^{-1} \\ &= -\log_{10}[\text{H}^+], \end{aligned}$$

where $[\text{H}^+]$ *represents a numerical value for the concentration of hydrogen ions in aqueous solution in moles per liter.*

Example Calculate the pH of a solution with hydrogen ion concentration 3.7×10^{-6}.

Solution Substitute 3.7×10^{-6} for $[\text{H}^+]$ in the relationship $\text{pH} = \log_{10}\frac{1}{[\text{H}^+]}$.

$$\begin{aligned} \text{pH} &= \log_{10}\frac{1}{3.7 \times 10^{-6}} \\ &= \log_{10}\left(\frac{1}{3.7} \times 10^{6}\right) \\ &= \log_{10} 1 - \log_{10} 3.7 + \log_{10} 10^{6} \\ &= 0 - 0.5682 + 6 = 5.4 \end{aligned}$$

23. Calculate the pH of a solution with hydrogen ion concentration 2.0×10^{-8}.

24. Calculate the pH of a solution with hydrogen ion concentration 6.3×10^{-7}.

25. Calculate the hydrogen ion concentration of a solution with pH 5.6.

26. Calculate the hydrogen ion concentration of a solution with pH 7.2.

27. The atmospheric pressure p, in inches of mercury, is given approximately by $p = 30.0(10)^{-0.09a}$, where a is the altitude in miles above sea level. What is the atmospheric pressure at sea level? At 3 miles above sea level?

28. Using the information in Problem 27, find the atmospheric pressure 6 miles above sea level.

B

29. Find a value for $\log_3 18$ using a table of logarithms to the base 10. [*Hint:* Let $y = \log_3 18$ and write in exponential form.]

30. Find a value for $\log_2 3$ using a table of logarithms to the base 10. [*Hint:* Let $y = \log_2 3$ and write in exponential form.]

9.8

CHANGING THE BASE OF A LOGARITHM—$\log_e$

The number e ($e \approx 2.7182818$) mentioned in Section 9.4 is of great mathematical interest. It has certain very useful properties of wide application in more advanced mathematics and in many practical situations. For our purposes, we wish only to find a way to change from $\log_{10} x$ to $\log_e x$, or, in general, from $\log_a x$ to $\log_b x$.

First note that for $a, x > 0$ and $a \neq 1$,

$$x = a^{\log_a x}. \tag{1}$$

By applying assumption L-1 from Section 9.6, we can equate the logarithms to the base b ($b > 0$, $b \neq 1$) of each member of (1), which gives

$$\log_b x = \log_b a^{\log_a x}.$$

By the third law of logarithms,

$$\log_b x = \log_a x \cdot \log_b a, \tag{2}$$

from which

$$\boldsymbol{\log_a x = \frac{\log_b x}{\log_b a}}. \tag{3}$$

Equation (3) gives us a means of finding $\log_a x$ when we have a table of logarithms to the base b. In particular, if $a = e$ and $b = 10$, we have

$$\log_e x = \frac{\log_{10} x}{\log_{10} e}. \tag{4}$$

Since $e \approx 2.718$, we can use the table of logarithms and interpolation to obtain $\log_{10} e \approx 0.4343$. Hence, (4) can be written as

$$\log_e x = \frac{\log_{10} x}{0.4343} = \frac{1}{0.4343} \cdot \log_{10} x \tag{5}$$

or

$$\log_e x = 2.303 \log_{10} x. \tag{6}$$

Equation (5) can also be written in the form

$$\log_{10} x = 0.4343 \log_e x. \tag{7}$$

[*Note:* In (5), (6), and (7) we are again using = for $\approx$.]

Another useful relationship can be obtained from (3) by setting $x = b$. This yields

$$\log_a b = \frac{\log_b b}{\log_b a}. \tag{8}$$

But, from the definition of a logarithm,

$$\log_b b = 1,$$

so that (8) becomes

$$\log_a b = \frac{1}{\log_b a}.$$

EXERCISE 9.8

A

Find each logarithm to the nearest hundredth using the table of logarithms to the base 10.

Examples

a. $\log_e 14$

b. $\log_3 7$

Solutions

a. Represent the number to the base 10.

$$\log_e 14 = \frac{\log_{10} 14}{\log_{10} e}$$

Use equation (6).

$$\log_e 14 = 2.303 \log_{10} 14$$
$$= (2.303)(1.1461) = 2.64$$

b. Represent the number to the base 10.

$$\log_3 7 = \frac{\log_{10} 7}{\log_{10} 3}$$

Use the table of logarithms.

$$\log_3 7 = \frac{0.8451}{0.4771} = 1.77$$

1. $\log_e 3$ **2.** $\log_e 8$ **3.** $\log_e 17$ **4.** $\log_e 98$

5. $\log_e 327$ **6.** $\log_e 107$ **7.** $\log_e 24$ **8.** $\log_e 14$

9. $\log_2 10$ **10.** $\log_2 5$ **11.** $\log_5 240$ **12.** $\log_3 18$

13. $\log_7 8.1$ **14.** $\log_5 60$ **15.** $\log_{15} 8.1$ **16.** $\log_{20} 200$

17. $\log_{100} 38$ **18.** $\log_{1000} 240$

Solve without using the table of logarithms.

19. If $\log_{10} 4 = 0.6021$, find $\log_4 10$.

20. If $\log_{10} 2 = 0.3010$, find $\log_2 10$.

21. If $\log_{10} 3 = 0.4771$, find $\log_3 10$.

22. If $\log_{10} e = 0.4343$, find $\log_e 10$.

23. If $\log_{10} 5 = 0.6990$, find $\log_5 100$.

24. If $\log_{10} 3 = 0.4771$, find $\log_3 100$.

B

Without using the table of logarithms, show that each is a true statement.

25. $\log_9 7 = \frac{1}{2} \log_3 7$

26. $\log_8 5 = \frac{1}{3} \log_2 5$

27. $\log_3 x = 2 \log_9 x$

28. $\log_b x = 3 \log_{b^3} x$

29. $(\log_{10} 4 - \log_{10} 2)\log_2 10 = 1$

30. $(2 \log_2 3)(\log_9 2 + \log_9 4) = 3$

31. The voltage V across a capacitor in a certain circuit is given by

$$V = 100(1 - e^{-0.5t}),$$

where t is the time in seconds. How much time must elapse for the voltage to reach 75 volts?

32. The amount of a radioactive element present at any time (t) is given by $y = y_0 e^{-0.4t}$, where t is measured in seconds and y_0 is the amount present initially. How much of the element would remain after 3 seconds if 40 grams were present initially? Use $e \approx 2.718$.

33. The number N of bacteria present in a culture is related to time by the formula $N = N_0 e^{0.04t}$, where N_0 is the number of bacteria present at time $t = 0$, and t is time in hours. If 10,000 bacteria are present 10 hours after the beginning of the experiment, how many were present when $t = 0$?

34. The intensity I (in lumens) of a light beam after passing through a thickness t (in centimeters) of a medium having an absorption coefficient of 0.1 is given by $I = 1000e^{-0.1t}$. How many centimeters of the material would reduce the illumination to 800 lumens?

CHAPTER SUMMARY

[9.1] A function defined by an equation of the form

$$f(x) = b^x \qquad (b > 0),$$

where $b \neq 1$, is called an **exponential function**. If $b > 1$, it is an increasing function; if $0 < b < 1$, it is a decreasing function.

[9.2] The inverse of an exponential function is called a **logarithmic function**, and is defined by an equation of the form

$$x = b^y \qquad \text{or} \qquad y = \log_b x \qquad (b > 0, b \neq 1).$$

These equations are equivalent to the statement

$$b^{\log_b x} = x \qquad (b > 0, b \neq 1).$$

The domain of a logarithmic function is the set of positive real numbers.

[9.3] The properties of logarithms given below follow from the definition of a logarithm and the properties of exponents developed in Chapter 4:

$$\log_b(x_1 x_2) = \log_b x_1 + \log_b x_2,$$

$$\log_b \frac{x_2}{x_1} = \log_b x_2 - \log_b x_1,$$

$$\log_b(x_1)^m = m \log_b x_1.$$

[9.4] By definition of $\log_{10} x$, the values of $\log_{10} 10^n$, $n \in J$, are integers and can be obtained directly from the definition. Values of $\log_{10} x$, $1 < x < 10$, are between 0 and 1, and can be obtained directly from the table of logarithms. Values of $\log_{10} x$, $x > 10$, are greater than 1 and can be obtained from the logarithmic table in conjunction with the first law of logarithms. The integral part of a logarithm to the base 10 is the **characteristic** of the logarithm and the decimal part is the **mantissa**.

In the expression $\log_{10} x$, the number x is called the **antilogarithm** (**antilog**$_{10}$) of $\log_{10} x$.

[9.5] To determine the mantissa of a logarithm (or the antilogarithm of a mantissa) that is not an entry in the table of logarithms, we can use **linear interpolation**.

[9.6–9.7] Logarithms are useful in computations involving multiplication, division, powers, and roots, and in solving exponential equations.

[9.8] Given a table of logarithms to a base b, we can find logarithms to any other base a by using the property

$$\log_a x = \frac{\log_b x}{\log_b a}.$$

The symbols introduced in this chapter are listed on the inside of the front cover.

REVIEW EXERCISES

[9.1] **1.** Sketch the graph of each equation.

a. $y = 5^x$ **b.** $y = 5^{-x}$

[9.2] **2.** Write each statement in logarithmic notation.

a. $9^{3/2} = 27$ **b.** $\left(\frac{4}{9}\right)^{1/2} = \frac{2}{3}$

3. Write each statement in exponential notation.

a. $\log_5 625 = 4$ **b.** $\log_{10} 0.0001 = -4$

4. Find a value for each variable.

a. $\log_3 x = 3$ **b.** $\log_b 3 = 3$

[9.3] **5.** Express each as the sum or difference of simpler logarithmic quantities.

a. $\log_b 3x^2y$ **b.** $\log_b \frac{y\sqrt{x}}{z^2}$

6. Write each expression as a single logarithm with a coefficient of 1.

a. $2\log_b x - 3\log_b y$ **b.** $\frac{1}{2}(\log_b x - 3\log_b y - \log_b z)$

[9.4] **7.** Find a value for each logarithm.

a. $\log_{10} 0.713$ **b.** $\log_{10} 1810$

8. Find a value for each antilogarithm.

a. $\text{antilog}_{10} 8.1761 - 10$ **b.** $\text{antilog}_{10} 3.7235$

[9.5] **9.** Use linear interpolation to find a value for each expression.

a. $\log_{10} 27.93$ **b.** $\text{antilog}_{10} 8.5800 - 10$

[9.6] *Compute by means of logarithms.*

10. a. $(3.17)(8.23)$ **b.** $\frac{\sqrt{18.72}}{3.12}$

11. a. $\frac{(2.12)^2(3.42)^3}{147}$ **b.** $\frac{\sqrt{(10.6)(2.3)^3}}{(20^9)(7)}$

12. a. $\frac{2.04}{7.31}$ **b.** $\frac{(31.2)(0.02)^3}{0.003}$

[9.7] *Solve.*

13. a. $3^x = 15$ **b.** $2^{x-4} = 10$

14. a. $N = N_0 e^{-kt}$ for t **b.** $E = ke^{1-t}$ for t

15. Find n to the nearest year, given that $(1 + 0.04)^n = 1.60$.

16. Find $[H^+]$, given that $\log_{10} \frac{1}{[H^+]} = 7.4$.

[9.8] *Find the value of each logarithm.*

17. a. $\log_e 7$ **b.** $\log_3 5$

18. Given that $y = y_0 e^{-0.2t}$, find t if $y = 4$ and $y_0 = 16$.

10

SYSTEMS OF EQUATIONS

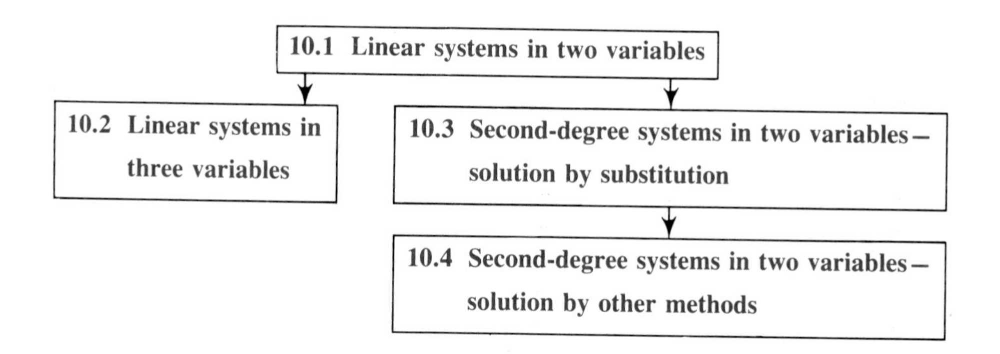

10.1

LINEAR SYSTEMS IN TWO VARIABLES

In Chapters 7 and 8 we observed that the solution set of an open sentence in two variables, such as

$$ax + by = c,$$

$$y = ax^2 + bx + c,$$

or

$$ax^2 + by^2 = c,$$

might contain infinitely many ordered pairs of numbers. It is often necessary to consider pairs of such sentences and to inquire whether or not the solution sets of the sentences contain ordered pairs in common. More specifically, we are interested in determining the members of the intersection of their solution sets.

We shall begin by considering the system

$$a_1x + b_1y = c_1 \qquad (a_1, b_1 \text{ not both } 0)$$

$$a_2x + b_2y = c_2 \qquad (a_2, b_2 \text{ not both } 0)$$

and studying $A \cap B$, where

$$A = \{(x, y) \,|\, a_1x + b_1y = c_1\},$$

$$B = \{(x, y) \,|\, a_2x + b_2y = c_2\}.$$

In a geometric sense, because the graphs of both A and B are straight lines, we are confronted with three possibilities, as illustrated in Figure 10.1 on page 304:

a. The graphs are the same line.

b. The graphs are parallel but distinct lines.

c. The graphs intersect in one and only one point.

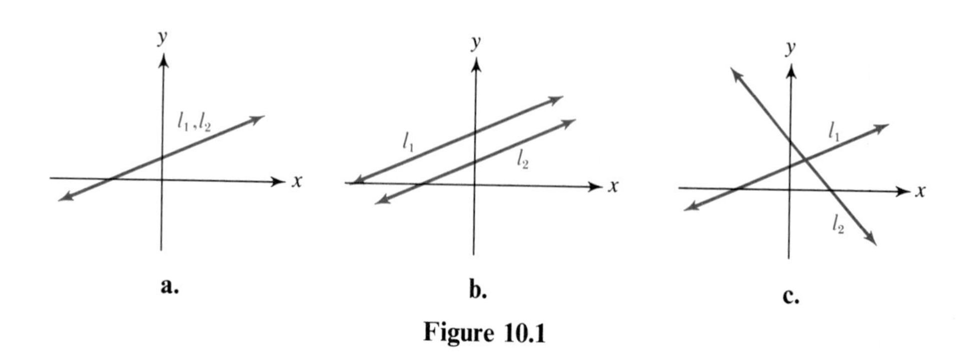

Figure 10.1

These possibilities lead, correspondingly, to the conclusion that one and only one of the following is true for any given system of two such linear equations in x and y:

a. The solution sets of the equations are equal, and their intersection contains all (an infinite number of) ordered pairs found in either one of the given solution sets.

b. The intersection of the two solution sets is the null set.

c. The intersection of the two solution sets contains one and only one ordered pair.

In case a, the left-hand members (or the equations themselves) of the two linear equations in x and y are said to be **dependent**, and in case b, the equations are said to be **inconsistent**.

Consider the system

$$x + y = 5$$
$$x - y = 1.$$

From the graph of the system in Figure 10.2, it is evident that the ordered pair $(3, 2)$ is the only ordered pair common to the solution sets of both equations. That is,

$$\{(x, y) \mid x + y = 5\} \cap \{(x, y) \mid x - y = 1\} = \{(3, 2)\}.$$

This can be verified by substituting $(3, 2)$ into each equation and observing that a true statement results in each case.

As another example, consider the system

$$x + y = 5$$
$$2x + 2y = 3,$$

and the graphs of these equations in Figure 10.3, where the lines appear to be parallel. We conclude from this that the solution set of this system is $\varnothing$. That is,

$$\{(x, y) \mid x + y = 5\} \cap \{(x, y) \mid 2x + 2y = 3\} = \varnothing.$$

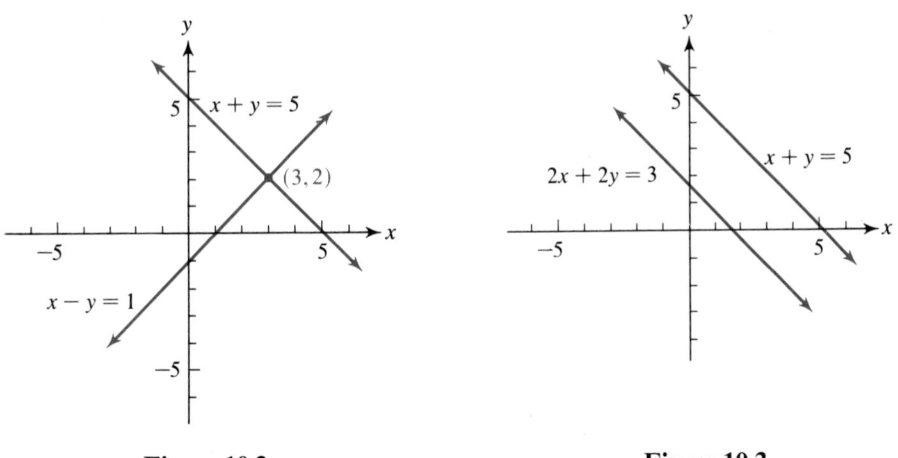

Figure 10.2

Figure 10.3

Because graphing equations is a time-consuming process, and, more importantly, because graphic results are not always precise, solutions to systems of linear equations are usually sought by analytic methods. One such method depends on the following theorem:

► *Any ordered pair* (x, y) *that satisfies the equations*

(1) $$a_1x + b_1y = c_1$$

(2) $$a_2x + b_2y = c_2$$

will also satisfy the equation

(3) $$A(a_1x + b_1y) + B(a_2x + b_2y) = Ac_1 + Bc_2$$

for all real numbers A and B.

We can see the validity of this theorem if we first rewrite the system of equations (1) and (2) as

(1a) $$a_1x + b_1y - c_1 = 0$$

(2a) $$a_2x + b_2y - c_2 = 0,$$

and equation (3) as

(3a) $$A(a_1x + b_1y - c_1) + B(a_2x + b_2y - c_2) = 0.$$

Note that the replacement of the variables in (3a) with the components of any ordered pair (x, y) that satisfies both (1a) and (2a) results in

$$A(0) + B(0) = 0,$$

which is clearly true for any values of A and B. The left-hand member of (3a) is called a **linear combination** of the left-hand members of (1a) and (2a). The theorem stated above asserts that any ordered pair satisfying both (1) and (2) must also satisfy the sum of any real number multiples of (1) and (2). This fact can be used to identify any such ordered pairs.

Consider the system

$$(4) \qquad 2x + 3y = 8$$

$$(5) \qquad 3x - 4y = -5.$$

We can multiply each member of (4) by 3 and each member of (5) by -2 to obtain

$$(4a) \qquad 6x + 9y = 24$$

$$(5a) \qquad -6x + 8y = 10.$$

Then, adding the corresponding members of 4a and 5a, we obtain

$$(6) \qquad 17y = 34$$

$$(6a) \qquad y = 2,$$

which must contain in its solution set in $R \times R$ any solution common to the solution sets of the two original equations. But any solution of (6) or (6a) is of the form $(x, 2)$—that is, it has y component 2. Now, substituting 2 for y in either (4) or (5), we can determine the x component for the ordered pair $(x, 2)$ that satisfies both (4) and (5). If (4) is used, we have

$$2x + 3(2) = 8$$
$$x = 1;$$

and if (5) is used, we have

$$3x - 4(2) = -5$$
$$x = 1.$$

Since the ordered pair $(1, 2)$ satisfies both (4) and (5), the required solution set is $\{(1, 2)\}$. That is,

$$\{(x, y) \mid 2x + 3y = 8\} \cap \{(x, y) \mid 3x - 4y = -5\} = \{(1, 2)\}.$$

Notice that $(1, 2)$ also satisfies (6) and (6a).

Observe how the concept of a linear combination is used. Multipliers A and B are chosen so that the coefficients of *one* of the variables, x or y, are additive inverses and the resulting equation is free of *one* of the variables—which one is immaterial.

If the coefficients of the variables in one equation in a system are proportional to the coefficients in the other equation, the equations are either *dependent* and there are an infinite number of solutions, or *inconsistent* and there are no solutions. For example, consider the system

$$(7) \qquad 2x + 3y = 2$$

$$(8) \qquad 4x + 6y = 7.$$

If we isolate the coefficients of x and the coefficients of y in (7) and (8) and set up proportions, we find that

$$\frac{2}{4}=\frac{3}{6}.$$

Because of this relationship, any attempt to form a linear combination of the left-hand members of these equations free of *one* variable will result in an equation free of *both* variables. Thus, if we multiply -2 times equation (7) and 1 times equation (8), we have

$$-4x-6y=-4$$
$$4x+6y=7,$$

which, upon adding left-hand members and right-hand members, yields

$$0x+0y=3.$$

Since this equation is not true for any values of x and y, the solution set of the system is $\varnothing$ and the equations are inconsistent. Now, consider the system

$$2x+3y=2 \tag{9}$$

$$4x+6y=4, \tag{10}$$

where the coefficients of the variables and the constant terms are proportional. That is,

$$\frac{2}{4}=\frac{3}{6}=\frac{2}{4}.$$

The result of adding -2 times equation (9) to 1 times equation (10) is

$$0x+0y=0.$$

Because this equation is true for any values of x and y, the solution sets of the given equations must be identical (there are an infinite number of solutions) and the equations are dependent.

The foregoing examples suggest that any system of the form

$$a_1x+b_1y=c_1$$
$$a_2x+b_2y=c_2$$

has one and only one solution if

$$\frac{a_1}{a_2}\neq\frac{b_1}{b_2},$$

has no solution if

$$\frac{a_1}{a_2}=\frac{b_1}{b_2}\neq\frac{c_1}{c_2},$$

and has an infinite number of solutions if

$$\frac{a_1}{a_2} = \frac{b_1}{b_2} = \frac{c_1}{c_2}.$$

Systems of equations are quite useful in expressing relationships in practical applications. By the assignment of separate variables to represent separate physical quantities, the difficulty encountered in symbolically representing these relationships can usually be decreased. In writing systems of equations, we must be careful that the conditions giving rise to one equation are independent of the conditions giving rise to any other equation.

EXERCISE 10.1

A

Find the solution set for each system by analytic methods. Check your solutions of Problems 1–10 by graphic methods. If the equations are inconsistent or dependent, so state.

Example (1) $$\frac{2}{3}x - y = 2$$

(2) $$x + \frac{1}{2}y = 7$$

Solution Multiply each member of equation (1) by 3 and each member of equation (2) by 2.

(1a) $$2x - 3y = 6$$

(2a) $$2x + y = 14$$

Add -1 times equation (1a) to 1 times equation (2a), and solve for y.

$$4y = 8$$
$$y = 2$$

Substitute 2 for y in (1), (2), (1a), or (2a), and solve for x. In this example, (2) is used.

$$x + \frac{1}{2}(2) = 7$$
$$x = 6$$

The solution set is $\{(6, 2)\}$.

Check

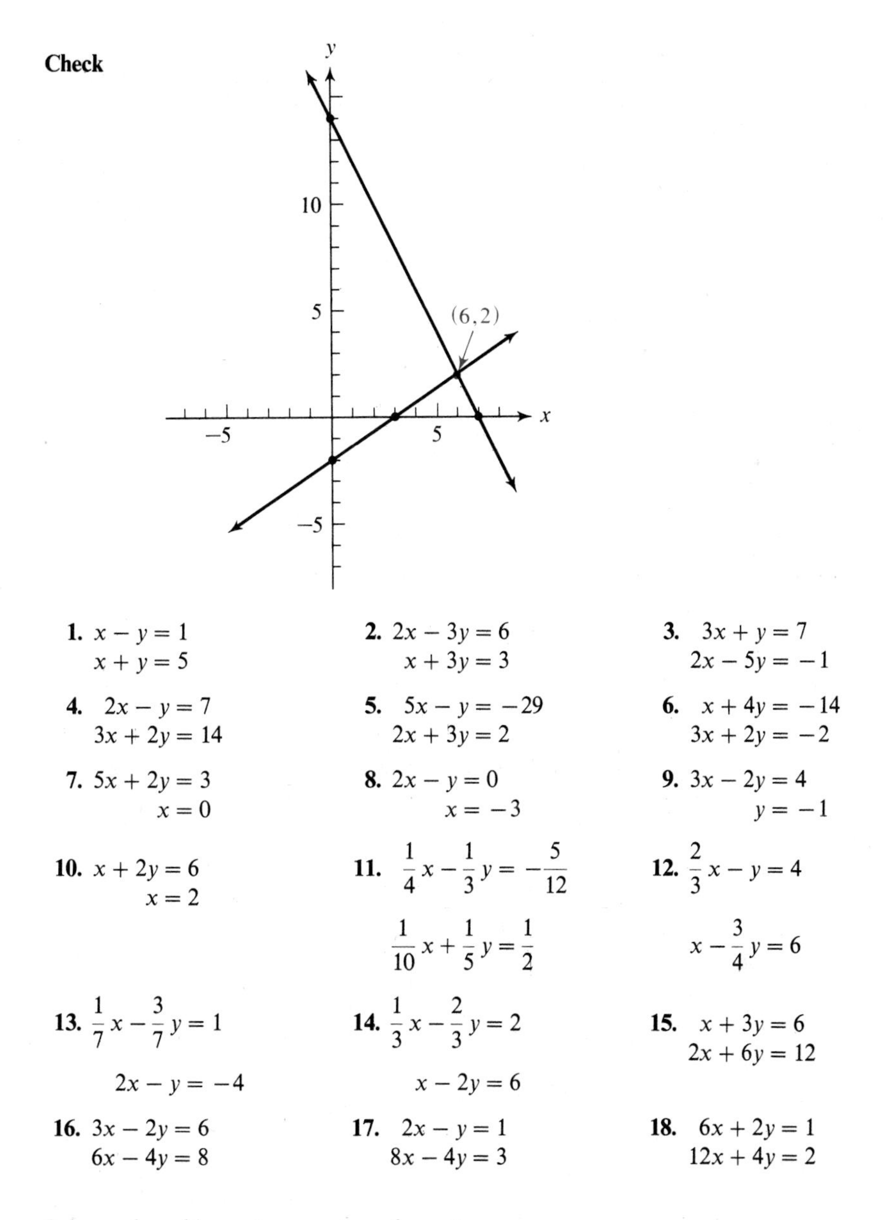

1. $x - y = 1$
$x + y = 5$

2. $2x - 3y = 6$
$x + 3y = 3$

3. $3x + y = 7$
$2x - 5y = -1$

4. $2x - y = 7$
$3x + 2y = 14$

5. $5x - y = -29$
$2x + 3y = 2$

6. $x + 4y = -14$
$3x + 2y = -2$

7. $5x + 2y = 3$
$x = 0$

8. $2x - y = 0$
$x = -3$

9. $3x - 2y = 4$
$y = -1$

10. $x + 2y = 6$
$x = 2$

11. $\frac{1}{4}x - \frac{1}{3}y = -\frac{5}{12}$
$\frac{1}{10}x + \frac{1}{5}y = \frac{1}{2}$

12. $\frac{2}{3}x - y = 4$
$x - \frac{3}{4}y = 6$

13. $\frac{1}{7}x - \frac{3}{7}y = 1$
$2x - y = -4$

14. $\frac{1}{3}x - \frac{2}{3}y = 2$
$x - 2y = 6$

15. $x + 3y = 6$
$2x + 6y = 12$

16. $3x - 2y = 6$
$6x - 4y = 8$

17. $2x - y = 1$
$8x - 4y = 3$

18. $6x + 2y = 1$
$12x + 4y = 2$

Solve each problem using a system of equations.

Example The sum of two numbers is 17 and one of the numbers is 4 less than 2 times the other. Find the numbers.

Solution Represent each number by a separate variable: Let x and y represent the numbers.

Represent the two independent conditions stated in the problem by two equations.

$$x + y = 17$$
$$x = 2y - 4$$

(*solution continued*)

Rewrite the equations in the form

$$x + y = 17$$
$$x - 2y = -4.$$

Solve the resulting system.

The numbers are 10 and 7.

Check Does $(10) + (7) = 17$? Yes.
Does $(10) = 2(7) - 4$? Yes.

19. The sum of two numbers is 24 and one of the numbers is 6 less than the other. Find the numbers.

20. The difference of two numbers is 14 and one of the numbers is 1 more than 2 times the other. Find the numbers.

21. If $\frac{1}{3}$ of an integer is added to $\frac{1}{2}$ the next consecutive integer, the sum is 33. Find the integers.

22. If $\frac{1}{2}$ of an integer is added to $\frac{1}{5}$ the next consecutive integer, the sum is 17. Find the integers.

23. The admission at a baseball game was \$1.50 for adults and \$.85 for children. The receipts were \$93.10 for 82 paid admissions. How many adults and how many children attended the game?

24. In an election, 7179 votes were cast for 2 candidates. If 6 votes had switched from the winner to the loser, the loser would have won by 1 vote. How many votes were cast for each candidate?

B

In Problems 25–30, set $u = \frac{1}{x}$ *and* $v = \frac{1}{y}$*, solve for u and v, and then solve for x and y.*

Example

$$\frac{4}{x} + \frac{3}{y} = 1$$

$$\frac{2}{x} - \frac{3}{y} = 2$$

Solution Substituting u for $\frac{1}{x}$ and v for $\frac{1}{y}$ in each equation, we obtain

$$4u + 3v = 1$$
$$2u - 3v = 2.$$

Solving for u and v, we have

$$6u = 3$$

$$u = \frac{1}{2},$$

and then we obtain

$$v = -\frac{1}{3}.$$

Since $u = \frac{1}{x}$ and $v = \frac{1}{y}$, we get $x = 2$ and $y = -3$.

Hence, the solution set is $\{(2, -3)\}$.

25. $\frac{1}{x} + \frac{1}{y} = 7$

$\frac{2}{x} + \frac{3}{y} = 16$

26. $\frac{1}{x} + \frac{2}{y} = -\frac{11}{12}$

$\frac{1}{x} + \frac{1}{y} = -\frac{7}{12}$

27. $\frac{5}{x} - \frac{6}{y} = -3$

$\frac{10}{x} + \frac{9}{y} = 1$

28. $\frac{1}{x} + \frac{2}{y} = 11$

$\frac{1}{x} - \frac{2}{y} = -1$

29. $\frac{1}{x} - \frac{1}{y} = 4$

$\frac{2}{x} - \frac{1}{2y} = 11$

30. $\frac{2}{3x} + \frac{3}{4y} = \frac{7}{12}$

$\frac{4}{x} - \frac{3}{4y} = \frac{7}{4}$

Solve each problem using a system of equations.

31. A sum of \$2000 is invested, part at 6% and the remainder at 8%. Find the amount invested at each rate if the yearly income from the two investments is \$132.

32. A man has \$1200 invested in two stocks, one of which returns 4% per year and the other 6% per year. How much has he invested in each stock if the income from the 4% stock is \$3 more than the income from the 6% stock?

33. In 1975 a vintner has a white wine that is 4 years older than a certain red wine. In 1965, the white wine was 2 times as old as the red wine. In what years were the wines produced?

34. In 1970 Mr. Evans died leaving a will saying that when his son was 2 times as old as his daughter, both would receive the funds from a trust. If the boy was 3 times as old as his sister in 1970 and if they received their funds in 1975, how old was each when their father died?

35. Find a and b so that the graph of $ax + by + 3 = 0$ passes through the points $(-1, 2)$ and $(-3, 0)$. [*Hint*: If the graph of the equation passes through the points, the components of each ordered pair must be valid replacements for x and y. Substitute -1 for x and 2 for y, and -3 for x and 0 for y to obtain a system of equations in a and b.]

36. Find a and b so that the solution set of the system

$$ax + by = 4$$
$$bx - ay = -3$$

is $\{(1, 2)\}$.

Solve for x and y in terms of a and b.

37. $x + y = b$
$x - y = a$

38. $ax - by = 2$
$bx - ay = 1$

By algebraic methods, determine whether or not the graphs of the three equations have any point in common.

39. $2x + 1 = y$
$x - 2y = 5$
$y = 3x$

40. $2y + 3x = 8$
$x - 3y = -5$
$y = 2x$

41. Solve the system

$$a_1x + b_1y = c_1$$
$$a_2x + b_2y = c_2$$

for x and y in terms of the coefficients and the constants c_1 and c_2.

42. Use the results of Problem 41 to find the solution set of the system of Problem 3.

10.2

LINEAR SYSTEMS IN THREE VARIABLES

A solution of an equation in three variables, such as

$$x + 2y - 3z = -4,$$

is an ordered triple of numbers (x, y, z), because all three of the variables must be replaced by numerals before we can decide whether the result is an equality. Thus, $(0, -2, 0)$ and $(-1, 0, 1)$ are solutions of this equation, while $(1, 1, 1)$ is not. There are, of course, infinitely many members in the solution set.

The solution set of a system of three linear equations in three variables, such as

(1) $$x + 2y - 3z = -4$$

(2) $$2x - y + z = 3$$

(3) $$3x + 2y + z = 10,$$

is the intersection of the solution sets of all three equations in the system. That is,

$$S = \{(x, y, z) \mid x + 2y - 3z = -4\} \cap \{(x, y, z) \mid 2x - y + z = 3\} \cap \{(x, y, z) \mid 3x + 2y + z = 10\}.$$

We seek solution sets of systems such as (1), (2), and (3) by methods analogous to those used in solving linear systems in two variables. Since graphic treatments would be three-dimensional (linear equations in three variables can be represented by planes, and their common point(s) of intersection, if any, would represent the solution set), we shall consider analytic solutions only. In the system presented here, we might begin by multiplying equation (1) by -2 and adding the result to 1 times equation (2), to produce

$$(4) \qquad -5y + 7z = 11,$$

which is satisfied by any ordered triple (x, y, z) that satisfies (1) and (2). Similarly, we can add -3 times equation (1) to 1 times equation (3) to obtain

$$(5) \qquad -4y + 10z = 22,$$

which is satisfied by any ordered triple (x, y, z) that satisfies both (1) and (3). We can now argue that any ordered triple satisfying the system (1), (2), and (3) will also satisfy the system

$$(4) \qquad -5y + 7z = 11$$

$$(5) \qquad -4y + 10z = 22.$$

Since the system (4) and (5) does not depend on x, the problem has been reduced to one of finding only the y and z components of the solution. The system (4) and (5) can be solved by the method of Section 10.1, which leads to the values $y = 2$ and $z = 3$. Since now any solution of (1), (2), and (3) must be of the form $(x, 2, 3)$, 2 can be substituted for y and 3 for z in (1) to obtain $x = 1$, so that the desired solution set is

$$\{(1, 2, 3)\}.$$

If at any step in this procedure the resulting linear combination vanishes or yields a contradiction, the system contains dependent equations or else two or three inconsistent equations, and it has either an infinite number of members or no members in its solution set.

The process of solving a system of equations can be reduced to a series of mechanical procedures, as illustrated by the first example in Exercise 10.2.

EXERCISE 10.2

A

Solve. If the system does not have a unique (one and only one) solution, so state.

Example

$$(1) \qquad x + 2y - z = -3$$

$$(2) \qquad x - 3y + z = 6$$

$$(3) \qquad 2x + y + 2z = 5$$

(*solution on page* 314)

Solution Multiply equation (1) by -1 and add the result to 1 times equation (2) to get (4); multiply equation (1) by -2 and add the result to 1 times equation (3) to get (5).

(4) $$-5y + 2z = 9$$

(5) $$-3y + 4z = 11$$

Multiply equation (4) by -2 and add the result to 1 times equation (5).

$$7y = -7$$
$$y = -1$$

Substitute -1 for y in either (4) or (5)—we will use (4)—and solve for z.

$$-5(-1) + 2z = 9$$
$$z = 2$$

Substitute -1 for y and 2 for z in (1), (2), or (3)—we will use (1)—and solve for x.

$$x + 2(-1) - 2 = -3$$
$$x = 1$$

The solution set is $\{(1, -1, 2)\}$.

1. $\begin{aligned} x + y + z &= 2 \\ 2x - y + z &= -1 \\ x - y - z &= 0 \end{aligned}$

2. $\begin{aligned} x + y + z &= 1 \\ 2x - y + 3z &= 2 \\ 2x - y - z &= 2 \end{aligned}$

3. $\begin{aligned} x + y + 2z &= 0 \\ 2x - 2y + z &= 8 \\ 3x + 2y + z &= 2 \end{aligned}$

4. $\begin{aligned} x - 2y + 4z &= -3 \\ 3x + y - 2z &= 12 \\ 2x + y - 3z &= 11 \end{aligned}$

5. $\begin{aligned} x - 2y + z &= -1 \\ 2x + y - 3z &= 3 \\ 3x + 3y - 2z &= 10 \end{aligned}$

6. $\begin{aligned} x + 5y - z &= 2 \\ 3x - 9y + 3z &= 6 \\ x - 3y + z &= 4 \end{aligned}$

7. $\begin{aligned} x - 2y + 3z &= 4 \\ 2x - y + z &= 1 \\ 3x - 3y + 4z &= 5 \end{aligned}$

8. $\begin{aligned} 2x - 3y + z &= 3 \\ x - y - 2z &= -1 \\ -x + 2y - 3z &= -4 \end{aligned}$

9. $\begin{aligned} 2x + z &= 7 \\ y - z &= -2 \\ x + y &= 2 \end{aligned}$

10. $\begin{aligned} 5y - 8z &= -19 \\ 5x - 8z &= 6 \\ 3x - 2y &= 12 \end{aligned}$

11. $\begin{aligned} x - \frac{1}{2}y - \frac{1}{2}z &= 4 \\ x - \frac{3}{2}y - 2z &= 3 \\ \frac{1}{4}x + \frac{1}{4}y - \frac{1}{4}z &= 0 \end{aligned}$

12. $\begin{aligned} x + 2y + \frac{1}{2}z &= 0 \\ x + \frac{3}{5}y - \frac{2}{5}z &= \frac{1}{5} \\ 4x - 7y - 7z &= 6 \end{aligned}$

Example The sum of three numbers is 12. Twice the first number is equal to the second, and the third is equal to the sum of the other two. Find the numbers.

Solution Represent each number by a separate variable: Let x, y, and z represent numbers.

Write the three conditions stated in the problem as three equations:

$$x + y + z = 12$$
$$2x = y$$
$$x + y = z$$

Rewrite the equations in the form

$$\begin{aligned} &(1) & x + y + z &= 12 \\ &(2) & 2x - y \quad &= 0 \\ &(3) & x + y - z &= 0. \end{aligned}$$

Multiply equation (1) by 1 and add the result to 1 times equation (3) to obtain

$$2x + 2y = 12.$$

Multiply this equation by $\frac{1}{2}$ and add the result to 1 times equation (2).

$$3x = 6$$

Solve this equation to get $x = 2$, and substitute the x value in (2) to obtain $y = 4$. Substitute these values in (1), (2), or (3) to obtain $z = 6$.

The numbers are 2, 4, and 6.

13. The sum of three numbers is 15. The second equals 2 times the first and the third equals the second. Find the numbers.

14. The sum of three numbers is 2. The first number is equal to the sum of the other two, and the third number is the result of subtracting the first from the second. Find the numbers.

15. A box contains $6.25 in nickels, dimes, and quarters. There are 85 coins in all with 3 times as many nickels as dimes. How many coins of each kind are there?

16. The perimeter of a triangle is 155 inches. Side x is 20 inches shorter than side y, and side y is 5 inches longer than side z. Find the lengths of the sides of the triangle.

B

17. The equation for a circle can be written $x^2 + y^2 + ax + by + c = 0$. Find the equation of the circle whose graph contains the points (2, 3), (3, 2), and $(-4, -5)$.

18. Find values for a, b, and c such that the graph of $x^2 + y^2 + ax + by + c = 0$ will contain the points $(-2, 3)$, (1, 6), and (2, 4).

19. Find values for a, b, and c such that the graph of $y = ax^2 + bx + c$ will contain the points $(-1, 2)$, (1, 6), and (2, 11).

20. Show that the system

$$\begin{aligned} x + y + 2z &= 2 \\ 2x - y + z &= 3 \end{aligned}$$

has an infinite number of members in its solution set. [*Hint*: Express x in terms of z alone; then express y in terms of z alone.] List two ordered triples that are solutions.

10.3

SECOND-DEGREE SYSTEMS IN TWO VARIABLES— SOLUTION BY SUBSTITUTION

Approximate solutions of systems of equations in two variables, where one or both of the equations are quadratic, can often be found by graphing both equations and estimating the coordinates of any points they have in common. For example, to find the solution set of the system

(1) $$x^2 + y^2 = 26$$

(2) $$x + y = 6,$$

we graph the equations on the same set of axes, as shown in Figure 10.4, and observe that the graphs appear to intersect at (1, 5) and (5, 1). The solution set of the system (1) and (2) is, in fact,

$$\{(1, 5), (5, 1)\}.$$

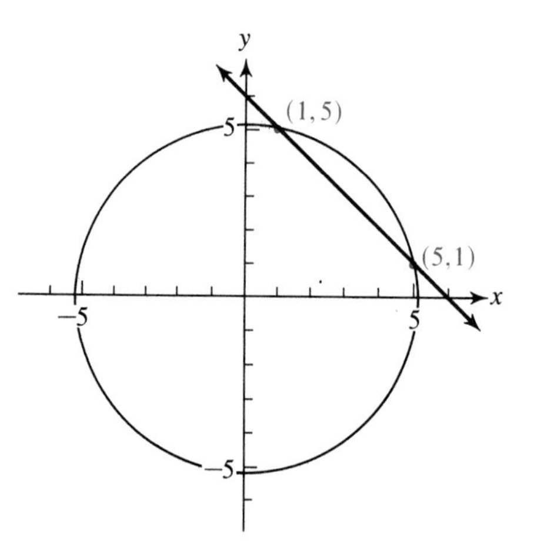

Figure 10.4

However, solving second-degree systems graphically on the real plane may produce only approximations to real solutions, and we cannot expect to locate solutions among the complex numbers. It is therefore more practical to concentrate on analytic methods of solution, since the results are exact and we can obtain complex solutions. It is suggested, however, that whenever feasible you sketch the graphs of the equations as a rough check on an analytic solution.

One of the most useful techniques available for finding solution sets for systems of equations is **substitution**. Consider the system above,

$$(1) \qquad x^2 + y^2 = 26$$

$$(2) \qquad x + y = 6.$$

Equation (2) can be written in the form

$$(3) \qquad y = 6 - x,$$

and we can argue that for any ordered pair (x, y) in the solution set of both (1) and (2), x and y in (1) represent the same numbers as x and y in (3), and hence the substitution axiom can be invoked to replace y in (1) by its equal $(6 - x)$ from (3). This will produce

$$(4) \qquad x^2 + (6 - x)^2 = 26,$$

which will have as a solution set those values of x for which the ordered pair (x, y) is a common solution of (1) and (2). Rewriting (4) equivalently, we have

$$\begin{aligned} x^2 + 36 - 12x + x^2 &= 26 \\ 2x^2 - 12x + 10 &= 0 \\ x^2 - 6x + 5 &= 0 \\ (x - 5)(x - 1) &= 0, \end{aligned}$$

from which x is either 1 or 5. Now, by replacing x in (3) with each of these numbers, we have

$$y = 6 - (1) = 5 \qquad \text{and} \qquad y = 6 - (5) = 1,$$

so that the solution set of the system (1) and (2) is $\{(1, 5), (5, 1)\}$.

Check this solution and notice that these ordered pairs are also solutions of (1). Using (1) rather than (2) or (3) to obtain values for the y component, we would have

$$\begin{aligned} (1)^2 + y^2 &= 26 \qquad & (5)^2 + y^2 &= 26 \\ y &= \pm 5 & y &= \pm 1 \end{aligned}$$

and the solutions obtained are $(1, 5)$, $(1, -5)$, $(5, 1)$, and $(5, -1)$. However, $(1, -5)$ and $(5, -1)$ are not solutions of (2). Therefore, the solution set is again $\{(1, 5), (5, 1)\}$. This example suggests that if the degrees of equations differ, one component of a solution should be substituted in the equation of lower degree in order to find *only* those ordered pairs that are solutions of *both* equations.

In the foregoing example, the components of each solution are real numbers. Therefore, the solutions are elements in the Cartesian product $R \times R$, and their graphs are the points of intersection of the graphs of each equation, as shown in Figure 10.4. If one or more of the components of the solutions of a system are imaginary numbers, we can find these solutions in $C \times C$, the Cartesian product

$\{(x, y) | x \in C \text{ and } y \in C\}$. However, in such cases, the graphs in the real plane do not have points of intersection. For example, consider the system

$$(5) \qquad x^2 + y^2 = 26$$

$$(6) \qquad x + y = 8.$$

Solving (6) for y, we have

$$(6a) \qquad y = 8 - x.$$

Substituting $8 - x$ for y in (5) and simplifying yields

$$\begin{aligned} x^2 + (8 - x)^2 &= 26 \\ x^2 + 64 - 16x + x^2 &= 26 \\ 2x^2 - 16x + 38 &= 0 \\ x^2 - 8x + 19 &= 0. \end{aligned}$$

Using the quadratic formula to solve for x, we obtain

$$\begin{aligned} x &= \frac{8 \pm \sqrt{64 - 76}}{2(1)} = \frac{8 \pm \sqrt{-12}}{2} \\ &= \frac{8 \pm 2i\sqrt{3}}{2} = \frac{2(4 \pm i\sqrt{3})}{2} = 4 \pm i\sqrt{3}. \end{aligned}$$

Then, substituting $4 + i\sqrt{3}$ for x in (6a) gives $y = 4 - i\sqrt{3}$, and substituting $4 - i\sqrt{3}$ for x in (6a) gives $y = 4 + i\sqrt{3}$. Hence, the solution set is $\{(4 + i\sqrt{3}, 4 - i\sqrt{3}), (4 - i\sqrt{3}, 4 + i\sqrt{3})\}$; the graphs of the equations are as shown in Figure 10.5.

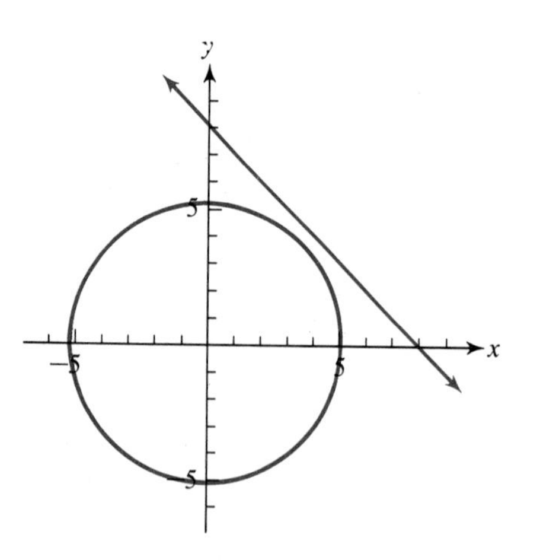

Figure 10.5

The technique of solution by substitution can be used very easily with systems containing two first-degree equations, or one first-degree and one higher-degree equation, but is less satisfactory for systems where both equations are of degree greater than one in both variables.

EXERCISE 10.3

A

Solve by the method of substitution. In Problems 1–12, check the solutions by sketching the graphs of the equations and estimating the coordinates of any points of intersection.

Example (1) $y = x^2 + 2x + 1$

(2) $y - x = 3$

Solution Solve (2) explicitly for y.

(2a) $y = x + 3$

Substitute $(x + 3)$ for y in (1).

$$x + 3 = x^2 + 2x + 1$$

Solve for x.

$$x^2 + x - 2 = 0$$

$$(x + 2)(x - 1) = 0$$

$$x = -2 \qquad x = 1$$

Substitute each of these values in (2a) to determine values for y.

If $x = -2$, then $y = 1$.

If $x = 1$, then $y = 4$.

The solution set is $\{(-2, 1), (1, 4)\}$.

Check

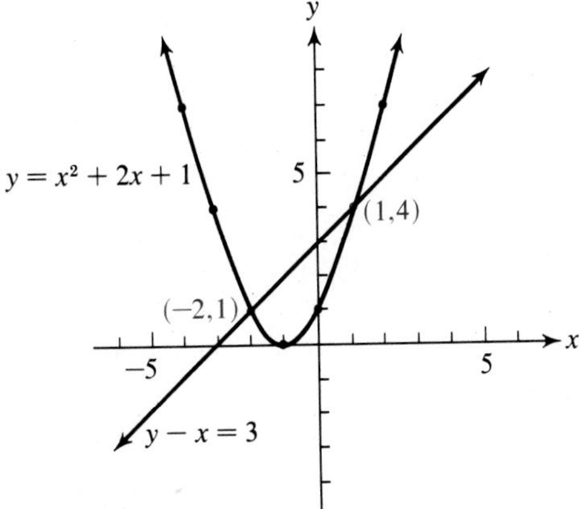

1. $y = x^2 - 5$
$y = 4x$

2. $y = x^2 - 2x + 1$
$y + x = 3$

3. $x^2 + y^2 = 13$
$x + y = 5$

4. $x^2 + 2y^2 = 12$
$2x - y = 2$

5. $x + y = 1$
$xy = -12$

6. $2x - y = 9$
$xy = -4$

7. $xy = 4$
$x^2 + y^2 = 8$

8. $x^2 - y^2 = 35$
$xy = 6$

9. $x^2 + y^2 = 9$
$y = 4$

10. $2x^2 - 4y^2 = 12$
$x = 4$

11. $x^2 + 2y^2 = 6$
$x + y = 10$

12. $x^2 + 9y^2 = 36$
$x - 2y = -8$

13. $x^2 - xy - 2y^2 = 4$
$x - y = 2$

14. $x^2 - 2x + y^2 = 3$
$2x + y = 4$

15. $2x^2 - 5xy + 2y^2 = 5$
$2x - y = 1$

16. $2x^2 + xy + y^2 = 9$
$-x + 3y = 9$

17. The sum of the squares of two positive numbers is 13. If 2 times the first number is added to the second, the sum is 7. Find the numbers.

18. The sum of two numbers is 6 and their product is $35/4$. Find the numbers.

19. The perimeter of a rectangle is 26 inches and the area is 12 square inches. Find the dimensions of the rectangle.

20. The area of a rectangle is 216 square feet. If the perimeter is 60 feet, find the dimensions of the rectangle.

21. A rectangle has a perimeter of 18 feet. If the length is decreased by 5 feet and the width is increased by 12 feet, the area is doubled. Find the dimensions of the original rectangle.

22. The annual income from an investment is $32. If the amount invested were $200 more and the rate $\frac{1}{2}\%$ less, the annual income would be $35. What is the amount and rate of the investment?

B

23. At a constant temperature the pressure (P) and volume (V) of a gas are related by the equation $PV = K$, where K is a constant. The product of the pressure (in pounds per square inch) and the volume (in cubic inches) of a certain gas is 30 inch-pounds. If the temperature remains constant as the pressure is increased 4 pounds per square inch, the volume is decreased by 2 cubic inches. Find the original pressure and volume of the gas.

24. What relationship must exist between the numbers a and b so that the solution set of the system

$$x^2 + y^2 = 25$$
$$y = ax + b$$

will have two ordered pairs of real numbers? One ordered pair of real numbers? No ordered pairs of real numbers? [*Hint*: Use substitution and consider the nature of the roots of the resulting quadratic equation.]

25. Consider the system

(1) $$x^2 + y^2 = 8$$
(2) $$xy = 4.$$

We can solve this system by substituting $4/x$ for y in (1) to obtain

$$x^2 + \frac{16}{x^2} = 8,$$

from which we have $x = 2$ or $x = -2$. Now if we obtain the y components of the solution from (2), we find that for $x = 2$, $y = 2$, and for $x = -2$, $y = -2$. But if we seek y components from (1), we have that for $x = 2$, $y = \pm 2$ and for $x = -2$, $y = \pm 2$. Discuss the fact that we seem to obtain two more solutions from (1) than from (2). What is the solution set of the system?

26. Graph equations (1) and (2) of Problem 25 and relate your discussion of Problem 25 to these graphs.

10.4

SECOND-DEGREE SYSTEMS IN TWO VARIABLES— SOLUTION BY OTHER METHODS

If both the equations in a system are second-degree in both variables, the use of linear combinations of members of the equations often provides a simpler means of solution than does substitution. For example, consider the system

$$(1) \qquad 4x^2 + y^2 = 25$$

$$(2) \qquad x^2 - y^2 = -5.$$

By forming a linear combination using 1 times equation (1) and 1 times equation (2), we have

$$5x^2 = 20,$$

from which

$$x = 2 \qquad \text{or} \qquad x = -2,$$

and we have the x components of the members of the solution set of the system (1) and (2). Substituting 2 for x in either (1) or (2)—we will use (1)—we obtain

$$4(2)^2 + y^2 = 25$$
$$y^2 = 25 - 16$$
$$y^2 = 9,$$

from which

$$y = 3 \qquad \text{or} \qquad y = -3.$$

Thus, the ordered pairs (2, 3) and (2, −3) are in the solution set of the system. Substituting −2 for x in (1) or (2)—this time we shall use (2)—gives us

$$(-2)^2 - y^2 = -5$$
$$-y^2 = -5 - 4$$
$$y^2 = 9,$$

so that

$$y = 3 \qquad \text{or} \qquad y = -3.$$

Thus, the ordered pairs $(-2, 3)$ and $(-2, -3)$ are also solutions of the system, and the complete solution set is

$$\{(2, 3), (2, -3), (-2, 3), (-2, -3)\}.$$

For an example of a slightly different procedure, consider the system

$$x^2 + y^2 = 5 \tag{3}$$

$$x^2 - 2xy + y^2 = 1. \tag{4}$$

By forming a linear combination using 1 times equation (3) and -1 times equation (4), we have

$$2xy = 4$$

$$xy = 2, \tag{5}$$

which has a solution set containing all the ordered pairs that satisfy both (3) and (4). Therefore, forming the new system

$$x^2 + y^2 = 5 \tag{3}$$

$$xy = 2, \tag{5}$$

we can be sure that the solution set of this system is the same as the solution set of the system (3) and (4). This latter system can be solved by substitution. We have, from (5),

$$y = \frac{2}{x}.$$

Replacing y in (3) by $\frac{2}{x}$, we find

$$x^2 + \left(\frac{2}{x}\right)^2 = 5,$$

from which

$$x^2 + \frac{4}{x^2} = 5. \tag{6}$$

Multiplying each member by x^2, we have

$$x^4 + 4 = 5x^2 \tag{7}$$

$$x^4 - 5x^2 + 4 = 0, \tag{7a}$$

which is quadratic in x^2. Factoring the left-hand member of (7a), we obtain

$$(x^2 - 1)(x^2 - 4) = 0,$$

from which

$$x^2 - 1 = 0 \qquad \text{or} \qquad x^2 - 4 = 0,$$

and

$$x = 1, \quad x = -1, \qquad x = 2, \quad x = -2.$$

Since the step from (6) to (7) was a nonelementary transformation, we are careful to note that these all satisfy (6). [What value for x would have to be excluded if it appeared as a solution of (7)?] Now substituting 1, -1, 2, and -2 for x in (5), we have

$$\begin{aligned} &\text{for } x = 1, && y = 2; \\ &\text{for } x = -1, && y = -2; \\ &\text{for } x = 2, && y = 1; \\ &\text{for } x = -2, && y = -1; \end{aligned}$$

and the solution set of either system (3) and (5) or system (3) and (4) is

$$\{(1, 2), (-1, -2), (2, 1), (-2, -1)\}.$$

There are other techniques involving substitution in conjunction with linear combinations that are useful in handling systems of higher-degree equations, but they are all similar to those illustrated. Each system should be scrutinized for some means of finding an equivalent system that will lend itself to solution by linear combination or substitution.

EXERCISE 10.4

A

Solve each system.

Example (1) $\quad 3x^2 + y^2 = 15$

(2) $\quad 11x^2 - 2y^2 = 4$

Solution Obtain a linear combination using 2 times equation (1) and 1 times equation (2), and solve for x.

$$6x^2 + 2y^2 = 30$$
$$11x^2 - 2y^2 = 4$$

$$17x^2 = 34$$
$$x^2 = 2$$
$$x = \sqrt{2} \quad \text{or} \quad x = -\sqrt{2}$$

(solution continued)

Substitute values for x in (1) or (2)—we will use (1)—to obtain associated values for y.

$$3(\sqrt{2})^2 + y^2 = 15 \qquad 3(-\sqrt{2})^2 + y^2 = 15$$
$$6 + y^2 = 15 \qquad 6 + y^2 = 15$$
$$y^2 = 9 \qquad y^2 = 9$$
$$y = \pm 3 \qquad y = \pm 3$$

The solution set is $\{(\sqrt{2}, 3), (\sqrt{2}, -3), (-\sqrt{2}, 3), (-\sqrt{2}, -3)\}$.

1. $x^2 + y^2 = 10$
$9x^2 + y^2 = 18$

2. $x^2 + 4y^2 = 52$
$x^2 + y^2 = 25$

3. $x^2 + 4y^2 = 17$
$3x^2 - y^2 = -1$

4. $9x^2 + 16y^2 = 100$
$x^2 + y^2 = 8$

5. $x^2 - y^2 = 7$
$2x^2 + 3y^2 = 24$

6. $x^2 + 4y^2 = 25$
$4x^2 + y^2 = 25$

7. $3x^2 + 4y^2 = 16$
$x^2 - y^2 = 3$

8. $4x^2 + 3y^2 = 12$
$x^2 + 3y^2 = 12$

9. $4x^2 - 9y^2 + 132 = 0$
$x^2 + 4y^2 - 67 = 0$

10. $16y^2 + 5x^2 - 26 = 0$
$25y^2 - 4x^2 - 17 = 0$

11. $2x^2 + xy - 4y^2 = -12$
$x^2 - 2y^2 = -4$

12. $x^2 + 2xy - y^2 = 14$
$x^2 - y^2 = 8$

13. $x^2 + 3xy - y^2 = -3$
$x^2 - xy - y^2 = 1$

14. $2x^2 + xy - 2y^2 = 16$
$x^2 + 2xy - y^2 = 17$

B

15. $3x^2 + 3xy - y^2 = 35$
$x^2 - xy - 6y^2 = 0$
[*Hint*: Factor $x^2 - xy - 6y^2$.]

16. $x^2 - xy + y^2 = 21$
$x^2 + 2xy - 8y^2 = 0$

17. $\dfrac{1}{x^2} + \dfrac{4}{y^2} = 15$

$\dfrac{2}{x^2} - \dfrac{3}{y^2} = -14$

18. $\dfrac{1}{x^2} + \dfrac{3}{y^2} = 7$

$\dfrac{2}{x^2} - \dfrac{5}{y^2} = 3$

19. How many *real* solutions are possible for systems of *independent* equations that consist of (support your answers with sketches):

a. two linear equations in two variables?

b. one linear equation and one quadratic equation in two variables?

c. two quadratic equations in two variables?

CHAPTER SUMMARY

[10.1] The system of linear equations

$$a_1x + b_1y = c_1$$
$$a_2x + b_2y = c_2$$

has no solution if the equations are **inconsistent**, infinitely many solutions if the equations are **dependent**, and otherwise exactly one solution.

The system has exactly one solution if

$$\frac{a_1}{a_2} \neq \frac{b_1}{b_2};$$

the equations are inconsistent if

$$\frac{a_1}{a_2} = \frac{b_1}{b_2} \neq \frac{c_1}{c_2};$$

and they are dependent if

$$\frac{a_1}{a_2} = \frac{b_1}{b_2} = \frac{c_1}{c_2}.$$

The unique solution of two linear equations in two variables (if such a solution exists) can be obtained by using linear combinations of the members of the equations. Any ordered pair (x, y) that satisfies the equations

$$a_1x + b_1y = c_1$$
$$a_2x + b_2y = c_2$$

also satisfies the equation

$$A(a_1x + b_1y) + B(a_2x + b_2y) = Ac_1 + Bc_2$$

for all real numbers A and B.

Systems of equations can be used to solve practical problems.

[10.2] The solution of a system of three linear equations in three variables (if a solution exists) can be obtained by first using linear combinations to form a system of two equations in two variables with a solution containing components that are the respective components of the solution of the original system in three variables. The third component can be obtained by substituting these two values into any one of the equations of the original system.

[10.3–10.4] Systems of equations in two variables in which either or both equations are second-degree in one or both variables may have solutions with real components or with imaginary components, or solutions of both kinds. Such systems can be solved by using substitution methods or by using linear combinations of the members of the equations in the system.

REVIEW EXERCISES

[10.1] *Solve each system by linear combinations.*

1. $x + 5y = 18$
$x - y = -3$

2. $x + 5y = 11$
$2x + 3y = 8$

3. $\frac{2}{3}x - 3y = 8$
$x + \frac{3}{4}y = 12$

4. $\frac{3}{x} - \frac{1}{y} = \frac{7}{2}$
$\frac{2}{x} + \frac{3}{y} = 7$

State whether the equations in each system are dependent, inconsistent, or neither dependent nor inconsistent.

5. a. $2x - 3y = 4$
$x + 2y = 7$

b. $2x - 3y = 4$
$6x - 9y = 12$

6. a. $2x - 3y = 4$
$6x - 9y = 4$

b. $x - y = 6$
$x + y = 6$

[10.2] *Solve each system by linear combinations.*

7. $x + 3y - z = 3$
$2x - y + 3z = 1$
$3x + 2y + z = 5$

8. $x + y + z = 2$
$3x - y + z = 4$
$2x + y + 2z = 3$

9. $\frac{1}{5}x + \frac{1}{5}z = 1$
$\frac{1}{8}y - \frac{1}{8}z = -1$
$\frac{2}{7}x + \frac{1}{7}z = 1$

10. $\frac{1}{4}x + y + z = -5$
$x - \frac{2}{3}y + \frac{1}{3}z = -\frac{4}{3}$
$\frac{1}{2}x - y + \frac{1}{4}z = -1$

11. $\frac{1}{2}x + y + z = 3$
$x - 2y - \frac{1}{3}z = -5$
$\frac{1}{2}x - 3y - \frac{2}{3}z = -6$

12. $\frac{3}{4}x - \frac{1}{2}y + 6z = 2$
$\frac{1}{2}x + y - \frac{3}{4}z = 0$
$\frac{1}{4}x + \frac{1}{2}y - \frac{1}{2}z = 0$

[10.3–10.4] *Solve each system by substitution or linear combination.*

13. $x + 3y^2 = 4$
$x = 3$

14. $x^2 + 2y^2 = -8$
$y = -2$

15. $x^2 + y = 3$
$5x + y = 7$

16. $x^2 + 3xy + x = -12$
$2x - y = 7$

17. $6x^2 - y^2 = 1$
$3x^2 + 2y^2 = 13$

18. $2x^2 + 5y^2 - 53 = 0$
$4x^2 + 3y^2 - 43 = 0$

11

NATURAL NUMBER FUNCTIONS

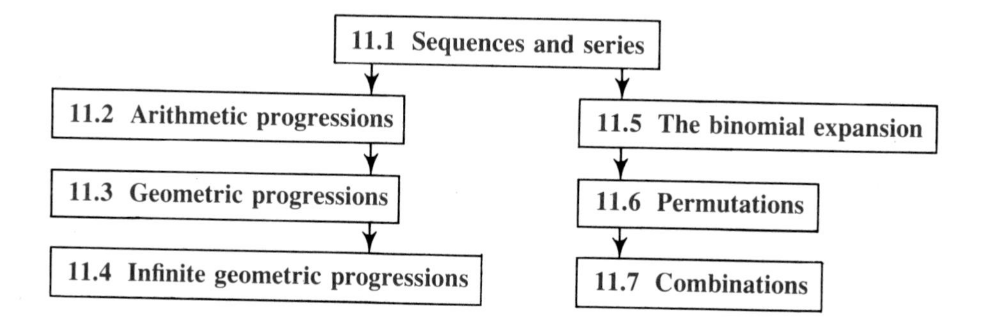

11.1

SEQUENCES AND SERIES

A function whose domain is a set of successive positive integers, for example, a function defined by an equation such as

$$(1) \qquad s(n) = 2n - 1 \qquad (n \in \{3, 4, 5\})$$

or

$$(2) \qquad s(n) = n + 3 \qquad (n \in \{1, 2, 3, \ldots\}),$$

is called a **sequence function**. The function defined by (1) is called a **finite sequence**, and the function defined by (2) is called an **infinite sequence**. The elements in the range of such functions arranged in the order

$$s(3), s(4), s(5) \qquad \text{or} \qquad s(1), s(2), s(3), \ldots$$

are said to form a **sequence**, and the elements are referred to as the **terms** of the sequence. Thus, the sequence associated with (2) is found by successively substituting the numbers 1, 2, 3, ... for n:

$$\begin{aligned} s(1) &= (1) + 3 = 4 \\ s(2) &= (2) + 3 = 5 \\ s(3) &= (3) + 3 = 6 \\ s(4) &= (4) + 3 = 7, \end{aligned}$$

and the first four terms are 4, 5, 6, and 7. The ***n*th** term, or **general term**, is $n + 3$. As another example, the first five terms of the sequence defined by the equation

$$s(n) = \frac{3}{2n - 1} \qquad (n \in \{1, 2, 3, \ldots\})$$

are $3/1$, $3/3$, $3/5$, $3/7$, and $3/9$, and the twenty-fifth term is

$$s(25) = \frac{3}{2(25) - 1} = \frac{3}{49}.$$

The notation ordinarily used for the terms in a sequence is not function notation as such; rather, it is customary to denote a term in a sequence by means of a subscript. Thus, we will use s_n rather than $s(n)$, and the sequence $s(1), s(2), s(3), \ldots$ will appear as $s_1, s_2, s_3, \ldots$.

Associated with any sequence is a **series**; the series is defined as the sum of the terms in the sequence. Thus, associated with the finite sequence

$$4, 7, 10, \ldots, 3n + 1 \tag{3}$$

is the finite series

$$S_n = 4 + 7 + 10 + \cdots + (3n + 1), \tag{4}$$

and associated with the sequence

$$x, x^2, x^3, x^4, \ldots, x^n$$

is the series

$$S_n = x + x^2 + x^3 + x^4 + \cdots + x^n.$$

Since the terms in the series are the same as those in the corresponding sequence, we can refer to the first term or the second term or the general term of a series in the same manner as we do for a sequence.

A series with a general term that is known can be represented in a very convenient, compact way by means of the symbol $\sum$ (**sigma** or **summation**) in conjunction with the general term; this denotes the sum of all the terms in the series. For example, series (4) can be written

$$S_n = \sum_{i=1}^{n} (3i + 1),$$

where we understand that S_n is the series with terms obtained by successively replacing i in the expression $3i + 1$ with the numbers, 1, 2, 3, ..., n. Thus,

$$S_6 = \sum_{i=1}^{6} (3i + 1)$$

appears in expanded form as

$$S_6 = 4 + 7 + 10 + 13 + 16 + 19.$$

The variable used in conjunction with summation notation (in this case, i) is called the **index of summation**; the set of integers over which we sum in this case is $\{1, 2, 3, 4, 5, 6\}$. The use of the symbol i as an index of summation should not be confused with its use as an imaginary unit in the set of complex numbers;

alternatively, the summation index can be any letter such as j, k, l, etc., and clearly,

$$\sum_{i=1}^{6}(3i+1)=\sum_{j=1}^{6}(3j+1)=\sum_{k=1}^{6}(3k+1)=\cdots$$

The first member of the replacement set for the index of summation is not necessarily 1. For example,

$$\sum_{i=3}^{6}(3i+1)=[3(3)+1]+[3(4)+1]+[3(5)+1]+[3(6)+1]$$
$$=10+13+16+19,$$

where the first replacement for i is 3 and the set of integers over which we sum is $\{3,4,5,6\}$. Note also that the series contains four terms, which is one more than the difference between the last and the first replacement for i. In general, the series $\sum_{i=a}^{b} s_i$ contains $b-a+1$ terms.

To show that a series has an infinite number of terms—that is, has no last term—we adopt a special notation. For example,

$$S_\infty=\sum_{i=4}^{\infty}(3i+1)$$

denotes the series which appears in expanded form as

$$S_\infty=13+16+19+22+\cdots, \tag{5}$$

where in this case i has been replaced by 4, 5, 6, 7, ...

A series can be represented in sigma notation by various general terms and different ranges. For example, both

$$\sum_{i=5}^{\infty}(3i-2) \qquad \text{and} \qquad \sum_{i=6}^{\infty}(3i-5)$$

also represent (5). This can be verified by writing the first few terms in each series.

EXERCISE 11.1

A

Find the first four terms in a sequence with the general term as given.

Examples a. $s_n = \dfrac{n(n+1)}{2}$ b. $s_n = (-1)^n 2^n$

Solutions a. $s_1 = \dfrac{1(1+1)}{2} = 1$

$s_2 = \dfrac{2(2+1)}{2} = 3$

$s_3 = \dfrac{3(3+1)}{2} = 6$

$s_4 = \dfrac{4(4+1)}{2} = 10$

The first four terms are 1, 3, 6, 10.

b. $s_1 = (-1)^1 2^1 = -2$

$s_2 = (-1)^2 2^2 = 4$

$s_3 = (-1)^3 2^3 = -8$

$s_4 = (-1)^4 2^4 = 16$

The first four terms are $-2, 4, -8, 16$.

1. $s_n = n - 5$ **2.** $s_n = 2n - 3$ **3.** $s_n = \dfrac{n^2 - 2}{2}$

4. $s_n = \dfrac{3}{n^2 + 1}$ **5.** $s_n = 1 + \dfrac{1}{n}$ **6.** $s_n = \dfrac{n}{2n - 1}$

7. $s_n = \dfrac{n(n-1)}{2}$ **8.** $s_n = \dfrac{5}{n(n+1)}$ **9.** $s_n = (-1)^n$

10. $s_n = (-1)^{n+1}$ **11.** $s_n = \dfrac{(-1)^n(n-2)}{n}$ **12.** $s_n = (-1)^{n-1}3^{n+1}$

Write in expanded form.

Examples a. $\sum_{i=2}^{4} (i^2 + 1)$ b. $\sum_{k=1}^{\infty} (-1)^k 2^{k+1}$

Solutions a. i takes values 2, 3, 4

$i = 2, \quad (2)^2 + 1 = 5$
$i = 3, \quad (3)^2 + 1 = 10$
$i = 4, \quad (4)^2 + 1 = 17$

Expanded form: $5 + 10 + 17$.

b. k takes values $1, 2, 3, \ldots$

$k = 1, \quad (-1)^1 2^{1+1} = (-1)(4) = -4$
$k = 2, \quad (-1)^2 2^{2+1} = (1)(8) = 8$
$k = 3, \quad (-1)^3 2^{3+1} = (-1)(16) = -16$

Expanded form: $-4 + 8 - 16 + \cdots$.

13. $\sum_{i=1}^{4} i^2$ **14.** $\sum_{i=1}^{3} (3i - 2)$ **15.** $\sum_{j=5}^{7} (j - 2)$

16. $\sum_{j=2}^{6} (j^2 + 1)$ **17.** $\sum_{k=1}^{4} k(k+1)$ **18.** $\sum_{i=2}^{6} \dfrac{i}{2}(i + 1)$

19. $\sum_{i=1}^{4} \frac{(-1)^i}{2^i}$

20. $\sum_{i=3}^{5} \frac{(-1)^{i+1}}{i-2}$

21. $\sum_{i=1}^{\infty} (2i-1)$

22. $\sum_{j=1}^{\infty} \frac{1}{j}$

23. $\sum_{k=0}^{\infty} \frac{1}{2^k}$

24. $\sum_{k=0}^{\infty} \frac{k}{1+k}$

B

Write in sigma notation. (There are no unique solutions.)

Examples **a.** $5 + 8 + 11 + 14$ **b.** $x^2 + x^4 + x^6 + \cdots + x^{2n}$

Solutions Find a general term.

a. $3i + 2$

$$\sum_{i=1}^{4} (3i + 2)$$

b. x^{2i}

$$\sum_{i=1}^{n} x^{2i}$$

25. $1 + 2 + 3 + 4$

26. $2 + 4 + 6 + 8$

27. $x + x^3 + x^5 + x^7$

28. $x^3 + x^5 + x^7 + x^9 + x^{11}$

29. $1 + 4 + 9 + 16 + 25$

30. $1 + 8 + 27 + 64 + 125$

Examples **a.** $3 + 6 + 9 + 12 + \cdots$ **b.** $\frac{3}{5} + \frac{5}{7} + \frac{7}{9} + \frac{9}{11} + \cdots$

Solutions Find a general term.

a. $3i$

$$\sum_{i=1}^{\infty} 3i$$

b. $\frac{2i+1}{2i+3}$

$$\sum_{i=1}^{\infty} \frac{2i+1}{2i+3}$$

31. $\frac{1}{2} + \frac{2}{3} + \frac{3}{4} + \frac{4}{5} + \cdots$

32. $\frac{2}{1} + \frac{3}{2} + \frac{4}{3} + \frac{5}{4} + \cdots$

33. $\frac{1}{1} + \frac{2}{3} + \frac{3}{5} + \frac{4}{7} + \cdots$

34. $\frac{3}{1} + \frac{5}{3} + \frac{7}{5} + \frac{9}{7} + \cdots$

35. $\frac{1}{1} + \frac{2}{2} + \frac{4}{3} + \frac{8}{4} + \cdots$

36. $\frac{1}{2} + \frac{3}{4} + \frac{9}{6} + \frac{27}{8} + \cdots$

37. A culture of bacteria doubles every hour. If there were 10 bacteria in the culture originally, how many are there after 2 hours? 4 hours? n hours?

38. A ball rebounds one-half of the distance it falls. When dropped from 8 feet, how high does it rebound on the first bounce? On the second bounce? On the nth bounce?

39. A certain radioactive substance has a half-life of 2400 years (50% of the original material is present at the end of 2400 years). If 100 grams were produced today, how many grams would be present in 4800 years? In 9600 years?

11.2

ARITHMETIC PROGRESSIONS

Any sequence with a general term that is linear in n has the property that each term except the first can be obtained from the preceding term by adding a common number called the **common difference**. A sequence with this property is called an **arithmetic progression** or **arithmetic sequence**, and we can state the definition for such a sequence symbolically:

$$s_1 = a,$$
$$s_{n+1} = s_n + d,$$

where d is the common difference. Definitions of this sort are called **recursive definitions**. It is customary to denote the first term in such a sequence by the letter a, the common difference between successive terms by d, the number of terms in the sequence (when finite) by n, and the nth term by s_n.

We can verify that a finite sequence is an arithmetic progression simply by subtracting each term from its successor and noting that the difference in each case is the same. For example,

$$7, 18, 29, 40$$

is an arithmetic progression, because

$$18 - 7 = 11, \qquad 29 - 18 = 11, \qquad 40 - 29 = 11.$$

If at least two consecutive terms in an arithmetic progression are known, we can determine the common difference and generate as many terms as we wish. Furthermore, a linear expression can be found for the nth term. Consider the general arithmetic progression with first term a and common difference d. The

first term is	$a,$
second term is	$a + d,$
third term is	$a + d + d = a + 2d,$
fourth term is	$a + d + d + d = a + 3d,$
⋮	⋮
nth term is	$a + d + d + \cdots + d = a + (n - 1)d.$

Thus,

$$s_n = a + (n - 1)d. \tag{1}$$

Here we have used an informal inductive process to obtain equation (1). We shall assume its validity for all natural numbers n. For example, the twenty-third term of the sequence

$$4, 7, 10, \ldots,$$

which has first term 4 and common difference 3, is given by

$$s_{23} = 4 + (23 - 1)3 = 70.$$

The problem of finding an explicit representation for the sum of n terms of a sequence in terms of n is, in general, very difficult; however, we can obtain such a representation for the sum of n terms in an arithmetic progression. Consider the series of n terms associated with the general arithmetic progression

$$a, (a + d), (a + 2d), \ldots, a + (n - 1)d.$$

That is,

$$S_n = a + (a + d) + (a + 2d) + \cdots + [a + (n - 1)d], \tag{2}$$

and then consider the same series written

$$S_n = s_n + (s_n - d) + (s_n - 2d) + \cdots + [s_n - (n - 1)d], \tag{3}$$

where the terms are written in reverse order. Adding (2) and (3) term-by-term, we have

$$S_n + S_n = (a + s_n) + (a + s_n) + (a + s_n) + \cdots + (a + s_n),$$

where the term $(a + s_n)$ occurs n times. Then,

$$2S_n = n(a + s_n),$$

from which

$$\mathbf{S_n = \frac{n}{2}(a + s_n).} \tag{4}$$

If (4) is rewritten as

$$S_n = n\left(\frac{a + s_n}{2}\right),$$

we observe that the sum is given by the product of the number of terms in the series and the average of the first and last terms.

An alternative form for (4) is obtained by substituting in (4) the value for s_n equal to $a + (n - 1)d$ as given by (1) to obtain

$$S_n = \frac{n}{2}(a + [a + (n - 1)d])$$

or

$$\mathbf{S_n = \frac{n}{2}[2a + (n - 1)d],} \tag{5}$$

where the sum is now expressed in terms of a, n, and d.

EXERCISE 11.2

A

Write the next three terms in each arithmetic progression. Find an expression for the general term.

Examples **a.** $5, 9, \ldots$ **b.** $x, x - a, \ldots$

Solutions **a.** Find the common difference and then continue the sequence.

$$d = 9 - 5 = 4$$

$$13, 17, 21$$

Use $s_n = a + (n - 1)d$ to find an expression for the general term.

$$s_n = 5 + (n - 1)4$$

$$s_n = 4n + 1$$

b. Find the common difference and then continue the sequence.

$$d = (x - a) - x = -a$$

$$x - 2a, x - 3a, x - 4a$$

Use $s_n = a + (n - 1)d$ to find an expression for the general term.

$$s_n = x + (n - 1)(-a)$$

$$s_n = x - a(n - 1)$$

1. $3, 7, \ldots$ **2.** $-6, -1, \ldots$ **3.** $-1, -5, \ldots$ **4.** $-10, -20, \ldots$

5. $x, x + 1, \ldots$ **6.** $a, a + 5, \ldots$ **7.** $x + a, x + 3a, \ldots$ **8.** $y - 2b, y, \ldots$

9. $2x + 1, 2x + 4, \ldots$ **10.** $a + 2b, a - 2b, \ldots$ **11.** $x, 2x, \ldots$ **12.** $3a, 5a, \ldots$

Example Find the fourteenth term of the arithmetic progression $-6, -1, 4, \ldots$.

Solution Find the common difference.

$$d = -1 - (-6) = 5$$

Use $s_n = a + (n - 1)d$.

$$s_{14} = -6 + (13)5 = 59$$

13. Find the seventh term in the arithmetic progression $7, 11, 15, \ldots$.

14. Find the tenth term in the arithmetic progression $-3, -12, -21, \ldots$.

15. Find the twelfth term in the arithmetic progression $2, \frac{5}{2}, 3, \ldots$.

16. Find the seventeenth term in the arithmetic progression $-5, -2, 1, \ldots$.

17. Find the twentieth term in the arithmetic progression $3, -2, -7, \ldots$.

18. Find the tenth term in the arithmetic progression $\frac{3}{4}, 2, \frac{13}{4}, \ldots$.

Example Find the first term in an arithmetic progression in which the third term is 7 and the eleventh term is 55.

Solution A diagram of the situation is helpful here.

$$n: \ 1, 2, 3, 4, 5, 6, 7, 8, 9, 10, 11$$

$$s_n: \ \underline{?}, \underline{\quad}, \underline{7}, \underline{\quad}, \underline{\quad}, \underline{\quad}, \underline{\quad}, \underline{\quad}, \underline{\quad}, \underline{\quad}, \underline{55}$$

Find a common difference by considering an arithmetic progression with first term 7 and ninth term 55. Use $s_n = a + (n - 1)d$.

$$s_9 = 7 + (9 - 1)d$$
$$55 = 7 + 8d$$
$$d = 6$$

Use this difference to find the first term in an arithmetic progression in which the third term is 7. Use $s_n = a + (n - 1)d$.

$$s_3 = a + (3 - 1)6$$
$$7 = a + 12$$
$$a = -5$$

The first term is -5.

Alternative Solution Use $s_n = a + (n - 1)d$, with $s_3 = 7$.

(1) $$7 = a + (3 - 1)d$$
$$7 = a + 2d$$

Use $s_n = a + (n - 1)d$ with $s_{11} = 55$.

(2) $$55 = a + (11 - 1)d$$
$$55 = a + 10d$$

Solve the system (1) and (2) to obtain $a = -5$.

19. If the third term in an arithmetic progression is 7 and the eighth term is 17, find the common difference. What is the first term? What is the twentieth term?

20. If the fifth term of an arithmetic progression is -16 and the twentieth term is -46, what is the twelfth term?

21. What term in the arithmetic progression 4, 1, -2, ... is -77?

22. What term in the arithmetic progression 7, 3, -1, ... is -81?

Terms between given terms in an arithmetic progression are called ***arithmetic means*** *of the given terms. Insert the given number of arithmetic means between the given two numbers.*

Example Three between 4 and -8.

Solution If there are three terms between 4 and -8, then the difference between 4 and -8 must be 4 times the common difference (d), as suggested by

$$\underset{\leftarrow d \rightarrow}{4 \quad ?} \underset{\leftarrow d \rightarrow}{\quad ?} \underset{\leftarrow d \rightarrow}{\quad ?} \underset{\leftarrow d \rightarrow}{\quad -8}.$$

Therefore,

$$4d = (-8) - (4) = -12$$
$$d = -3.$$

(*solution continued*)

The three requested arithmetic means can then be obtained by successive additions of -3. We obtain 1, -2, and -5.

23. Two between -6 and 15.

24. Four between 10 and 65.

25. One between 12 and 20.

26. One between -11 and 7.

27. Three between 24 and 4.

28. Six between -12 and 23.

Find the sum of each finite series.

Example $\sum_{i=1}^{12} (4i + 1)$

Solution Write the first two or three terms in expanded form.

$$5 + 9 + 13 + \cdots$$

By inspection, the first term is 5 and the common difference is 4.

Use $S_n = \frac{n}{2}[2a + (n - 1)d]$ with $n = 12$.

$$S_{12} = \frac{12}{2}[2(5) + (12 - 1)4] = 324$$

29. $\sum_{i=1}^{7} (2i + 1)$

30. $\sum_{i=1}^{21} (3i - 2)$

31. $\sum_{j=3}^{15} (7j - 1)$

32. $\sum_{j=10}^{20} (2j - 3)$

33. $\sum_{k=1}^{8} \left(\frac{1}{2}k - 3\right)$

34. $\sum_{k=1}^{100} k$

B

35. Find the sum of all even integers n, where $13 < n < 89$.

36. Find the sum of all integral multiples of 7 between 8 and 110.

37. How many bricks will there be in a pile one brick thick if there are 27 bricks in the first row, 25 in the second row, ..., and 1 in the top row?

38. If there is a total of 256 bricks in a pile arranged in the manner of those in Problem 37, how many bricks are there in the third row from the bottom of the pile?

39. Find three numbers that form an arithmetic sequence such that their sum is 21 and their product is 168.

40. Find three numbers that form an arithmetic sequence such that their sum is 21 and their product is 231.

41. Find k if $\sum_{j=1}^{5} kj = 14$.

42. Find p and q if $\sum_{i=1}^{4} (pi + q) = 28$ and $\sum_{i=2}^{5} (pi + q) = 44$.

43. Show that the sum of the first n odd natural numbers is n^2.

44. Show that the sum of the first n even natural numbers is $n^2 + n$.

11.3

GEOMETRIC PROGRESSIONS

Any sequence in which each term except the first is obtained by multiplying the preceding term by a common multiplier is called a **geometric progression** or **geometric sequence**, and is defined by the recursive equations

$$\mathbf{s_1 = a,}$$
$$\mathbf{s_{n+1} = rs_n,}$$

where r is called the **common ratio**. Thus,

$$2, 6, 18, 54, \ldots$$

is a geometric progression in which each term except the first is obtained by multiplying the preceding term by 3. The effect of multiplying the terms in this way is to produce a fixed ratio between any two successive terms—hence, the name "common ratio" for r.

If the first term is designated by a, the

second term is	ar,
third term is	$ar \cdot r \;= ar^2$,
fourth term is	$ar^2 \cdot r = ar^3$,

and it appears that the nth term will take the form

(1) $$\mathbf{s_n = ar^{n-1}.}$$

We assume the validity of this expression for all natural numbers n. The general geometric progression will appear as

$$a, ar, ar^2, ar^3, ar^4, \ldots, ar^{n-1}, \ldots.$$

For example, consider the geometric progression

$$2, 6, 18, \ldots.$$

By writing the ratio of any term to its predecessor, say $^{18}/_{6}$, we find that $r = 3$. A representation for s_n of this sequence can now be written in terms of n by substituting 2 for a and 3 for r in (1). Thus,

$$s_n = 2(3)^{n-1}.$$

The sequence function that generates a geometric progression is, of course, an exponential function, because the variable n is an exponent in the defining equation.

To find an explicit representation for the sum of a given number of terms in a geometric progression in terms of a, r, and n, we employ a device somewhat similar to the one used in finding the sum of an arithmetic progression. Consider the geometric series containing n terms (2), and the series (3) obtained by multiplying both members of (2) by r:

$$S_n = a + ar + ar^2 + ar^3 + \cdots + ar^{n-2} + ar^{n-1}, \tag{2}$$

$$rS_n = ar + ar^2 + ar^3 + ar^4 + \cdots + ar^{n-1} + ar^n. \tag{3}$$

Subtracting (3) from (2), we find that all terms in the right-hand member vanish except the first term in (2) and the last term in (3), and therefore

$$S_n - rS_n = a - ar^n.$$

Factoring S_n from the left-hand member yields

$$(1 - r)S_n = a - ar^n,$$

from which

$$S_n = \frac{a - ar^n}{1 - r} \quad (r \neq 1), \tag{4}$$

and we have a formula for the sum of n terms of a geometric progression. For the special case $r = 1$, the sum $S_n = na$.

An alternative expression for (4) can be obtained by noting that (4) can be written

$$S_n = \frac{a - r(ar^{n-1})}{1 - r}$$

and, since $s_n = ar^{n-1}$,

$$S_n = \frac{a - rs_n}{1 - r} \quad (r \neq 1), \tag{5}$$

where the sum is now given in terms of a, s_n, and r.

EXERCISE 11.3

A

Write the next three terms in each geometric progression. Find the general term.

Examples **a.** 3, 6, 12, ...

b. $x, 2, \frac{4}{x}, \ldots$

Solutions **a.** Find the common ratio.

$$r = \frac{6}{3} = 2$$

Multiply successively by r to determine the following terms:

$$24, 48, 96.$$

Use $s_n = ar^{n-1}$ to find the general term.

$$s_n = 3(2)^{n-1}$$

b. Find the common ratio.

$$r = \frac{2}{x}$$

Multiply successively by r to determine the following terms:

$$\frac{8}{x^2}, \frac{16}{x^3}, \frac{32}{x^4}$$

Use $s_n = ar^{n-1}$ to find the general term.

$$s_n = x\left(\frac{2}{x}\right)^{n-1} = \frac{2^{n-1}}{x^{n-2}}$$

1. 2, 8, 32, ...

2. 4, 8, 16, ...

3. $\frac{2}{3}, \frac{4}{3}, \frac{8}{3}, \ldots$

4. $6, 3, \frac{3}{2}, \ldots$

5. $4, -2, 1, \ldots$

6. $\frac{1}{2}, -\frac{3}{2}, \frac{9}{2}, \ldots$

7. $\frac{a}{x}, -1, \frac{x}{a}, \ldots$

8. $\frac{a}{b}, \frac{a}{bc}, \frac{a}{bc^2}, \ldots$

Example Find the ninth term of the geometric progression $-24, 12, -6, \ldots$.

Solution Find the common ratio.

$$r = \frac{12}{-24} = -\frac{1}{2}$$

Use $s_n = ar^{n-1}$.

$$s_9 = -24\left(-\frac{1}{2}\right)^8 = -\frac{3}{32}$$

9. Find the sixth term in the geometric progression 48, 96, 192, ...

10. Find the eighth term in the geometric progression $-3, \frac{3}{2}, -\frac{3}{4}, \ldots$.

11. Find the seventh term in the geometric progression $-\frac{1}{3}a^2, a^5, -3a^8, \ldots$.

12. Find the ninth term in the geometric progression $-81, -27, -9, \ldots$.

13. Find the first term of a geometric progression with fifth term 48 and ratio 2.

14. Find the first term of a geometric progression with fifth term 1 and ratio $-\frac{1}{2}$.

Terms between two given terms in a geometric progression are called **geometric means**. *Insert the given number of geometric means between the two given numbers.*

Example Two between 3 and 24.

Solution Since there are three multiplications by the common ratio between 3 and 24, the quotient when 24 is divided by 3 must be the third power of the common ratio as suggested by

$$3 \underset{\times r}{\longrightarrow} ? \underset{\times r}{\longrightarrow} ? \underset{\times r}{\longrightarrow} 24.$$

Hence, $r^3 = 24/3 = 8$, so that $r = \sqrt[3]{8} = 2$. Therefore, the missing terms can be determined by successive multiplications by 2, and we have $3 \times 2 = 6$ and $6 \times 2 = 12$. Thus, the requested geometric means are 6 and 12.

15. Two between 1 and 27.

16. Two between -4 and -32.

17. One between 36 and 9. (Two answers are possible.)

18. One between -12 and $-\frac{1}{12}$. (Two answers are possible.)

19. Three between 32 and 2. (Two answers are possible.)

20. Three between -25 and $-\frac{1}{25}$. (Two answers are possible.)

Find each sum.

Example $\sum_{i=2}^{7} \left(\frac{1}{3}\right)^i$

Solution Write the first two or three terms in expanded form.

$$\left(\frac{1}{3}\right)^2 + \left(\frac{1}{3}\right)^3 + \cdots$$

By inspection, the first term is $\frac{1}{9}$, the ratio is $\frac{1}{3}$, and $n = 6$ (recall from page 331 that the series $\sum_{i=a}^{b}$ contains $b - a + 1$ terms).

Use $S_n = \dfrac{a - ar^n}{1 - r}$.

$$S_6 = \frac{\frac{1}{9} - \frac{1}{9}\left(\frac{1}{3}\right)^6}{1 - \frac{1}{3}} = \frac{\frac{1}{9}\left(1 - \frac{1}{729}\right)}{\frac{2}{3}} = \frac{1}{9} \cdot \frac{728}{729} \cdot \frac{3}{2} = \frac{364}{2187}$$

21. $\sum_{i=1}^{6} 3^i$

22. $\sum_{j=1}^{4} (-2)^j$

23. $\sum_{k=3}^{7} \left(\frac{1}{2}\right)^{k-2}$

24. $\sum_{i=3}^{12} (2)^{i-5}$

25. $\sum_{j=1}^{6} \left(\frac{1}{3}\right)^j$

26. $\sum_{k=1}^{5} \left(\frac{1}{4}\right)^k$

B

27. Find $\sum_{i=1}^{n} \left(\frac{1}{2}\right)^i$ for $n = 2, 3, 4, 5$. What number do you think $\sum_{i=1}^{n} \left(\frac{1}{2}\right)^i$ approaches as n becomes larger and larger?

11.4

INFINITE GEOMETRIC PROGRESSIONS

Consider the infinite geometric series

$$\frac{1}{2} + \frac{1}{4} + \frac{1}{8} + \frac{1}{16} + \cdots$$

and the partial sums of terms of the series,

$$S_1 = \frac{1}{2}$$

$$S_2 = \frac{1}{2} + \frac{1}{4} = \frac{3}{4}$$

$$S_3 = \frac{1}{2} + \frac{1}{4} + \frac{1}{8} = \frac{7}{8}$$

$$S_4 = \frac{1}{2} + \frac{1}{4} + \frac{1}{8} + \frac{1}{16} = \frac{15}{16}$$

$$\vdots$$

Note that the nth term of the sequence of partial sums

$$S_1, S_2, S_3, S_4, \ldots, S_n, \ldots \qquad \text{or} \qquad \frac{1}{2}, \frac{3}{4}, \frac{7}{8}, \frac{15}{16}, \ldots, S_n, \ldots$$

appears to be "approaching" 1. That is, as the number n becomes very large, S_n is very close to 1. In fact, we can make the difference between S_n and 1 as small as we like by using a sufficiently large value for n.

Recall from Section 11.3 that the sum of n terms of a geometric progression is given by

(1) $$S_n = \frac{a - ar^n}{1 - r}.$$

If $|r| < 1$, that is, if $-1 < r < 1$, then r^n becomes smaller and smaller for increasingly large n. For example, if $r = \frac{1}{2}$,

$$r^2 = \left(\frac{1}{2}\right)^2 = \frac{1}{4},$$

$$r^3 = \left(\frac{1}{2}\right)^3 = \frac{1}{8},$$

$$r^4 = \left(\frac{1}{2}\right)^4 = \frac{1}{16},$$

etc., and we can make $(\frac{1}{2})^n$ as small as we please by taking n sufficiently large. Writing (1) in the form

(2) $$S_n = \frac{a}{1 - r}(1 - r^n),$$

we see that the value of the factor $(1 - r^n)$ can be made as close as we please to 1, providing $|r| < 1$ and n is taken large enough. Since this asserts that the sum (2) can be made to approximate

$$\frac{a}{1 - r}$$

as close as we please, we define the sum of an infinite geometric progression with $|r| < 1$ to be

(3) $$\mathbf{S_\infty = \frac{a}{1 - r}}.$$

A sequence that has an nth term (and all terms after the nth) that can be made to approximate a fixed number L as closely as desired by simply taking n large enough is said to approach the limit L as n increases without bound. We can indicate this in terms of symbols,

$$\lim_{n \to \infty} S_n = L,$$

where S_n is the nth term of the sequence of sums $S_1, S_2, S_3, \ldots, S_n, \ldots$. Thus, (3) might be written

$$S_\infty = \lim_{n \to \infty} S_n = \frac{a}{1 - r}.$$

If $|r| \geq 1$, then r^n in (2) does not approach 0 and $\lim_{n \to \infty} S_n$ does not exist.

An interesting application of this sum arises in connection with repeating decimals—that is, decimal numerals that, after a finite number of decimal places, have endlessly repeating groups of digits. For example,

$$0.2121\overline{21}, \qquad 0.333\overline{3}, \qquad 0.138512512\overline{512},$$

are repeating decimals, where in each case the bar indicates the repeating digits. Consider the problem of expressing such a decimal numeral as a fraction. We illustrate the process involved with the first example above:

$$0.2121\overline{21}. \tag{4}$$

This decimal can be written either as

$$0.21 + 0.0021 + 0.000021 + \cdots, \tag{5}$$

or as

$$\frac{21}{100} + \frac{21}{10{,}000} + \frac{21}{1{,}000{,}000} + \cdots, \tag{6}$$

which are sums with terms that form a geometric progression with ratio 0.01 (or $1/100$). Since the ratio is less than 1 in absolute value, we can use (3) to find the sum of an infinite number of terms of (6). Thus,

$$S_\infty = \frac{a}{1-r} = \frac{\dfrac{21}{100}}{1 - \dfrac{1}{100}} = \frac{\dfrac{21}{100}}{\dfrac{99}{100}} = \frac{21}{99} = \frac{7}{33},$$

and the given decimal numeral, $0.2121\overline{21}$, is equivalent to $7/33$.

EXERCISE 11.4

A

Find the sum of each infinite geometric series. If the series has no sum, so state.

Examples

a. $3 + 2 + \dfrac{4}{3} + \cdots$

b. $\dfrac{1}{81} - \dfrac{1}{54} + \dfrac{1}{36} + \cdots$

Solutions

a. $r = \dfrac{2}{3}$; the series has a sum, since $|r| < 1$.

$$S_\infty = \frac{a}{1-r} = \frac{3}{1 - \dfrac{2}{3}} = 9$$

b. $r = -\dfrac{1}{54} \div \dfrac{1}{81} = -\dfrac{3}{2}$; the series does not have a sum, since $|r| > 1$.

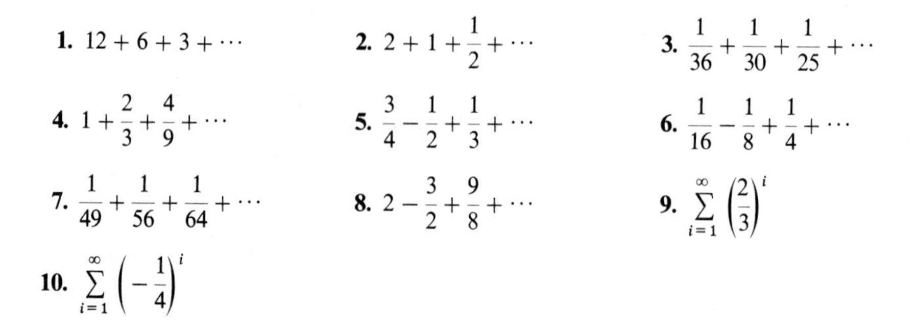

1. $12 + 6 + 3 + \cdots$

2. $2 + 1 + \frac{1}{2} + \cdots$

3. $\frac{1}{36} + \frac{1}{30} + \frac{1}{25} + \cdots$

4. $1 + \frac{2}{3} + \frac{4}{9} + \cdots$

5. $\frac{3}{4} - \frac{1}{2} + \frac{1}{3} + \cdots$

6. $\frac{1}{16} - \frac{1}{8} + \frac{1}{4} + \cdots$

7. $\frac{1}{49} + \frac{1}{56} + \frac{1}{64} + \cdots$

8. $2 - \frac{3}{2} + \frac{9}{8} + \cdots$

9. $\sum_{i=1}^{\infty} \left(\frac{2}{3}\right)^i$

10. $\sum_{i=1}^{\infty} \left(-\frac{1}{4}\right)^i$

Find a fraction equivalent to each of the given decimal numerals.

Example $2.045045\overline{045}$

Solution Rewrite as a series.

$$2 + \frac{45}{1000} + \frac{45}{1{,}000{,}000} + \cdots$$

Find the common ratio: $r = \frac{1}{1000}$ for series beginning with $\frac{45}{1000}$.

Use $S_\infty = \frac{a}{1-r}$.

$$S_\infty = \frac{\frac{45}{1000}}{1 - \frac{1}{1000}} = \frac{\frac{45}{1000}}{\frac{999}{1000}} = \frac{45}{999} = \frac{5}{111}$$

Hence, $2.045045\overline{045} = 2\frac{5}{111} = \frac{227}{111}$.

11. $0.3333\overline{3}$ **12.** $0.6666\overline{6}$ **13.** $0.3131\overline{31}$ **14.** $0.4545\overline{45}$

15. $2.410\overline{410}$ **16.** $3.027\overline{027}$ **17.** $0.12888\overline{8}$ **18.** $0.8333\overline{3}$

B

19. A force is applied to a particle moving in a straight line in such a fashion that each second it moves only one-half of the distance it moved the preceding second. If the particle moves 10 centimeters the first second, approximately how far will it move before coming to rest?

20. The arc length through which the bob on a pendulum moves is nine-tenths of its preceding arc length. Approximately how far will the bob move before coming to rest if the first arc length is 12 inches?

21. A ball returns two-thirds of its preceding height on each bounce. If the ball is dropped from a height of 6 feet, approximately what is the total distance the ball travels before coming to rest?

11.5

THE BINOMIAL EXPANSION

Sometimes it is necessary to write the product of consecutive positive integers. To do this, we use a special symbol, $n!$ (read "n factorial" or "factorial n"), which is defined by

$$\boldsymbol{n! = n(n-1)(n-2)\cdots(1).}$$

Thus,

$$5! = 5 \cdot 4 \cdot 3 \cdot 2 \cdot 1$$

and

$$8! = 8 \cdot 7 \cdot 6 \cdot 5 \cdot 4 \cdot 3 \cdot 2 \cdot 1.$$

Factorial notation can also be used to represent the products of consecutive positive integers, beginning with integers different from 1. For example,

$$8 \cdot 7 \cdot 6 \cdot 5 = \frac{8!}{4!},$$

because

$$\frac{8!}{4!} = \frac{8 \cdot 7 \cdot 6 \cdot 5 \cdot 4 \cdot 3 \cdot 2 \cdot 1}{4 \cdot 3 \cdot 2 \cdot 1} = 8 \cdot 7 \cdot 6 \cdot 5.$$

Since

$$n! = n(n-1)(n-2)(n-3)\cdots 5 \cdot 4 \cdot 3 \cdot 2 \cdot 1$$

and

$$(n-1)! = (n-1)(n-2)(n-3)\cdots 5 \cdot 4 \cdot 3 \cdot 2 \cdot 1,$$

we can, for $n > 1$, write the recursive relationship

$$\boldsymbol{n! = n(n-1)!.}$$

For example,

$$7! = 7 \cdot 6!,$$
$$27! = 27 \cdot 26!,$$
$$(n+2)! = (n+2)(n+1)!.$$

Then, if $n = 1$, we should have

$$1! = 1 \cdot (1-1)!$$
$$1! = 1 \cdot 0!.$$

Therefore, for consistency, we shall define

$$\boldsymbol{0! = 1.}$$

The series obtained by expanding a binomial of the form

$$(a+b)^n \qquad (n \in N)$$

is particularly useful in certain branches of mathematics. Starting with familiar examples where n takes the value 1, 2, 3, 4, 5, we can show by direct multiplication that

$$
\begin{aligned}
(a+b)^1 &= a+b,\\
(a+b)^2 &= a^2+2ab+b^2,\\
(a+b)^3 &= a^3+3a^2b+3ab^2+b^3,\\
(a+b)^4 &= a^4+4a^3b+6a^2b^2+4ab^3+b^4,\\
(a+b)^5 &= a^5+5a^4b+10a^3b^2+10a^2b^3+5ab^4+b^5.
\end{aligned}
$$

We observe that in each case where there is a sufficient number of terms:

1. The first term is a^n.
2. The variable factors of the second term are $a^{n-1}b^1$, and the coefficient is n, which can be written in the form
$$\frac{n}{1!}.$$
3. The variable factors of the third term are $a^{n-2}b^2$, and the coefficient can be written in the form
$$\frac{n(n-1)}{2!}.$$
4. The variable factors of the fourth term are $a^{n-3}b^3$, and the coefficient can be written in the form
$$\frac{n(n-1)(n-2)}{3!}.$$

The above results can be generalized to obtain the **binomial expansion**,

$$
\begin{aligned}
(1)\qquad (a+b)^n &= a^n+\frac{n}{1!}a^{n-1}b+\frac{n(n-1)}{2!}a^{n-2}b^2+\frac{n(n-1)(n-2)}{3!}a^{n-3}b^3\\
&\quad+\cdots+\frac{n(n-1)(n-2)\cdots(n-r+2)}{(r-1)!}a^{n-r+1}b^{r-1}+\cdots+b^n,
\end{aligned}
$$

where r is the number of the term. For example,

$$
\begin{aligned}
&(x-2)^4\\
&\quad= x^4+\frac{4}{1!}x^3(-2)^1+\frac{4\cdot 3}{2!}x^2(-2)^2+\frac{4\cdot 3\cdot 2}{3!}x(-2)^3+\frac{4\cdot 3\cdot 2\cdot 1}{4!}(-2)^4\\
&\quad= x^4-8x^3+24x^2-32x+16.
\end{aligned}
$$

In this case, $a = x$ and $b = -2$ in the binomial expansion.

Note that the rth term in a binomial expansion is given by

$$\frac{n(n-1)(n-2)\cdots(n-r+2)}{(r-1)!}a^{n-r+1}b^{r-1}. \tag{2}$$

For example, we have for the fifth term in the expression $(x-2)^{10}$,

$$\frac{10\cdot 9\cdot 8\cdot 7}{4\cdot 3\cdot 2\cdot 1}x^6(-2)^4 = 3360x^6.$$

EXERCISE 11.5

A

1. Write $(2n)!$ in expanded form for $n = 4$.
2. Write $(3n)!$ in expanded form for $n = 4$.
3. Write $2n!$ in expanded form for $n = 4$.
4. Write $3n!$ in expanded form for $n = 4$.
5. Write $n(n-1)!$ in expanded form for $n = 6$.
6. Write $2n(2n-1)!$ in expanded form for $n = 2$.

Write in expanded form and simplify.

Examples **a.** $\frac{7!}{4!}$ **b.** $\frac{4!6!}{8!}$

Solutions **a.** $\frac{7!}{4!} = \frac{7\cdot 6\cdot 5\cdot 4!}{4!} = 210$ **b.** $\frac{4!6!}{8!} = \frac{4\cdot 3\cdot 2\cdot 1\cdot 6!}{8\cdot 7\cdot 6!} = \frac{3}{7}$

7. $5!$ **8.** $7!$ **9.** $\frac{9!}{7!}$ **10.** $\frac{12!}{11!}$

11. $\frac{5!7!}{8!}$ **12.** $\frac{12!8!}{16!}$ **13.** $\frac{8!}{2!(8-2)!}$ **14.** $\frac{10!}{4!(10-4)!}$

Write each product in factorial notation.

Examples **a.** $1\cdot 2\cdot 3\cdot 4\cdot 5\cdot 6$ **b.** $11\cdot 12\cdot 13\cdot 14$ **c.** 150

Solutions **a.** $1\cdot 2\cdot 3\cdot 4\cdot 5\cdot 6 = 6!$ **b.** $11\cdot 12\cdot 13\cdot 14 = \frac{14!}{10!}$ **c.** $150 = \frac{150!}{149!}$

15. $1 \cdot 2 \cdot 3$

16. $1 \cdot 2 \cdot 3 \cdot 4 \cdot 5$

17. $3 \cdot 4 \cdot 5 \cdot 6$

18. 7

19. $8 \cdot 7 \cdot 6$

20. $28 \cdot 27 \cdot 26 \cdot 25 \cdot 24$

Write each expression in factored form and show the first three factors and the last three factors.

Example $(2n + 1)!$

Solution $(2n + 1)! = (2n + 1)(2n)(2n - 1)\cdots 3 \cdot 2 \cdot 1$

21. $n!$

22. $(n + 4)!$

23. $(3n)!$

24. $3n!$

25. $(n - 2)!$

26. $(3n - 2)!$

Expand.

Example $(a - 3b)^4$

Solution From the binomial expansion (1) on page 348,

$$(a - 3b)^4 = a^4 + \frac{4}{1!}a^3(-3b) + \frac{4 \cdot 3}{2!}a^2(-3b)^2 + \frac{4 \cdot 3 \cdot 2}{3!}a(-3b)^3 + \frac{4 \cdot 3 \cdot 2 \cdot 1}{4!}(-3b)^4$$

$$= a^4 - 12a^3b + 54a^2b^2 - 108ab^3 + 81b^4.$$

27. $(x + 3)^5$

28. $(2x + y)^4$

29. $(x - 3)^4$

30. $(2x - 1)^5$

31. $\left(2x - \frac{y}{2}\right)^3$

32. $\left(\frac{x}{3} + 3\right)^5$

33. $\left(\frac{x}{2} + 2\right)^6$

34. $\left(\frac{2}{3} - a^2\right)^4$

Write the first four terms in each expansion. Do not simplify the terms.

Example $(x + 2y)^{15}$

Solution $(x + 2y)^{15} = x^{15} + \frac{15}{1!}x^{14}(2y) + \frac{15 \cdot 14}{2!}x^{13}(2y)^2 + \frac{15 \cdot 14 \cdot 13}{3!}x^{12}(2y)^3$

35. $(x + y)^{20}$

36. $(x - y)^{15}$

37. $(a - 2b)^{12}$

38. $(2a - b)^{12}$

39. $(x - \sqrt{2})^{10}$

40. $\left(\frac{x}{2} + 2\right)^8$

Find each specified term.

Example $(x - 2y)^{12}$, seventh term

Solution Use (2) on page 349. Set $n = 12$ and $r = 7$.

$$\frac{12 \cdot 11 \cdot 10 \cdot 9 \cdot 8 \cdot 7}{6 \cdot 5 \cdot 4 \cdot 3 \cdot 2 \cdot 1} x^6(-2y)^6 = 59{,}136x^6y^6$$

41. $(a - b)^{15}$, sixth term

42. $(x + 2)^{12}$, fifth term

43. $(x - 2y)^{10}$, fifth term

44. $(a^3 - b)^9$, seventh term

B

45. Given that the binomial formula holds for $(1 + x)^n$, where n is a negative integer:

a. Write the first four terms of $(1 + x)^{-1}$.

b. Find the first four terms of the quotient $\dfrac{1}{(1 + x)}$ by dividing $(1 + x)$ into 1.

Compare the results of parts a and b.

46. Given that the binomial formula holds as an infinite "sum" for $(1 + x)^n$, where n is a noninteger rational number and $|x| < 1$, find to two decimal places:

a. $\sqrt{1.02}$

b. $\sqrt{0.99}$

11.6

PERMUTATIONS

An arrangement of the elements of a set in specified order is called a **permutation** of the elements. For example, the possible permutations of the elements of $\{a, b, c\}$ are

$$abc, \quad bac, \quad cab, \quad acb, \quad bca, \quad cba.$$

To count the permutations of a given set containing n elements you can think of placing them in order as follows:

1. Select one member to be first. There are n possible selections.
2. Having selected a first element, select a second. Since one element has been placed in the first position, there remain $n - 1$ elements to be placed in the second position. The total number of possible first and second selections is then given by the product $n(n - 1)$.
3. Select a third element. Since one has been used for the first position and one for the second, there are $n - 2$ possibilities for the third. The total number of possible first, second, and third selections is $n(n - 1)(n - 2)$.

4. Continue this process until the last element is put in the last position. There are then $n(n-1)(n-2)\cdots 3 \cdot 2 \cdot 1$ possible permutations of n elements.

Formally, we represent the number of permutations of n things by means of the symbol $P(n, n)$ (read "the number of permutations of n things taken n at a time"). From the foregoing discussion, it follows that

$$P(n, n) = n! \tag{1}$$

For example, suppose we wish to determine how many different signals can be formed with 5 different signal flags on a pole by altering the location of the flags. A helpful way to picture the situation is to first draw 5 dashes to represent the positions of the flags:

$$\_\ \ \_\ \ \_\ \ \_\ \ \_.$$

Next, since we can place any of the 5 flags in the first position, we write 5 on the first dash:

$$\underline{5}\ \ \_\ \ \_\ \ \_\ \ \_.$$

Having selected 1 flag for the first position, we have a choice from among 4 flags for the second position.

$$\underline{5}\ \ \underline{4}\ \ \_\ \ \_\ \ \_$$

Similarly, 3 flags remain for the third position, 2 for the fourth, and 1 for the fifth. Thus, we fill the dashes:

$$\underline{5}\ \ \underline{4}\ \ \underline{3}\ \ \underline{2}\ \ \underline{1}.$$

Then, the total number of different signals is

$$5 \cdot 4 \cdot 3 \cdot 2 \cdot 1 = 120.$$

Counting permutations of distinct things in this way makes use of a more general counting principle, namely:

If one thing can be done m ways, another thing n ways, another p ways, etc., then the number of ways all the things can be done is

$$m \cdot n \cdot p \cdots.$$

For example, the number of different license plates containing four-digit numerals using the digits 1, 2, 3, 4, 5, 6, 7, 8, and 9, is given by

$$9 \cdot 9 \cdot 9 \cdot 9 = 6561,$$

since nothing prevents the use of the same digit in two, three, or four positions.

Frequently, it is important to be able to compute the number of possible permutations of subsets of a given set. Thus, we might wish to know the number of possible permutations of n different things taken r ($r \leq n$) at a time. The symbol representing this number is $P(n, r)$, and, by reasoning in exactly the same way as we did to compute $P(n, n)$, we find that

$$P(n, r) = n(n-1)(n-2)\cdots(n-r+1), \tag{2}$$

or

$$P(n, r) = \frac{n!}{(n-r)!}. \tag{3}$$

For example, to find the number of permutations of 5 things taken 3 at a time, we would have

$$P(5, 3) = \frac{5!}{2!} = \frac{5 \cdot 4 \cdot 3 \cdot 2 \cdot 1}{2 \cdot 1} = 5 \cdot 4 \cdot 3 = 60.$$

Again, if we wish to use dashes to help visualize the situation, we could first draw 3 dashes,

$$\_\ \_\ \_,$$

and then fill each dash in order,

$$\underline{5}\ \underline{4}\ \underline{3}.$$

The product of these numbers then gives us the total number of possibilities for permuting 5 things taken 3 at a time.

Sometimes the elements with which we wish to form permutations are not all different. Thus, to find the number of distinguishable permutations of the letters in the word *toast*, we must take into consideration the fact that we cannot distinguish between the 2 t's in any permutation. The number of permutations of the 5 letters in the word is clearly $5!$, but since in any one of these the 2 t's can be permuted in $2!$ ways without producing a different result, the number of *distinguishable* permutations P is given by

$$2!\,P = 5!.$$

This leads to the result

$$P = \frac{5!}{2!} = \frac{5 \cdot 4 \cdot 3 \cdot 2 \cdot 1}{2 \cdot 1} = 60.$$

As another example, consider the letters in the word *pineapple*. Here there is a total of $9!$ permutations. However, there are 3 p's and 2 e's which can be permuted $3!$ and $2!$ ways, respectively, in each permutation of the 9 letters without altering the result. Accordingly, the number of distinguishable permutations P is given by

$$3!\,2!\,P = 9!,$$

from which

$$P = \frac{9!}{3!\,2!} = \frac{9 \cdot 8 \cdot 7 \cdot 6 \cdot 5 \cdot 4 \cdot 3 \cdot 2 \cdot 1}{3 \cdot 2 \cdot 1 \cdot 2 \cdot 1}$$

$$= 30{,}240.$$

EXERCISE 11.6

Example In how many ways can a jury of 12 persons be seated in a jury box containing 12 chairs?

Solution What is wanted is $P(12, 12)$.

$$P(12, 12) = 12! = 12 \cdot 11 \cdot 10 \cdot 9 \cdot 8 \cdot 7 \cdot 6 \cdot 5 \cdot 4 \cdot 3 \cdot 2 \cdot 1$$
$$= 479{,}001{,}600.$$

1. In how many different ways can 4 books be arranged between bookends?

2. In how many ways can 7 students be seated at 7 desks?

3. In how many ways can 9 players be assigned to the 9 positions on a baseball team?

4. In how many ways can 10 floats be arranged for a parade?

Examples How many different three-digit numerals for whole numbers can be formed from the digits 0, 1, 2, 3, 4, 5, 6, if:

a. No restrictions are placed on the repetition of digits?

b. No digit can be used more than once?

c. The number named is odd and no digit can be used more than once?

Solutions **a.** Since the numeral must contain 3 digits, the first digit cannot be 0. Drawing 3 dashes, we see that the first place can be filled with any 1 of the digits from 1 to 6, a total of 6 digits:

$$\underline{6}\ \_\ \_.$$

The remaining places can be filled with any of the 7 digits, so that we have

$$\underline{6}\ \underline{7}\ \underline{7},$$

and the answer to the original question is

$$6 \cdot 7 \cdot 7 = 294.$$

b. If no digit can be used more than once, then we can select any one of the digits from 1 to 6 for the first digit, but, having fixed the first digit, we have 6 possibilities for the second, and 5 for the third. Our diagram appears as follows:

$$\underline{6}\ \underline{6}\ \underline{5}.$$

The answer to the original question then is

$$6 \cdot 6 \cdot 5 = 180.$$

c. If the number named must be odd, we first must look at the last digit. We may use any one of 1, 3, or 5 for this digit, so our diagram begins like this:

$$\_\ \_\ \underline{3}.$$

Next, having fixed one digit, and being unable to use 0 as a first digit, we have

$$\underline{5}\ \underline{5}\ \underline{3}.$$

The answer to the original question then is

$$5 \cdot 5 \cdot 3 = 75.$$

How many different four-digit numerals for whole numbers can be formed from the digits 0, 1, 2, 3, 4, 5, 6, *if:*

5. No restrictions are placed on the repetition of digits?

6. No restrictions are placed on the repetition of digits and the number named is odd?

7. No digit can be used more than once?

8. No digit may be used more than once, and the number named is odd?

9. No restrictions are placed on the repetition of digits and the number named is even?

10. No digit may be used more than once and the number named is even?

Repeat the problem cited for three-digit numerals, using the digits 0, 1, 2, 3, 4, 5, 6, 7, 8.

11. Problem 5 **12.** Problem 6 **13.** Problem 7

14. Problem 8 **15.** Problem 9 **16.** Problem 10

17. How many three-letter words can be formed from the twenty-six-letter English alphabet?

18. How many five-letter words can be formed from the twenty-six-letter English alphabet?

19. How many license numbers can be formed if each license contains 2 letters of the alphabet followed by 3 digits?

20. How many license numbers can be formed if each license contains 1 digit followed by 1 letter of the alphabet followed by 2 digits?

How many distinguishable permutations are there in the letters of the given word?

Example Seventeen

Solution There are 9 letters in the word; 2 *n*'s and 4 *e*'s are repeated. Hence, the number of distinguishable permutations is given by

$$2!\,4!\,P = 9!$$

$$P = \frac{9!}{2!\,4!} = \frac{9 \cdot 8 \cdot 7 \cdot 6 \cdot 5 \cdot 4 \cdot 3 \cdot 2 \cdot 1}{2 \cdot 4 \cdot 3 \cdot 2}$$

$$= 7560.$$

21. Hurry **22.** Sonnet **23.** Word **24.** Between

25. Committee **26.** Consists **27.** Selected **28.** Permutation

29. Banana **30.** Tennessee **31.** Mississippi **32.** Tallahassee

11.7

COMBINATIONS

If the order in which the elements of a set are considered is unimportant, then the set is called a **combination**. Thus, a, b and b, a are two permutations of the elements of $\{a, b\}$; however, the set constitutes a single combination. In general, we reserve the name *combination* to refer to any of the r-element subsets of an n-element set. The *number* of such combinations is designated by the symbol $\binom{n}{r}$ or sometimes by $C(n, r)$, which are both read "the number of combinations of n things taken r at a time." For example, for $\{a, b, c\}$ we can list the combinations of 3 things taken 2 at a time as follows: a, b; a, c; b, c. Therefore,

$$\binom{3}{2} = 3.$$

In counting combinations we can use permutations. Let S be an n-element set, and consider $\binom{n}{r}$ the number of r-element subsets of S. Each of these subsets can be permuted in $P(r, r)$, or $r!$, ways. Then the total number of permutations of n things taken r at a time can be computed by multiplying $P(r, r)$ by $\binom{n}{r}$. That is,

$$P(r, r)\binom{n}{r} = P(n, r).$$

Multiplying both members by $1/P(r, r)$, we have

$$(1) \qquad \binom{n}{r} = \frac{P(n, r)}{P(r, r)}.$$

Since $P(n, r) = n(n-1)(n-2)\cdots(n-r+1)$ and $P(r, r) = r!$, it follows that

$$(2) \qquad \binom{n}{r} = \frac{n(n-1)(n-2)\cdots(n-r+1)}{r!}.$$

For example, the number of different committees of 3 people that could be appointed in a club having 8 members is the number of combinations of 8 things taken 3 at a time, rather than the number of permutations of 8 things taken 3 at a time, because the order in which the members of a committee are considered is of no consequence. Thus, from (2) we would have

$$\binom{8}{3} = \frac{8 \cdot 7 \cdot 6}{3 \cdot 2 \cdot 1} = 56,$$

and 56 such committees could be appointed in the club. Notice that in computing $\binom{n}{r}$, it is easier to first write $r!$ in the denominator, because we can then simply write the same number of factors in the numerator, rather than actually determining $(n - r + 1)$ as the final factor of the numerator.

In Section 11.6 we observed that

$$P(n,r) = \frac{n!}{(n-r)!}$$

Substituting the right-hand member of this equation for $P(n,r)$ in (1) above, we have

$$\binom{n}{r} = \frac{P(n,r)}{P(r,r)} = \frac{\frac{n!}{(n-r)!}}{r!}$$

from which we have the formula

$$(3) \qquad \binom{n}{r} = \frac{n!}{r!(n-r)!}$$

which is sometimes easier to use than (2).

EXERCISE 11.7

A

1. How many different committees of 4 people can be appointed in a club containing 15 members?
2. How many different committees of 5 people can be appointed from a group containing 9 people?

Example How many different amounts of money can be formed from a penny, a nickel, a dime, and a quarter?

Solution Using 1 coin, we can form $\binom{4}{1} = 4$ different amounts; using 2 coins, we obtain $\binom{4}{2} = \frac{4 \cdot 3}{1 \cdot 2} = 6$; using 3 coins, we have $\binom{4}{3} = \frac{4 \cdot 3 \cdot 2}{1 \cdot 2 \cdot 3} = 4$; using all the coins we obtain $\binom{4}{4} = \frac{4 \cdot 3 \cdot 2 \cdot 1}{1 \cdot 2 \cdot 3 \cdot 4} = 1$, so that the total number of different amounts of money is

$$\binom{4}{1} + \binom{4}{2} + \binom{4}{3} + \binom{4}{4} = 4 + 6 + 4 + 1 = 15.$$

3. How many different amounts of money can be formed from a nickel, a dime, and a quarter?
4. How many different amounts of money can be formed from a nickel, a dime, a quarter, a penny, and a half-dollar?

5. In how many different ways can a set of 5 cards be selected from a deck of 52 cards?
6. In how many different ways can a set of 13 cards be selected from a deck of 52 cards?
7. How many straight lines are determined by 6 points, no 3 of which are collinear?
8. How many diagonals does a regular octagon (8 sides) have?
9. In how many ways can a person draw 3 marbles from an urn containing 9 marbles?
10. In how many ways can 2 white and 4 red marbles be drawn from an urn containing 5 white and 5 red marbles? [*Hint*: 2 white marbles can be drawn in $\binom{5}{2}$ ways, and 4 red marbles can be drawn in $\binom{5}{4}$ ways.]
11. In how many ways can a bridge hand consisting of 6 hearts, 3 spades, and 4 diamonds be selected from a deck of 52 cards?
12. In how many ways can a hand consisting of 4 aces and 1 card that is not an ace be selected from a deck of 52 cards?
13. In how many ways can a committee of 2 men and 3 women be selected from a club with 10 men and 11 women members?
14. In how many ways can a committee containing 8 men and 8 women be selected from the club in Problem 13?

B

15. Use equation (2) or (3) on pages 356 and 357 to show that $\binom{n}{r} = \binom{n}{n-r}$.
16. Use the results of Problem 15 to compute $\binom{100}{98}$.
17. Given $\binom{n}{3} = \binom{n}{4}$, find n.
18. Given $\binom{n}{7} = \binom{n}{5}$, find n.
19. Show that the coefficients of the terms in the binomial expansion of $(a+b)^4$ are $\binom{4}{0}, \binom{4}{1}, \binom{4}{2}, \binom{4}{3}$, and $\binom{4}{4}$.
20. Show that the coefficients of the first five terms in the binomial expansion of $(a+b)^n$ are $\binom{n}{0}, \binom{n}{1}, \binom{n}{2}, \binom{n}{3}$, and $\binom{n}{4}$.

CHAPTER SUMMARY

[11.1] The elements in the range of a function whose domain is a set of successive positive integers form a **sequence**. The sequence is **finite** if it has a last member, otherwise it is **infinite**.

A **series** is the indicated sum of the terms in a sequence. We can use **sigma** or **summation** notation to represent a series.

[11.2] A sequence in which each term after the first is obtained by adding a constant to the preceding term is an **arithmetic progression**. The constant is called the **common difference** of the terms.

The nth term of an arithmetic sequence is given by

$$s_n = a + (n - 1)d,$$

and the sum of n terms is given by

$$S_n = \frac{n}{2}(a + s_n) = \frac{n}{2}[2a + (n - 1)d],$$

where a is the first term, n is the number of terms, and d is the common difference.

[11.3] A sequence in which each term after the first is obtained by multiplying its predecessor by a constant is called a **geometric progression**. The constant is called the **common ratio**.

The nth term of a geometric sequence is given by

$$s_n = ar^{n-1},$$

and the sum of n terms is given by

$$S_n = \frac{a - ar^n}{1 - r} = \frac{a - rs_n}{1 - r} \qquad (r \neq 1),$$

where a is the first term, n is the number of terms, and r is the common ratio.

[11.4] An infinite geometric series has a sum if the common ratio has an absolute value less than 1. This sum is given by

$$S_\infty = \lim_{n \to \infty} S_n = \frac{a}{1 - r}.$$

[11.5] Factorial notation is convenient to represent special kinds of products. For $n \in N$,

$$\begin{aligned} n! &= n(n - 1)(n - 2) \cdots (3)(2)(1) \\ &= n(n - 1)! \end{aligned}$$

The binomial power $(a + b)^n$ can be expanded into a series containing $n + 1$ terms. For $n \in N$,

$$(a + b)^n = a^n + \frac{n}{1!}a^{n-1}b + \frac{n(n - 1)}{2!}a^{n-2}b^2 + \frac{n(n - 1)(n - 2)}{3!}a^{n-3}b^3 + \cdots + \frac{n(n - 1)(n - 2) \cdots (n - r + 2)}{(r - 1)!}a^{n-r+1}b^{r-1} + \cdots + b^n,$$

where r is the number of the term.

[11.6] An arrangement of the elements in a set is called a **permutation** of the elements. To count permutations, we can use the general counting principle:

If one thing can be done m ways, another thing n ways, another thing p ways, etc., then the number of ways all the things can be done is $m \cdot n \cdot p \cdots$.

For $n, r \in N$ $(r \leq n)$,

$$P(n,n) = n!$$
$$P(n,r) = n(n-1)(n-2)\cdots(n-r+1)$$
$$= \frac{n!}{(n-r)!}$$

[11.7] Any r-element subset of an n-element set is called a **combination**. For $n, r \in N$,

$$\binom{n}{r} = \frac{n(n-1)(n-2)\cdots(n-r+1)}{r!}$$
$$= \frac{n!}{r!(n-r)!}$$

The symbols introduced in this chapter are listed on the inside of the front cover.

REVIEW EXERCISES

[11.1]

1. Find the first four terms in a sequence with the general term

$$s_n = \frac{(-1)^{n-1}}{n}.$$

2. Write $\sum_{k=2}^{5} k(k-1)$ in expanded form.

[11.2]

3. Given that 5, 9 are the first two terms of an arithmetic progression, find an expression for the general term.

4. **a.** Find the twenty-third term of the arithmetic progression $-82, -74, -66, \ldots$.

   b. Find the sum of the first twenty-three terms.

5. The first term of an arithmetic progression is 8 and the twenty-eighth term is 89. Find the twenty-first term.

[11.3]

6. Given that 5, 9 are the first two terms of a geometric progression, find an expression for the general term.

7. **a.** Find the eighth term of the geometric progression $\frac{16}{27}, -\frac{8}{9}, \frac{4}{3}, \ldots$.

   b. Find the sum of the first eight terms.

8. The second term of a geometric progression is 3 and the fifth term is $\frac{81}{8}$. Find the seventh term.

9. Find $\sum_{j=1}^{5} \left(\frac{1}{3}\right)^j$.

[11.4] 10. Find $\sum_{i=1}^{\infty} \left(\frac{1}{3}\right)^i$.

11. Find a fraction equivalent to 0.444.

[11.5] 12. Simplify $\frac{8!}{3!5!}$

13. Write the first four terms of the binomial expansion of $(x - 2y)^{10}$.

14. Find the eighth term in the expansion of $(x - 2y)^{10}$.

[11.6] 15. How many different three-digit numerals for whole numbers can be formed from the digits 1, 2, 3, 4, 5 when:

a. No restrictions are placed on the repetition of digits?

b. No digit can be used more than once?

16. In how many ways can 6 men be assigned to a 4 man relay team?

17. How many distinguishable permutations are there in the letters of the word *fifteen*?

[11.7] 18. How many different doubles combinations could be formed from a tennis team containing 6 members?

19. In how many ways can 3 aces and 2 kings be selected from a deck of 52 cards?

20. In how many ways can a bridge hand consisting of 4 hearts, 5 spades, 2 diamonds, and 2 clubs be selected from a deck of 52 cards?

APPENDIX A

SYNTHETIC DIVISION; POLYNOMIAL FUNCTIONS

A.1 Synthetic division

↓

A.2 Graphing polynomial functions

A.1

SYNTHETIC DIVISION

In Section 3.3, we rewrote quotients of polynomials of the form $P(x)/D(x)$ using a long division algorithm (process). If the divisor $D(x)$ is of the form $x - a$, this algorithm can be simplified by a procedure known as **synthetic division**. Consider the quotient

$$\frac{x^4 + x^2 + 2x - 1}{x + 3}.$$

The division can be accomplished as follows:

$$
\begin{array}{r|rrrrr}
\multicolumn{1}{r}{} & x^3 & -\,3x^2 & +\,10x & -\,28 & \\ \cline{2-6}
x + 3 & x^4 & +\,0x^3 & +\,x^2 & +\,2x & -\,1 \\
\multicolumn{1}{r}{} & x^4 & +\,3x^3 & & & \\ \cline{2-3}
\multicolumn{1}{r}{} & & -\,3x^3 & +\,x^2 & & \\
\multicolumn{1}{r}{} & & -\,3x^3 & -\,9x^2 & & \\ \cline{3-4}
\multicolumn{1}{r}{} & & & 10x^2 & +\,2x & \\
\multicolumn{1}{r}{} & & & 10x^2 & +\,30x & \\ \cline{4-5}
\multicolumn{1}{r}{} & & & & -\,28x & -\,1 \\
\multicolumn{1}{r}{} & & & & -\,28x & -\,84 \\ \cline{5-6}
\multicolumn{1}{r}{} & & & & & 83 \text{ (remainder).}
\end{array}
$$

We see that

$$\frac{x^4 + x^2 + 2x - 1}{x + 3} = x^3 - 3x^2 + 10x - 28 + \frac{83}{x + 3} \qquad (x \neq -3).$$

If we omit the variables, writing only the coefficients of the terms, and use 0 for the coefficient of any missing power, we have

$$
\begin{array}{r|l}
 & 1 - 3 + 10 \;\; - 28 \\
\hline
1 + 3 & 1 + 0 + \;1 \;+\; 2 \;-\; 1 \\
 & \underline{1 + 3} \\
 & \quad - 3 + (1) \\
 & \quad \underline{- 3 - \;9} \\
 & \qquad\qquad 10 \;+ (2) \\
 & \qquad\qquad \underline{10 \;+ 30} \\
 & \qquad\qquad\quad - 28 \;-\; (1) \\
 & \qquad\qquad\quad \underline{- 28 \;-\; 84} \\
 & \qquad\qquad\qquad\qquad 83 \text{ (remainder).}
\end{array}
$$

Now, observe that the numbers in color are repetitions of the numbers written immediately above and are also repetitions of the coefficients of the associated variable in the quotient; the numbers in parentheses are repetitions of the coefficients of the dividend. Therefore, the whole process can be written in compact form as

$$
\begin{array}{lrrrrrrl}
(1) & \underline{3 \;|} & 1 & 0 & 1 & 2 & -1 & \\
(2) & & & 3 & -9 & 30 & -84 & \\
\cline{3-7}
(3) & & 1 & -3 & 10 & -28 & 83 & \text{(remainder: 83),}
\end{array}
$$

where the repetitions are omitted and where 1, the coefficient of x in the divisor, has also been omitted.

The numbers in line (3), which are the coefficients of the variables in the quotient and the remainder, have been obtained by *subtracting* the **detached coefficients** in line (2) from the detached coefficients of terms of the same degree in line (1). We could obtain the same result by replacing 3 with -3 in the divisor and *adding* instead of subtracting at each step, and this is what is done in the *synthetic division* process. The final form then is

$$
\begin{array}{lrrrrrrl}
(1) & \underline{-3 \;|} & 1 & 0 & 1 & 2 & -1 & \\
(2) & & & -3 & 9 & -30 & 84 & \\
\cline{3-7}
(3) & & 1 & -3 & 10 & -28 & 83 & \text{(remainder: 83).}
\end{array}
$$

Note that the numbers in line (2) can be obtained by multiplying the preceding number to the left in line (3) by -3.

Comparing the results of using synthetic division with the same process using long division, we observe that the numbers in line (3) are the coefficients of the polynomial

$$x^3 - 3x^2 + 10x - 28,$$

and that there is a remainder of 83.

As another example, let us write the quotient

$$\frac{3x^3 - 4x - 1}{x - 2}$$

in the form $Q(x) + r/(x - a)$. Using synthetic division, we begin by writing

$$\begin{array}{r|rrrr} 2 & 3 & 0 & -4 & -1 \\ \hline \end{array}$$

where 0 has been inserted in the position corresponding to the coefficient of a second-degree term. The divisor is the negative of -2, or 2. Then, we have

$$\begin{array}{lr|rrrrl} (1) & 2 & 3 & 0 & -4 & -1 & \\ (2) & & & 6 & 12 & 16 & \\ \hline (3) & & 3 & 6 & 8 & 15 & \text{(remainder: 15).} \end{array}$$

This process employs these steps:

1. 3 is "brought down" from line (1) to line (3)
2. 6, the product of 2 and 3, is written in the next position on line (2)
3. 6, the sum of 0 and 6, is written on line (3)
4. 12, the product of 2 and 6, is written in the next position on line (2)
5. 8, the sum of -4 and 12, is written on line (3)
6. 16, the product of 2 and 8, is written in the next position on line (2)
7. 15, the sum of -1 and 16, is written on line (3)

We can use the first three numbers on line (3) as coefficients to write a polynomial of degree one less than the degree of the dividend. This polynomial is the quotient lacking the remainder. The last number is the remainder. Thus, for $x - 2 \neq 0$, the quotient of $3x^3 - 4x - 1$ divided by $x - 2$ is

$$3x^2 + 6x + 8$$

with a remainder of 15; that is,

$$\frac{3x^3 - 4x - 1}{x - 2} = 3x^2 + 6x + 8 + \frac{15}{x - 2} \qquad (x \neq 2).$$

EXERCISE A.1

Use synthetic division to write each quotient $P(x)/(x - a)$ in the form $Q(x)$ or $Q(x) + r/(x - a)$, where r is a constant.

Examples **a.** $\dfrac{2x^4 + x^3 - 1}{x + 2}$ **b.** $\dfrac{x^3 - 1}{x - 1}$

(*solution on page* 366)

Solutions

a.
$$\begin{array}{r|rrrrr} -2 & 2 & 1 & 0 & 0 & -1 \\ & & -4 & 6 & -12 & 24 \\ \hline & 2 & -3 & 6 & -12 & 23 \end{array}$$

$$2x^3 - 3x^2 + 6x - 12 + \frac{23}{x+2} \qquad (x \neq -2)$$

b.
$$\begin{array}{r|rrrr} 1 & 1 & 0 & 0 & -1 \\ & & 1 & 1 & 1 \\ \hline & 1 & 1 & 1 & 0 \end{array}$$

$$x^2 + x + 1 \qquad (x \neq 1)$$

1. $\dfrac{x^2 - 8x + 12}{x - 6}$ **2.** $\dfrac{a^2 + a - 6}{a + 3}$ **3.** $\dfrac{x^2 + 4x + 4}{x + 2}$

4. $\dfrac{x^2 + 6x + 9}{x + 3}$ **5.** $\dfrac{x^4 - 3x^3 + 2x^2 - 1}{x - 2}$ **6.** $\dfrac{x^4 + 2x^2 - 3x + 5}{x - 3}$

7. $\dfrac{2x^3 + x - 5}{x + 1}$ **8.** $\dfrac{3x^3 + x^2 - 7}{x + 2}$ **9.** $\dfrac{2x^4 - x + 6}{x - 5}$

10. $\dfrac{3x^4 - x^2 + 1}{x - 4}$ **11.** $\dfrac{x^3 + 4x^2 + x - 2}{x + 2}$ **12.** $\dfrac{x^3 - 7x^2 - x + 3}{x + 3}$

13. $\dfrac{x^6 + x^4 - x}{x - 1}$ **14.** $\dfrac{x^6 + 3x^3 - 2x - 1}{x - 2}$ **15.** $\dfrac{x^5 - 1}{x - 1}$

16. $\dfrac{x^5 + 1}{x + 1}$ **17.** $\dfrac{x^6 - 1}{x - 1}$ **18.** $\dfrac{x^6 + 1}{x + 1}$

A.2

GRAPHING POLYNOMIAL FUNCTIONS

In Section 3.3, we observed that the quotient

$$\frac{x^2 + 2x + 2}{x + 1}$$

can be expressed as

$$x + 1 + \frac{1}{x + 1} \qquad (x \neq -1).$$

In general, we have that

$$\frac{P(x)}{x - a} = Q(x) + \frac{r}{x - a}, \tag{1}$$

as indicated in Section A.1, or

$$P(x) = (x - a)Q(x) + r, \tag{2}$$

where $P(x)$ is a polynomial with real coefficients of degree $n \geq 1$, $Q(x)$ is a polynomial with real coefficients of degree $n - 1$, and r is a real number.

Now in (2), if $x = a$, we have

$$P(a) = (a - a)Q(a) + r$$
$$P(a) = 0 \cdot Q(a) + r$$
$$P(a) = r.$$

Hence, the remainder, where $P(x)$ is divided by $x - a$, is the value $P(a)$. This is sometimes called the **remainder theorem**. Since synthetic division offers a means of finding values of $r = P(a)$, we can sometimes find such values more quickly by synthetic division than by direct substitution. For example, if we set $P(x) = 2x^3 - 3x^2 + 2x + 1$, then we can find $P(2)$ by synthetically dividing $2x^3 - 3x^2 + 2x + 1$ by $x - 2$:

$$\begin{array}{r|rrrr} 2 & 2 & -3 & 2 & 1 \\ & & 4 & 2 & 8 \\ \hline & 2 & 1 & 4 & 9 \end{array}$$

By inspection we note that $r = P(2) = 9$.

In Section 7.3, we graphed linear functions defined by

$$f(x) = a_0 x + a_1,$$

and in Sections 8.1 and 8.2, we graphed quadratic functions defined by

$$f(x) = a_0 x^2 + a_1 x + a_2.$$

We can graph any polynomial function defined by

$$f(x) = a_0 x^n + a_1 x^{n-1} + \cdots + a_n,$$

where $a_0, a_1, \ldots, a_n$, and x are real numbers, by obtaining a number of solutions (ordered pairs) sufficient to determine the behavior of its graph. We can obtain the ordered pairs $(x, f(x))$ by direct substitution of arbitrary values of x, as we did earlier in this book, or by synthetic division. In this appendix we will use synthetic division. For example, we can obtain solutions of

$$P(x) = x^3 - 2x^2 - 5x + 6 \tag{3}$$

for selected values of x, say $-3, -2, -1, 0, 1, 2, 3, 4$, by dividing $x^3 - 2x^2 - 5x + 6$ by $x + 3$, $x + 2$, etc. Dividing synthetically by $x + 3$, we have

$$\begin{array}{r|rrrr} -3 & 1 & -2 & -5 & 6 \\ & & -3 & 15 & -30 \\ \hline & 1 & -5 & 10 & -24 \end{array}$$

and $(-3, -24)$ is a solution of (3). Dividing by $x + 2$, we have

$$\begin{array}{r|rrrr} -2 & 1 & -2 & -5 & 6 \\ & & -2 & 8 & -6 \\ \hline & 1 & -4 & 3 & 0 \end{array}$$

and $(-2, 0)$ is a solution of (3). Dividing by $x + 1$, we have

$$\begin{array}{r|rrrr} -1 & 1 & -2 & -5 & 6 \\ & & -1 & 3 & 2 \\ \hline & 1 & -3 & -2 & 8 \end{array}$$

and $(-1, 8)$ is a solution of (3). Similarly, we find that $(0, 6)$, $(1, 0)$, $(2, -4)$, $(3, 0)$, and $(4, 18)$ are solutions, and their graphs are on the graph of the function. These points, shown in part a of Figure A.1, make the general appearance of the graph clear and enable us to complete the graph as shown in part b of Figure A.1. Any additional values of x less than -3 and greater than 4 would not change the general appearance of the graph. Note that the graph of this

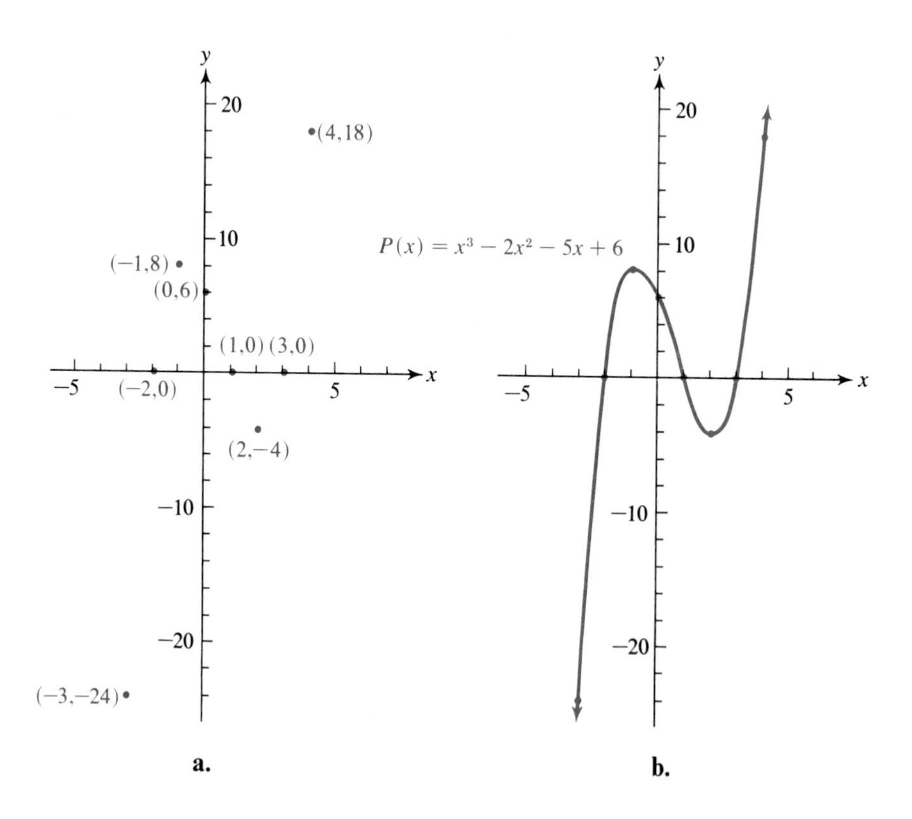

Figure A.1

third-degree polynomial function changes direction twice. In fact, it can be shown, although we will not do so, that the maximum number of direction changes of the graph of a polynomial function is one less than the degree of the polynomial.

We observe from (2) on page 366 that if $r = P(a) = 0$, then

$$P(x) = (x - a) \cdot Q(x) \tag{4}$$

and $x - a$ is a factor of $P(x)$. This is sometimes called the **factor theorem**. In the above example we established that $x + 2$, $x - 1$, and $x - 3$ are factors of $x^3 - 2x^2 - 5x + 6$, because in each synthetic division by these factors the remainder was 0.

From (4) we note that a is a solution of $P(x) = 0$ if and only if $x - a$ is a factor of $P(x)$. This suggests that an equation $P(x) = 0$, where $P(x)$ is of nth degree, has n solutions. This is indeed the case. For example, we note that the third-degree polynomial $x^3 - 2x^2 - 5x + 6$ is equivalent to $(x + 2)(x - 1)(x - 3)$, and that there are exactly three solutions of

$$x^3 - 2x^2 - 5x + 6 = (x + 2)(x - 1)(x - 3) = 0,$$

namely, -2, 1, and 3.

Of course, it may be that one or more factors of such an expression are the same. When this happens, we count the solution as many times as the factor involved occurs. Thus, because

$$P(x) = x^4 + 2x^3 - 2x - 1 = (x + 1)(x + 1)(x + 1)(x - 1),$$

we can see that 1 and -1 are the only solutions of $P(x) = 0$, but we say that -1 is a solution of multiplicity three.

EXERCISE A.2

A

Find the designated value.

Example In $P(x) = 4x^4 - 2x^3 + 3x - 2$, find $P(-1)$.

Solution

$$\begin{array}{r|rrrrr} -1 & 4 & -2 & 0 & 3 & -2 \\ & & -4 & 6 & -6 & 3 \\ \hline & 4 & -6 & 6 & -3 & 1 \end{array}$$

$P(-1) = 1$

1. If $P(x) = 3x^3 - 2x^2 + 5x - 4$, find $P(3)$ and $P(-2)$.
2. If $P(x) = 4x^4 - 2x^3 + 3x^2 - 5$, find $P(1)$ and $P(-1)$.
3. If $P(x) = 2x^5 - 3x^3 + x^2 - x + 2$, find $P(-1)$ and $P(2)$.
4. If $P(x) = x^4 - 10x^3 + 5x^2 - 3x + 6$, find $P(-2)$ and $P(3)$.

Use synthetic division and the remainder theorem to find solutions to each equation, and then graph the equation.

5. $y = x^3 + x^2 - 6x$
6. $y = x^3 + 5x^2 + 4x$
7. $y = x^3 - 2x^2 + 1$
8. $y = x^3 - 4x^2 + 3x$
9. $y = 2x^3 + 9x^2 + 7x - 6$
10. $y = x^3 - 3x^2 - 6x + 8$
11. $y = x^4 - 4x^2$
12. $y = x^4 - x^3 - 4x^2 + 4x$

Use the remainder in synthetic division to determine whether or not the given binomial is a factor of the given polynomial.

13. $x - 2$; $x^3 - 3x^2 + 2x + 2$
14. $x - 1$; $2x^3 - 5x^2 + 4x - 1$
15. $x + 3$; $3x^3 + 11x^2 + x - 15$
16. $x + 1$; $2x^3 - 5x^2 + 3x + 3$

B

17. Verify that 1 is a solution of $x^3 + 2x^2 - x - 2 = 0$, and find the other solutions.
18. Verify that 3 is a solution of $x^3 - 6x^2 - x + 30 = 0$, and find the other solutions.
19. Verify that -3 is a solution of $x^4 - 3x^3 - 10x^2 + 24x = 0$, and find the other solutions.
20. Verify that -5 is a solution of $x^4 + 5x^3 - x^2 - 5x = 0$, and find the other solutions.

APPENDIX A SUMMARY

[A.1] **Synthetic division** is a condensation of the division algorithm using only coefficients.

[A.2] A quotient of the form $P(x)/(x - a)$, where $P(x)$ is a polynomial with real coefficients of degree $n \geq 1$, can be expressed in the form $Q(x) + r/(x - a)$, where $Q(x)$ is a polynomial with real coefficients of degree $n - 1$ and r is a real number. Furthermore, $r = P(a)$ (**remainder theorem**).

Synthetic division can be used to find sufficient values $P(a)$ of a polynomial $P(x)$ in order to determine the behavior of its graph.

If $P(x)/(x - a)$ yields a remainder $r = P(a) = 0$, then $x - a$ is a factor of $P(x)$ (**factor theorem**).

An equation of the form $P(x) = 0$, where $P(x)$ is of nth degree, has n solutions.

REVIEW EXERCISES

[A.1] **1.** Use synthetic division to write the quotient below as a polynomial in simple form.

$$\frac{y^3 + 3y^2 - 2y - 4}{y + 1}$$

2. Use synthetic division to write the quotient below as a polynomial in simple form.

$$\frac{y^7 - 1}{y - 1}$$

[A.2] **3.** If $P(x) = 2x^3 - x^2 + 3x + 1$, find $P(2)$ and $P(-2)$.

4. If $P(x) = x^4 + 3x^2 - 2x + 2$, find $P(1)$ and $P(-1)$.

Use synthetic division and the remainder theorem to find solutions to each equation, and then graph the equation.

5. $y = x^3 + x^2 - 2x$

6. $y = x^4 + 2x^3 - 5x^2 - 6x$

7. Is $x + 2$ a factor of $3x^3 - 2x^2 - x + 4$?

8. Verify that $x - 2$ is a factor of $x^3 - 4x^2 + x + 6$, and find the other factors.

APPENDIX B

MATRICES AND DETERMINANTS

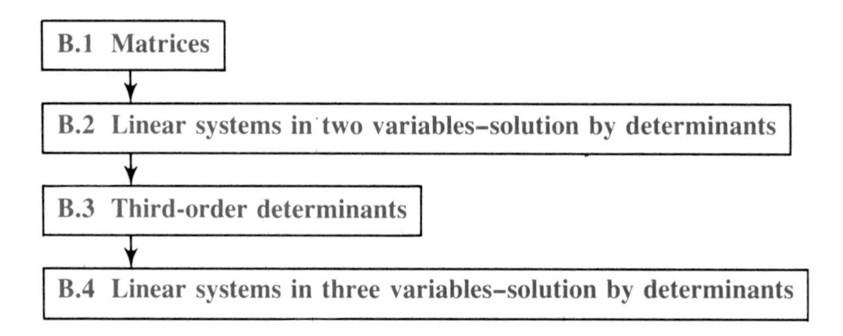

B.1

MATRICES

A **matrix** is a rectangular array of elements or **entries** (in this book, real numbers). These entries are ordinarily displayed using brackets or parentheses (we shall use brackets). Thus,

$$\begin{bmatrix} 1 & 2 & 3 \\ 4 & 5 & 6 \\ 7 & 8 & 9 \end{bmatrix}, \quad \begin{bmatrix} 2 & -1 & 3 \\ 4 & 0 & 2 \end{bmatrix}, \quad \begin{bmatrix} 4 \\ 5 \\ 6 \end{bmatrix}$$

are matrices with real number elements. The **order**, or **dimension**, of a matrix is the ordered pair having as first component the number of (horizontal) rows and as second component the number of (vertical) columns in the matrix. Thus, the matrices above are 3×3 (read "three-by-three"), 2×3 (read "two-by-three"), and 3×1 (read "three-by-one"), respectively.

We can transform one matrix into another in a variety of ways. However, here we are only concerned with the following kinds of transformations:

1. Multiplying the entries of any row by a nonzero real number.
2. Interchanging two rows.
3. Multiplying the entries of any row by a real number and adding the results to the corresponding elements of another row.

Such transformations are called **elementary transformations**, and if a matrix A is transformed into a matrix B by a finite succession of such transformations, then we say that A and B are **row-equivalent**. We represent this by writing $A \sim B$ (read "A is row-equivalent to B"). For example,

$$A = \begin{bmatrix} 1 & 3 & -1 \\ 2 & 1 & 4 \\ 6 & 2 & -1 \end{bmatrix} \quad \text{and} \quad B = \begin{bmatrix} 1 & 3 & -1 \\ 6 & 3 & 12 \\ 6 & 2 & -1 \end{bmatrix}$$

are equivalent, because we can multiply each entry in row 2 of A by 3 to obtain B;

$$A = \begin{bmatrix} 3 & -1 & 2 \\ 2 & 1 & 4 \\ 3 & 1 & 9 \end{bmatrix} \quad \text{and} \quad B = \begin{bmatrix} 3 & 1 & 9 \\ 2 & 1 & 4 \\ 3 & -1 & 2 \end{bmatrix}$$

are equivalent, because we can interchange rows 1 and 3 of A to obtain B; and

$$A = \begin{bmatrix} 1 & 2 & 1 \\ 2 & 0 & -1 \\ 3 & 1 & 2 \end{bmatrix} \quad \text{and} \quad B = \begin{bmatrix} 1 & 2 & 1 \\ 0 & -4 & -3 \\ 3 & 1 & 2 \end{bmatrix}$$

are equivalent, because we can multiply each entry in row 1 of A by -2 and add the results to the corresponding entries in row 2 of A to obtain B.

It is often convenient to perform more than one elementary transformation on a given matrix. For example, if in the matrices

$$\begin{bmatrix} 1 & -2 & 1 \\ 2 & 1 & 3 \\ -3 & 0 & 0 \end{bmatrix} \sim \begin{bmatrix} 1 & -2 & 1 \\ 0 & 5 & 1 \\ 0 & -6 & 3 \end{bmatrix}$$

we add -2 times row 1 to row 2, and 3 times row 1 to row 3 in the left-hand matrix, we obtain the row-equivalent right-hand matrix.

In a system of linear equations of the form

$$\begin{aligned} a_1 x + b_1 y + c_1 z &= d_1 \\ a_2 x + b_2 y + c_2 z &= d_2 \\ a_3 x + b_3 y + c_3 z &= d_3, \end{aligned}$$

the matrices

$$\begin{bmatrix} a_1 & b_1 & c_1 \\ a_2 & b_2 & c_2 \\ a_3 & b_3 & c_3 \end{bmatrix} \quad \text{and} \quad \begin{bmatrix} a_1 & b_1 & c_1 & d_1 \\ a_2 & b_2 & c_2 & d_2 \\ a_3 & b_3 & c_3 & d_3 \end{bmatrix}$$

are called the **coefficient matrix** and the **augmented matrix**, respectively. By performing elementary transformations on the augmented matrix of a system of equations, we can obtain a matrix from which the solution set of the system is readily determined. The validity of the method, as illustrated in the examples below, stems from the fact that performing elementary row transformations on the augmented matrix of a system corresponds to forming equivalent systems of equations. For example, the augmented matrix of

$$\begin{aligned} x - 3y + z &= -2 \\ 3x + y - z &= 8 \\ 2x - 2y + 3z &= -1 \end{aligned} \quad \text{is} \quad \begin{bmatrix} 1 & -3 & 1 & -2 \\ 3 & 1 & -1 & 8 \\ 2 & -2 & 3 & -1 \end{bmatrix}.$$

Then, performing elementary transformations (each time on the matrix directly above the step), we have:

$$\begin{array}{r} \\ \text{row } 2 + [-3 \times \text{row } 1] \rightarrow \\ \text{row } 3 + [-2 \times \text{row } 1] \rightarrow \end{array} \begin{bmatrix} 1 & -3 & 1 & -2 \\ 0 & 10 & -4 & 14 \\ 0 & 4 & 1 & 3 \end{bmatrix} \qquad \begin{aligned} x - 3y + z &= -2 \\ 0x + 10y - 4z &= 14 \\ 0x + 4y + z &= 3 \end{aligned}$$

$$\frac{1}{2} \times \text{row } 2 \rightarrow \begin{bmatrix} 1 & -3 & 1 & -2 \\ 0 & 5 & -2 & 7 \\ 0 & 4 & 1 & 3 \end{bmatrix} \qquad \begin{aligned} x - 3y + z &= -2 \\ 0x + 5y - 2z &= 7 \\ 0x + 4y + z &= 3 \end{aligned}$$

$$\text{row } 3 + \left[-\frac{4}{5} \times \text{row } 2\right] \rightarrow \begin{bmatrix} 1 & -3 & 1 & -2 \\ 0 & 5 & -2 & 7 \\ 0 & 0 & \frac{13}{5} & -\frac{13}{5} \end{bmatrix} \qquad \begin{aligned} x - 3y + z &= -2 \\ 0x + 5y - 2z &= 7 \\ 0x + 0y + \frac{13}{5}z &= -\frac{13}{5} \end{aligned}$$

$$5 \times \text{row } 3 \rightarrow \begin{bmatrix} 1 & -3 & 1 & -2 \\ 0 & 5 & -2 & 7 \\ 0 & 0 & 13 & -13 \end{bmatrix} \qquad \begin{aligned} x - 3y + z &= -2 \\ 0x + 5y - 2z &= 7 \\ 0x + 0y + 13z &= -13 \end{aligned}$$

Note that the last matrix corresponds to the system

$$\begin{aligned} x - 3y + z &= -2 \\ 5y - 2z &= 7 \\ 13z &= -13. \end{aligned}$$

Since, from the last equation, $z = -1$, we can substitute -1 for z in the second equation to obtain $y = 1$. Finally, substituting -1 for z and 1 for y in the first equation, we have $x = 2$, so that the solution set is $\{(2, 1, -1)\}$.

EXERCISE B.1

Use row transformations on the augmented matrix to solve each system.

1. $x + 3y = 11$
$2x - y = 1$

2. $x - 5y = 11$
$2x + 3y = -4$

3. $x - 4y = -6$
$3x + y = -5$

4. $x + 6y = -14$
$5x - 3y = -4$

5. $2x + y = 5$
$3x - 5y = 14$

6. $3x - 2y = 16$
$4x + 2y = 12$

7. $x + 3y - z = 5$
$3x - y + 2z = 5$
$x + y + 2z = 7$

8. $x - 2y + 3z = -11$
$2x + 3y - z = 6$
$3x - y - z = 3$

9. $2x - y + z = 8$
$x - 2y - 3z = 4$
$3x + 3y - z = -4$

10. $\begin{aligned} x - 2y - 2z &= 4 \\ 2x + y - 3z &= 7 \\ x - y - z &= 3 \end{aligned}$

11. $\begin{aligned} 2x - y - z &= -4 \\ x + y + z &= -5 \\ x + 3y - 4z &= 12 \end{aligned}$

12. $\begin{aligned} x - 2y - 5z &= 2 \\ 2x + 3y + z &= 11 \\ 3x - y - z &= 11 \end{aligned}$

B.2

LINEAR SYSTEMS IN TWO VARIABLES— SOLUTION BY DETERMINANTS

Associated with each square matrix A that has real number entries is a real number called the **determinant** of A, which is denoted by δA or $\delta(A)$ and read "the determinant of A." The determinant is customarily displayed in the same form as the matrix, but with vertical bars instead of brackets. In this section, we consider 2×2 determinants with values defined as follows:

▶ *If* $A = \begin{bmatrix} a_1 & b_1 \\ a_2 & b_2 \end{bmatrix}$, *then*

$$\delta(A) = \begin{vmatrix} a_1 & b_1 \\ a_2 & b_2 \end{vmatrix} = a_1 b_2 - a_2 b_1.$$

This value is obtained by multiplying the elements on the diagonals and adding the negative of the second product to the first product. This process can be shown schematically as

$$\begin{vmatrix} a_1 & b_1 \\ a_2 & b_2 \end{vmatrix} = a_1 b_2 - a_2 b_1.$$

For example,

$$\begin{vmatrix} 1 & 2 \\ -1 & 3 \end{vmatrix} = 3 - (-2) = 5 \quad \text{and} \quad \begin{vmatrix} 0 & -1 \\ -1 & 7 \end{vmatrix} = 0 - 1 = -1.$$

A determinant, therefore, is simply another way to represent a single number.

Determinants can be used to solve linear systems. In this section, we shall confine our attention to linear systems of two equations in two variables of the form

(1) $$a_1 x + b_1 y = c_1$$

(2) $$a_2 x + b_2 y = c_2.$$

If this system is solved by means of a linear combination, we have, upon multiplication of equation (1) by $-a_2$ and equation (2) by a_1, the equations

(1a) $$-a_1 a_2 x - a_2 b_1 y = -a_2 c_1$$

(2a) $$a_1 a_2 x + a_1 b_2 y = a_1 c_2.$$

The sum of the members of (1a) and (2a) is

$$a_1b_2y - a_2b_1y = a_1c_2 - a_2c_1.$$

Now, factoring y from each term in the left-hand member, we have

$$(a_1b_2 - a_2b_1)y = a_1c_2 - a_2c_1,$$

from which

$$y = \frac{a_1c_2 - a_2c_1}{a_1b_2 - a_2b_1} \qquad (a_1b_2 - a_2b_1 \neq 0). \tag{3}$$

But the numerator of (3) is just the value of the determinant

$$\begin{vmatrix} a_1 & c_1 \\ a_2 & c_2 \end{vmatrix},$$

which we designate as D_y, and the denominator is the value of the determinant

$$\begin{vmatrix} a_1 & b_1 \\ a_2 & b_2 \end{vmatrix},$$

which we designate as D, so that (3) can be written

$$y = \frac{D_y}{D} = \frac{\begin{vmatrix} a_1 & c_1 \\ a_2 & c_2 \end{vmatrix}}{\begin{vmatrix} a_1 & b_1 \\ a_2 & b_2 \end{vmatrix}}. \tag{4}$$

The elements of the determinant in the denominator of (4) are the coefficients of the variables in (1) and (2). The elements of the determinant in the numerator of (4) are identical to those in the denominator, except that *the elements in the column containing the coefficients of y have been replaced by c_1 and c_2*, the constant terms of (1) and (2).

By exactly the same procedure, we can show that

$$x = \frac{D_x}{D} = \frac{\begin{vmatrix} c_1 & b_1 \\ c_2 & b_2 \end{vmatrix}}{\begin{vmatrix} a_1 & b_1 \\ a_2 & b_2 \end{vmatrix}}, \tag{5}$$

and (4) and (5) together yield the components of the ordered pair in the solution set of the system. The use of determinants in this way is known as **Cramer's rule** for the solution of a system of linear equations. As an example of the use of Cramer's rule, consider the system

$$\begin{aligned} 2x + y &= 4 \\ x - 3y &= -5. \end{aligned}$$

We have

$$D = \begin{vmatrix} 2 & 1 \\ 1 & -3 \end{vmatrix} = -6 - 1 = -7$$

$$D_x = \begin{vmatrix} 4 & 1 \\ -5 & -3 \end{vmatrix} = -12 + 5 = -7$$

$$D_y = \begin{vmatrix} 2 & 4 \\ 1 & -5 \end{vmatrix} = -10 - 4 = -14.$$

Therefore,

$$x = \frac{D_x}{D} = \frac{-7}{-7} = 1, \qquad y = \frac{D_y}{D} = \frac{-14}{-7} = 2,$$

and the solution set is $\{(1, 2)\}$.

If $D = 0$ in this procedure, the equations in the system are either dependent or inconsistent depending upon whether or not D_y and D_x are both 0. This follows from the discussion on page 307 where these conditions are considered in terms of the coefficients a_1, b_1, a_2, b_2, and the constant terms c_1 and c_2.

EXERCISE B.2

A

Evaluate.

Example $\begin{vmatrix} 2 & -3 \\ 1 & 4 \end{vmatrix}$

Solution $\begin{vmatrix} 2 & -3 \\ 1 & 4 \end{vmatrix} = (2)(4) - (1)(-3) = 11$

1. $\begin{vmatrix} 1 & 0 \\ 2 & 1 \end{vmatrix}$ **2.** $\begin{vmatrix} 3 & -2 \\ 4 & 1 \end{vmatrix}$ **3.** $\begin{vmatrix} -5 & -1 \\ 3 & 3 \end{vmatrix}$ **4.** $\begin{vmatrix} 1 & -2 \\ -1 & 2 \end{vmatrix}$

5. $\begin{vmatrix} -1 & 6 \\ 0 & -2 \end{vmatrix}$ **6.** $\begin{vmatrix} 20 & 3 \\ -20 & -2 \end{vmatrix}$ **7.** $\begin{vmatrix} -2 & -1 \\ -3 & -4 \end{vmatrix}$ **8.** $\begin{vmatrix} -1 & -5 \\ -2 & -6 \end{vmatrix}$

Find the solution set of each system by Cramer's rule.

Example $2x - 3y = 6$
$2x + y = 14$

Solution

$$D = \begin{vmatrix} 2 & -3 \\ 2 & 1 \end{vmatrix} = (2)(1) - (2)(-3) = 8$$

The elements in D_x are obtained from the elements in D by replacing the elements in the column containing the coefficients of x with the corresponding constants 6 and 14.

$$D_x = \begin{vmatrix} 6 & -3 \\ 14 & 1 \end{vmatrix} = (6)(1) - (14)(-3) = 48$$

The elements in D_y are obtained from the elements in D by replacing the elements in the column containing the coefficients of y with the corresponding constants 6 and 14.

$$D_y = \begin{vmatrix} 2 & 6 \\ 2 & 14 \end{vmatrix} = (2)(14) - (2)(6) = 16$$

Values for x and y can now be determined by Cramer's rule.

$$x = \frac{D_x}{D} = \frac{48}{8} = 6 \qquad y = \frac{D_y}{D} = \frac{16}{8} = 2$$

The solution set is $\{(6, 2)\}$.

9. $2x - 3y = -1$, $x + 4y = 5$

10. $3x - 4y = -2$, $x - 2y = 0$

11. $3x - 4y = -2$, $6x + 12y = 36$

12. $2x - 4y = 7$, $x - 2y = 1$

13. $\frac{1}{3}x - \frac{1}{2}y = 0$, $\frac{1}{2}x + \frac{1}{4}y = 4$

14. $\frac{2}{3}x + y = 1$, $x - \frac{4}{3}y = 0$

15. $x - 2y = 5$, $\frac{2}{3}x - \frac{4}{3}y = 6$

16. $\frac{1}{2}x + y = 3$, $-\frac{1}{4}x - y = -3$

17. $x - 3y = 1$, $y = 1$

18. $2x - 3y = 12$, $x = 4$

19. $ax + by = 1$, $bx + ay = 1$

20. $x + y = a$, $x - y = b$

B

Show that each statement is true for every real value of each variable.

21. $\begin{vmatrix} a & a \\ b & b \end{vmatrix} = 0$

22. $\begin{vmatrix} a_1 & b_1 \\ a_2 & b_2 \end{vmatrix} = -\begin{vmatrix} a_2 & b_2 \\ a_1 & b_1 \end{vmatrix}$

23. $\begin{vmatrix} a_1 & b_1 \\ a_2 & b_2 \end{vmatrix} = -\begin{vmatrix} b_1 & a_1 \\ b_2 & a_2 \end{vmatrix}$

24. $\begin{vmatrix} ka_1 & b_1 \\ ka_2 & b_2 \end{vmatrix} = k\begin{vmatrix} a_1 & b_1 \\ a_2 & b_2 \end{vmatrix}$

25. $\begin{vmatrix} ka & a \\ kb & b \end{vmatrix} = 0$

26. $\begin{vmatrix} a_1 + ka_2 & b_1 + kb_2 \\ a_2 & b_2 \end{vmatrix} = \begin{vmatrix} a_1 & b_1 \\ a_2 & b_2 \end{vmatrix}$

27. Show that if both $D_y = 0$ and $D_x = 0$, it follows that $D = 0$ when c_1 and c_2 are not both 0, and the equations in the system

$$a_1x + b_1y = c_1$$
$$a_2x + b_2y = c_2$$

are dependent. [*Hint*: Show that the first two determinant equations imply that $a_1c_2 = a_2c_1$ and $b_1c_2 = b_2c_1$ and that the rest follows from the formation of a proportion with these equations.

28. Show that for the system given in Problem 27, if $D = 0$ and $D_x = 0$, then $D_y = 0$.

B.3

THIRD-ORDER DETERMINANTS

Associated with each 3×3 matrix A that has real number entries is the determinant $\delta(A)$ with the value defined as follows:

$$\textit{If } A = \begin{bmatrix} a_1 & b_1 & c_1 \\ a_2 & b_2 & c_2 \\ a_3 & b_3 & c_3 \end{bmatrix}, \textit{ then}$$

$$(1) \qquad \delta(A) = \begin{vmatrix} a_1 & b_1 & c_1 \\ a_2 & b_2 & c_2 \\ a_3 & b_3 & c_3 \end{vmatrix}$$
$$= a_1b_2c_3 - a_1b_3c_2 + a_3b_1c_2 - a_2b_1c_3 + a_2b_3c_1 - a_3b_2c_1.$$

Again, we note that a 3×3 determinant is simply a number, namely, that number represented by the expression in the right-hand member of (1).

The **minor** of an element in a determinant is defined as the determinant that remains after deleting the row and column in which the element appears. In the determinant (1):

the minor of the element a_1 is $\begin{vmatrix} b_2 & c_2 \\ b_3 & c_3 \end{vmatrix}$;

the minor of the element b_1 is $\begin{vmatrix} a_2 & c_2 \\ a_3 & c_3 \end{vmatrix}$;

the minor of the element c_1 is $\begin{vmatrix} a_2 & b_2 \\ a_3 & b_3 \end{vmatrix}$; *etc.*

If, by suitably factoring pairs of terms in the right-hand member, (1) is rewritten in the form

$$(2) \quad \begin{vmatrix} a_1 & b_1 & c_1 \\ a_2 & b_2 & c_2 \\ a_3 & b_3 & c_3 \end{vmatrix} = a_1(b_2c_3 - b_3c_2) - b_1(a_2c_3 - a_3c_2) + c_1(a_2b_3 - a_3b_2),$$

we observe that the sums enclosed in parentheses in the right-hand member of (2) are the respective minors (second-order determinants) of the elements a_1, b_1, and c_1. Therefore, (2) can be written

$$(3)\quad \begin{vmatrix} a_1 & b_1 & c_1 \\ a_2 & b_2 & c_2 \\ a_3 & b_3 & c_3 \end{vmatrix} = a_1\begin{vmatrix} b_2 & c_2 \\ b_3 & c_3 \end{vmatrix} - b_1\begin{vmatrix} a_2 & c_2 \\ a_3 & c_3 \end{vmatrix} + c_1\begin{vmatrix} a_2 & b_2 \\ a_3 & b_3 \end{vmatrix}.$$

The right-hand member of (3) is called the **expansion** of the determinant by minors about the *first row*.

Suppose, instead of factoring the right-hand member of (1) into the right-hand member of (2), we factor it as

$$(4)\quad \begin{vmatrix} a_1 & b_1 & c_1 \\ a_2 & b_2 & c_2 \\ a_3 & b_3 & c_3 \end{vmatrix} = a_1(b_2 c_3 - b_3 c_2) - a_2(b_1 c_3 - b_3 c_1) + a_3(b_1 c_2 - b_2 c_1).$$

Then we have the expansion of the determinant by minors about the *first column*,

$$(5)\quad \begin{vmatrix} a_1 & b_1 & c_1 \\ a_2 & b_2 & c_2 \\ a_3 & b_3 & c_3 \end{vmatrix} = a_1\begin{vmatrix} b_2 & c_2 \\ b_3 & c_3 \end{vmatrix} - a_2\begin{vmatrix} b_1 & c_1 \\ b_3 & c_3 \end{vmatrix} + a_3\begin{vmatrix} b_1 & c_1 \\ b_2 & c_2 \end{vmatrix}.$$

With the proper use of signs it is possible to expand a determinant by minors about *any* row or *any* column and obtain an expression equivalent to a factored form of the right-hand member of (1). A helpful device for determining the signs of the terms in an expansion of a third-order determinant by minors is the array of alternating signs

$$\begin{matrix} + & - & + \\ - & + & - \\ + & - & + \end{matrix}$$

which we will call the **sign array** for the determinant. To obtain an expansion of (1) about a given row or column, the appropriate sign from the sign array is prefixed to each term in the expansion.

As an example, let us first expand the determinant

$$\begin{vmatrix} 1 & 2 & -3 \\ 0 & 2 & -1 \\ 1 & 1 & 0 \end{vmatrix}$$

about the second row. We have

$$\begin{aligned} \begin{vmatrix} 1 & 2 & -3 \\ 0 & 2 & -1 \\ 1 & 1 & 0 \end{vmatrix} &= -0\begin{vmatrix} 2 & -3 \\ 1 & 0 \end{vmatrix} + 2\begin{vmatrix} 1 & -3 \\ 1 & 0 \end{vmatrix} - (-1)\begin{vmatrix} 1 & 2 \\ 1 & 1 \end{vmatrix} \\ &= 0 + 2(0 + 3) + 1(1 - 2) \\ &= 6 - 1 = 5. \end{aligned}$$

If we expand about the third row, we have

$$\begin{vmatrix} 1 & 2 & -3 \\ 0 & 2 & -1 \\ 1 & 1 & 0 \end{vmatrix} = 1\begin{vmatrix} 2 & -3 \\ 2 & -1 \end{vmatrix} - 1\begin{vmatrix} 1 & -3 \\ 0 & -1 \end{vmatrix} + 0\begin{vmatrix} 1 & 2 \\ 0 & 2 \end{vmatrix}$$

$$= 1(-2+6) - 1(-1-0) + 0$$

$$= 4 + 1 = 5.$$

You should expand this determinant about the first row and about each column to verify that the result is the same in each expansion.

The expansion of a higher-order determinant by minors can be accomplished in the same way. By continuing the pattern of alternating signs used for third-order determinants, the sign array extends to higher-order determinants. The determinants in each term in the expansion will be of order one less than the order of the original determinant.

EXERCISE B.3

A

Evaluate.

Example $\begin{vmatrix} 1 & 2 & 0 \\ 3 & -1 & 4 \\ -2 & 1 & 3 \end{vmatrix}$

Solution Expand about any row or column; the first row is used here.

$$\begin{vmatrix} 1 & 2 & 0 \\ 3 & -1 & 4 \\ -2 & 1 & 3 \end{vmatrix} = 1\begin{vmatrix} -1 & 4 \\ 1 & 3 \end{vmatrix} - 2\begin{vmatrix} 3 & 4 \\ -2 & 3 \end{vmatrix} + 0\begin{vmatrix} 3 & -1 \\ -2 & 1 \end{vmatrix}$$

$$= 1[(-1)(3) - (1)(4)] - 2[(3)(3) - (-2)(4)] + 0$$

$$= (-3 - 4) - 2(9 + 8)$$

$$= -7 - 34 = -41$$

1. $\begin{vmatrix} 2 & 0 & 1 \\ 1 & 1 & 2 \\ -1 & 0 & 1 \end{vmatrix}$ **2.** $\begin{vmatrix} 1 & 3 & 1 \\ -1 & 2 & 1 \\ 0 & 2 & 0 \end{vmatrix}$ **3.** $\begin{vmatrix} 2 & -1 & 0 \\ -3 & 1 & 2 \\ 1 & -3 & 1 \end{vmatrix}$

4. $\begin{vmatrix} 2 & 4 & -1 \\ -1 & 3 & 2 \\ 4 & 0 & 2 \end{vmatrix}$ **5.** $\begin{vmatrix} 1 & 2 & 3 \\ 3 & -1 & 2 \\ 2 & 0 & 2 \end{vmatrix}$ **6.** $\begin{vmatrix} 1 & 0 & 0 \\ 0 & 1 & 2 \\ 0 & 3 & 4 \end{vmatrix}$

7. $\begin{vmatrix} -1 & 0 & 2 \\ -2 & 1 & 0 \\ 0 & 1 & -3 \end{vmatrix}$ **8.** $\begin{vmatrix} 2 & 1 & 4 \\ 3 & 2 & 6 \\ 5 & -3 & 10 \end{vmatrix}$ **9.** $\begin{vmatrix} 2 & 5 & -1 \\ 1 & 0 & 2 \\ 0 & 0 & 1 \end{vmatrix}$

10. $\begin{vmatrix} 2 & 3 & 1 \\ 0 & 1 & 0 \\ -4 & 2 & 1 \end{vmatrix}$ **11.** $\begin{vmatrix} a & b & 1 \\ a & b & 1 \\ 1 & 1 & 1 \end{vmatrix}$ **12.** $\begin{vmatrix} a & a & a \\ 1 & 2 & 3 \\ 4 & 5 & 6 \end{vmatrix}$

13. $\begin{vmatrix} x & 0 & 0 \\ 0 & x & 0 \\ 0 & 0 & x \end{vmatrix}$ **14.** $\begin{vmatrix} 0 & 0 & x \\ 0 & x & 0 \\ x & 0 & 0 \end{vmatrix}$ **15.** $\begin{vmatrix} x & y & 0 \\ x & y & 0 \\ 0 & 0 & 1 \end{vmatrix}$

16. $\begin{vmatrix} 0 & a & b \\ a & 0 & a \\ b & a & 0 \end{vmatrix}$ **17.** $\begin{vmatrix} a & b & 0 \\ b & 0 & b \\ 0 & b & a \end{vmatrix}$ **18.** $\begin{vmatrix} 0 & b & 0 \\ b & a & b \\ 0 & b & 0 \end{vmatrix}$

Solve for x.

19. $\begin{vmatrix} x & 0 & 0 \\ 2 & 1 & 3 \\ 0 & 1 & 4 \end{vmatrix} = 3$ **20.** $\begin{vmatrix} x^2 & 0 & 1 \\ 2 & -1 & 3 \\ 3 & 2 & 0 \end{vmatrix} = 1$

21. $\begin{vmatrix} x^2 & x & 1 \\ 0 & 2 & 1 \\ 3 & 1 & 4 \end{vmatrix} = 28$ **22.** $\begin{vmatrix} x & 1 & 1 \\ 0 & x & 1 \\ 0 & x & 0 \end{vmatrix} = -4$

B

23. Show that for all values of x, y, and z,

$$\begin{vmatrix} x & x & a \\ y & y & b \\ z & z & c \end{vmatrix} = 0.$$

[*Hint*: Expand about the elements of the third column.] Make a conjecture about determinants containing two identical columns.

24. Show that

$$\begin{vmatrix} 0 & 0 & 0 \\ a & b & c \\ d & e & f \end{vmatrix} = 0$$

for all values of a, b, c, d, e, and f. Make a conjecture about determinants containing a row of 0 elements.

25. Show that

$$\begin{vmatrix} 1 & 2 & 3 \\ 4 & 5 & 6 \\ 0 & 0 & 1 \end{vmatrix} = -\begin{vmatrix} 4 & 5 & 6 \\ 1 & 2 & 3 \\ 0 & 0 & 1 \end{vmatrix}.$$

Make a conjecture about the result of interchanging any two rows of a determinant.

26. Show that

$$\begin{vmatrix} 2 & 0 & 1 \\ 4 & 1 & -2 \\ 6 & 1 & 1 \end{vmatrix} = 2\begin{vmatrix} 1 & 0 & 1 \\ 2 & 1 & -2 \\ 3 & 1 & 1 \end{vmatrix}.$$

Make a conjecture about the result of factoring a common factor from each element of a column in a determinant.

B.4

LINEAR SYSTEMS IN THREE VARIABLES— SOLUTION BY DETERMINANTS

Consider the linear system in three variables

(1) $$a_1 x + b_1 y + c_1 z = d_1$$

(2) $$a_2 x + b_2 y + c_2 z = d_2$$

(3) $$a_3 x + b_3 y + c_3 z = d_3.$$

By solving this system using the methods of Section 10.2, it can be shown that Cramer's rule is applicable to such systems and, in fact, to all similar systems as well as to linear systems in two variables. That is,

$$x = \frac{D_x}{D}, \qquad y = \frac{D_y}{D}, \qquad z = \frac{D_z}{D},$$

where

$$D = \begin{vmatrix} a_1 & b_1 & c_1 \\ a_2 & b_2 & c_2 \\ a_3 & b_3 & c_3 \end{vmatrix}, \qquad D_x = \begin{vmatrix} d_1 & b_1 & c_1 \\ d_2 & b_2 & c_2 \\ d_3 & b_3 & c_3 \end{vmatrix},$$

$$D_y = \begin{vmatrix} a_1 & d_1 & c_1 \\ a_2 & d_2 & c_2 \\ a_3 & d_3 & c_3 \end{vmatrix}, \qquad D_z = \begin{vmatrix} a_1 & b_1 & d_1 \\ a_2 & b_2 & d_2 \\ a_3 & b_3 & d_3 \end{vmatrix}.$$

Note that the elements of the determinant D in each denominator are the coefficients of the variables in (1), (2), and (3), and that the numerators are formed from D by replacing the elements in the x, y, or z column, respectively, by d_1, d_2, and d_3. We illustrate the application of Cramer's rule by considering the system of equations used in the example on page 312;

$$x + 2y - 3z = -4$$
$$2x - y + z = 3$$
$$3x + 2y + z = 10.$$

The determinant D, with elements that are the coefficients of the variables, is given by

$$D = \begin{vmatrix} 1 & 2 & -3 \\ 2 & -1 & 1 \\ 3 & 2 & 1 \end{vmatrix}.$$

We can expand the determinant about the first column, and get

$$D = \begin{vmatrix} 1 & 2 & -3 \\ 2 & -1 & 1 \\ 3 & 2 & 1 \end{vmatrix} = 1\begin{vmatrix} -1 & 1 \\ 2 & 1 \end{vmatrix} - 2\begin{vmatrix} 2 & -3 \\ 2 & 1 \end{vmatrix} + 3\begin{vmatrix} 2 & -3 \\ -1 & 1 \end{vmatrix}$$

$$= -3 - 16 - 3 = -22.$$

Replacing the first column in D with -4, 3, and 10, we obtain

$$D_x = \begin{vmatrix} -4 & 2 & -3 \\ 3 & -1 & 1 \\ 10 & 2 & 1 \end{vmatrix}.$$

Expanding D_x about the third column, we have

$$D_x = \begin{vmatrix} -4 & 2 & -3 \\ 3 & -1 & 1 \\ 10 & 2 & 1 \end{vmatrix} = 3\begin{vmatrix} 3 & -1 \\ 10 & 2 \end{vmatrix} - 1\begin{vmatrix} -4 & 2 \\ 10 & 2 \end{vmatrix} + 1\begin{vmatrix} -4 & 2 \\ 3 & -1 \end{vmatrix}$$

$$= -48 + 28 - 2 = -22.$$

In a similar fashion, we can compute D_y and D_z:

$$D_y = \begin{vmatrix} 1 & -4 & -3 \\ 2 & 3 & 1 \\ 3 & 10 & 1 \end{vmatrix} = -44, \qquad D_z = \begin{vmatrix} 1 & 2 & -4 \\ 2 & -1 & 3 \\ 3 & 2 & 10 \end{vmatrix} = -66.$$

We then have

$$x = \frac{D_x}{D} = \frac{-22}{-22} = 1, \qquad y = \frac{D_y}{D} = \frac{-44}{-22} = 2, \qquad z = \frac{D_z}{D} = \frac{-66}{-22} = 3,$$

and the solution set of the system is $\{(1, 2, 3)\}$.

As noted on page 378 for a linear system in two variables, if $D = 0$ for a linear system in three variables, the system does not have a unique solution.

EXERCISE B.4

Solve by Cramer's rule. If a unique solution does not exist ($D = 0$), so state.

Example

$$\begin{aligned} 4x + 10y - z &= 2 \\ 2x + 8y + z &= 4 \\ x - 3y - 2z &= 3 \end{aligned}$$

(*solution on page* 386)

Solution Determine values for D, D_x, D_y, and D_z. The elements of D are the coefficients of the variables in the order they occur. For D_x, D_y, and D_z, the respective column of elements in D is replaced by the constants 2, 4, and 3.

$$D = \begin{vmatrix} 4 & 10 & -1 \\ 2 & 8 & 1 \\ 1 & -3 & -2 \end{vmatrix} = 12 \qquad D_x = \begin{vmatrix} 2 & 10 & -1 \\ 4 & 8 & 1 \\ 3 & -3 & -2 \end{vmatrix} = 120$$

$$D_y = \begin{vmatrix} 4 & 2 & -1 \\ 2 & 4 & 1 \\ 1 & 3 & -2 \end{vmatrix} = -36 \qquad D_z = \begin{vmatrix} 4 & 10 & 2 \\ 2 & 8 & 4 \\ 1 & -3 & 3 \end{vmatrix} = 96$$

Use Cramer's rule to determine x, y, and z.

$$x = \frac{D_x}{D} = \frac{120}{12} = 10 \qquad y = \frac{D_y}{D} = \frac{-36}{12} = -3 \qquad z = \frac{D_z}{D} = \frac{96}{12} = 8$$

The solution set is $\{(10, -3, 8)\}$.

1. $x + y = 2$
$2x - z = 1$
$2y - 3z = -1$

2. $2x - 6y + 3z = -12$
$3x - 2y + 5z = -4$
$4x + 5y - 2z = 10$

3. $x - 2y + z = -1$
$3x + y - 2z = 4$
$y - z = 1$

4. $2x + 5z = 9$
$4x + 3y = -1$
$3y - 4z = -13$

5. $2x + 2y + z = 1$
$x - y + 6z = 21$
$3x + 2y - z = -4$

6. $4x + 8y + z = -6$
$2x - 3y + 2z = 0$
$x + 7y - 3z = -8$

7. $x + y + z = 0$
$2x - y - 4z = 15$
$x - 2y - z = 7$

8. $x + y - 2z = 3$
$3x - y + z = 5$
$3x + 3y - 6z = 9$

9. $x - 2y + 2z = 3$
$2x - 4y + 4z = 1$
$3x - 3y - 3z = 4$

10. $3x - 2y + 5z = 6$
$4x - 4y + 3z = 0$
$5x - 4y + z = -5$

11. $\frac{1}{4}x - z = -\frac{1}{4}$
$x + y = \frac{2}{3}$
$3x + 4z = 5$

12. $2x - \frac{2}{3}y + z = 2$
$\frac{1}{2}x - \frac{1}{3}y - \frac{1}{4}z = 0$
$4x + 5y - 3z = -1$

13. $x + 4z = 3$
$y + 3z = 9$
$2x + 5y - 5z = -5$

14. $2x + y = 18$
$y + z = -1$
$3x - 2y - 5z = 38$

15. $2x + z = 2$
$3y - 2z = 22$
$2x - y = 13$

APPENDIX B SUMMARY

[B.1] A **matrix** is a rectangular array of elements or **entries**. Two matrices are said to be **row-equivalent** if one matrix can be transformed to the other by one or more of the following transformations:

1. Multiplying the entries of any row by a nonzero real number.
2. Interchanging two rows.
3. Multiplying the entries of any row by a real number and adding the results to the corresponding elements of another row.

We can use these transformations to solve linear systems of equations.

[B.2] The order of a **determinant** is the number of rows or columns in the determinant. A second-order determinant is defined by

$$\begin{vmatrix} a_1 & b_1 \\ a_2 & b_2 \end{vmatrix} = a_1b_2 - a_2b_1.$$

Cramer's rule uses determinants to solve systems of two linear equations in two variables.

If

$$a_1x + b_1y = c_1$$
$$a_2x + b_2y = c_2,$$

then, for $D \neq 0$,

$$x = \frac{D_x}{D} = \frac{\begin{vmatrix} c_1 & b_1 \\ c_2 & b_2 \end{vmatrix}}{\begin{vmatrix} a_1 & b_1 \\ a_2 & b_2 \end{vmatrix}} \quad \text{and} \quad y = \frac{D_y}{D} = \frac{\begin{vmatrix} a_1 & c_1 \\ a_2 & c_2 \end{vmatrix}}{\begin{vmatrix} a_1 & b_1 \\ a_2 & b_2 \end{vmatrix}}.$$

[B.3] A third-order determinant is defined by

$$\begin{vmatrix} a_1 & b_1 & c_1 \\ a_2 & b_2 & c_2 \\ a_3 & b_3 & c_3 \end{vmatrix} = a_1b_2c_3 - a_1b_3c_2 + a_3b_1c_2 - a_2b_1c_3 + a_2b_3c_1 - a_3b_2c_1.$$

The **minor** of an element in a determinant D is the determinant that results when the row and column containing the element are deleted. A determinant of an order greater than or equal to three can be evaluated by expansion by minors of the elements of a row or column.

[B.4] Cramer's rule can also be used to solve systems of three linear equations in three variables.

If

$$a_1x + b_1y + c_1z = d_1$$
$$a_2x + b_2y + c_2z = d_2$$
$$a_3x + b_3y + c_3z = d_3,$$

then, for $D \neq 0$,

$$x = \frac{D_x}{D}, \qquad y = \frac{D_y}{D}, \qquad z = \frac{D_z}{D}.$$

REVIEW EXERCISES

[B.1] *Use row transformations to solve each system.*

1. $x - 2y = 5$
$2x + y = 5$

2. $x + 2y - z = -3$
$2x - 3y + 2z = 2$
$x - y + 4z = 7$

[B.2] **3.** Evaluate $\begin{vmatrix} 3 & -2 \\ 1 & -5 \end{vmatrix}$.

4. Solve the system by Cramer's rule.

$$2x + 3y = -2$$
$$x - 8y = -39$$

[B.3] **5.** Evaluate $\begin{vmatrix} 2 & 1 & 3 \\ 0 & 4 & -1 \\ 2 & 0 & 3 \end{vmatrix}$.

[B.4] **6.** Solve the system by Cramer's rule.

$$2x + 3y - z = -2$$
$$x - y + z = 6$$
$$3x - y + z = 10$$

TABLES

FORMULAS FROM GEOMETRY

Plane Figures

1. Triangle ABC with sides of lengths a and c, base b, and altitude h.
Perimeter: $P = a + b + c$

Area: $A = \frac{1}{2}bh$

$\angle A + \angle B + \angle C = 180^\circ$

a. Isosceles triangle.
Two sides of equal length.
Two angles with equal measure.

b. Equilateral triangle.
Three sides of equal length.
Three angles with equal measure.

c. Right triangle with hypotenuse c.
$c^2 = a^2 + b^2$

2. Square with side of length s.
Perimeter: $P = 4s$
Area: $A = s^2$

3. Rectangle with length l and width w.
Perimeter: $P = l + l + w + w$
$P = 2l + 2w$
Area: $A = lw$

4. Circle with radius r.
Diameter: $d = 2r$
Circumference: $C = 2\pi r$ or πd
Area: $A = \pi r^2$

Solid Figures

1. Right circular cylinder with height h and radius r of the base.
Volume: $V = \pi r^2 h$
Lateral area: $S = 2\pi rh$

2. Rectangular prism with length l, width w, and height h.
Volume: $V = lwh$

TABLE OF SQUARES, SQUARE ROOTS, AND PRIME FACTORS

Number	Square	Square root	Prime factors
1	1	1.000	
2	4	1.414	2
3	9	1.732	3
4	16	2.000	2^2
5	25	2.236	5
6	36	2.449	$2 \cdot 3$
7	49	2.646	7
8	64	2.828	2^3
9	81	3.000	3^2
10	100	3.162	$2 \cdot 5$
11	121	3.317	11
12	144	3.464	$2^2 \cdot 3$
13	169	3.606	13
14	196	3.742	$2 \cdot 7$
15	225	3.873	$3 \cdot 5$
16	256	4.000	2^4
17	289	4.123	17
18	324	4.243	$2 \cdot 3^2$
19	361	4.359	19
20	400	4.472	$2^2 \cdot 5$
21	441	4.583	$3 \cdot 7$
22	484	4.690	$2 \cdot 11$
23	529	4.796	23
24	576	4.899	$2^3 \cdot 3$
25	625	5.000	5^2
26	676	5.099	$2 \cdot 13$
27	729	5.196	3^3
28	784	5.292	$2^2 \cdot 7$
29	841	5.385	29
30	900	5.477	$2 \cdot 3 \cdot 5$
31	961	5.568	31
32	1,024	5.657	2^5
33	1,089	5.745	$3 \cdot 11$
34	1,156	5.831	$2 \cdot 17$
35	1,225	5.916	$5 \cdot 7$
36	1,296	6.000	$2^2 \cdot 3^2$
37	1,369	6.083	37
38	1,444	6.164	$2 \cdot 19$
39	1,521	6.245	$3 \cdot 13$
40	1,600	6.325	$2^3 \cdot 5$
41	1,681	6.403	41
42	1,764	6.481	$2 \cdot 3 \cdot 7$
43	1,849	6.557	43
44	1,936	6.633	$2^2 \cdot 11$
45	2,025	6.708	$3^2 \cdot 5$
46	2,116	6.782	$2 \cdot 23$
47	2,209	6.856	47
48	2,304	6.928	$2^4 \cdot 3$
49	2,401	7.000	7^2
50	2,500	7.071	$2 \cdot 5^2$
51	2,601	7.141	$3 \cdot 17$
52	2,704	7.211	$2^2 \cdot 13$
53	2,809	7.280	53
54	2,916	7.348	$2 \cdot 3^3$
55	3,025	7.416	$5 \cdot 11$
56	3,136	7.483	$2^3 \cdot 7$
57	3,249	7.550	$3 \cdot 19$
58	3,364	7.616	$2 \cdot 29$
59	3,481	7.681	59
60	3,600	7.746	$2^2 \cdot 3 \cdot 5$
61	3,721	7.810	61
62	3,844	7.874	$2 \cdot 31$
63	3,969	7.937	$3^2 \cdot 7$
64	4,096	8.000	2^6
65	4,225	8.062	$5 \cdot 13$
66	4,356	8.124	$2 \cdot 3 \cdot 11$
67	4,489	8.185	67
68	4,624	8.246	$2^2 \cdot 17$
69	4,761	8.307	$3 \cdot 23$
70	4,900	8.367	$2 \cdot 5 \cdot 7$
71	5,041	8.426	71
72	5,184	8.485	$2^3 \cdot 3^2$
73	5,329	8.544	73
74	5,476	8.602	$2 \cdot 37$
75	5,625	8.660	$3 \cdot 5^2$
76	5,776	8.718	$2^2 \cdot 19$
77	5,929	8.775	$7 \cdot 11$
78	6,084	8.832	$2 \cdot 3 \cdot 13$
79	6,241	8.888	79
80	6,400	8.944	$2^4 \cdot 5$
81	6,561	9.000	3^4
82	6,724	9.055	$2 \cdot 41$
83	6,889	9.110	83
84	7,056	9.165	$2^2 \cdot 3 \cdot 7$
85	7,225	9.220	$5 \cdot 17$
86	7,396	9.274	$2 \cdot 43$
87	7,569	9.327	$3 \cdot 29$
88	7,744	9.381	$2^3 \cdot 11$
89	7,921	9.434	89
90	8,100	9.487	$2 \cdot 3^2 \cdot 5$
91	8,281	9.539	$7 \cdot 13$
92	8,464	9.592	$2^2 \cdot 23$
93	8,649	9.644	$3 \cdot 31$
94	8,836	9.695	$2 \cdot 47$
95	9,025	9.747	$5 \cdot 19$
96	9,216	9.798	$2^5 \cdot 3$
97	9,409	9.849	97
98	9,604	9.899	$2 \cdot 7^2$
99	9,801	9.950	$3^2 \cdot 11$
100	10,000	10.000	$2^2 \cdot 5^2$

ODD-NUMBERED ANSWERS

Exercise 1.1 [page 6]

1. $\{2,3,4,5,6\}$ **3.** $\{1,2,3\}$ **5.** $\{4,8,12,\ldots\}$ **7.** $\{-2,0,2\}$

9. {natural number multiples of 15} **11.** {odd integers between 8 and 14}

13. {first three whole numbers} **15.** $\{0,8\}$ **17.** $\{-\sqrt{15}\}$

19. $\{-5,-\sqrt{15},-3.44,\ldots,-2/3\}$ **21.** finite **23.** finite **25.** infinite

27. variable **29.** constant **31.** constant **33.** $\in$ **35.** $\subset$ **37.** $\subset$ **39.** $\not\subset$

41. $\subset$ **43.** $\in$ **45.** $\in$ **47.** $\frac{1}{2}\notin N$ **49.** $Q\not\subset J$ **51.** $N\subset R$ **53.** $W\subset R$

55. no **57.** no **59.** yes **61.** yes **63.** yes **65.** $M=N$

67. $\{1\}$, $\{2\}$, $\{3\}$, $\{1,2\}$, $\{1,3\}$, $\{2,3\}$, $\{1,2,3\}$, $\varnothing$ **69.** 5

Exercise 1.2 [page 9]

1. $\{1,3,5\}$ **3.** $\{2,4,6,7,8,9,10\}$ **5.** $\{6,8,10\}$ **7.** $\{1,2,3,4,5,6,7,8,9,10\}$

9. $\{2,4,6,8,10\}$ **11.** $\{1,2,3,4,5\}$ **13.** $\varnothing$ **15.** $\{1,3,5,7,9\}$ **17.** $\{1,2,3,4,5\}$

19. $\{6,8,10\}$ **21.** $\{1,2,3,4,5,6,7,8,9,10\}$ **23.** $G=\varnothing$ and $H=\varnothing$

25. $G\subset H$ **27.** $G=\varnothing$

29. 31.

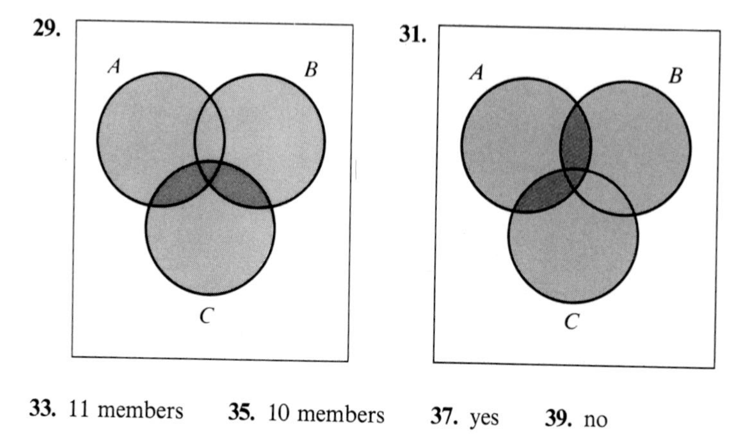

33. 11 members **35.** 10 members **37.** yes **39.** no

Exercise 1.3 [page 14]

1. $3r$ **3.** 6 **5.** $n; t$ **7.** r **9.** $6 + x$ **11.** $4 > -2$ **13.** $-4 < -2$

15. $x + 3 > 0$ **17.** $2n - 1 \geq 0$ **19.** $0 < z < 1$ **21.** $1 \leq x < 7$ **23.** $-2 < 8$

25. $-7 > -13$ **27.** $-6 < -3$ **29.** $1\frac{1}{2} = \frac{3}{2}$ **31.** $3 < 5 < 7$ **33.** $-7 < 0 < 2$

35. $2 < 5$ **37.** $7 < 8$ **39.** $x \geq y$ **41.** $x \nleq 0; x > 0$ **43.** $x \nless 0; x \geq 0$

45. **47.** **49.** **51.** **53.** **55.** **57.** **59.** **61.** **63.**

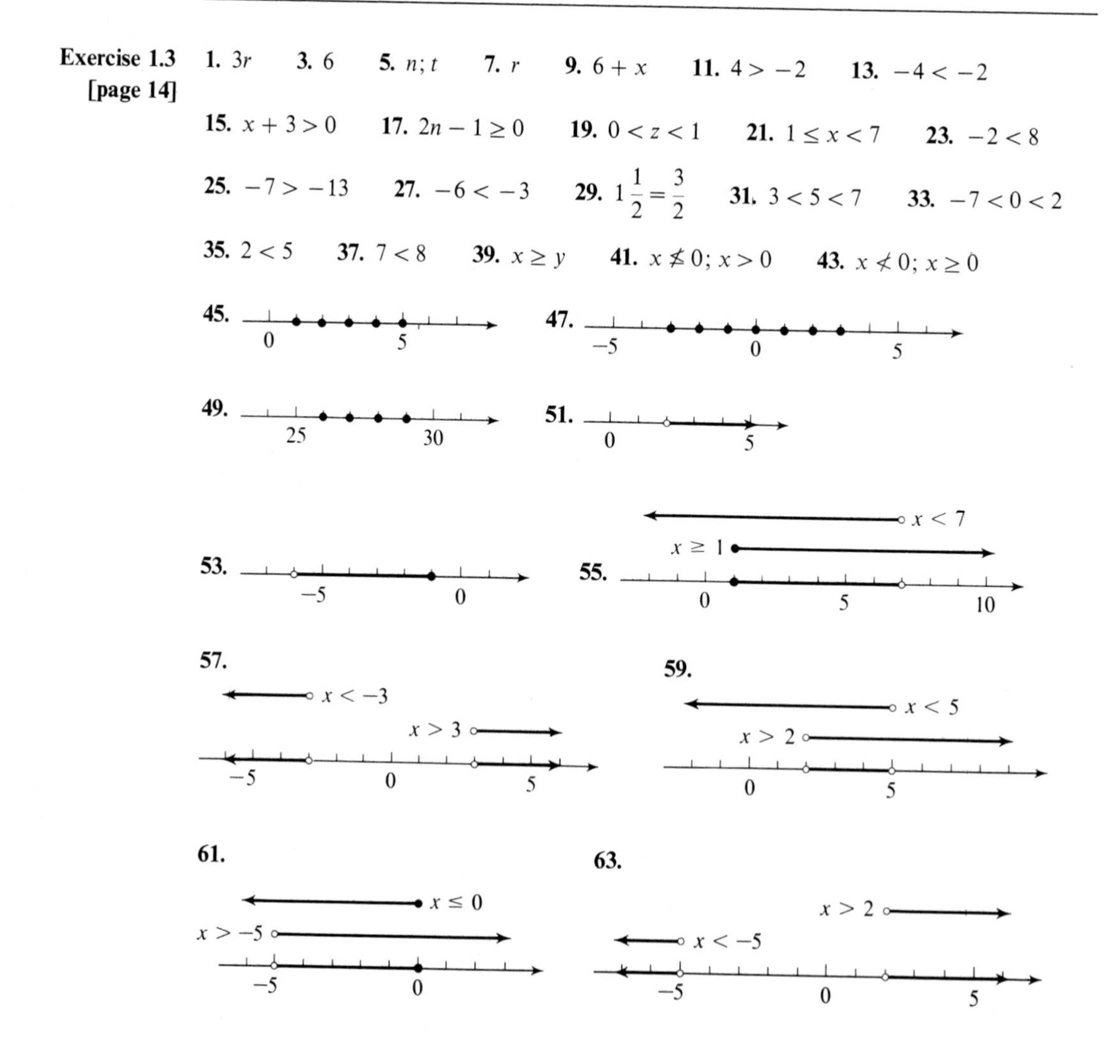

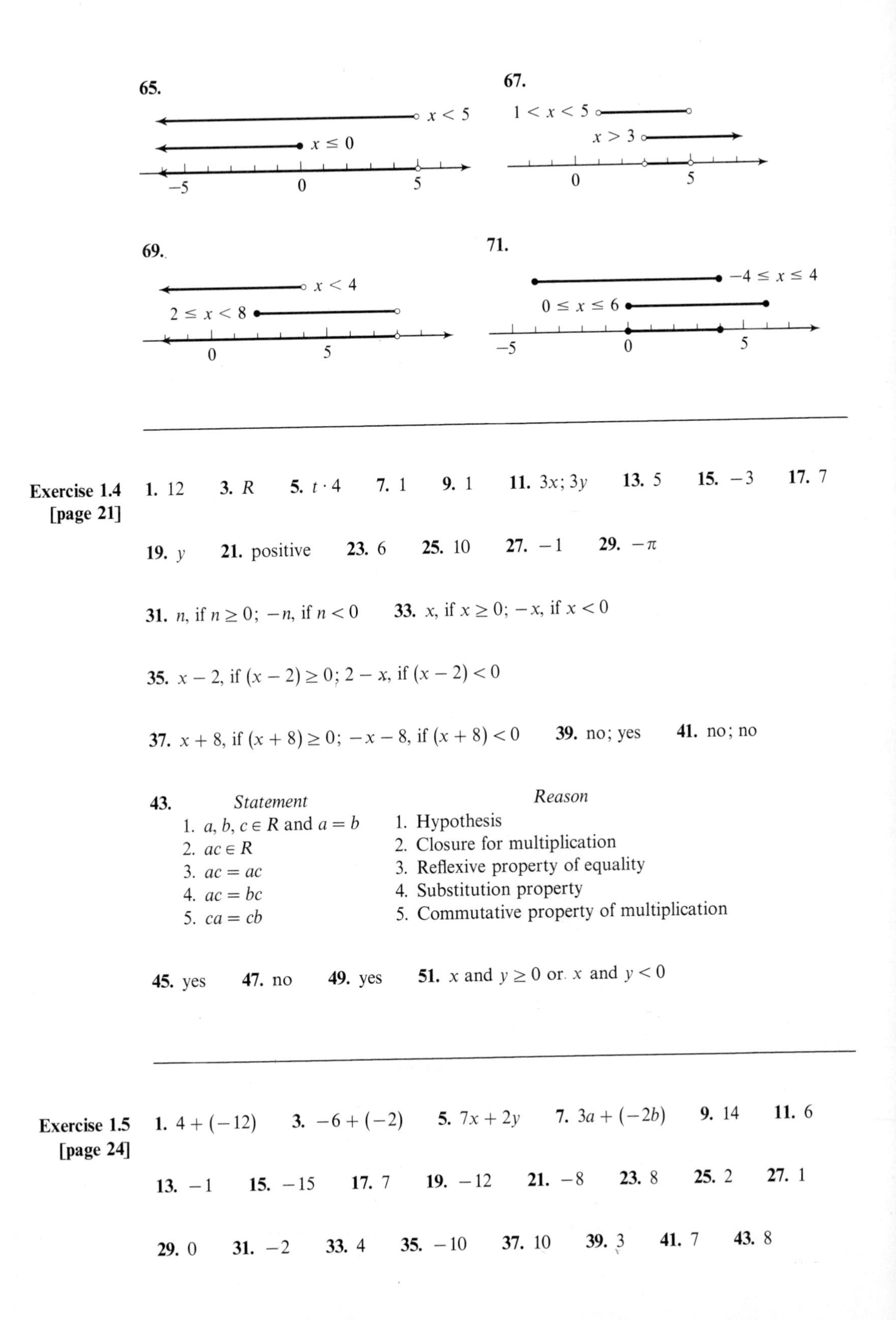

Exercise 1.4 [page 21]

1. 12 **3.** R **5.** $t \cdot 4$ **7.** 1 **9.** 1 **11.** $3x; 3y$ **13.** 5 **15.** -3 **17.** 7

19. y **21.** positive **23.** 6 **25.** 10 **27.** -1 **29.** $-\pi$

31. n, if $n \geq 0$; $-n$, if $n < 0$ **33.** x, if $x \geq 0$; $-x$, if $x < 0$

35. $x - 2$, if $(x - 2) \geq 0$; $2 - x$, if $(x - 2) < 0$

37. $x + 8$, if $(x + 8) \geq 0$; $-x - 8$, if $(x + 8) < 0$ **39.** no; yes **41.** no; no

43.

Statement	*Reason*
1. $a, b, c \in R$ and $a = b$	1. Hypothesis
2. $ac \in R$	2. Closure for multiplication
3. $ac = ac$	3. Reflexive property of equality
4. $ac = bc$	4. Substitution property
5. $ca = cb$	5. Commutative property of multiplication

45. yes **47.** no **49.** yes **51.** x and $y \geq 0$ or x and $y < 0$

Exercise 1.5 [page 24]

1. $4 + (-12)$ **3.** $-6 + (-2)$ **5.** $7x + 2y$ **7.** $3a + (-2b)$ **9.** 14 **11.** 6

13. -1 **15.** -15 **17.** 7 **19.** -12 **21.** -8 **23.** 8 **25.** 2 **27.** 1

29. 0 **31.** -2 **33.** 4 **35.** -10 **37.** 10 **39.** 3 **41.** 7 **43.** 8

45. $a - b = d$ implies that $b + d = a$. It follows that

$(b + d) + (-b) = a + (-b)$	Addition property of equality
$(d + b) + (-b) = a + (-b)$	Commutative property of addition
$d + [b + (-b)] = a + (-b)$	Associative property of addition
$d + 0 = a + (-b)$	Additive inverse property
$d = a + (-b)$	Identity element for addition

Hence,

$a - b = a + (-b)$	Transitive property of equality

47. Substituting 5 and 3 for a and b, we get

$$5 - 3 = 3 - 5$$
$$2 \neq -2$$

Exercise 1.6 [page 28]

1. -30 **3.** 42 **5.** -8 **7.** 30 **9.** -70 **11.** 0 **13.** 6 **15.** -30

17. -4 **19.** -13 **21.** 3 **23.** 0 **25.** undefined **27.** 4 **29.** $15 = -3(-5)$

31. $-38 = 19(-2)$ **33.** $-52 = -4(13)$ **35.** $0 = -7(0)$ **37.** $7\left(\frac{1}{8}\right)$ **39.** $3\left(\frac{1}{8}\right)$

41. $82\left(\frac{1}{11}\right)$ **43.** $7\left(\frac{1}{100}\right)$ **45.** $\frac{3}{2}$ **47.** $\frac{2}{7}$ **49.** $\frac{5}{8}$ **51.** $\frac{9}{2}$ **53.** $(2)(2)(2)$

55. $(7)(7)$ **57.** prime **59.** $-1(2)(2)(3)$ **61.** $(2)(2)(2)(7)$ **63.** $-1(2)(19)$

65. $(2)(2)(5)$ **67.** $(2)(53)$ **69.** $x = 0, y \neq 0$ **71.** $x, y > 0$ or $x, y < 0$

73. positive; negative

75.

Statement	*Reason*
1. $a, b \in R, b \neq 0$	1. Hypothesis
2. $\frac{a}{b} = q$ implies $bq = a$.	2. Definition of quotient
3. $\frac{1}{b}(bq) = \frac{1}{b} \cdot a$	3. Multiplication property
4. $\left(\frac{1}{b} \cdot b\right)q = a \cdot \frac{1}{b}$	4. Associative and commutative properties of multiplication
5. $1 \cdot q = a \cdot \frac{1}{b}$	5. Reciprocal axiom
6. $q = a \cdot \frac{1}{b}$	6. Identity element for multiplication
7. $\frac{a}{b} = a \cdot \frac{1}{b}$	7. Substitution of $\frac{a}{b}$ for q

77. Substituting 8 and 4 for a and b, we get $8 \div 4 = 4 \div 8$; $2 \neq \frac{1}{2}$

Exercise 1.7 [page 31]

1. 20 **3.** 0 **5.** -17 **7.** -13 **9.** -2 **11.** 6 **13.** -44 **15.** 5

17. -2 **19.** 100 **21.** 4 **23.** 20 **25.** 10 **27.** 2 **29.** undefined **31.** 1

33. -12

Review Exercises [page 34]

1. $\{-3, 0, 1\}$ **2.** $\{1,2\}, \{1,2,3\}, \{1,2,4\}, \{1,2,3,4\}$

3. $N \subset W, W \subset J, J \subset Q, Q \subset R$, and $H \subset R$ **4.** $\{-1, 2, 3, 5, 6, 7, 9\}$ **5.** $\{9\}$

6. disjoint **7.** $<$ **8.** $a < b < c$ or $c < b < a$

9. 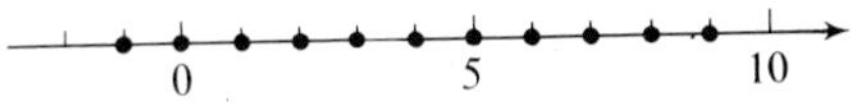

10. 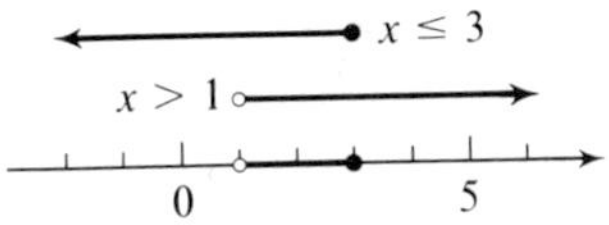

11. 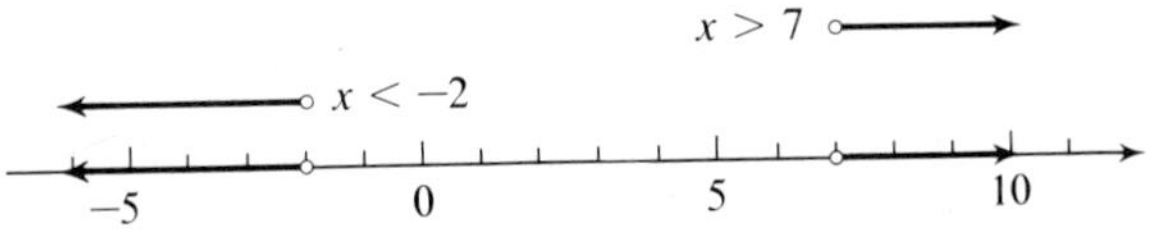

12. $2; x$ **13.** $x + 2$ **14.** $18 + t$ **15.** $\frac{1}{3}$ **16.** 5 **17.** 1 **18.** 0

19. a. 14 **b.** 8 **20. a.** -6 **b.** 5 **21. a.** -36 **b.** 0 **c.** 21 **d.** 24

22. a. 12 **b.** 0 **c.** -8 **d.** -8 **23. a.** $-7\left(\frac{1}{5}\right)$ **b.** $24\left(\frac{1}{7}\right)$

24. (2)(2)(2)(2)(2)(3) **25.** (2)(3)(23) **26.** (2)(2)(3)(17) **27. a.** 3 **b.** 6

28. a. 5 **b.** 28

Exercise 2.1 [page 39]

1. trinomial; degree 4 **3.** binomial; degree 3 **5.** monomial; degree 4 **7.** one

9. two **11.** two **13.** 36 **15.** 21 **17.** 11 **19.** -2 **21.** -5 **23.** -64

25. 7 **27.** 13 **29.** 5 **31.** 0 **33.** 1 **35.** 26 **37.** 64 **39.** 8 **41.** 54

43. -1; -21; 1 **45.** -4; 60; 16 **47.** 19; 7 **49.** 37; 11 **51.** 14 **53.** -4

55. 4 **57.** 8 **59.** closure for multiplication; closure for both addition and multiplication

---

Exercise 2.2 [page 44]

1. a^2 **3.** $6t$ **5.** $6n$ **7.** $x^2y + 2x$ **9.** $4n^2 - 3n$ **11.** $2x^2y - 4xy^2$

13. $4k^2 + u^2 - 3k - 3u + 3$ **15.** $-x^2 - 6x + 12$ **17.** $2x^2 - 4x + 11$

19. $3a^2 + 4a + 1$ **21.** $-5xy - xy^2$ **23.** $3x^3 + 2x^2 - 4x - 1$ **25.** $-x^2 - 4x + 7$

27. $-t^3 - 4t^2 + t + 1$ **29.** $3x^3 - 5x^2y + 3xy^2$ **31.** $-2x^2 + x - 3$

33. $b^2 + 3b + 1$ **35.** $2x^2 + 3x + 2$ **37.** $-2y - 1$ **39.** $2 - x$ **41.** $x - 1$

43. $-x^2 - 3x - 1$ **45.** $x - 2y$ **47.** $2x - y$ **49.** $-4x - 5$ **51.** $-6x + y$

53. $2x - 1$ **55.** -1 **57.** closure for addition; closure for multiplication

---

Exercise 2.3 [page 48]

1. $-14t^3$ **3.** $-40a^3b^3c$ **5.** $44x^3y^4z^2$ **7.** $6x^5y^5$ **9.** $-2r^6s^4t^2$ **11.** $3x^2y^6z^4$

13. $2x^2y$ **15.** $2tr^2$ **17.** $-2c^2$ **19.** ab^6c^3 **21.** $x - 4$ **23.** -1 **25.** a^{3n-3}

27. x^n **29.** a^{3n+1} **31.** a^{2n} **33.** a^{2n+3} **35.** x^{n^2-2} **37.** a^nb^4

39. For $m, n \in N$ and $m > n$,

$$\frac{x^m}{x^n} = x^m\left(\frac{1}{x^n}\right) = x^{m-n} \cdot x^n\left(\frac{1}{x^n}\right) = x^{m-n}\left(x^n \cdot \frac{1}{x^n}\right)$$

Hence,

$$\frac{x^m}{x^n} = x^{m-n} \qquad (x \neq 0)$$

41. $$(xy)^n = \underbrace{(xy)(xy)(xy)\cdots(xy)}_{n \text{ factors}} = \underbrace{(x \cdot x \cdot x \cdots x)}_{n \text{ factors}} \cdot \underbrace{(y \cdot y \cdot y \cdots y)}_{n \text{ factors}} = x^ny^n$$

Exercise 2.4 [page 51]

1. $2x^2 + 4xy$ **3.** $6t^3 - 18t^2 + 6t$ **5.** $-3x^2 - 2x + 4$ **7.** $x^2 + 4x + 4$

9. $x^2 - 4x - 5$ **11.** $y^2 - 7y + 6$ **13.** $2z^2 - 5z - 3$ **15.** $8r^2 + 2r - 3$

17. $4x^2 - a^2$ **19.** $y^3 - y + 6$ **21.** $x^3 + 2x^2 - 21x + 18$ **23.** $x^3 - 7x + 6$

25. $z^3 - 7z - 6$ **27.** $6x^3 + x^2 - 8x + 6$ **29.** $6a^4 - 5a^3 - 5a^2 + 5a - 1$

31. $a^5 - a^4 - 6a^3 + 3a^2 + 5a - 2$ **33.** $a^4 - 2a^3 - 4a^2 + 7a - 2$ **35.** $a^4 - 2a^2 + 1$

37. 6 **39.** $-2a^2 - 2a$ **41.** $4a + 4$ **43.** $4a + 4$ **45.** $-4x^2 - 11x$

47. $2x^{2n} - x^n$ **49.** $a^{2n+1} - a^{n+1}$ **51.** $a^{3n+1} + a^{2n+2}$ **53.** $1 - a^{2n}$

55. $a^{6n} + a^{3n} - 2$ **57.** $2a^{2n} + 3a^nb^n - 2b^{2n}$

59. $$\begin{aligned}(x + a)(x + b) &= x(x + a) + b(x + a)\\ &= x^2 + ax + bx + ab\\ &= x^2 + (a + b)x + ab\end{aligned}$$

61. $$\begin{aligned}(x + a)(x - a) &= x(x + a) - a(x + a)\\ &= x^2 + ax - ax - a^2\\ &= x^2 - a^2\end{aligned}$$

63. $$\begin{aligned}(x + a)(x^2 - ax + a^2) &= x(x^2 - ax + a^2) + a(x^2 - ax + a^2)\\ &= (x^3 - ax^2 + a^2x) + (ax^2 - a^2x + a^3)\\ &= x^3 - ax^2 + ax^2 + a^2x - a^2x + a^3\\ &= x^3 + a^3\end{aligned}$$

Exercise 2.5 [page 55]

1. $2(x + 3)$ **3.** $4x(x + 2)$ **5.** $3x(x + y + 1)$ **7.** $6(4a^2 + 2a - 1)$

9. $2x(x^3 - 2x + 4)$ **11.** $xz(xy^2z + 2y - 1)$ **13.** $(a + b)(a + 3)$ **15.** $(2x - y)(x + 3)$

17. $(2y - x)(a + b)$ **19.** $-(r - 7)$ **21.** $-(b - 2a)$ **23.** $-2(x - 1)$

25. $-a(b + c)$ **27.** $-xy(1 + x)$ **29.** $-(-2x + 1)$ **31.** $-(-x + y - z)$

33. $x - (y - z)$ **35.** $x^n(x^n - 1)$ **37.** $a^n(a^{2n} - a^n - 1)$ **39.** $x^n(x^2 + 1)$

41. $-x^n(x^n + 1)$ **43.** $-x^a(x + 1)$

Exercise 2.6 [page 58]

1. $(x + 4)(x + 3)$ **3.** $(a - 3)(a + 2)$ **5.** $(x + 5y)(x + y)$ **7.** $(x + 2)(x - 2)$

9. $(2 + b)(2 - b)$ **11.** $(ab + 1)(ab - 1)$ **13.** $(x^2 + 3)(x^2 - 3)$

15. $(x + 4y)(x - 4y)$ **17.** $(x - 6y)(x + 6y)$ **19.** $(2x - 1)(x + 2)$

21. $(4x - 1)(x + 2)$ **23.** $(3x + 1)(x + 1)$ **25.** $(3x + 1)(3x - 8)$

27. $(3x - a)(x - 2a)$ **29.** $(3x - y)(3x + y)$ **31.** $(2x + 3)(2x + 3)$

33. $(1 - 4xy)(1 + 4xy)$ **35.** $(3xy + 1)(3xy + 1)$ **37.** $3(x + 2)(x + 2)$

39. $2a(a - 5)(a + 1)$ **41.** $4(a - b)(a - b)$ **43.** $4y(x - 3)(x + 3)$

45. $x(4 + x)(3 - x)$ **47.** $x^2y^2(x - 1)(x + 1)$ **49.** $(y^2 + 1)(y^2 + 2)$

51. $(3x^2 + 1)(x^2 + 2)$ **53.** $(x^2 + 4)(x - 1)(x + 1)$ **55.** $(x - 2)(x + 2)(x - 1)(x + 1)$

57. $(2a^2 + 1)(a - 1)(a + 1)$ **59.** $(x^2 + 2a^2)(x - a)(x + a)$

61. $(x^n - 1)(x^n + 1)(x^{2n} + 1)$ **63.** $(x^{2n} + y^{2n})(x^n + y^n)(x^n - y^n)$

65. $(3x^{2n} - 1)(x^{2n} - 3)$

Exercise 2.7 [page 61]

1. $(ax + 1)(x + 1)$ **3.** $(ax + 1)(x + a)$ **5.** $(x + a)(x + y)$ **7.** $(3a - c)(b - d)$

9. $(3x + y)(1 - 2x)$ **11.** $(a^2 + 2b^2)(a - 2b)$ **13.** $(x + 2y)(x - 1)$

15. $(2a^2 - 1)(b + 3)$ **17.** $(x^3 - 3)(y^2 + 1)$ **19.** $(x - 1)(x^2 + x + 1)$

21. $(2x + y)(4x^2 - 2xy + y^2)$ **23.** $(a - 2b)(a^2 + 2ab + 4b^2)$

25. $(xy - 1)(x^2y^2 + xy + 1)$ **27.** $(3a + 4b)(9a^2 - 12ab + 16b^2)$

29. $(2x - y)(x^2 - xy + y^2)$ **31.** $[(x + 1) - 1][(x + 1)^2 + (x + 1) + 1] = x(x^2 + 3x + 3)$

33. $[(x + 1) - (x - 1)][(x + 1)^2 + (x + 1)(x - 1) + (x - 1)^2] = 2(3x^2 + 1)$

35. $[(x + y) + (x - y)][(x + y)^2 - (x + y)(x - y) + (x - y)^2] = 2x(x^2 + 3y^2)$

37. $ac - ad + bd - bc = (ac - ad) - (bc - bd) = a(c - d) - b(c - d) = (a - b)(c - d)$;
$ac - ad + bd - bc = (bd - ad) - (bc - ac) = (b - a)d - (b - a)c = (b - a)(d - c)$

39.
$$\begin{aligned} x^4 + 3x^2y^2 + 4y^4 + x^2y^2 - x^2y^2 &= (x^4 + 4x^2y^2 + 4y^4) - x^2y^2 \\ &= (x^2 + 2y^2)^2 - (xy)^2 \\ &= (x^2 + 2y^2 - xy)(x^2 + 2y^2 + xy) \end{aligned}$$

41.
$$\begin{aligned} a^4 + 6a^2b^2 + 25b^4 + 4a^2b^2 - 4a^2b^2 &= (a^4 + 10a^2b^2 + 25b^4) - 4a^2b^2 \\ &= (a^2 + 5b^2)^2 - (2ab)^2 \\ &= (a^2 + 5b^2 - 2ab)(a^2 + 5b^2 + 2ab) \end{aligned}$$

Review Exercises [page 63]

1. a. binomial; degree 3 **b.** trinomial; degree 2

2. a. bionomial; degree 3, degree 2 in x, degree 3 in y

b. trinomial; degree 4, degree 2 in x, degree 3 in y **3.** $\frac{1}{5}$ **4.** 7

5. a. 8 **b.** 13 **6. a.** -2 **b.** -14 **7. a.** $x + y - z$ **b.** $2x^2 - 2z^2 - x + 3y$

8. a. $4x + 3$ **b.** $-x^2 - 2x$ **9. a.** $-6x^3y^4$ **b.** $-6x^2y^3z^3$

10. a. $4x^3y$ **b.** $\frac{3}{2}xy$ **11. a.** $2x^3 - 4x^2 + 2x$ **b.** $4x^2 + 10x - 6$

12. a. $y^3 - 3y^2 + 3y - 2$ **b.** $z^3 + 2z^2 - z - 2$

13. a. $4y^2(y - 2)$ **b.** $x(x - 5)(x + 2)$ **14. a.** $-(y - 3x)$ **b.** $-(-2x + y - z)$

15. a. $-x(x - 2)$ **b.** $-3xy(2x - 1 + y)$ **16. a.** $(a + 2)(x - y)$ **b.** $(x - y)(2a + b)$

17. a. $(x - 7)(x + 5)$ **b.** $(y + 8)(y - 4)$ **18. a.** $(xy - 6)(xy + 6)$ **b.** $(a - 7b)(a + 7b)$

19. a. $(3y - 1)(y + 4)$ **b.** $x(x + 5)(x - 2)$ **20. a.** $9(x - 2)(x + 2)$ **b.** $3(2x - y)(2x + y)$

21. a. $(2x - y)(x + 2y)$ **b.** $(3x + y)(2x - y)$ **22. a.** $(3a + 2b)(5a + 6b)$ **b.** $6(2a - b)(a - b)$

23. a. $(x + y)(2x + 1)$ **b.** $(y - 3)(x - 1)$ **24. a.** $(x + y)(a - 2b)$ **b.** $(x - 2y)(2a + b)$

25. a. $(2x - y)(4x^2 + 2xy + y^2)$ **b.** $(x + 4y)(x^2 - 4xy + 16y^2)$

26. a. $(3y + z)(9y^2 - 3yz + z^2)$ **b.** $(x - 2a)(x^2 + 2ax + 4a^2)$

Exercise 3.1 [page 70]

1. $\frac{-2}{3}$ **3.** $\frac{5}{7}$ **5.** $\frac{2}{7}$ **7.** $\frac{6}{7}$ **9.** $\frac{-3x}{y}$ $(y \neq 0)$ **11.** $\frac{-2x^2}{y^2}$ $(y \neq 0)$

13. $\frac{-y-3}{y}$ $(y \neq 0)$ **15.** $\frac{-(x-y)}{y}$ or $\frac{y-x}{y}$ $(y \neq 0)$ **17.** $\frac{4}{y-3}$ **19.** $\frac{-1}{y-x}$

21. $\frac{2-x}{x-3}$ **23.** $\frac{x}{y-x}$ **25.** $\frac{1}{y-x}$ **27.** $\frac{a}{3a+b}$

29. No. For example, $\frac{4}{y-3}$ does not represent a real number when $y = 3$.

Exercise 3.2 [page 73]

1. $\frac{a^2c^2}{b}$ **3.** $\frac{1}{3a^2bc}$ **5.** 3 **7.** -1 **9.** $b - a$ **11.** $\frac{-1}{y+3}$ **13.** $\frac{2x+3}{3}$

15. $\frac{3x-1}{3}$ **17.** $y - 1$ **19.** $a^2 - 3a + 2$ **21.** $y + 7$ **23.** $\frac{x-4}{x+1}$ **25.** $\frac{2y-3}{y-1}$

27. $\frac{x+2y}{x+y}$ **29.** $4y^2+6y+9$ **31.** $x+y$ **33.** $\frac{6}{9}$ **35.** $\frac{-30}{14}$ **37.** $\frac{20}{5}$

39. $\frac{6}{18x}$ **41.** $\frac{-a^2b}{b^3}$ **43.** $\frac{xy^2}{xy}$ **45.** $\frac{x+y}{3(x+y)}$ **47.** $\frac{3(a^2-1)}{6(a+1)}$ **49.** $\frac{3a+3b}{a^2-b^2}$

51. $\frac{3xy-9x}{y^2-y-6}$ **53.** $\frac{-2x-4}{x^2+3x+2}$ **55.** $\frac{-2b-2a}{b^2-a^2}$ **57.** $\frac{-x^2+x}{x^2-3x+2}$

59. $\frac{3(a^2-3a+9)}{a^3+27}$ **61.** $\frac{3a^2-2ab-5b^2}{2ax+2bx-3ay-3by}$ **63.** $\frac{9a^2-b^2}{-4by-7bx+12ay+21ax}$

Exercise 3.3 [page 78]

1. $2y+5$ **3.** $2t-1-\frac{6}{2t-1}$ **5.** $x^2+4x+9+\frac{19}{x-2}$

7. $a^3-3a^2+6a-16+\frac{47}{a+3}$ **9.** $x^3+2x^2+4x+8+\frac{15}{x-2}$

11. $4z^2+z-\frac{2}{2z+1}$ **13.** $4a^2+2a+\frac{1}{2}$ **15.** $y^2-2+\frac{3}{7y^2}$ **17.** $6rs-5+\frac{2}{rs}$

19. $4ax-2x+\frac{1}{2}$ **21.** $-5m^3+3-\frac{7}{5m^3}$ **23.** $8m^2-5+\frac{7}{5m}$

25. $x-1+\frac{-7x+12}{x^2-2x+7}$ **27.** $4a^2-9a+31+\frac{-104a+32}{a^2+3a-1}$

29. $t-1+\frac{-t^2-3t+3}{t^3-2t^2+t+2}$ **31.** $k=-2$

Exercise 3.4 [page 82]

1. 60 **3.** 120 **5.** 252 **7.** $6ab^2$ **9.** $24x^2y^2$ **11.** $a(a-b)^2$

13. $(a-b)(a+b)$ **15.** $(a+4)(a+1)^2$ **17.** $(x+4)(x-1)^2$ **19.** $x(x-1)^3$

21. $4(a+1)(a-1)^2$ **23.** $x^3(x-1)^2$ **25.** $\frac{x-3}{2}$ **27.** $\frac{a+b-c}{6}$ **29.** $\frac{2x-1}{2y}$

31. $\frac{1-2x}{x+2y}$ **33.** $\frac{6-2a}{a^2-2a+1}$ **35.** $\frac{2-2a}{ax}$ **37.** $\frac{-a-4}{6}$ **39.** $\frac{2x^2+xy+2y^2}{2xy}$

41. $\frac{-4}{15(x-2)}$ **43.** $\frac{-x+33}{2(x-3)(x+3)}$ **45.** $\frac{4}{r-3}$ **47.** $\frac{-2ax-3a}{(3x+2)(x-1)}$

49. $\frac{-8x^2+2xy+6y^2}{(3x+y)(2x-y)}$ **51.** $\frac{6}{(x-2)(x+1)(x+1)}$ **53.** $\frac{-6y-4}{(y+4)(y-4)(y-1)}$

55. $\frac{y^2+6y-3}{(y+3)(y-3)(y+7)}$ **57.** $\frac{x^3-2x^2+2x-2}{(x-1)(x-1)}$ **59.** $\frac{-1}{(z-4)(z-3)(z-2)}$ **61.** 0

63. $\dfrac{y^3 - y^2 + y - 1}{y^3 + 1}$ **65.** $\dfrac{-12y^3 - 18y^2 - 27y - 5}{8y^3 - 27}$

67. $x = 1, -\dfrac{2}{3}$; $a = -2, -3$; $3x = -y$, $2x = y$; $2x = y, x = 2y$; $x = 2, -1$;

$b = 1, -1$; $y = 4, -4, 1$

Exercise 3.5 [page 87]

1. $\dfrac{2}{3}$ **3.** $\dfrac{7}{10}$ **5.** $\dfrac{10}{3}$ **7.** $\dfrac{-b^2}{a}$ **9.** $\dfrac{3c}{35ab}$ **11.** $\dfrac{5}{ab}$ **13.** 5 **15.** $\dfrac{a(2a-1)}{a+4}$

17. $\dfrac{x+3}{x-5}$ **19.** $\dfrac{x-7}{x-5}$ **21.** $\dfrac{(x-2)(3x+1)}{(3x-1)(x-1)}$ **23.** $\dfrac{3-a}{a+1}$ **25.** $\dfrac{4}{3}$ **27.** $\dfrac{1}{ax^2y}$

29. $\dfrac{20ay}{3}$ **31.** $\dfrac{2}{9y}$ **33.** $\dfrac{a+1}{a-2}$ **35.** $\dfrac{x-5}{x+5}$ **37.** $\dfrac{3x-1}{x-2}$ **39.** $3(x^2 - xy + y^2)$

41. $(y-3)(x+2)$ **43.** $\dfrac{a-1}{a-3}$ **45.** $\dfrac{x+1}{x}$ **47.** $\dfrac{1}{x+1}$

49. $a = -5, 3, 2, -2, -1$; $a = 7, -5, 2, 3, -3$; $x = -3, 1, -5, -2, 5$; $x = -2, 1, 5, -7, 2$;

$x = \dfrac{3}{2}, -1, -\dfrac{1}{5}, 2, \dfrac{1}{3}$; $x = -\dfrac{2}{3}, -\dfrac{3}{2}, \dfrac{1}{4}, \dfrac{1}{3}$; $x = 0, -y$; $x = -y, y, 2x = y$;

$x = 2, -1, -2$; $x = -\dfrac{3}{2}, y = 1, -2$; $a = 3, -1, 0, -4, 4$; $y = 0, 3, -2, -1$

Exercise 3.6 [page 90]

1. $\dfrac{3}{2}$ **3.** $\dfrac{2}{21}$ **5.** $\dfrac{a}{bc}$ **7.** $\dfrac{4y}{3}$ **9.** $\dfrac{1}{5}$ **11.** $\dfrac{1}{10}$ **13.** $\dfrac{7}{2(5a+1)}$ **15.** x

17. $\dfrac{x}{x-1}$ **19.** $\dfrac{y}{y+2}$ **21.** $\dfrac{10}{7}$ **23.** $\dfrac{a(4a-3)}{4a+1}$ **25.** $\dfrac{-1}{y-3}$ **27.** $\dfrac{a-2b}{a+2b}$

29. $\dfrac{a+6}{a-1}$ **31.** 1 **33.** $\dfrac{-(a^2+b^2)}{4ab}$ **35.** $\dfrac{29}{12}$ **37.** $\dfrac{209}{56}$

Review Exercises [page 94]

1. a. $\dfrac{-1}{-(a-b)}, \dfrac{-1}{b-a}, -\dfrac{-1}{a-b}, -\dfrac{-1}{-(b-a)}, -\dfrac{1}{-(a-b)}, -\dfrac{1}{b-a}$

b. $\dfrac{1}{a-b}$ is a positive number if $a > b$ and a negative number if $a < b$.

2. $\dfrac{1}{a-1}, \dfrac{-1}{1-a}$ **3. a.** $\dfrac{2x}{5y}$ **b.** $x - 2$ **4. a.** $-2x - 1$ **b.** $\dfrac{4x^2 - 2xy + y^2}{2x - y}$

5. a. $\frac{-18}{24}$ b. $\frac{x^2y}{2xy^2}$ 6. a. $\frac{2x-6y}{x^2-9y^2}$ b. $\frac{y-y^2}{y^2-4y+3}$

7. a. $6x-3+\frac{3}{2x}$ b. $2y+1-\frac{1}{3y}$ 8. a. $y-3+\frac{10}{2y+3}$ b. $3x^3+1+\frac{3}{2x-1}$

9. a. x b. $\frac{-x-y}{4x}$ 10. a. $\frac{8y-15x+7}{20xy}$ b. $\frac{3y+4}{6y-18}$

11. a. $\frac{5x+1}{(x^2-1)(x-1)}$ b. $\frac{-6y-4}{(y-4)(y+4)(y-1)}$

12. a. $\frac{3x^2+2x-4}{x^2-4}$ b. $\frac{x+6y+3}{x^2-4y^2}$ 13. a. $\frac{x^2}{3}$ b. $\frac{2}{y}$ 14. a. 1 b. $\frac{2y^2-2y}{y+1}$

15. a. $\frac{5xy}{3}$ b. 1 16. a. $\frac{(y+3)(y+2)}{(y-2)(y+1)}$ b. $\frac{1}{1-x}$ 17. a. $\frac{5}{3}$ b. $\frac{2}{11}$

18. a. $\frac{6x+9}{3x-2}$ b. $\frac{y^2-1}{y^2+1}$ 19. a. $\frac{2x}{x^2-1}$ b. 12

20. a. $\frac{3y+5}{y+1}$ b. $\frac{(x^2+1)(x-1)}{x(x-2)}$

Exercise 4.1 [page 99]

1. x^5 3. a^8 5. x^2 7. xy^2 9. a^6 11. x^6 13. x^3y^6 15. $a^4b^4c^8$

17. $\frac{x^3}{y^6}$ 19. $\frac{8x^3}{y^6}$ 21. $\frac{-8x^3}{27y^6}$ 23. $-16x^4$ 25. $4a^7b^6$ 27. $\frac{16}{x}$ 29. $\frac{y^4}{x}$

31. x^4y 33. $\frac{8x}{9y^2}$ 35. $36y^2$ 37. $\frac{x^2s^{14}}{y^2t^2}$ 39. $\frac{-1}{x^3a^3b}$ 41. $\frac{a^2b^6}{x^8}$ 43. x^{2n}

45. x^{3n} 47. x^{2n} 49. $\frac{y^{n+2}}{x}$ 51. $x^{3n-3}y^{9n-6}$

Exercise 4.2 [page 102]

1. $\frac{1}{2}$ 3. 3 5. $\frac{-1}{8}$ 7. $\frac{1}{5}$ 9. $\frac{5}{3}$ 11. $\frac{9}{5}$ 13. $\frac{82}{9}$ 15. $\frac{3}{16}$ 17. $\frac{x^2}{y^3}$

19. $\frac{1}{x^6y^3}$ 21. $\frac{x^2}{y^6}$ 23. $\frac{y^4}{x}$ 25. x^4 27. $\frac{1}{x^6}$ 29. $\frac{y}{x}$ 31. $4x^5y^2$ 33. $\frac{a^{18}b^{18}}{c^{18}}$

35. $\frac{b^2+a^2}{a^2b^2}$ 37. $\frac{r^2s^2+1}{rs}$ 39. $\frac{1}{(a-b)^2}$ 41. $\frac{y^2-x^2}{xy}$ 43. $\frac{a^2b+b}{a}$ 45. $\frac{y+x}{y-x}$

47. a^{3-n} 49. a^{2-2n} 51. $b^{-1}c^{-1}$ 53. x^{-1}

55. $\left(\frac{a}{b}\right)^{-n}=\frac{a^{-n}}{b^{-n}}=\frac{1}{a^n}\cdot b^n=\frac{b^n}{a^n}=\left(\frac{b}{a}\right)^n$

Exercise 4.3 [page 105]

1. 2.85×10^2 **3.** 2.1×10 **5.** 8.372×10^6 **7.** 2.4×10^{-2} **9.** 4.21×10^{-1}

11. 4×10^{-6} **13.** 240 **15.** 687,000 **17.** 0.0050 **19.** 0.0202 **21.** 12,270

23. 0.00235 **25.** 0.0005 **27.** 12.5 **29.** 0.00006 **31.** 10^{-5} or .00001

33. 10^0 or 1 **35.** 6×10^{-3} or .006 **37.** 1.8×10^6 or 1,800,000 **39.** 72×10 or 720

41. 4×10^{-1} or 0.4 **43.** 8 **45. a.** 3×10^8 **b.** 1.18×10^{10} inches per second

---

Exercise 4.4 [page 109]

1. 3 **3.** 2 **5.** -2 **7.** 9 **9.** 27 **11.** 16 **13.** $\frac{1}{4}$ **15.** $\frac{1}{8}$ **17.** $x^{2/3}$

19. $x^{1/3}$ **21.** $a^{3/2}$ **23.** $\frac{1}{x^{1/2}}$ **25.** $a^{1/3}b^{1/2}$ **27.** $\frac{a^4}{b^2}$ **29.** $\frac{yz}{x}$ **31.** $\frac{b^{1/3}}{a^{1/3}c}$

33. $x^{3/2} + x$ **35.** $x - x^{2/3}$ **37.** $x^{-1} + 1$ **39.** $x^{2/5}$ **41.** $x^{1/3}$ **43.** $x^{-2/3}$

45. $x^{1/2} + 1$ **47.** $x^{2/3} - x^{1/3}$ **49.** $x^{-1} + 1$ **51.** $x^{-1} + 1$ **53.** $x^{3n/2}$ **55.** $x^{3n/2}$

57. $x^{5n/2}y^{(3m+2)/2}$ **59.** $x^{n/3}y^n$ **61.** $\frac{x^3y^2}{z^{1/n}}$ **63.** 5 **65.** $2|x|$

67. $\frac{2}{|x|(x+5)^{1/2}}$ $(x > -5, x \neq 0)$ **69.** $16^{1/2} > 16^{1/4}$; $\left(\frac{1}{16}\right)^{1/4} > \left(\frac{1}{16}\right)^{1/2}$

Exercise 4.5 [page 113]

1. $\sqrt{3}$ **3.** $\sqrt{x^3}$ **5.** $3\sqrt[4]{y}$ **7.** $x\sqrt[3]{y}$ **9.** $\sqrt[3]{xy}$ **11.** $-3\sqrt[5]{x^3}$ **13.** $\sqrt{x+2y}$

15. $\sqrt[3]{(2x-y)^2}$ **17.** $\frac{1}{\sqrt[3]{4}}$ **19.** $\frac{1}{\sqrt[3]{x^2}}$ **21.** $5^{1/2}$ **23.** $x^{2/3}$ **25.** $(ab)^{1/2}$

27. $xy^{1/2}$ **29.** $2^{1/3}a^{1/3}b^{2/3}$ **31.** $(a-b)^{1/2}$ **33.** $a^{1/2} - 2b^{1/2}$ **35.** $a^{1/3}b^{1/2}$

37. $\frac{1}{x^{1/2}}$ **39.** $\frac{2}{(x+y)^{1/2}}$ **41.** 4 **43.** -5 **45.** 3 **47.** -4 **49.** x **51.** x^2

53. $2y^2$ **55.** $-x^2y^3$ **57.** $\frac{2}{3}xy^4$ **59.** $\frac{-2}{5}x$ **61.** $2xy^2$ **63.** $2a^2b^3$

65. 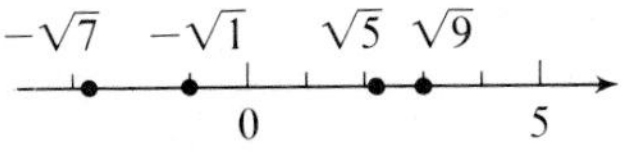

67. 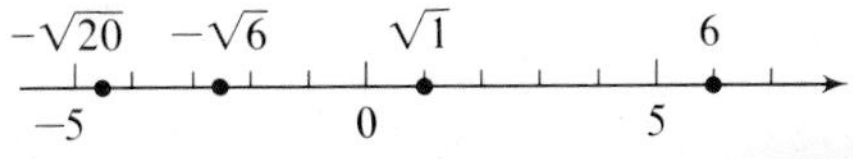

69. $2|x|$ **71.** $|x+1|$ **73.** $\frac{2}{|x+y|}$ $(x + y \neq 0)$ **75.** $\frac{5}{2x-y}$ $(2x - y \neq 0)$

77. $\sqrt{x^2} = x$, if $x \geq 0$; $\sqrt{x^2} = -x$, if $x < 0$

Exercise 4.6 [page 118]

1. $3\sqrt{2}$ **3.** $2\sqrt{5}$ **5.** $5\sqrt{3}$ **7.** $4\sqrt{10}$ **9.** x^2 **11.** $x\sqrt{x}$ **13.** $3x\sqrt{x}$

15. $2x^3\sqrt{2}$ **17.** x^2 **19.** $x\sqrt[4]{x}$ **21.** $xy\sqrt[4]{xy}$ **23.** $2y\sqrt[4]{xy^3}$ **25.** $xyz^2\sqrt[5]{x^2y^4z}$

27. $ab^2c^2\sqrt[6]{ac^3}$ **29.** $2a^3b^4c^4\sqrt[3]{2a^2c}$ **31.** $3abc\sqrt[7]{ab^2c^3}$ **33.** 6 **35.** x^3y **37.** 2

39. $x\sqrt[4]{x}$ **41.** $10\sqrt{3}$ **43.** $100\sqrt{6}$ **45.** $\frac{\sqrt{5}}{5}$ **47.** $\frac{-\sqrt{2}}{2}$ **49.** $\frac{\sqrt{2x}}{2}$

51. $\frac{-\sqrt{xy}}{x}$ **53.** $\sqrt{x}$ **55.** $-x\sqrt{y}$ **57.** $\frac{\sqrt[3]{4x^2y}}{2x}$ **59.** $\frac{\sqrt[4]{2}}{2}$ **61.** a^2b

63. $\frac{7\sqrt{2xy}}{y}$ **65.** $\frac{2b\sqrt[3]{b}}{a^2}$ **67.** $\frac{\sqrt[5]{a^3b}}{a}$ **69.** $\frac{1}{\sqrt{3}}$ **71.** $\frac{x}{\sqrt{xy}}$ **73.** $\sqrt{3}$ **75.** $\sqrt{3}$

77. $\sqrt[3]{9}$ **79.** $\sqrt{x}$ **81.** $\sqrt[3]{2x}$ **83.** $\sqrt{xy}$ **85.** $\sqrt[6]{72}$ **87.** $\sqrt[4]{20}$ **89.** $\sqrt[10]{y^7}$

91. $\sqrt[12]{16x^7y^9}$

93. Substitute two numbers such as 4 and 9 for a and b in the given equation and solve:

$$(\sqrt{a}+\sqrt{b})^2 \stackrel{?}{=} a+b$$
$$(\sqrt{4}+\sqrt{9})^2 \stackrel{?}{=} 4+9$$
$$(2+3)^2 \stackrel{?}{=} 13$$
$$25 \neq 13$$

95. Substitute a negative number such as -2 for a and solve the given equation:

$$\sqrt{a^2} \stackrel{?}{=} a$$
$$\sqrt{(-2)^2} \stackrel{?}{=} -2$$
$$\sqrt{4} \stackrel{?}{=} -2$$
$$2 \neq -2$$

Exercise 4.7 [page 122]

1. $5\sqrt{7}$ **3.** $\sqrt{3}$ **5.** $9\sqrt{2x}$ **7.** $-6y\sqrt{x}$ **9.** $15\sqrt{2a}$ **11.** $5\sqrt[3]{2}$

13. $6-2\sqrt{5}$ **15.** $3\sqrt{2}+\sqrt{6}$ **17.** $2\sqrt{3}+2\sqrt{5}$ **19.** $2\sqrt{3}+3\sqrt{2}$ **21.** $1-\sqrt{5}$

23. $x-9$ **25.** $-4+\sqrt{6}$ **27.** $7-2\sqrt{10}$ **29.** $4-2\sqrt[3]{2}$ **31.** $2(1+\sqrt{3})$

33. $6(\sqrt{3}+1)$ **35.** $4(1+\sqrt{y})$ **37.** $\sqrt{3}(y-x)$ **39.** $\sqrt{2}(1-\sqrt{3})$

41. $\sqrt{x}(1+\sqrt{3})$ **43.** $1+\sqrt{3}$ **45.** $1+\sqrt{2}$ **47.** $1-\sqrt{x}$ **49.** $x-y$

51. $2\sqrt{3}-2$ **53.** $\frac{2(\sqrt{7}+2)}{3}$ **55.** $\frac{4(1-\sqrt{x})}{1-x}$ **57.** $\frac{x(\sqrt{x}+3)}{x-9}$

59. $\dfrac{\sqrt{x}(\sqrt{x}+\sqrt{y})}{x-y}$ **61.** $\dfrac{x+2\sqrt{xy}+y}{x-y}$ **63.** $\dfrac{3\sqrt{2}+2\sqrt{3}}{6}$ **65.** $\dfrac{\sqrt{3}}{3}$

67. $3\sqrt{x}$ **69.** $\dfrac{\sqrt{x+1}}{x+1}$ **71.** $\dfrac{-\sqrt{x^2+1}}{x(x^2+1)}$ **73.** $\dfrac{(y^2-x+1)(\sqrt{y^2+1})}{y^2+1}$

75. $\dfrac{(x^2+3x+2)(\sqrt{x^2+2})}{3x^2+6}$ **77.** $\dfrac{-1}{2(1+\sqrt{2})}$ **79.** $\dfrac{x-1}{3(\sqrt{x}+1)}$ **81.** $\dfrac{x-y}{x(\sqrt{x}+\sqrt{y})}$

Review Exercises [page 127]

1. a. x^3y^2 **b.** $27x^6y^3$ **2. a.** $\dfrac{1}{x^3y}$ **b.** $\dfrac{y^6}{x^8}$ **3. a.** $5\dfrac{1}{2}$ **b.** $\dfrac{x^3y^2}{y^2-x^3}$

4. a. $\dfrac{2x}{4x^2+4x+1}$ **b.** $\dfrac{(y^2-x^2)(x-y)^2}{x^2y^2}$ **5. a.** 4.2×10^3 **b.** 3.07×10^{11}

6. a. 4.21×10^{-2} **b.** 2.3×10^{-11} **7. a.** 10^2 or 100 **b.** 1.5×10^{-3} or .0015

8. a. 9 **b.** $\dfrac{1}{2}$ **9. a.** x^2 **b.** x^2y^3 **10. a.** $x - x^{5/3}$ **b.** $y - y^0$ or $y - 1$

11. a. $x^{-1/2}(1+x^3)$ **b.** $y^{-1/4}(y-1)$

12. a. $\sqrt[3]{(1-x^2)^2}$ **b.** $\sqrt[3]{(1-x^2)^{-2}}$ or $\dfrac{1}{\sqrt[3]{(1-x^2)^2}}$

13. a. $(x^2y)^{1/3}$ or $x^{2/3}y^{1/3}$ **b.** $(a+b)^{-2/3}$ **14. a.** $2y$ **b.** $-2xy^2$

15. a. $\sqrt{2}|x|$ **b.** $|y-1|$ **16. a.** $6\sqrt{5}$ **b.** $2xy\sqrt[4]{2y}$ **17. a.** $\dfrac{\sqrt{xy}}{y}$ **b.** $\dfrac{\sqrt[3]{4}}{2}$

18. a. $x\sqrt{3xy}$ **b.** $\dfrac{2y\sqrt[3]{3y}}{x}$ **19. a.** $xy\sqrt[3]{x^2y}$ **b.** xy **20. a.** $x\sqrt[3]{y}$ **b.** $4\sqrt[3]{y}$

21. a. $18\sqrt{3}$ **b.** $19\sqrt{2x}$ **22. a.** $3y\sqrt{2x}$ **b.** $4x\sqrt{3x}$

23. a. $3\sqrt{2}-2\sqrt{3}$ **b.** $5\sqrt{2}-5$ **24. a.** $12-7\sqrt{3}$ **b.** $x-4$

25. a. $8+4\sqrt{3}$ **b.** $\dfrac{3-\sqrt{5}}{2}$ **26. a.** $\dfrac{y(\sqrt{y}+3)}{y-9}$ **b.** $\sqrt{x}-\sqrt{y}$

Exercise 5.1 [page 134]

11. $\{2\}$ **13.** $\{5\}$ **15.** $\{7\}$ **17.** $\left\{\dfrac{3}{2}\right\}$ **19.** $\{9\}$ **21.** $\{6\}$ **23.** $\{-4\}$ **25.** $\left\{\dfrac{2}{9}\right\}$

27. $\left\{\dfrac{7}{2}\right\}$ **29.** $\left\{\dfrac{-9}{2}\right\}$ **31.** $\{-1\}$ **33.** $\{8\}$ **35.** $\{-20\}$ **37.** $\left\{\dfrac{-27}{2}\right\}$ **39.** $\{3\}$

41. $\{-30\}$ **43.** $\{-7\}$ **45.** $\varnothing$ **47.** $\{13\}$ **49.** $\varnothing$ **51.** $k = -7$ **53.** $k = 11$

Exercise 5.2 [page 138]

17. $y \neq 0$ **19.** $x \neq 0$ **21.** $x \neq 4$ **23.** identity **25.** $\varnothing$ **27.** $\left\{\frac{5}{3}\right\}$ **29.** $\{7\}$

Exercise 5.3 [page 140]

1. $x = \frac{b + c}{a}$ **3.** $y = \frac{c + b}{3a}$ **5.** $x = \frac{b}{3a}$ **7.** $y = \frac{-bc}{a}$ **9.** $x = bc - a$

11. $y = \frac{c - ab}{2}$ **13.** $x = \frac{a}{a + 1}$ **15.** $x = a + b$ **17.** $x = \frac{5b + 8}{b + 2}$ **19.** $x = \frac{a}{3a - 1}$

21. $y = a$ **23.** $x = \frac{ab}{a + b}$ **25.** $k = v - gt$ **27.** $m = \frac{f}{a}$ **29.** $v = \frac{K}{p}$ **31.** $a = \frac{2s}{t^2}$

33. $t = \frac{v - k}{g}$ **35.** $h = \frac{V}{lw}$ **37.** $B = 180 - A - C$ **39.** $c = \frac{2A - hb}{h}$

41. $d = \frac{S - 5\pi D}{3\pi}$ **43.** $n = \frac{l - a + d}{d}$

45. Multiplying both sides by bd yields

$$\frac{a}{b}(bd) = \frac{c}{d}(bd); \quad \text{hence,} \quad \frac{abd}{b} = \frac{cbd}{d} \quad \text{and} \quad ad = bc.$$

47. Using the results of Problem 45, $ad = bc$, and multiplying both sides by $\frac{1}{cd}$ yields

$$ad\left(\frac{1}{cd}\right) = bc\left(\frac{1}{cd}\right); \quad \text{from which} \quad \frac{ad}{cd} = \frac{bc}{cd} \quad \text{and} \quad \frac{a}{c} = \frac{b}{d}.$$

49. Since $\frac{a}{b} = \frac{c}{d}$, adding 1 to both sides produces

$$\frac{a}{b} + 1 = \frac{c}{d} + 1; \quad \text{hence,} \quad \frac{a + b}{b} = \frac{c + d}{d}.$$

51. Multiplying each member by $x - a$ $(x \neq a)$ yields

$$\begin{aligned} x &= a + k(x - a) \\ x &= a + kx - ka \\ x - kx &= a - ka \\ x(1 - k) &= a(1 - k) \\ x &= a \qquad (k \neq 1). \end{aligned}$$

Since $x \neq a$, the equation has no solution.

Exercise 5.4 [page 143]

1. a. $x + (x + 24) = 110$ **b.** 43; 67 **3. a.** $x + (x + 122) = 584$ **b.** 231; 353

5. a. $x + (x + 1) + (x + 2) = 78$ **b.** 25; 26; 27 **7. a.** $7x = 6(x + 2)$ **b.** 12; 14

9. a. $0.04A + 0.05(8000 - A) = 350$ **b.** \$5000 at 4%; \$3000 at 5%

11. a. $0.02(3000) + 0.05A = 0.04(A + 3000)$ **b.** \$6000 at 5%

13. a. $5(s + 3) = 9s + 3$ **b.** 3 years; 27 years

15. a. $3(s + 4) = (60 - s) + 4$ **b.** 13 years; 47 years **17.** 40°C

19. 153 feet per second **21. a.** $x + 2x + (3x + 12) = 180$ **b.** 28°; 56°; 96°

23. a. $x + (3 + 2x) + (3 + 2x) = 86$ **b.** 16 centimeters; 35 centimeters, 35 centimeters

25. a. $24(x + 2) = 32x$ **b.** 24 grams at 8 centimeters; 32 grams at 6 centimeters

27. a. $200(6 - x) = 2200x$ **b.** 6 inches from 2200 pound weight

29. a. $10d + 25(d + 12) = 1245$ **b.** 27 dimes; 39 quarters

31. a. $80f + 64(42 - f) = 2880$ **b.** 12 first-class; 30 tourist

33. a. $0.30x + 0.12(40) = 0.20(x + 40)$ **b.** 32 liters

35. a. $x + 0.45(12) = 0.60(x + 12)$ **b.** 4.5 liters

37. a. $\dfrac{1260}{r + 120} = \dfrac{420}{r}$ **b.** auto, 60 miles per hour; plane, 180 miles per hour

39. a. $20t + 5 = 30t$ **b.** $\frac{1}{2}$ hour **41. a.** $\frac{1}{30}x + \frac{1}{45}x = 1$ **b.** 18 hours

43. a. $\left(\frac{1}{10}\right)6 + \left(\frac{1}{n}\right)6 = 1$ **b.** 15 hours

---

Exercise 5.5 [page 158]

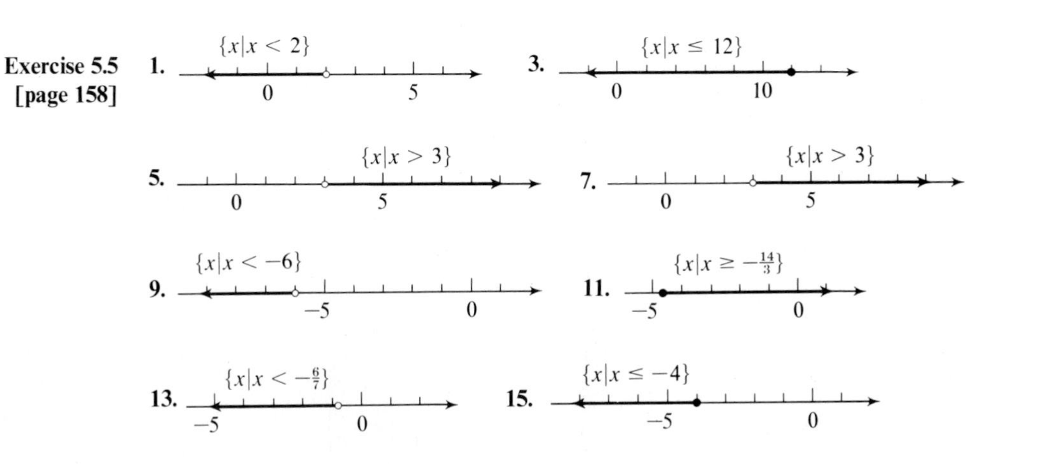

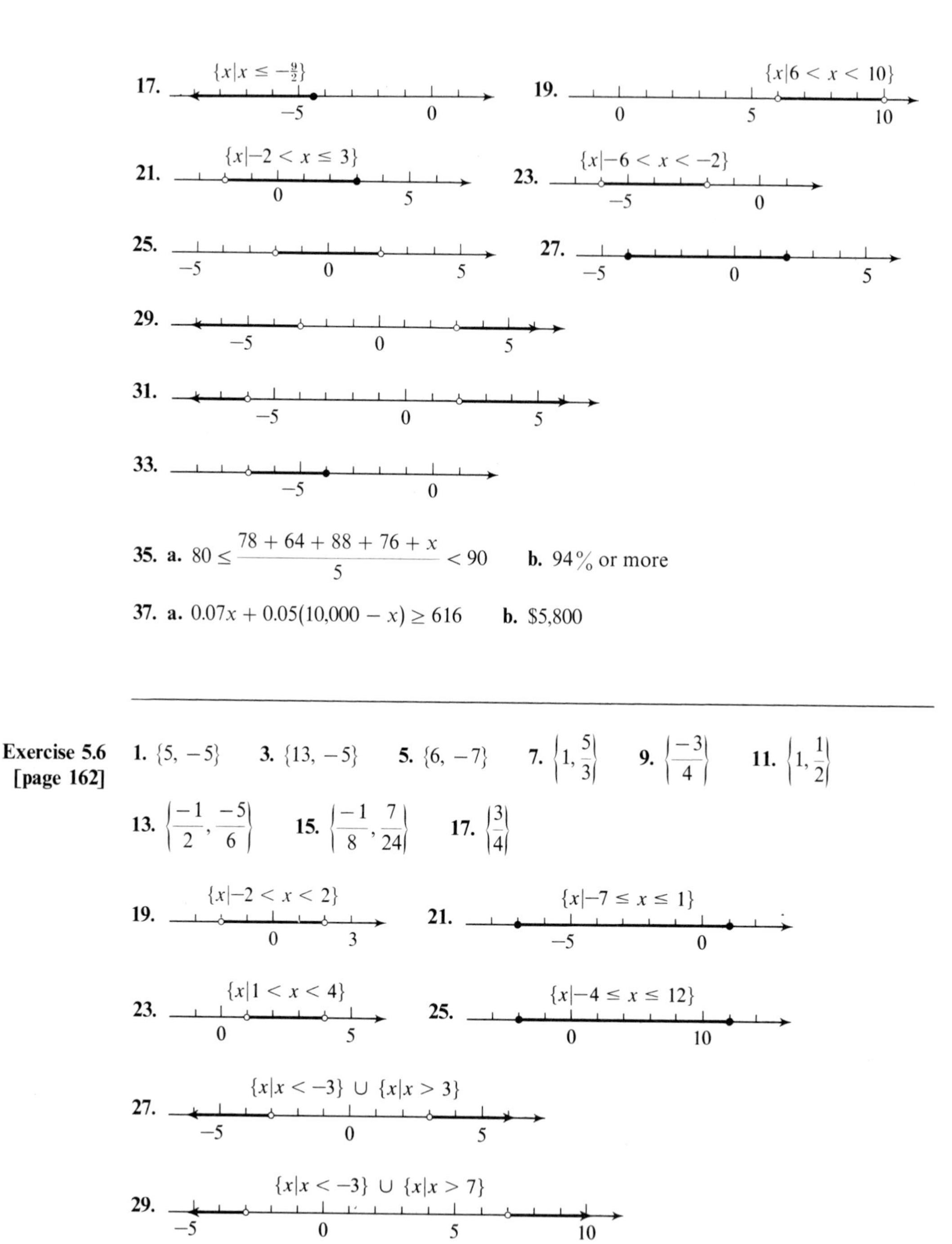

35. a. $80 \le \dfrac{78 + 64 + 88 + 76 + x}{5} < 90$ **b.** 94% or more

37. a. $0.07x + 0.05(10{,}000 - x) \ge 616$ **b.** \$5,800

Exercise 5.6 [page 162]

1. $\{5, -5\}$ **3.** $\{13, -5\}$ **5.** $\{6, -7\}$ **7.** $\left\{1, \frac{5}{3}\right\}$ **9.** $\left\{\frac{-3}{4}\right\}$ **11.** $\left\{1, \frac{1}{2}\right\}$

13. $\left\{\frac{-1}{2}, \frac{-5}{6}\right\}$ **15.** $\left\{\frac{-1}{8}, \frac{7}{24}\right\}$ **17.** $\left\{\frac{3}{4}\right\}$

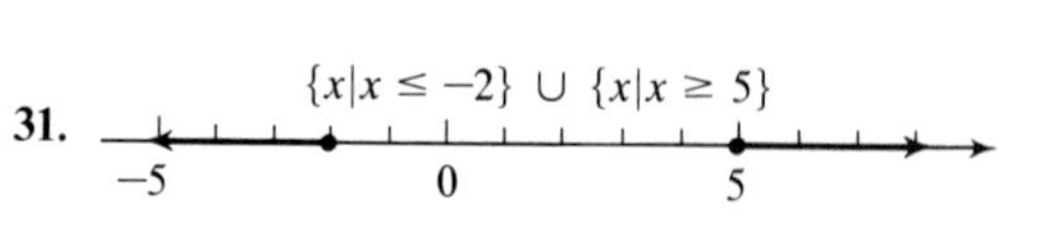

33. $\{x|x \ge 3\}$ **35.** $\{y|y \le -6\}$ **37.** $\left\{x \,|\, x \ge \frac{3}{2}\right\}$

Review Exercises [page 166]

1. a. $\{-4\}$ **b.** $\left\{-\frac{7}{2}\right\}$ **2. a.** $\left\{-\frac{26}{21}\right\}$ **b.** $\left\{\frac{5}{2}\right\}$ **4.** $y = -\frac{x}{5}$ **5.** $x = -5y$

6. $a = \frac{2s - 4t}{t^2}$ **7.** $d = \frac{a - l}{1 - n}$ **8.** 11; 18 **9.** 3 liters

10. \$1800 at 5%; \$5400 at $5\frac{1}{2}$%

11. a. $\{x|x \leq 27\}$

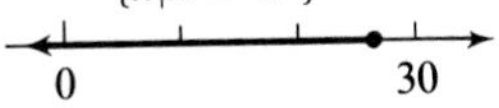

b. $\{x|x \leq \frac{3}{2}\}$

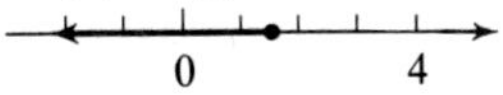

12. a. $\{x|2 < x \leq 5\}$

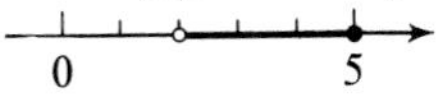

b. $\{x|1 < x < 3\}$

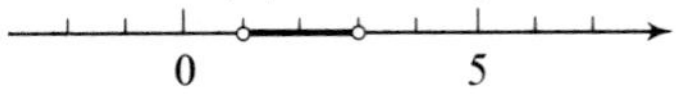

13. $\{x|\frac{1}{2} < x < 2\}$

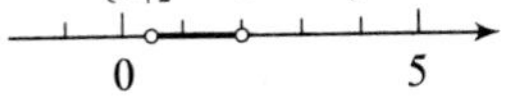

14. Must win between 14 and 23 games. **15. a.** $\{3, -4\}$ **b.** $\left\{-\frac{5}{4}, \frac{11}{4}\right\}$

16. a. $\{x|x \leq -3\} \cup \{x|x \geq 2\}$

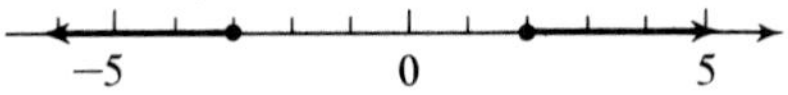

b.

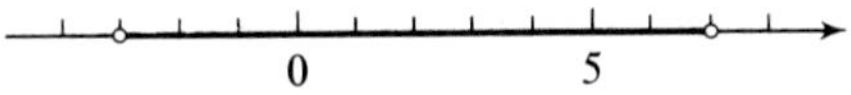

Exercise 6.1 [page 171]

1. $\{-2, 5\}$ **3.** $\left\{\frac{-5}{2}, 2\right\}$ **5.** $\left\{0, \frac{-1}{2}\right\}$ **7.** $\left\{6, \frac{-3}{2}\right\}$ **9.** $\left\{2, \frac{-1}{2}\right\}$ **11.** $\left\{\frac{5}{2}, \frac{-2}{3}\right\}$

13. $\{0, 3\}$ **15.** $\{0, 3\}$ **17.** $\{3, -3\}$ **19.** $\{3, -3\}$ **21.** $\left\{\frac{2}{3}, \frac{-2}{3}\right\}$ **23.** $\left\{\frac{3}{2}, \frac{-3}{2}\right\}$

25. $\{4, 1\}$ **27.** $\{7, -2\}$ **29.** $\{1\}$ **31.** $\left\{\frac{1}{2}, 1\right\}$ **33.** $\{3, -2\}$ **35.** $\{1, -6\}$

37. $\left\{\frac{3}{2}, -2\right\}$ **39.** $\{1, -3\}$ **41.** $\left\{1, -\frac{10}{3}\right\}$ **43.** $\{2, 13\}$ **45.** $x^2 + x - 2 = 0$

47. $x^2 + 5x = 0$ **49.** $4x^2 - 13x - 12 = 0$ **51.** $10x^2 - 11x + 3 = 0$ **53.** $\{2b, -2b\}$

55. $\{4a, -a\}$ **57.** $\{a, b\}$ **59.** $\left\{-a, -\frac{b}{2}\right\}$

Exercise 6.2 [page 176]

1. $\{10, -10\}$ **3.** $\left\{\frac{5}{3}, \frac{-5}{3}\right\}$ **5.** $\{\sqrt{7}, -\sqrt{7}\}$ **7.** $\{\sqrt{6}, -\sqrt{6}\}$ **9.** $\{\sqrt{6}, -\sqrt{6}\}$

11. $\left\{\frac{9}{2}, -\frac{9}{2}\right\}$ **13.** $\{5, -1\}$ **15.** $\left\{\frac{5}{2}, \frac{-3}{2}\right\}$ **17.** $\{-2+\sqrt{3}, -2-\sqrt{3}\}$

19. $\{2+2\sqrt{3}, 2-2\sqrt{3}\}$ **21.** $\{5, 9\}$ **23.** $\{1, 4\}$ **25. a.** 1 **b.** $(x+1)^2$

27. a. 9 **b.** $(x-3)^2$ **29. a.** $\frac{9}{4}$ **b.** $\left(x+\frac{3}{2}\right)^2$ **31. a.** $\frac{49}{4}$ **b.** $\left(x-\frac{7}{2}\right)^2$

33. a. $\frac{1}{4}$ **b.** $\left(x-\frac{1}{2}\right)^2$ **35. a.** $\frac{1}{16}$ **b.** $\left(x+\frac{1}{4}\right)^2$ **37.** $\{2, -6\}$ **39.** $\{1\}$

41. $\{-5, -4\}$ **43.** $\{1+\sqrt{2}, 1-\sqrt{2}\}$ **45.** $\left\{\frac{-3+\sqrt{41}}{4}, \frac{-3-\sqrt{41}}{4}\right\}$

47. $\left\{-1+\sqrt{\frac{5}{2}}, -1-\sqrt{\frac{5}{2}}\right\}$ **49.** $\{\sqrt{a}, -\sqrt{a}\}$ **51.** $\left\{\sqrt{\frac{bc}{a}}, -\sqrt{\frac{bc}{a}}\right\}$

53. $\{a+4, a-4\}$ **55.** $\left\{\frac{3-b}{a}, \frac{-3-b}{a}\right\}$ **57.** $\{\sqrt{c^2-a^2}, -\sqrt{c^2-a^2}\}$

59. $\left\{-1+\sqrt{\frac{A}{P}}, -1-\sqrt{\frac{A}{P}}\right\}$ **61.** $\left\{\frac{-b+\sqrt{b^2-4ac}}{2a}, \frac{-b-\sqrt{b^2-4ac}}{2a}\right\}$

Exercise 6.3 [page 182]

1. $2i$ **3.** $4i\sqrt{2}$ **5.** $6i\sqrt{2}$ **7.** $4+2i$ **9.** $2+15i\sqrt{2}$ **11.** $2+2i$

13. $5+5i$ **15.** $-2+i$ **17.** $-1-2i$ **19.** $8+i$ **21.** $-8-6i$ **23.** $-\frac{5}{2}i$

25. $1+i$ **27.** $4+2i$ **29.** $15+3i$ **31.** $-1+4i\sqrt{5}$ **33.** $-\frac{3}{2}i$ **35.** $\frac{6}{13}+\frac{9}{13}i$

37. $\frac{3}{5}-\frac{4}{5}i$ **39. a.** -1 **b.** 1 **c.** $-i$ **41.** $5+4i$ **43.** $x \geq 5; x < 5$

Exercise 6.4 [page 185]

1. $\{4, 1\}$ **3.** $\{1, -4\}$ **5.** $\left\{\frac{3+\sqrt{5}}{2}, \frac{3-\sqrt{5}}{2}\right\}$ **7.** $\left\{\frac{5+\sqrt{13}}{6}, \frac{5-\sqrt{13}}{6}\right\}$

9. $\left\{\frac{3}{2}, \frac{-2}{3}\right\}$ **11.** $\{0, 5\}$ **13.** $\{\sqrt{2}, -\sqrt{2}\}$ **15.** $\left\{\frac{3}{2}, -\frac{5}{2}\right\}$

17. $\{3, -4\}$ **19.** $\{1+\sqrt{3}, 1-\sqrt{3}\}$ **21.** 1; real and unequal

23. 24; real and unequal **25.** 0; one real **27.** $x = 2k; x = -k$

29. $x = \frac{1 \pm \sqrt{1-4ac}}{2a}$ **31.** $x = -1 \pm \sqrt{1+y}$ **33.** $x = \frac{1 \pm \sqrt{17-8y}}{4}$

35. $x = y \pm \sqrt{y^2+3y+2}$ **37.** $x = \frac{-y \pm \sqrt{24-11y^2}}{6}$ **39.** $y = \frac{-x \pm \sqrt{8-11x^2}}{2}$

41. $k = 4$ **43.** $\{k \mid k \leq -2\}$

45. If r_1 and r_2 are solutions of a quadratic equation, let

$$r_1 = \frac{-b + \sqrt{b^2 - 4ac}}{2a} \quad \text{and} \quad r_2 = \frac{-b - \sqrt{b^2 - 4ac}}{2a}.$$

Then,

$$r_1 + r_2 = \left(\frac{-b}{2a} + \frac{\sqrt{b^2 - 4ac}}{2a}\right) + \left(\frac{-b}{2a} - \frac{\sqrt{b^2 - 4ac}}{2a}\right) = \frac{-2b}{2a} = -\frac{b}{a}$$

and

$$r_1 \cdot r_2 = \left(\frac{-b}{2a} + \frac{\sqrt{b^2 - 4ac}}{2a}\right)\left(\frac{-b}{2a} - \frac{\sqrt{b^2 - 4ac}}{2a}\right) = \frac{b^2 - b^2 + 4ac}{4a^2} = \frac{c}{a}.$$

Exercise 6.5 [page 188]

1. 5, 8 **3.** 7 meters, 9 meters **5.** 4 or $\frac{1}{4}$ **7.** 1 second **9.** 2.5 seconds

11. 2.5 seconds **13.** numerator is 3; denominator is 4 **15.** 3 meters by 11 meters

17. -13; -12; -11 **19.** 15 **21.** 3 centimeters

23. 6 miles per hour going; 4 miles per hour returning **25.** 3 miles per hour

27. 20 miles per hour to the city; 30 miles per hour returning

---

Exercise 6.6 [page 190]

1. $\{64\}$ **3.** $\{-2\}$ **5.** $\left\{\frac{-1}{3}\right\}$ **7.** $\{4\}$ **9.** $\{-27\}$ **11.** $\{17\}$ **13.** $\{13\}$

15. $\{5\}$ **17.** $\{0\}$ **19.** $\{4\}$ **21.** $\{1, 3\}$ **23.** $A = \pi r^2$ **25.** $y = \frac{1}{x^3}$

27. $t = \pm\sqrt{r^2 + s^2}$ **29.** $x - 4$ cannot be negative **31.** $x > 0$; $y > 0$

Exercise 6.7 [page 193]

1. $\{1, -1, 2, -2\}$ **3.** $\left\{\frac{\sqrt{2}}{2}, \frac{-\sqrt{2}}{2}, 3i, -3i\right\}$ **5.** $\{25\}$ **7.** $\{3, -3\}$ **9.** $\{64, -8\}$

11. $\{1, 16\}$ **13.** $\{9, 36\}$ **15.** $\left\{\frac{1}{4}, 16\right\}$ **17.** $\left\{\frac{1}{4}, -\frac{1}{3}\right\}$ **19.** $\{626\}$ **21.** $\{9, 16\}$

23. $\{4, 81\}$

Exercise 6.8 [page 198]

1. $\{x|x < -1\} \cup \{x|x > 2\}$

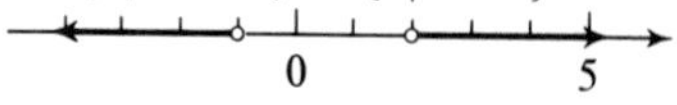

3. $\{x|0 \le x \le 2\}$

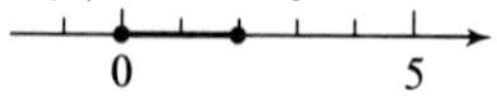

5. $\{x|x < -1\} \cup \{x|x > 4\}$

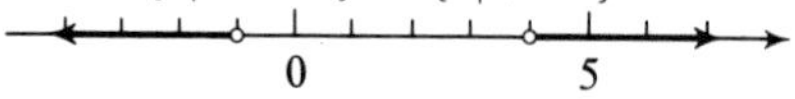

7. $\{x|-\sqrt{5} < x < \sqrt{5}\}$

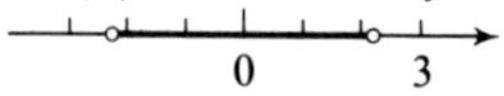

9. $\{x|x \in R\}$

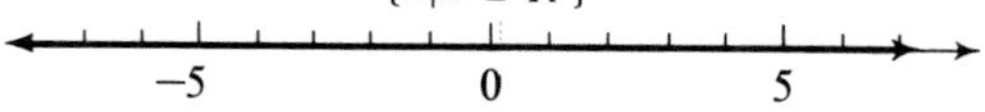

11. $\{x|x < 0\} \cup \{x|x \ge \frac{1}{2}\}$

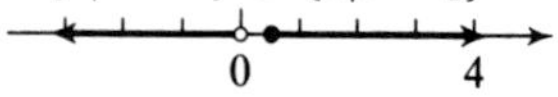

13. $\{x|-\frac{8}{3} < x < -2\}$

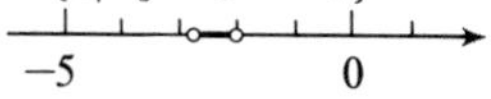

15. $\{x|x < -1\} \cup \{x|0 < x < 3\}$

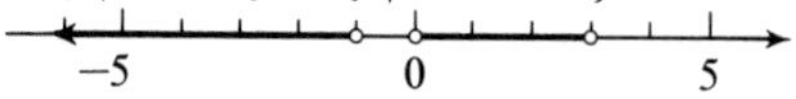

17. $\{x|x < 0\} \cup \{x|2 < x \le 4\}$

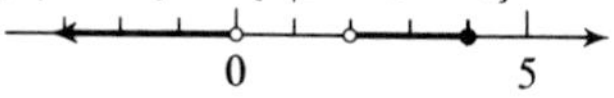

19. $\{x|-3 < x < -2\} \cup \{x|2 < x < 3\}$

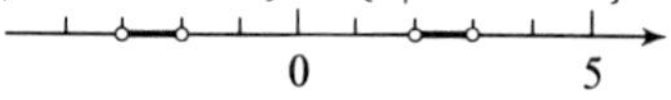

Review Exercises [page 200]

1. a. $\{0, 2\}$ **b.** $\{2, 3\}$ **2. a.** $\{5, -3\}$ **b.** $\{2, -3\}$

3. a. $\{2\}$ **b.** $\{1, 3\}$

4. a. $x^2 - 3x - 10 = 0$ **b.** $12x^2 - x - 1 = 0$ **5. a.** $\{5, -5\}$ **b.** $\left\{\sqrt{\frac{7}{3}}, -\sqrt{\frac{7}{3}}\right\}$

6. a. $\{2, -8\}$ **b.** $\{4 + \sqrt{15}, 4 - \sqrt{15}\}$

7. a. $\{2 + \sqrt{10}, 2 - \sqrt{10}\}$ **b.** $\left\{\frac{-3 + \sqrt{33}}{4}, \frac{-3 - \sqrt{33}}{4}\right\}$

8. a. $4 + 6i$ **b.** $5 - 6i\sqrt{3}$ **9. a.** $6 - i$ **b.** $5 + 3i$ **10. a.** $7 + 17i$ **b.** $4 + 3i$

11. a. $-\frac{4}{3}i$ **b.** $1 - i$ **12. a.** $\{1, 2\}$ **b.** $\left\{\frac{3 + i\sqrt{19}}{2}, \frac{3 - i\sqrt{19}}{2}\right\}$

13. a. $\left\{\frac{3 + \sqrt{5}}{2}, \frac{3 - \sqrt{5}}{2}\right\}$ **b.** $\left\{1, \frac{-3}{2}\right\}$ **14.** 9 inches; 4 inches

15. a. $\{1, 4\}$ **b.** $\{8\}$ **16. a.** $\{2, -2, i, -i\}$ **b.** $\{16\}$

17. $\{x|x < 0\} \cup \{x|x > 9\}$

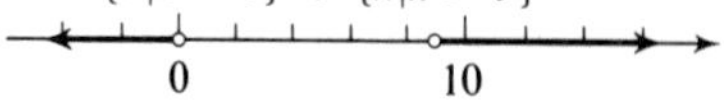

18. $\{x|-3 < x < -2\}$

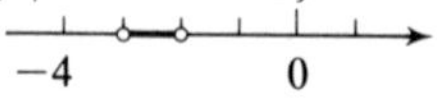

Exercise 7.1 [page 204]

1. a. $(0, 7)$ **b.** $(2, 9)$ **c.** $(-2, 5)$ **3. a.** $\left(0, \frac{-3}{2}\right)$ **b.** $(2, 0)$ **c.** $\left(-5, \frac{-21}{4}\right)$

5. $\{(-3, -7), (0, -4), (3, -1)\}$ **7.** $\{(1, 6), (2, 8), (3, 10)\}$ **9.** $\left\{\left(\frac{1}{2}, \frac{11}{2}\right), \left(\frac{3}{2}, \frac{9}{2}\right), \left(\frac{5}{2}, \frac{7}{2}\right)\right\}$

11. $k = \frac{7}{2}$ **13.** $y = \frac{x+2}{x} \quad (x \neq 0)$ **15.** $y = \frac{4}{x-1} \quad (x \neq 1)$ **17.** $y = \frac{x^2 - 7}{4}$

19. $y = \pm\frac{\sqrt{x+8}}{2}$ **21.** $y = \frac{3x^2 - 4}{4}$ **23.** $\{8\}$ **25.** $\{2, -2\}$ **27.** $\{1\}$ **29.** $\{3\}$

31. $\{8\}$ **33.** $\left\{\frac{1}{8}\right\}$

Exercise 7.2 [page 210]

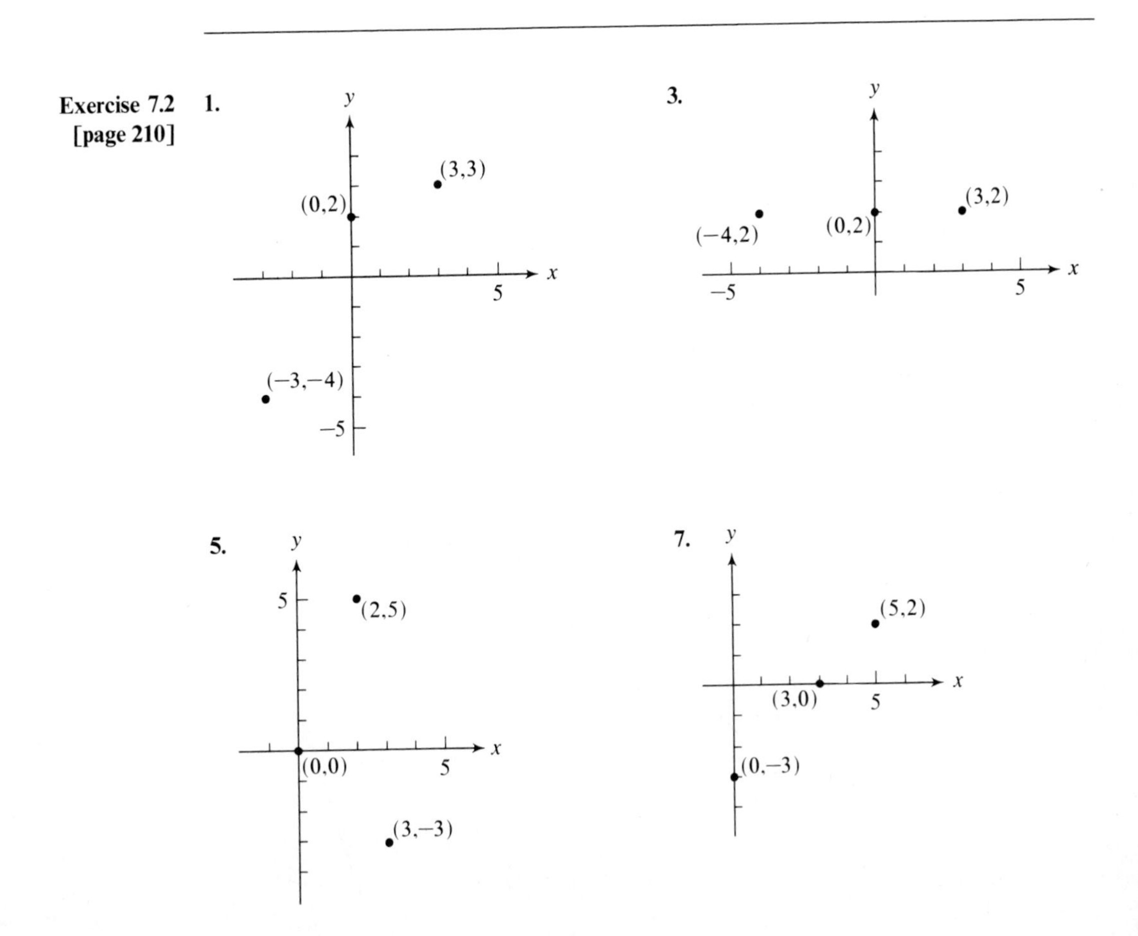

9.

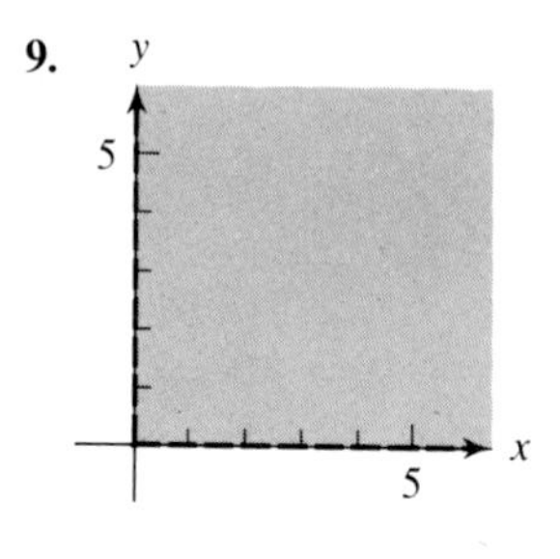

11.

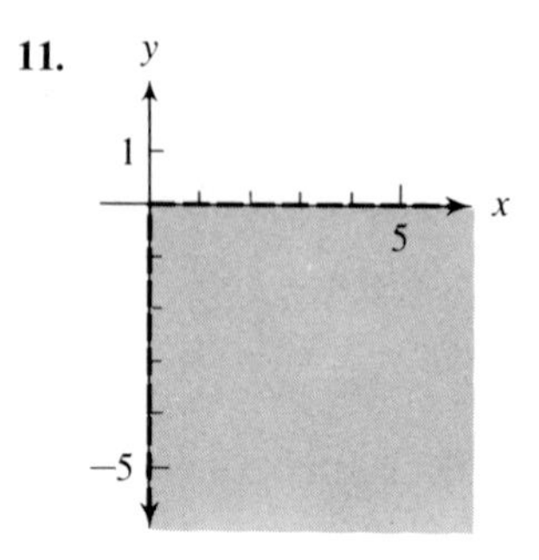

13.

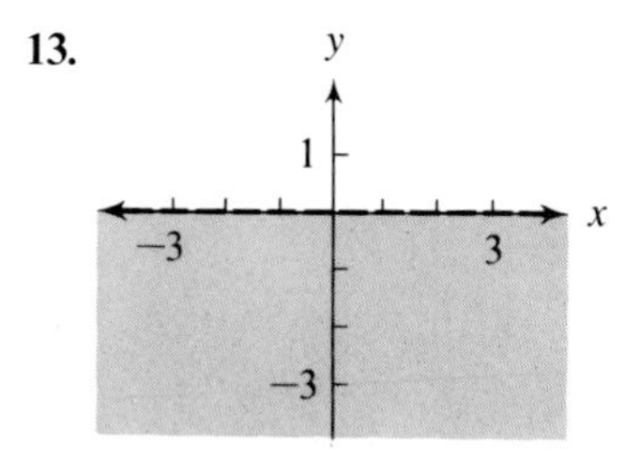

15.

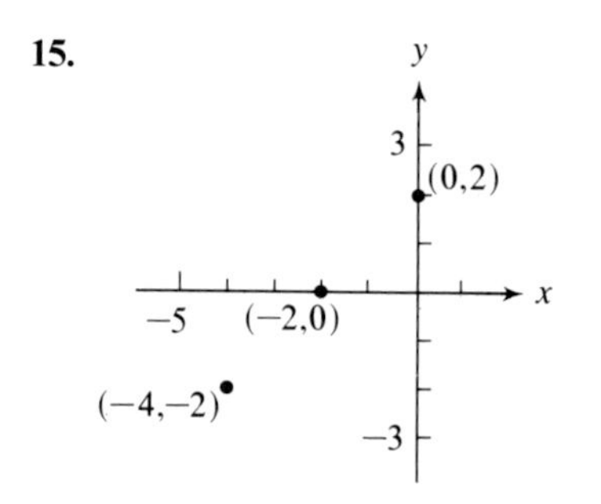

17.

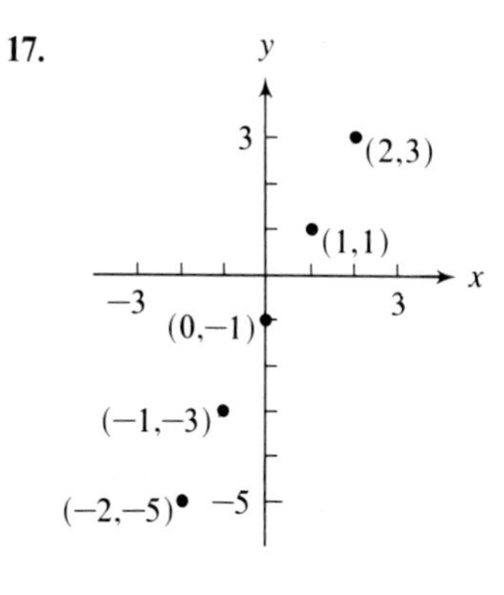

19.

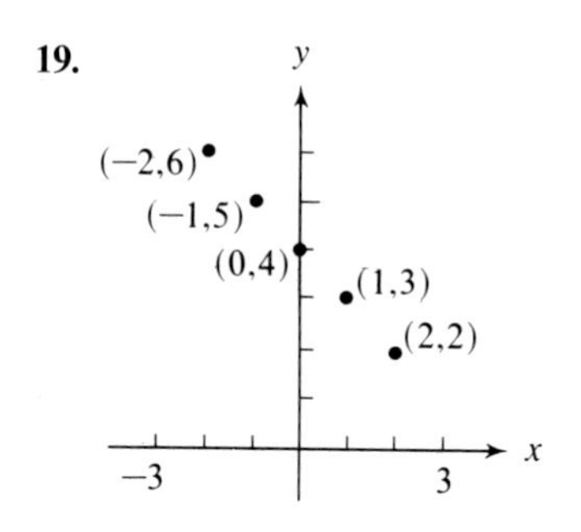

21.

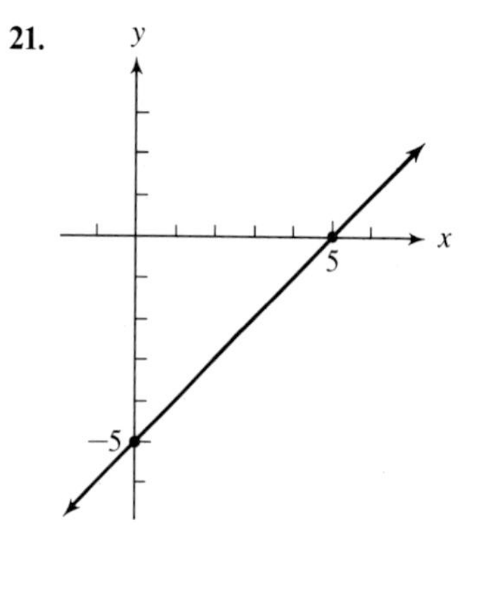

23.

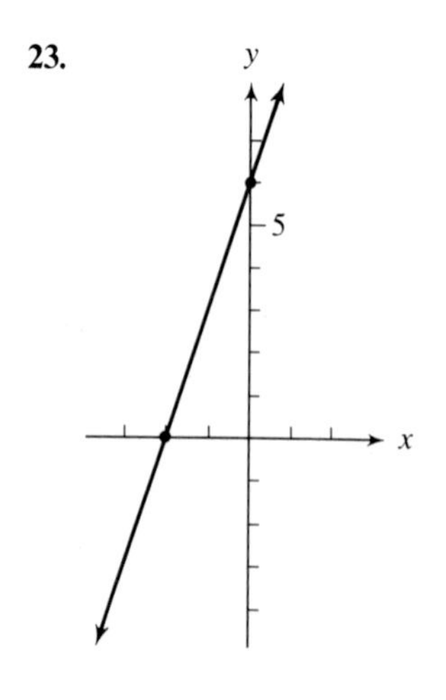

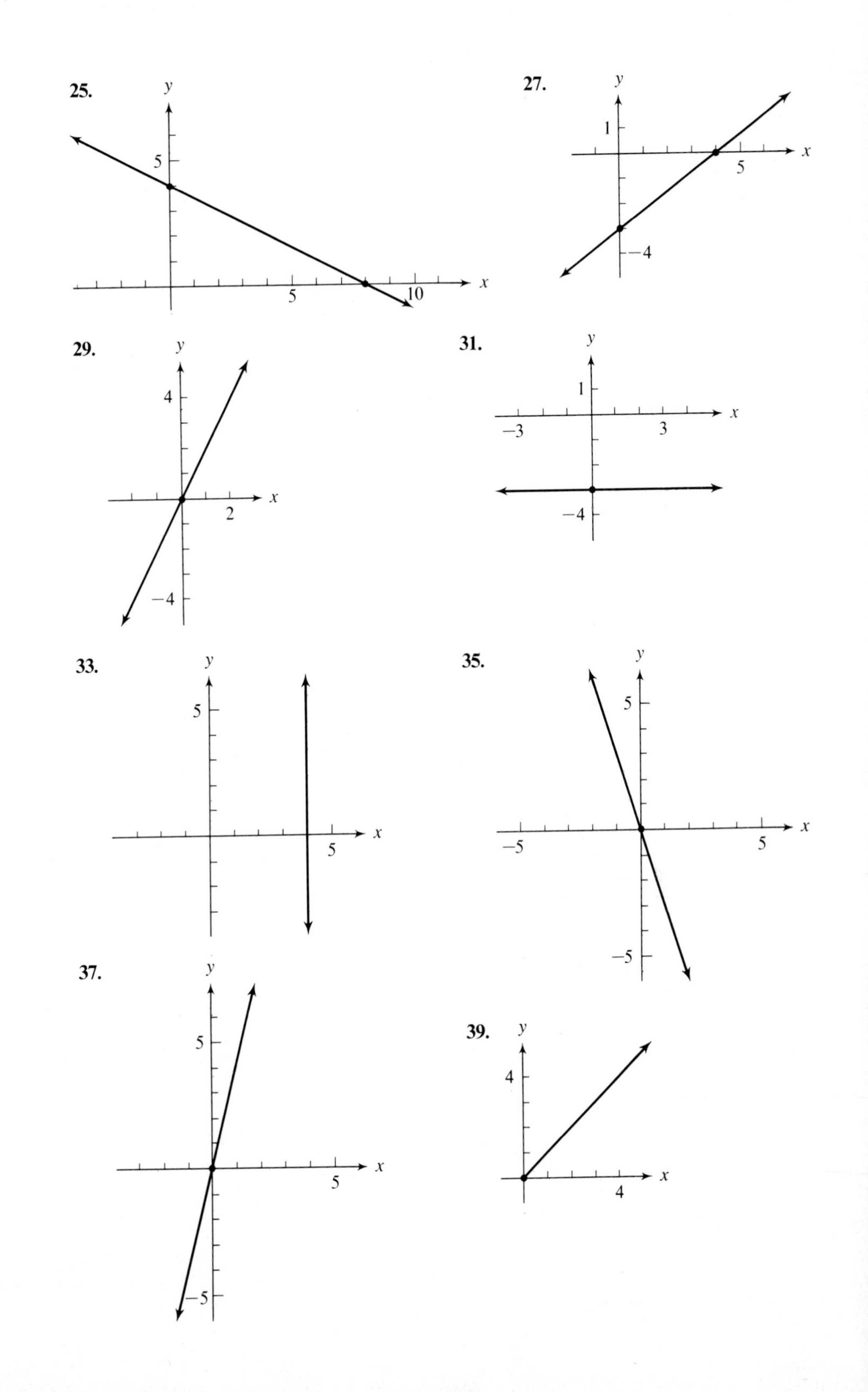
25.
y
5
5
10
x
27.
y
1
5
x
−4
29.
y
4
2
x
−4
31.
y
1
−3
3
x
−4
33.
y
5
5
x
35.
y
5
−5
5
x
−5
37.
y
5
5
x
−5
39.
y
4
4
x

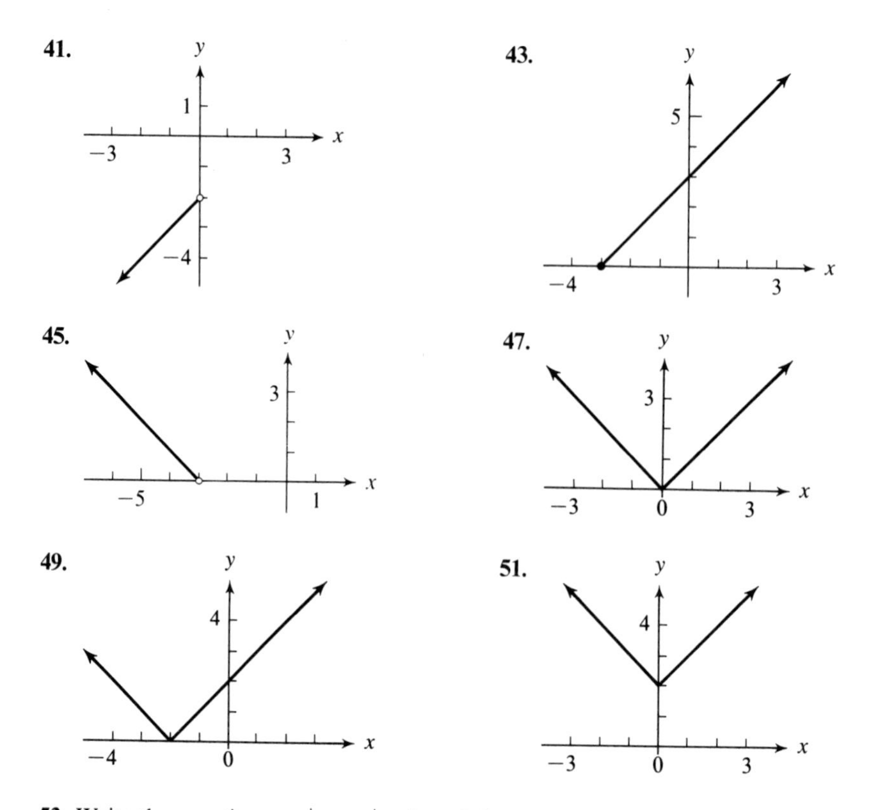

53. Write the equation $y = |x - a| + b$, and then graph the ordered pair (a, b)—this is the point where the two "branches" meet. Next, graph the ordered pairs $(a + 1, 1 + b)$ and $(a - 1, 1 + b)$ and sketch the branches.

Exercise 7.3
[page 215]

1. a. domain: $\{-2, -1, 0, 1\}$; range: $\{3, 4, 5, 6\}$ **b.** relation is a function

3. a. domain: $\{5, 6, 7, 8\}$; range $\{1\}$ **b.** relation is a function

5. a. domain: $\{2, 3\}$; range: $\{3, 4\}$ **b.** relation is not a function

7. a. domain: $\{0, 2, 3\}$; range: $\{0, 1, 2, 3\}$ **b.** relation is not a function

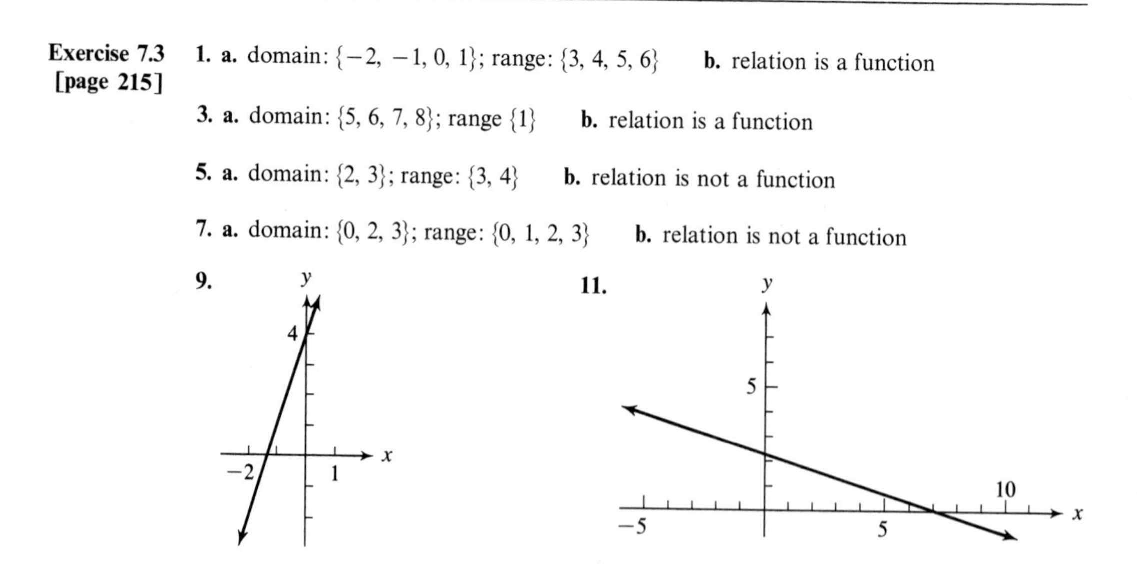

13.

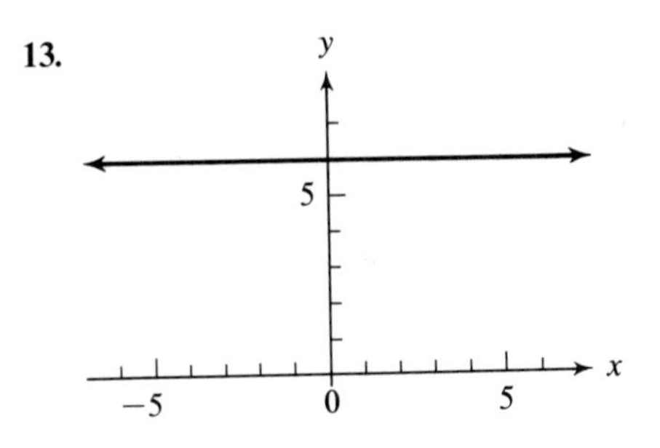

15. 9 **17.** 3 **19.** −6

21. **a.** $3x + 3h - 4$ **b.** $3h$ **c.** 3

23. **a.** $x^2 + 2hx + h^2 - 3x - 3h + 5$
b. $2hx + h^2 - 3h$
c. $2x + h - 3$

25. **a.** $x^3 + 3x^2h + 3xh^2 + h^3 + 2x + 2h - 1$
b. $3x^2h + 3xh^2 + h^3 + 2h$
c. $3x^2 + 3xh + h^2 + 2$

27.

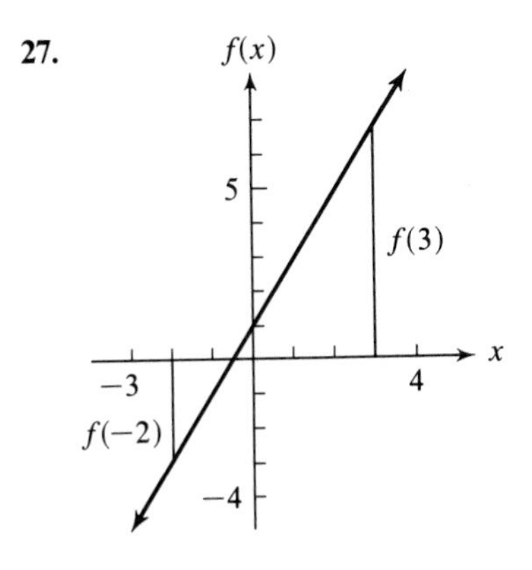

29.

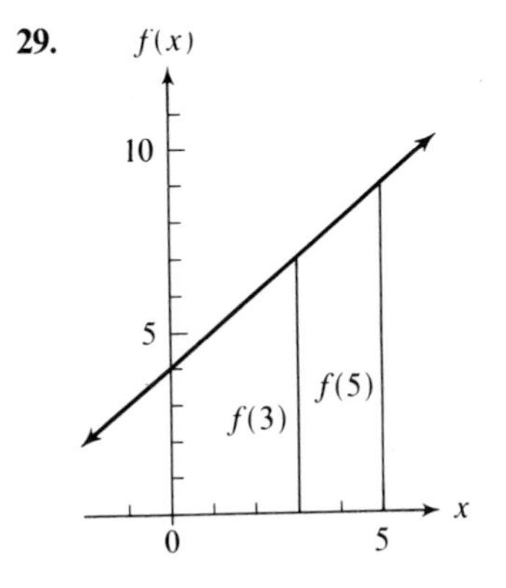

31. $f(x + h) - f(x)$

Exercise 7.4 [page 221]

1. distance: 5; slope: $\frac{4}{3}$

3. distance: 13; slope: $\frac{12}{5}$

5. distance: $\sqrt{5}$; slope: $\frac{-1}{2}$

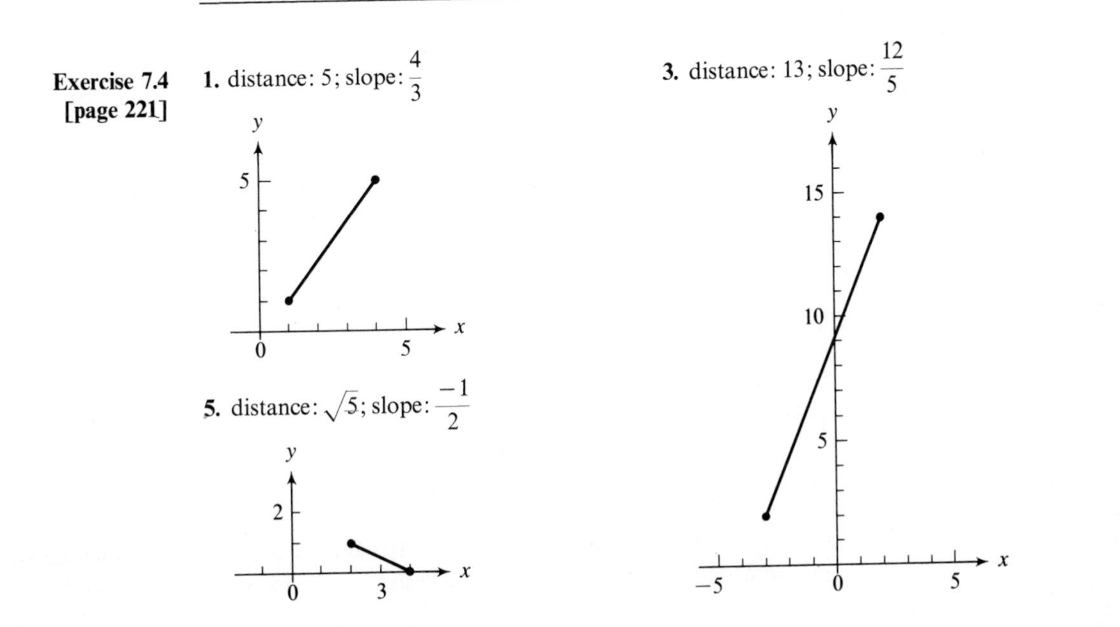

7. distance: $\sqrt{61}$; slope: $\frac{-5}{6}$

9. distance: 5; slope: 0

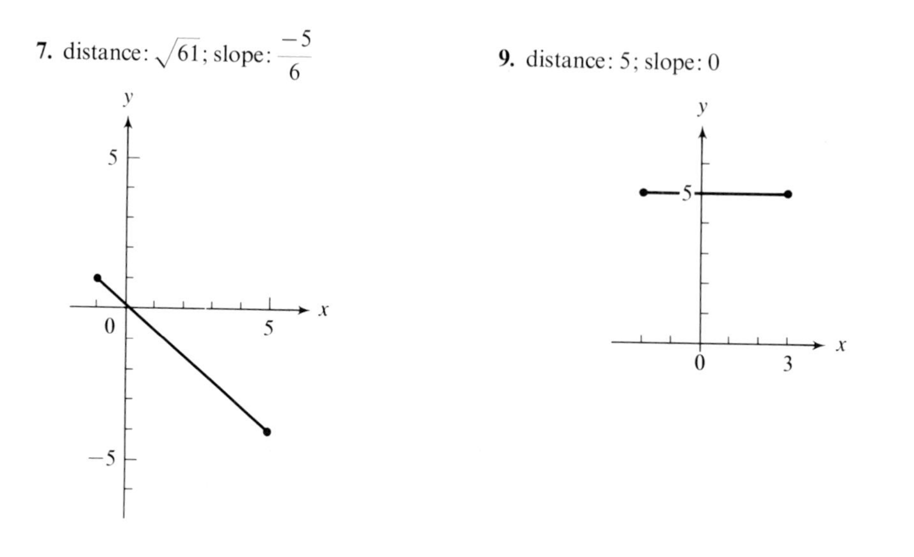

11. distance: 10; slope: not defined

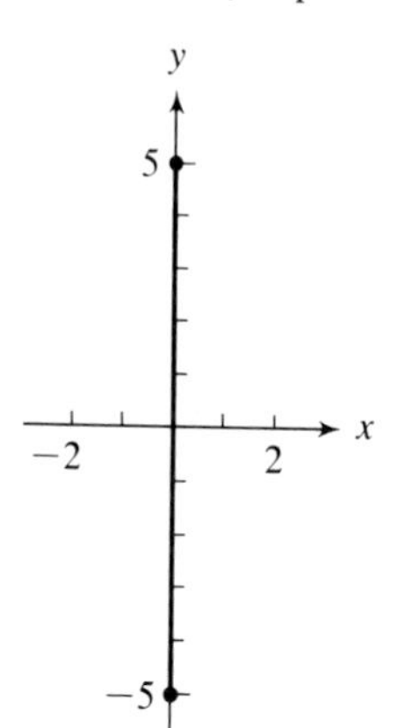

13. $15 + 9\sqrt{5}$

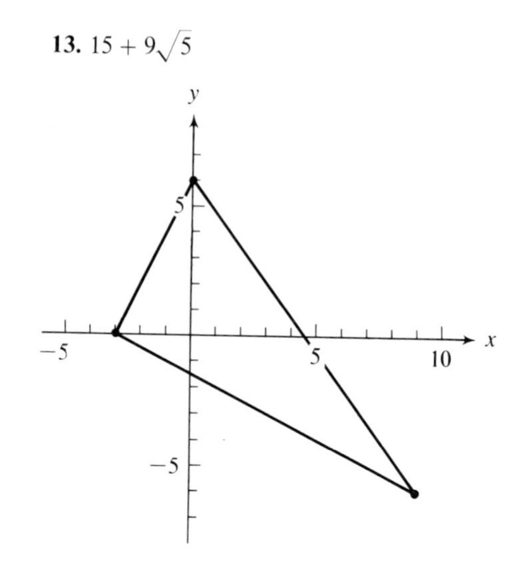

15. 48

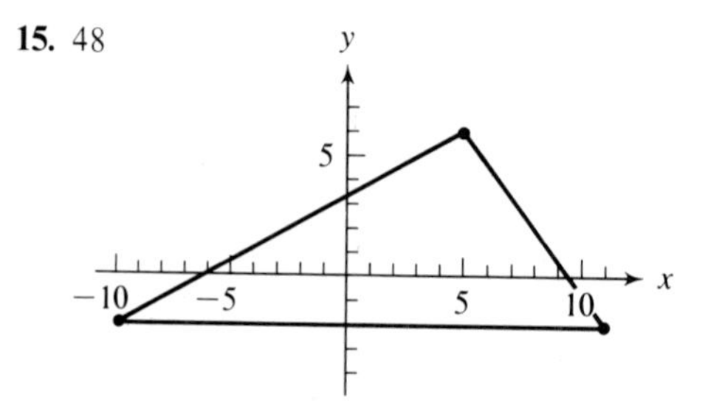

17. The slope of the segment with end points (5, 4) and (3, 0) is 2. The slope of the segment with end points (−1, 8) and (−4, 2) is 2. Therefore, the line segments are parallel.

19. Identify the given points as $A(0, -7)$, $B(8, -5)$, $C(5, 7)$, and $D(8, -5)$. Then, the slopes of AB and CD are

$$\text{slope } AB = \frac{-7-(-5)}{0-(8)} = \frac{1}{4}, \qquad \text{slope } CD = \frac{7-(-5)}{5-8} = -4,$$

and

$$\left(\frac{1}{4}\right)(-4) = -1.$$

21. Identify the vertices as $A(0, 6)$, $B(9, -6)$, and $C(-3, 0)$. Then, by the distance formula, $(AB)^2 = 225$ and $(BC)^2 + (AC)^2 = 180 + 45 = 225$; the triangle is a right triangle.

23. Let the given points be $A(2, 4)$, $B(3, 8)$, $C(5, 1)$, and $D(4, -3)$. Since AB and CD have equal slopes ($m = 4$) and BC and AD have equal slopes $\left(m = \frac{-7}{2}\right)$, there are two pairs of parallel sides. Hence, $ABCD$ is a parallelogram.

25. $k = -28$

Exercise 7.5 [page 224]

1. $4x - y - 7 = 0$

3. $x + y - 10 = 0$

5. $3x - y = 0$

7. $x + 2y + 2 = 0$

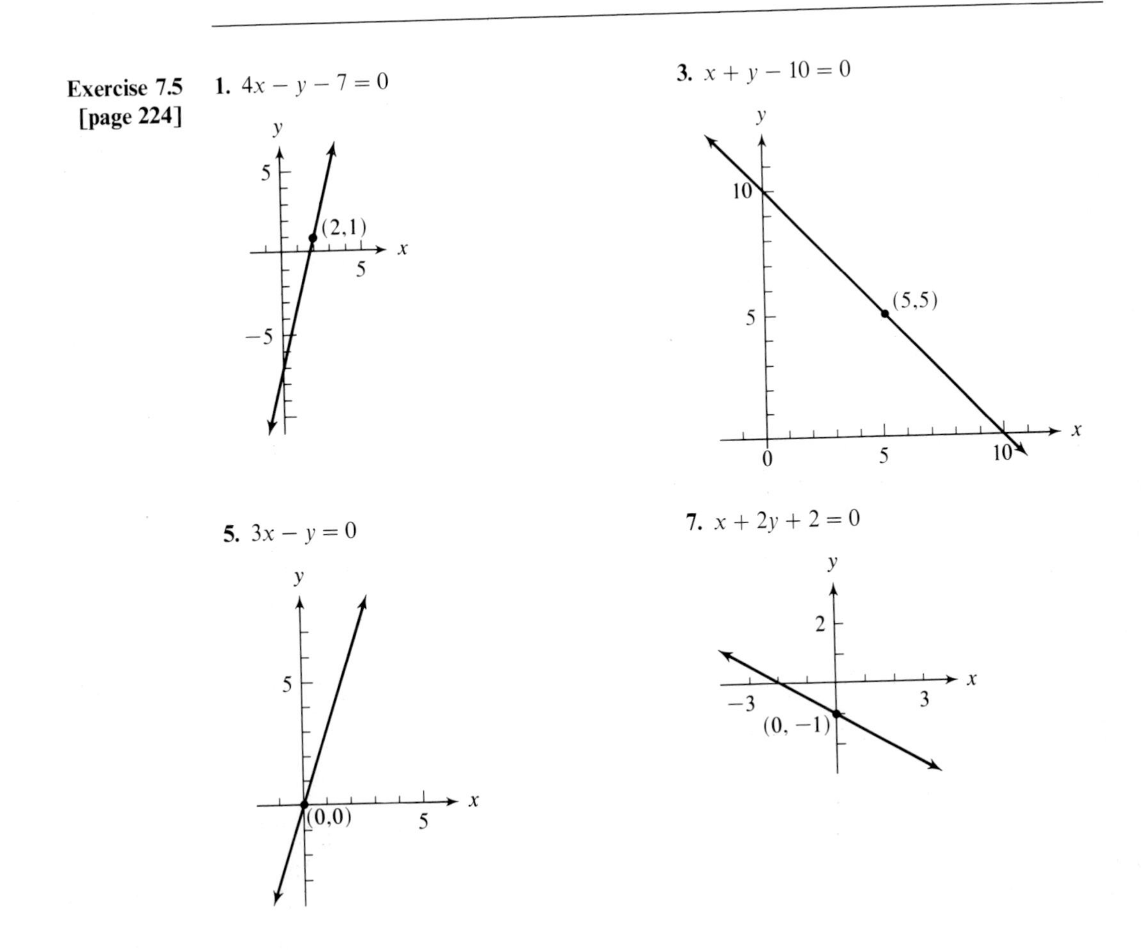

9. $3x + 4y + 14 = 0$

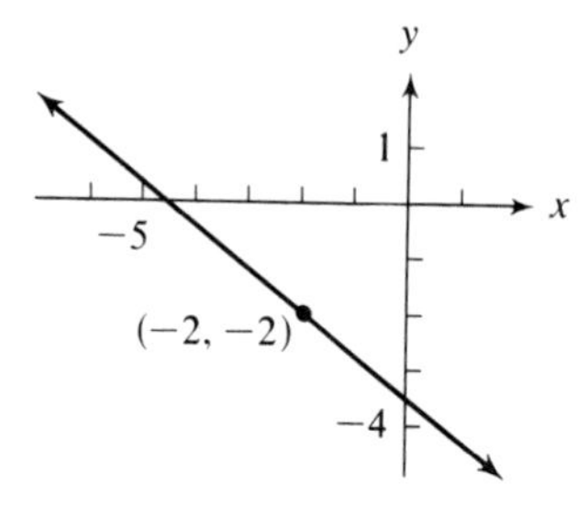

11. $y - 2 = 0$

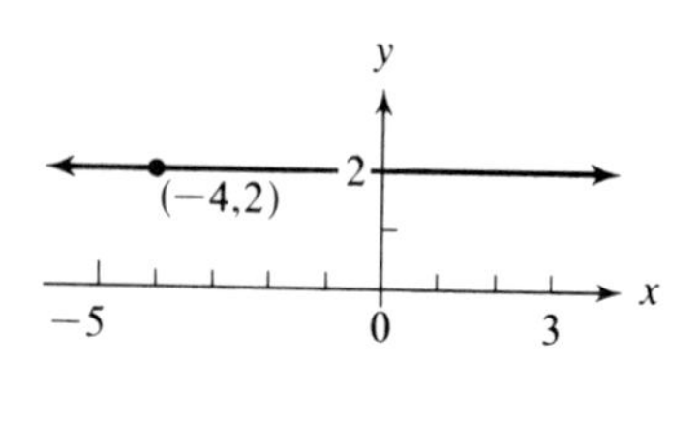

13. $y = -x + 3$; slope: -1; y intercept: 3

15. $y = \frac{-3}{2}x + \frac{1}{2}$; slope: $\frac{-3}{2}$; y intercept: $\frac{1}{2}$

17. $y = \frac{1}{3}x - \frac{2}{3}$; slope: $\frac{1}{3}$; y intercept: $\frac{-2}{3}$

19. $y = \frac{8}{3}x$; slope: $\frac{8}{3}$; y intercept: 0

21. $y = 0x - 2$; slope: 0; y intercept: -2

23. Since the graph is a line perpendicular to the x axis, there is no slope and no y intercept.

25. $x - 2y = 0$

27. $2x + y = 0$

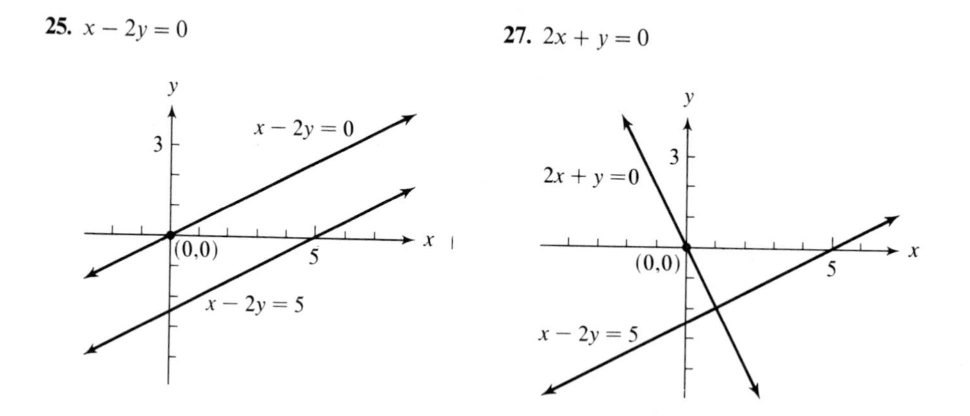

29. Since $y - y_1 = m(x - x_1)$ and $m = \frac{y_2 - y_1}{x_2 - x_1}$, by substitution, it follows that

$$y - y_1 = \left(\frac{y_2 - y_1}{x_2 - x_1}\right)(x - x_1).$$

31. $-\frac{a}{b}$ **33.** $\frac{b}{a}$

Exercise 7.6 [page 228]

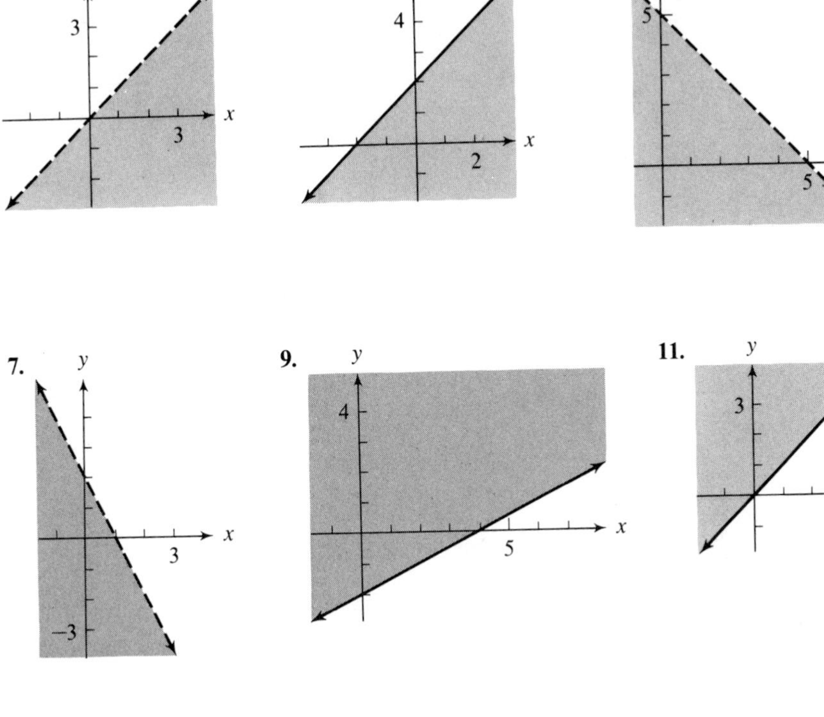

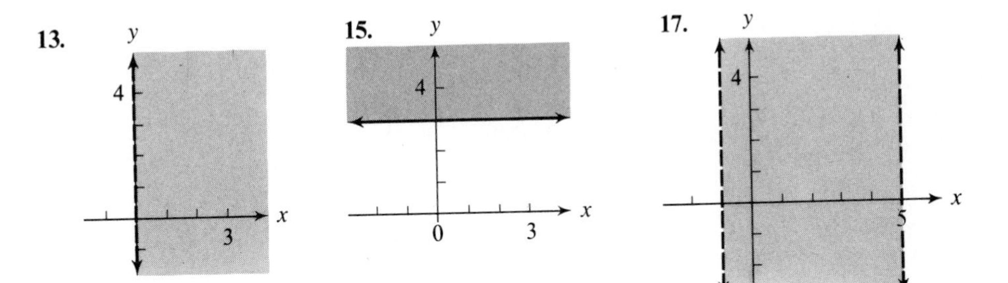

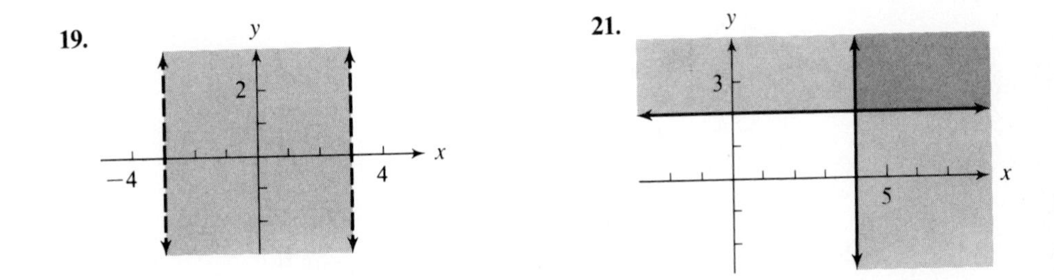

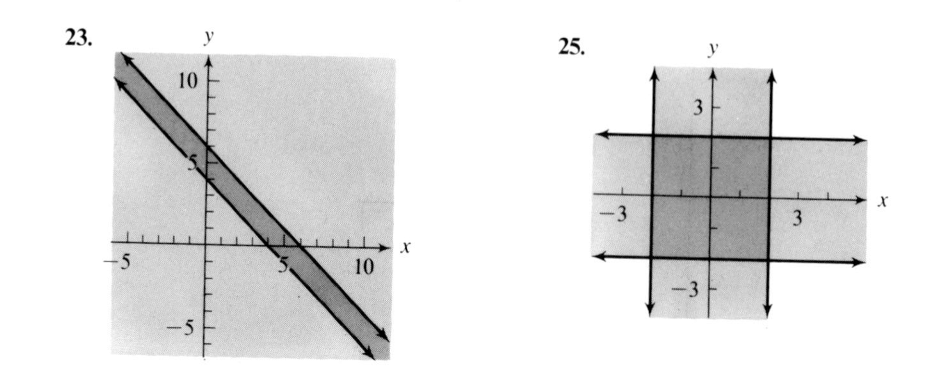

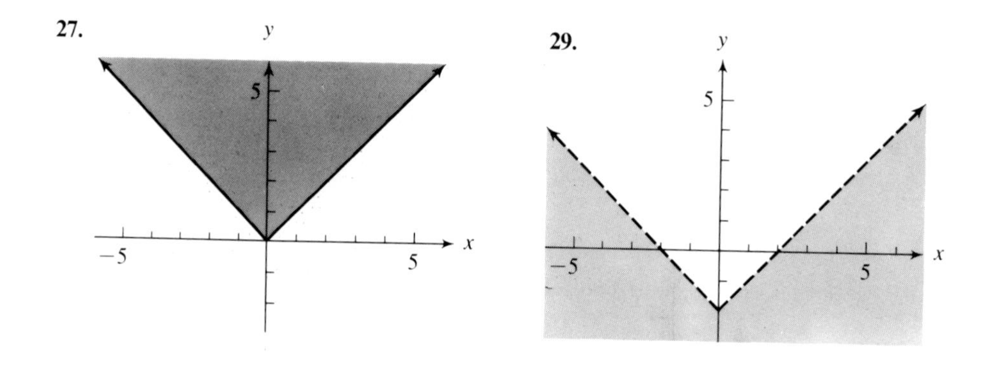

Review Exercises [page 231]

1. a. (0, −2) **b.** (6, 0) **c.** (3, −1) **2.** {(2, 1), (4, 5), (6, 9)} **3.** $k = 0$

4. $y = \dfrac{3}{x - 2x^2} \quad \left(x \neq 0, x \neq \dfrac{1}{2}\right)$

5. a. y 10 5 5 10 x

b. y 3 −5 5 x −5 −10

6. a. **b.**

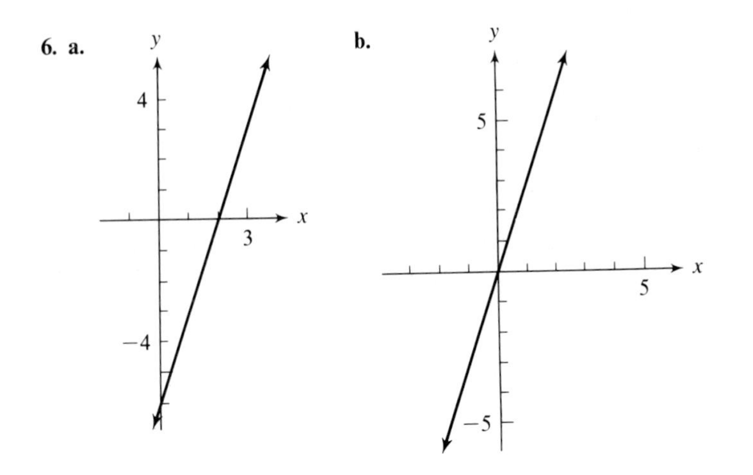

7. a. domain: {3, 4}; range: {5, 6}; relation is not a function
b. domain: {2, 3, 4, 5}; range: {3, 4, 5}; relation is a function

8. a. domain: {2, 4}; range: {4, 8} **b.** domain: {−2, 0, 2}; range: {−4, 2, 8}

9. a. 2 **b.** h

10. a. 10 **b.** $4x + 2h - 3$ **11. a.** $\sqrt{178}$ **b.** $\frac{13}{3}$

12. a. $3\sqrt{5}$ **b.** -2 **13. a.** $\sqrt{37} + \sqrt{41} + \sqrt{26}$ **b.** $\sqrt{113} + 5\sqrt{5}$

14. a. slope of $P_1P_2 = 1$; slope of $P_3P_4 = 1$, therefore P_1P_2 and P_3P_4 are parallel
b. slope of $P_1P_2 = 1$; slope of $P_1P_3 = -1$; since $1(-1) = -1$, the lines are perpendicular

15. a. $2x - y - 1 = 0$ **b.** $y = 2x - 1$

16. a. $y = -\frac{2}{3}x + 2$ **b.** slope: $-\frac{2}{3}$; y intercept: 2

17. a. $2x + 3y - 1 = 0$ **b.** $3x - 2y + 5 = 0$ **18. a.** $2x - 3y - 14 = 0$ **b.** $3x + 2y + 1 = 0$

19. a. **b.**

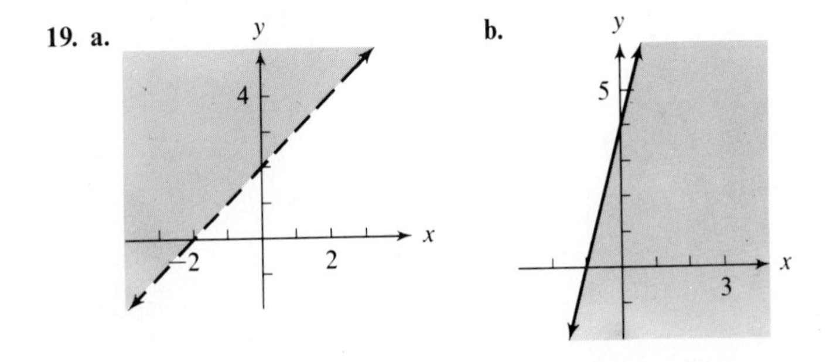

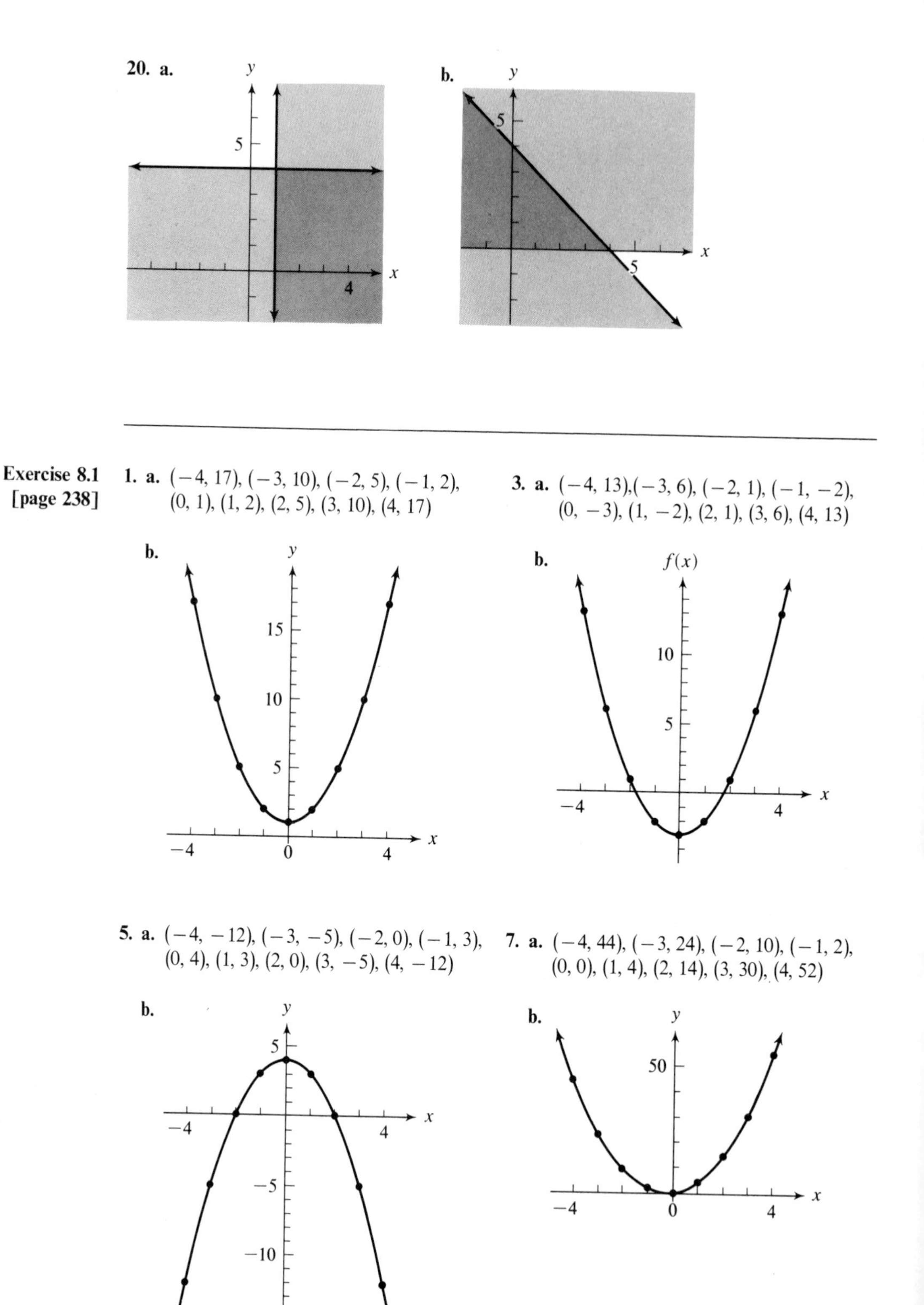

Exercise 8.1 [page 238]

1. a. $(-4, 17), (-3, 10), (-2, 5), (-1, 2), (0, 1), (1, 2), (2, 5), (3, 10), (4, 17)$

3. a. $(-4, 13), (-3, 6), (-2, 1), (-1, -2), (0, -3), (1, -2), (2, 1), (3, 6), (4, 13)$

5. a. $(-4, -12), (-3, -5), (-2, 0), (-1, 3), (0, 4), (1, 3), (2, 0), (3, -5), (4, -12)$

7. a. $(-4, 44), (-3, 24), (-2, 10), (-1, 2), (0, 0), (1, 4), (2, 14), (3, 30), (4, 52)$

9. a. $(-4, 9), (-3, 4), (-2, 1), (-1, 0), (0, 1), (1, 4), (2, 9), (3, 16), (4, 25)$

b.

11. a. $(-4, -39), (-3, -24), (-2, -13), (-1, -6), (0, -3), (1, -4), (2, -9), (3, -18), (4, -31)$

b.

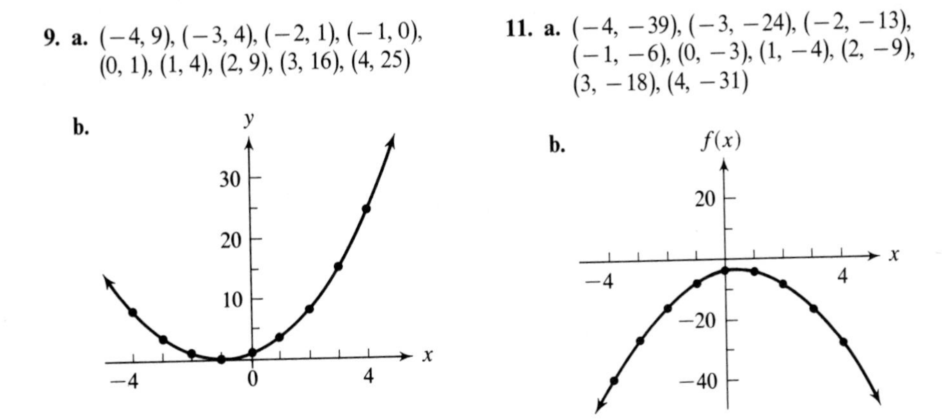

13. a. $(-4, -19), (-3, -11), (-2, -5), (-1, -1), (0, 1), (1, 1), (2, -1), (3, -5), (4, -11)$

b.

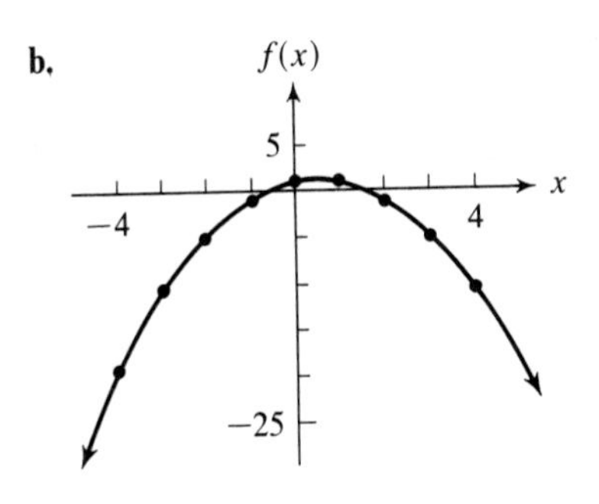

15. a. $(16, -4), (9, -3), (4, -2), (1, -1), (0, 0), (1, 1), (4, 2), (9, 3), (16, 4)$

b.

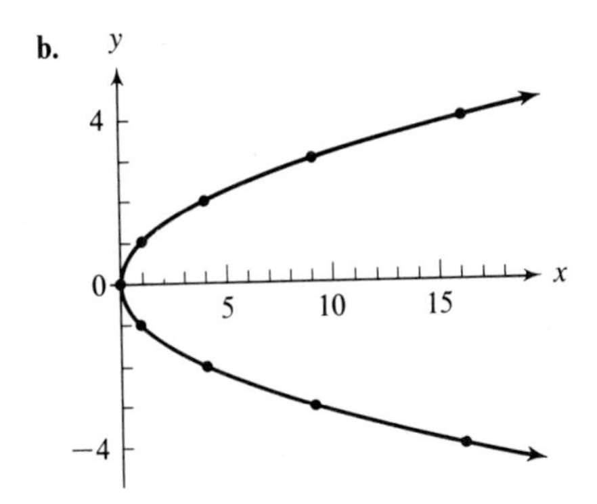

17. a. $(-60, -4), (-32, -3), (-12, -2), (0, -1), (4, 0), (0, 1), (-12, 2), (-32, 3), (-60, 4)$

b.

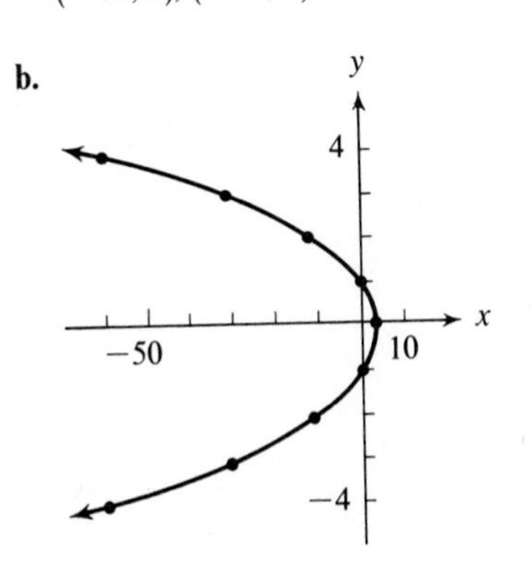

19. a. $(6, -4), (2, -3), (0, -2), (0, -1), (2, 0), (6, 1), (12, 2), (20, 3), (30, 4)$

b.

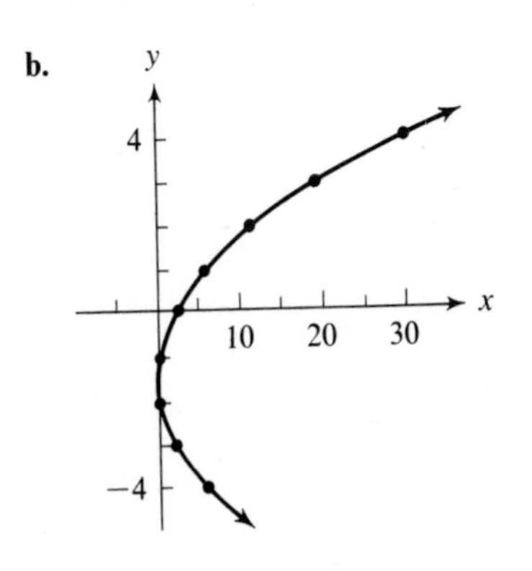

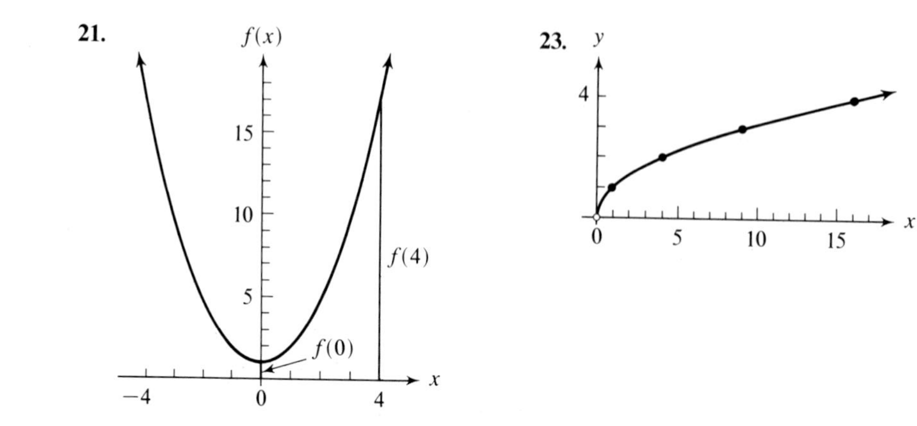
21.
f(x)
15
10
5
f(4)
f(0)
x
−4
0
4
23.
y
4
0
5
10
15
x

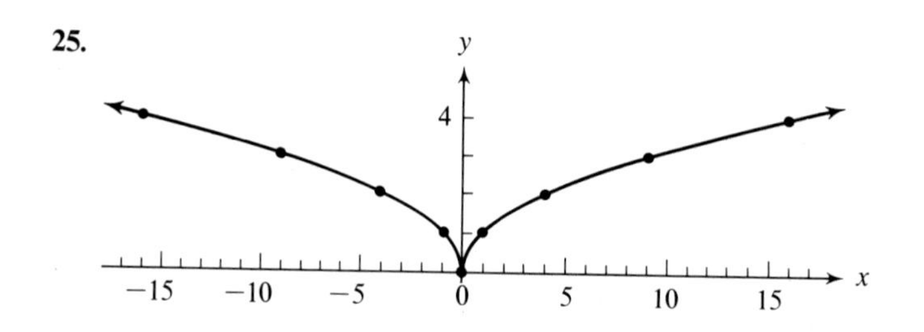
25.
y
4
−15
−10
−5
0
5
10
15
x

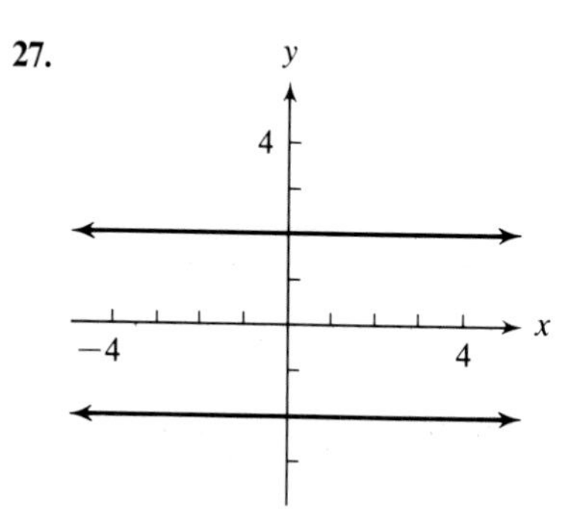
27.
y
4
x
−4
4

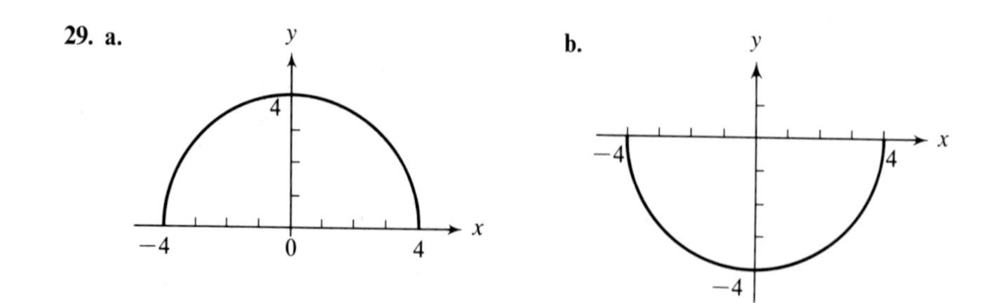
29. a.
y
4
x
−4
0
4
b.
y
x
−4
4
−4

c. Does not define a function, since some elements in the domain are associated with two elements in the range.

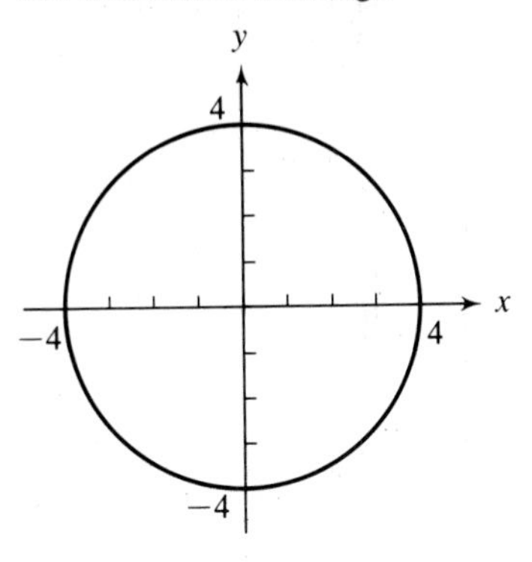

Exercise 8.2 [page 243]

1. x intercepts: 1 and 4; y intercept: 4;
minimum point: $\left(\frac{5}{2}, -\frac{9}{4}\right)$

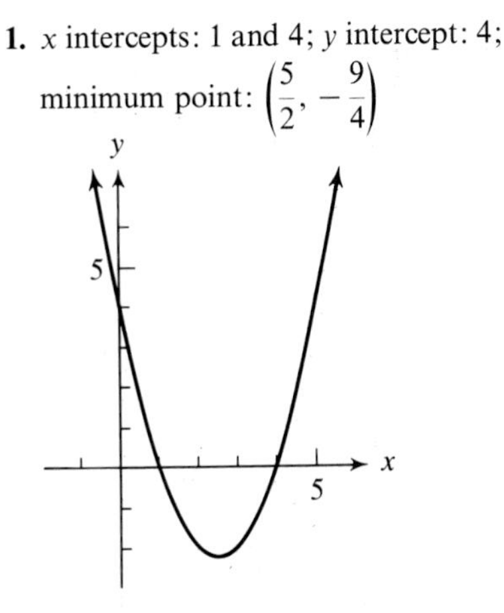

3. x intercepts: 0 and 4; $f(x)$ intercept: 0;
minimum point: $(2, -4)$

5. x intercepts: -5 and 1;
$g(x)$ intercept: -5;
minimum point: $(-2, -9)$

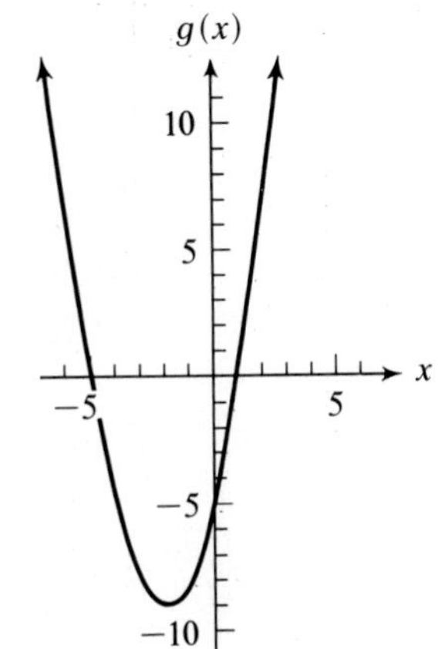

7. x intercepts: -3 and $\frac{3}{2}$; y intercept: -9;
minimum point: $\left(-\frac{3}{4}, -\frac{81}{8}\right)$

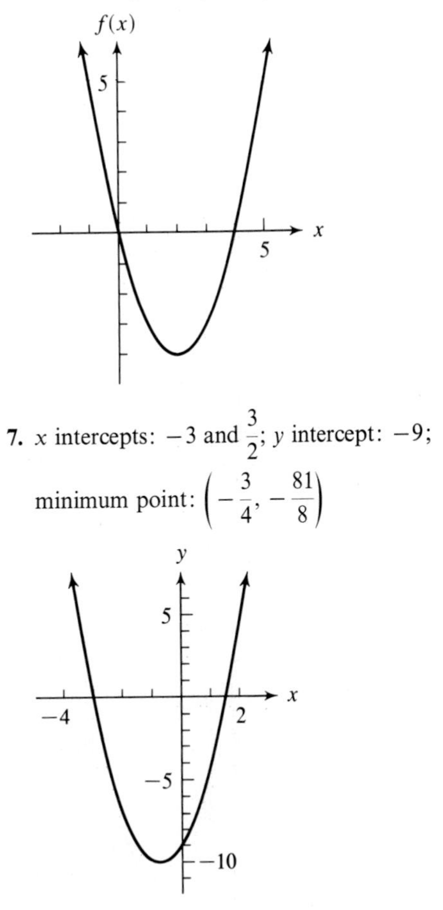

9. x intercepts: 1 and 6; $f(x)$ intercept: -6; maximum point: $\left(\frac{7}{2}, \frac{25}{4}\right)$

11. x intercepts: -3 and $-\frac{1}{2}$; y intercept: -3; maximum point: $\left(\frac{-7}{4}, \frac{25}{8}\right)$

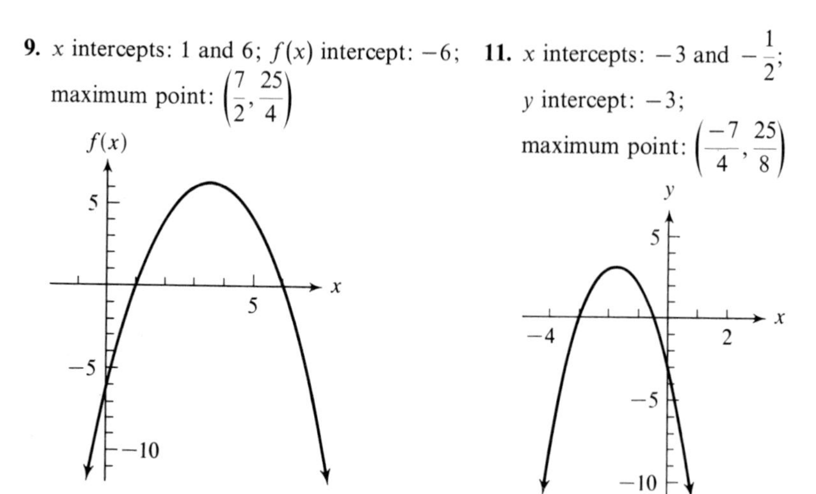

13. 6 and 6

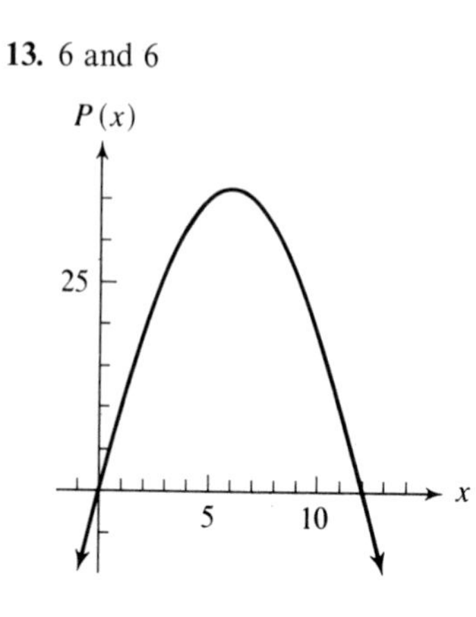

15.

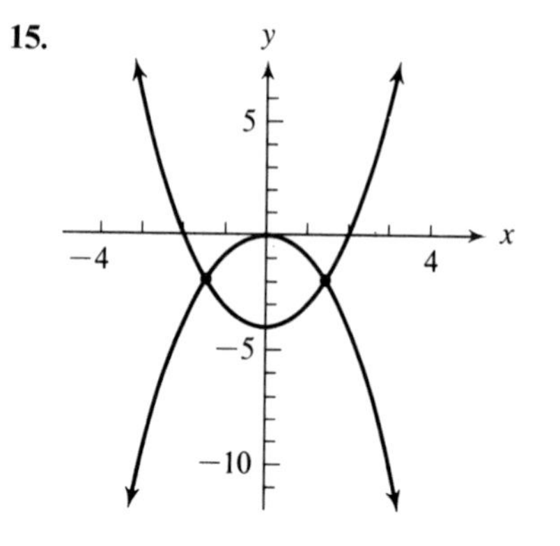

17.

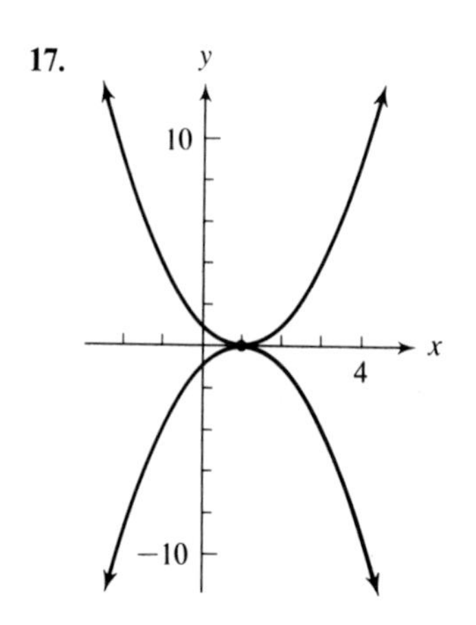

19.

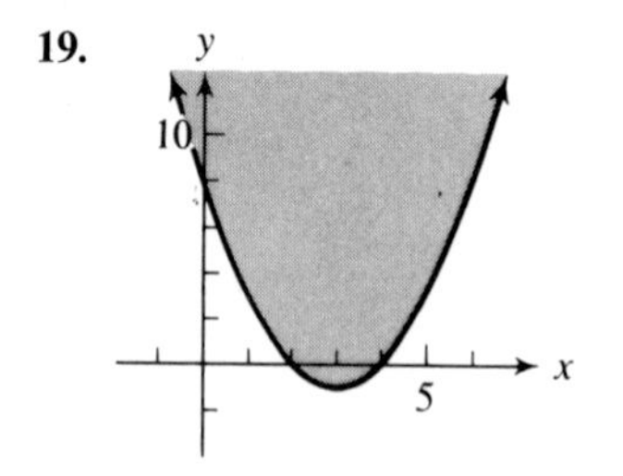

21.

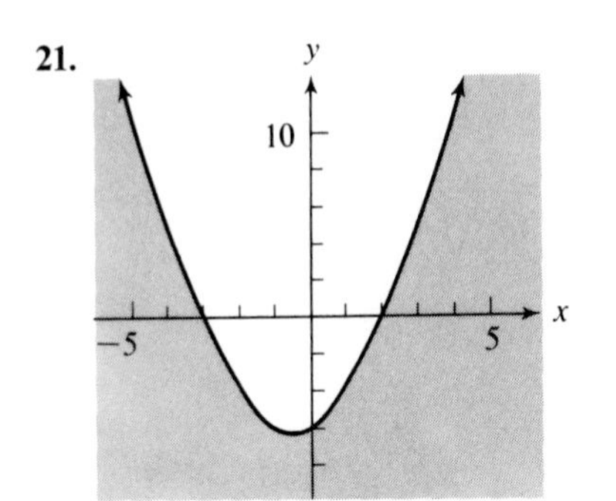

23.

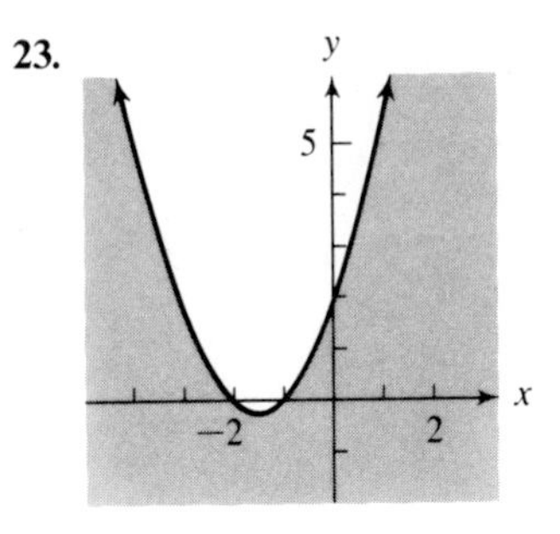

25.

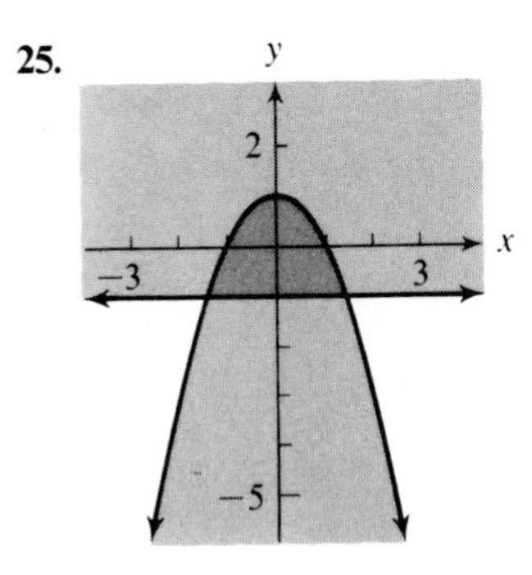

27.

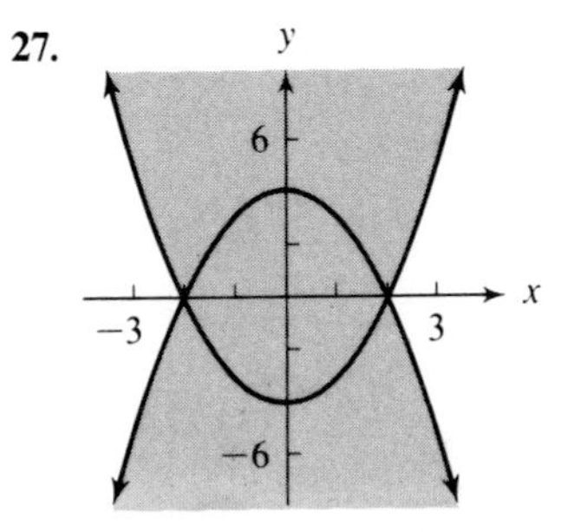

29.

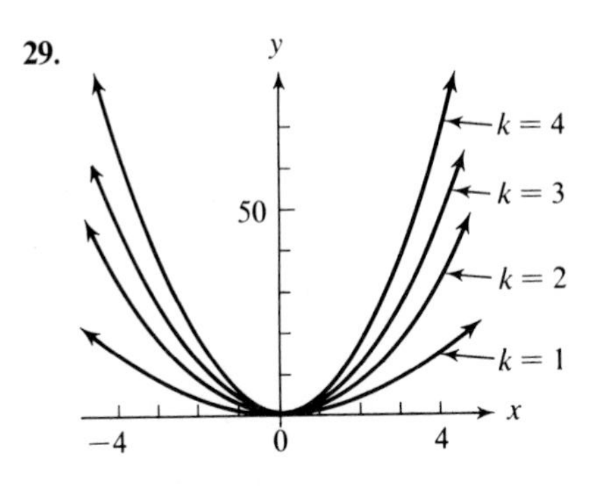

31.

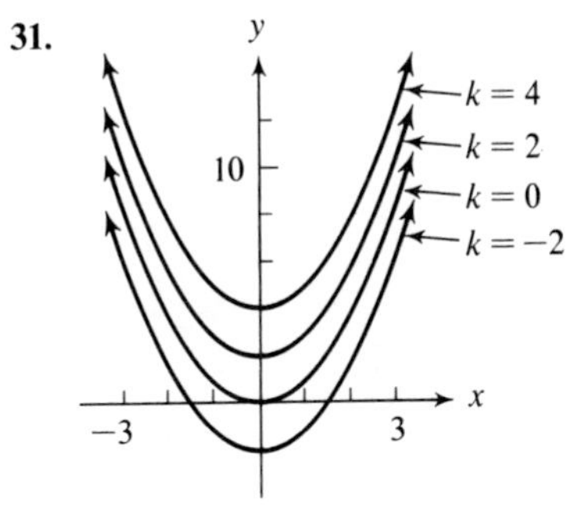

33. 2 seconds

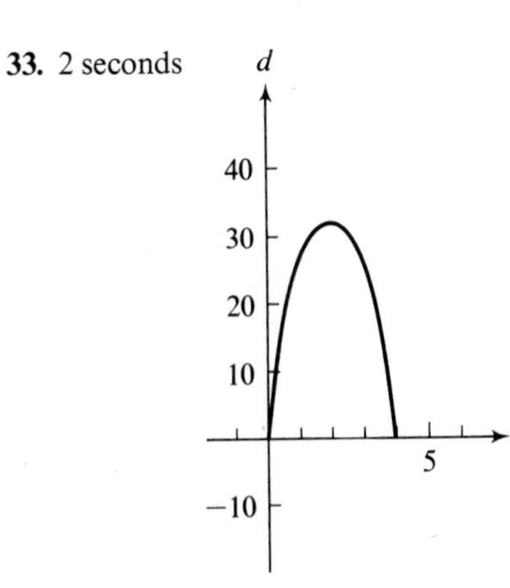

35.

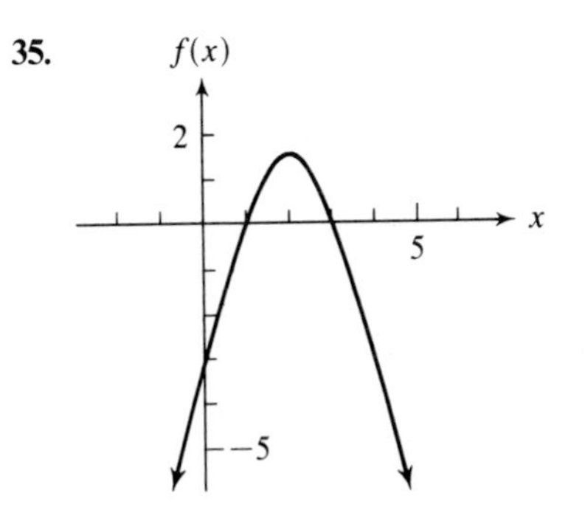

Exercise 8.3 [page 250]

1. **a.** $y = \pm\sqrt{4 - x^2}$ **b.** domain: $\{x \mid -2 \le x \le 2\}$

3. **a.** $y = \pm 3\sqrt{4 - x^2}$ **b.** domain: $\{x \mid -2 \le x \le 2\}$

5. **a.** $y = \pm\frac{1}{2}\sqrt{16 - x^2}$ **b.** domain: $\{x \mid -4 \le x \le 4\}$

7. **a.** $y = \pm\sqrt{x^2 - 1}$ **b.** domain: $\{x \mid x \le -1 \quad \text{or} \quad x \ge 1\}$

9. **a.** $y = \pm\sqrt{x^2 + 9}$ **b.** domain: $\{x \mid x \in R\}$

11. **a.** $y = \pm\sqrt{\dfrac{24 - 2x^2}{3}}$ **b.** domain: $\{x \mid -2\sqrt{3} \le x \le 2\sqrt{3}\}$

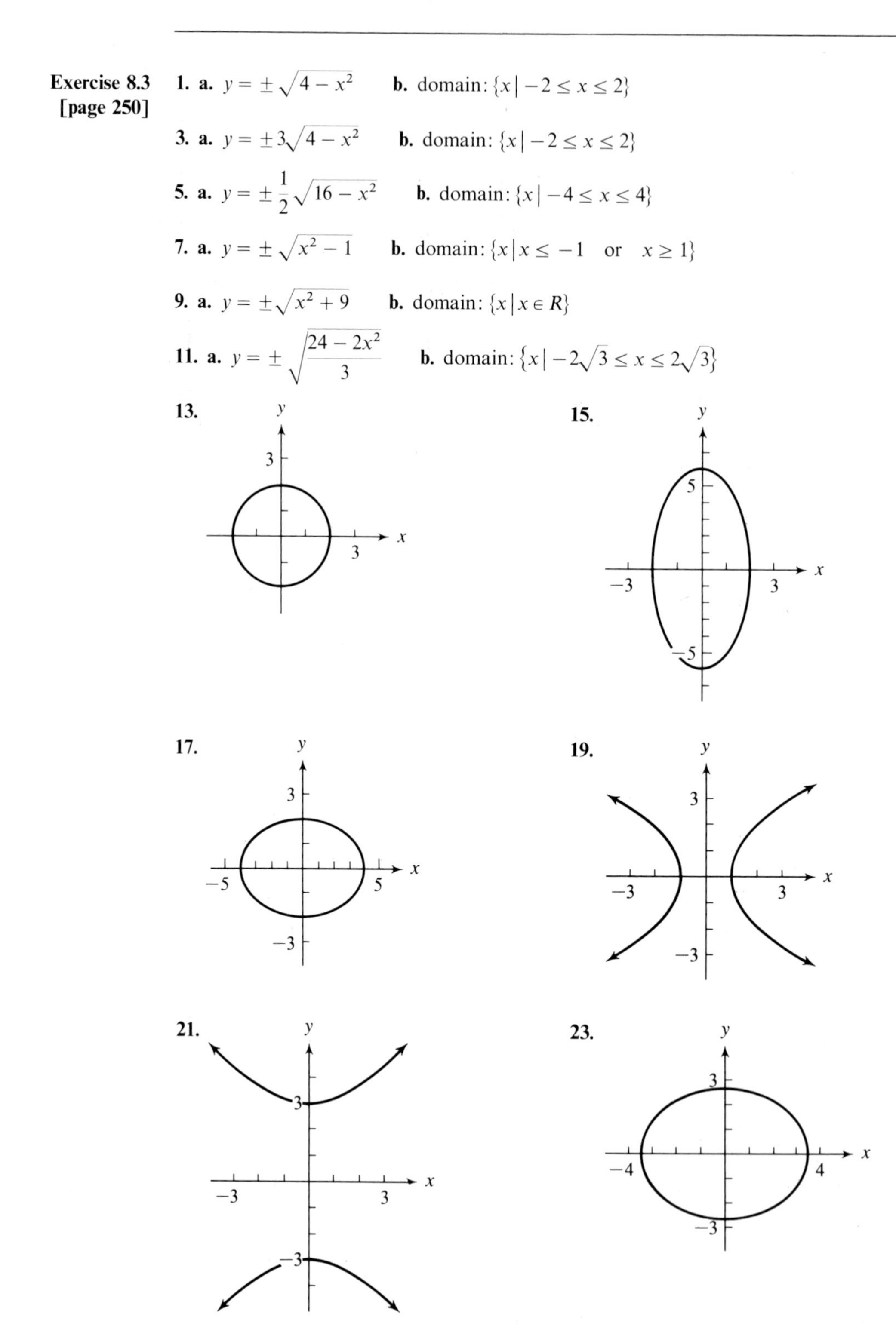

25.

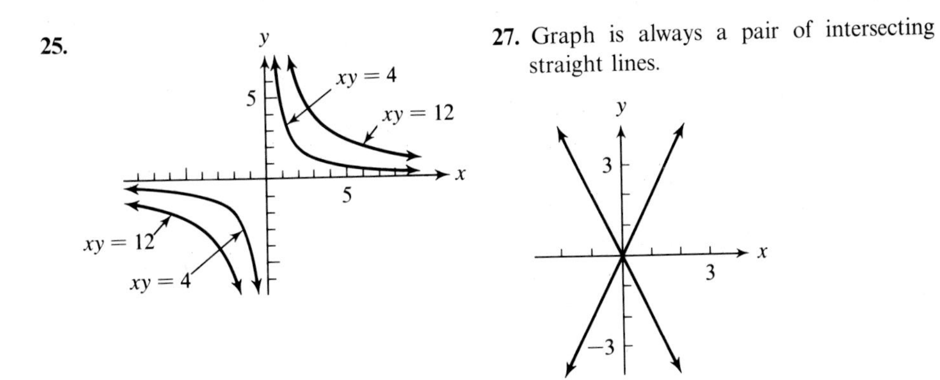

27. Graph is always a pair of intersecting straight lines.

29. Let (x, y) be any point such that its distance from the origin $(0, 0)$ is r. Then by the distance formula, $\sqrt{(x - 0)^2 + (y - 0)^2} = r$. Squaring both sides and simplifying the result yields $x^2 + y^2 = r^2$.

Exercise 8.4 [page 254]

1. circle

3. ellipse

5. hyperbola

7. two intersecting lines

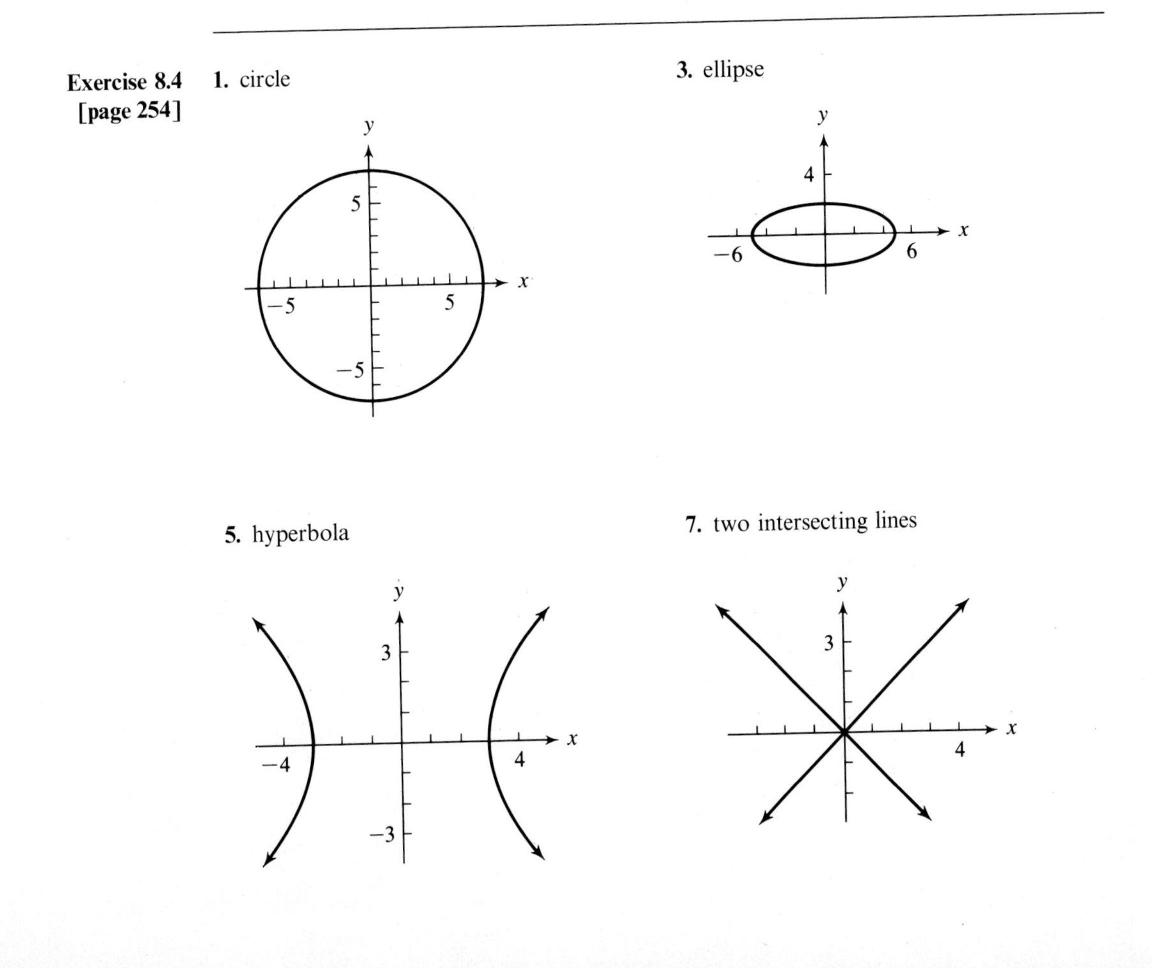

9. point

11. circle

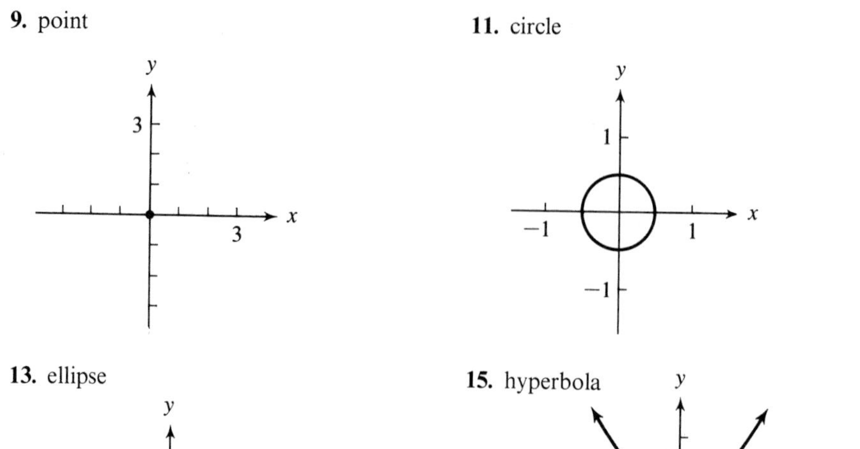

13. ellipse

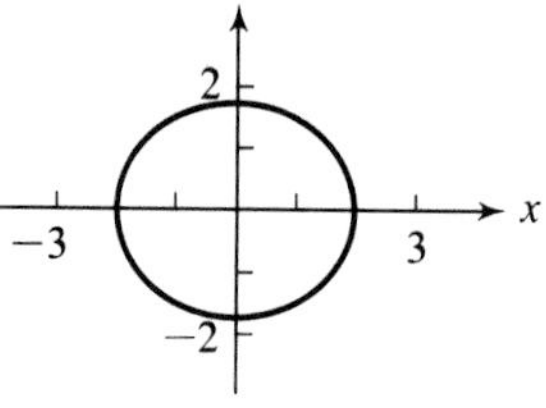

15. hyperbola

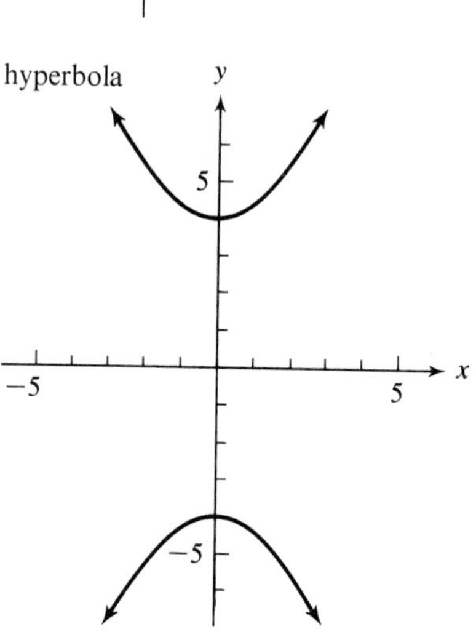

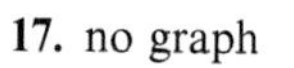

17. no graph

19.

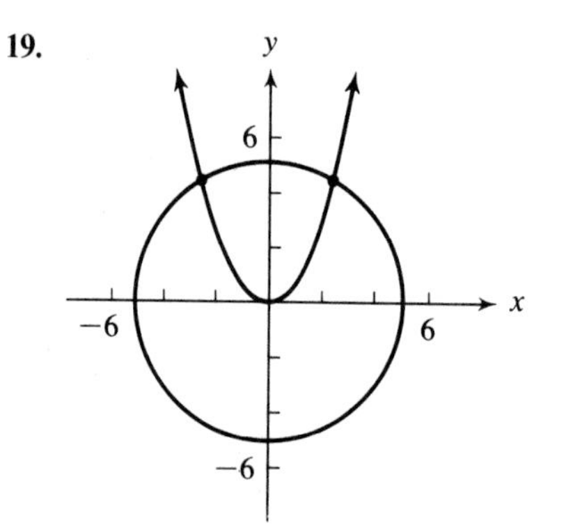

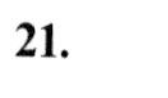

21.

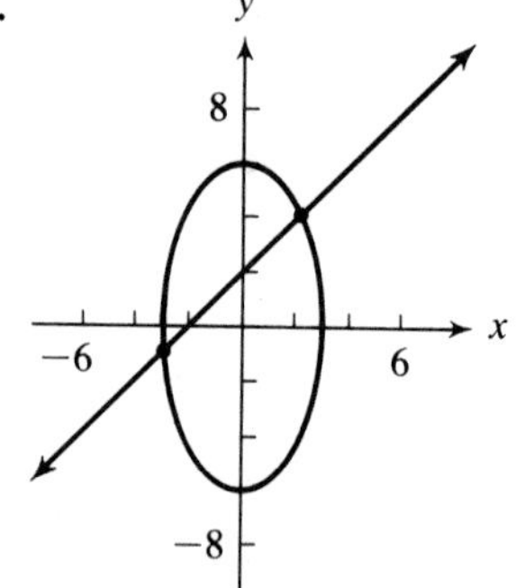

23.

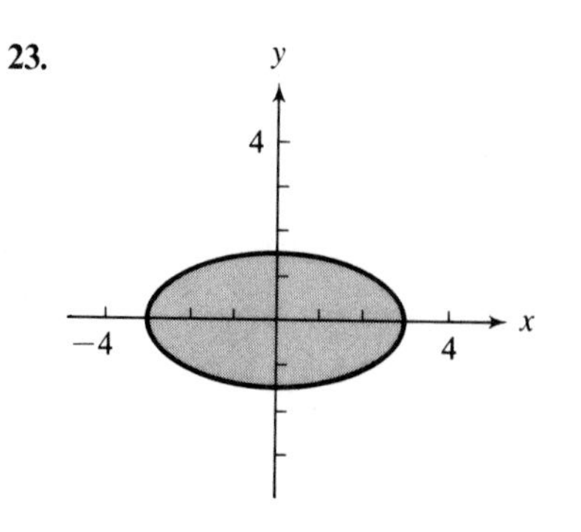

25.

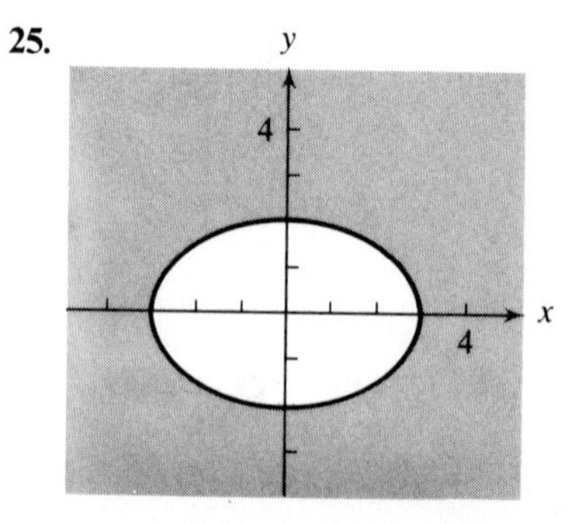

27. $by^2 = ax^2 - c$

$$y^2 = \frac{a}{b}x^2 - \frac{c}{b} = \frac{a}{b}x^2\left(1 - \frac{c}{ax^2}\right);$$

$$y = \pm\sqrt{\frac{a}{b}}\,x\sqrt{1 - \frac{c}{ax^2}}.$$

Exercise 8.5 [page 258]

1. $d = kt$ **3.** $I = \dfrac{k}{R}$ **5.** $V = klw$ **7.** 3 **9.** 200 **11.** 2 **13.** $\dfrac{32}{9}$ **15.** 264

17. 16 **19.** 400 feet **21.** 160 pounds per square foot

23. $\dfrac{3375}{2}$ pounds **25.** $\dfrac{9}{3^2} = \dfrac{y}{4^2}$; 16 **27.** $\dfrac{16}{2^2} = \dfrac{d}{10^2}$; 400 feet

29. $\dfrac{40}{10} = \dfrac{P}{40}$; 160 pounds per square foot **31.** $\dfrac{8 \cdot 750}{2 \cdot 4^2} = \dfrac{8 \cdot L}{2 \cdot 6^2}$; $\dfrac{3375}{2}$ pounds

33. Let D_1 and D_2 be the diameters and C_1 and C_2 be the corresponding circumferences of the two circles; then since $\pi = \dfrac{C}{D}$, it follows that $\dfrac{C_1}{D_1} = \pi = \dfrac{C_2}{D_2}$ and hence $\dfrac{C_1}{C_2} = \dfrac{D_1}{D_2}$.

35. The intensity will decrease by 1/4.

37. As k increases, the graph rises more steeply from left to right.

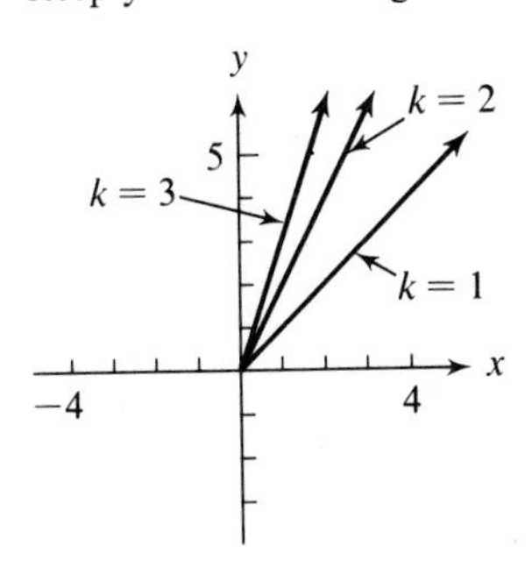

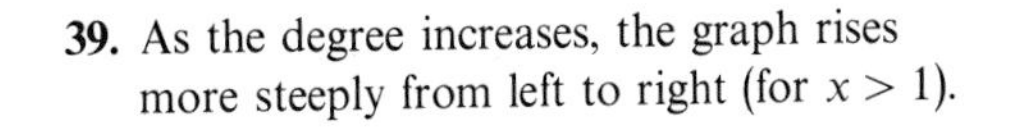

39. As the degree increases, the graph rises more steeply from left to right (for $x > 1$).

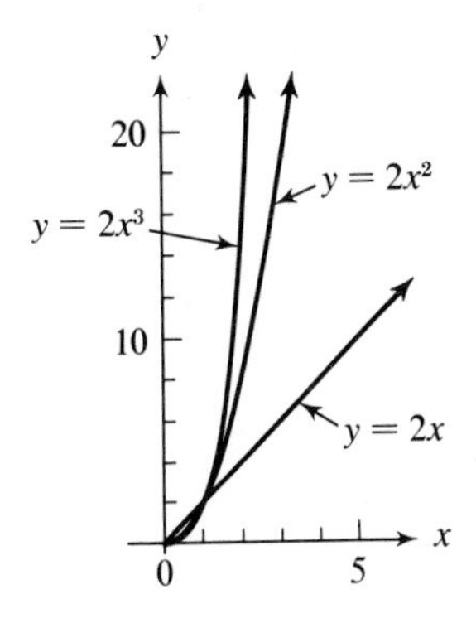

Exercise 8.6 [page 264]

1. a. $F^{-1} = \{(-2, -2), (2, 2)\}$ **b.** F^{-1} is a function

3. a. $Q^{-1} = \{(3, 1), (3, 2), (4, 3)\}$ **b.** Q^{-1} is not a function

5. a. $G^{-1} = \{(1, 1), (2, 2), (3, 3)\}$ **b.** G^{-1} is a function

7. $2y + 4x = 7$; defines a function

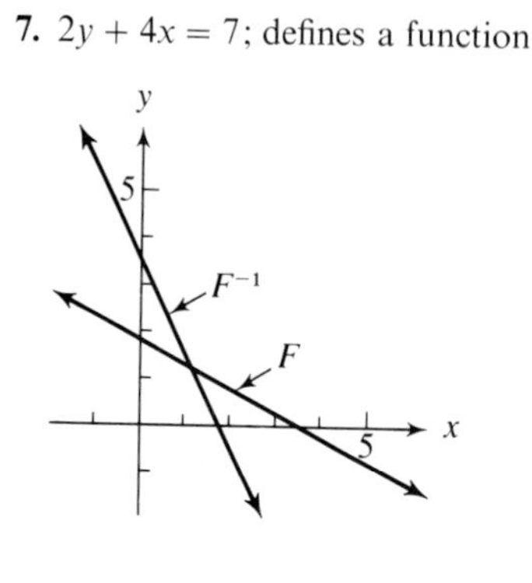

9. $x = y^2 - 4y$; does not define a function

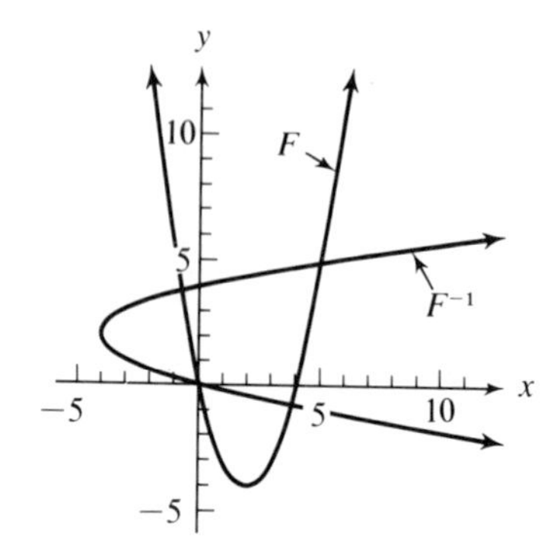

11. $y^2 + 4x^2 = 36$; does not define a function

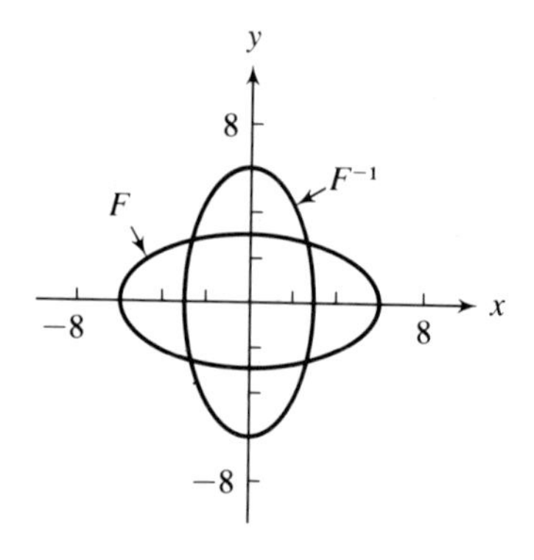

13. $y^2 - x^2 = 3$; does not define a function

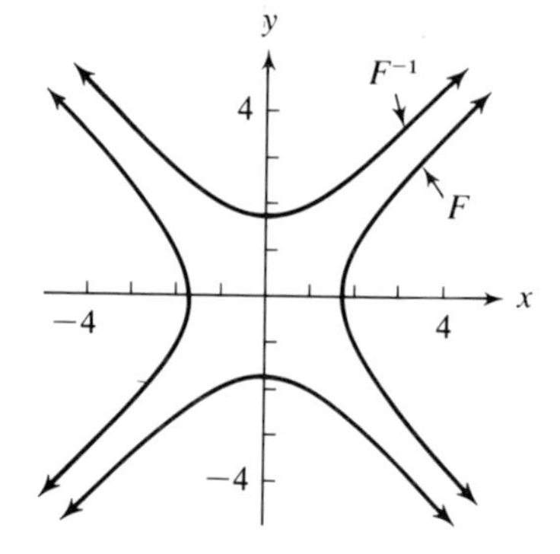

15. $x = \sqrt{4 + y^2}$; does not define a funetion

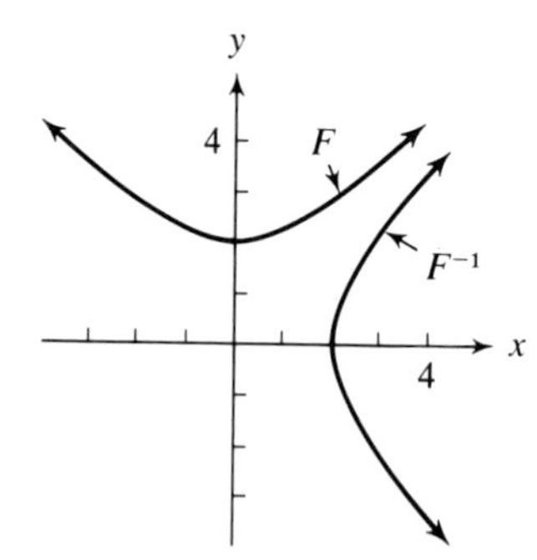

17. $x = |y|$; does not define a function

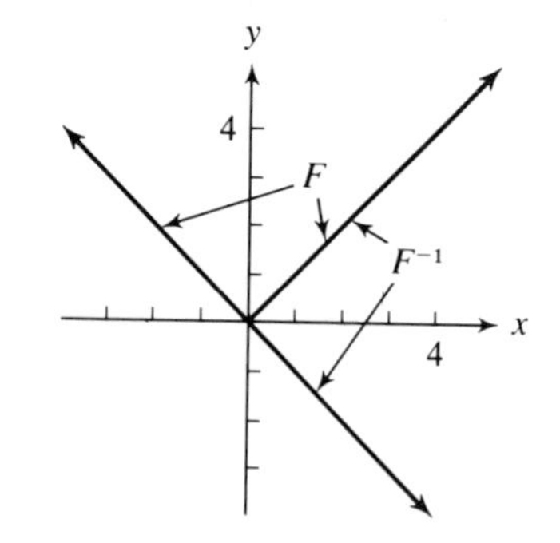

19. $F^{-1}: y = x; F[F^{-1}(x)] = F[x] = x; F^{-1}[F(x)] = F^{-1}[x] = x$

21. $F^{-1}: y = \dfrac{4 - x}{2}; F[F^{-1}(x)] = F\left[\dfrac{4 - x}{2}\right] = -2\left[\dfrac{4 - x}{2}\right] + 4 = -4 + x + 4 = x;$

$F^{-1}[F(x)] = F^{-1}[-2x + 4] = 2 - \dfrac{1}{2}[-2x + 4] = 2 + x - 2 = x$

23. $F^{-1}: y = \dfrac{4x+12}{3}$; $F[F^{-1}(x)] = F\left[\dfrac{4x+12}{3}\right] = \dfrac{3}{4}\left[\dfrac{4x+12}{3}\right] - 3 = x + 3 - 3 = x$;

$F^{-1}[F(x)] = F^{-1}\left[\dfrac{3x-12}{4}\right] = \dfrac{4}{3}\left[\dfrac{3x-12}{4}\right] + 4 = x - 4 + 4 = x$

25. Problems 7 and 8

27. a. $\{y \mid y \geq 0\}$ **b.** $F^{-1}: x = y^2 - 2y + 1$ or $y = 1 \pm \sqrt{x}$, $\{x \mid x \geq 0\}$ **c.** no

---

Review Exercises [page 267]

1. a. x intercepts, 1 and 5; y intercept, 5 **b.** $(3, -4)$

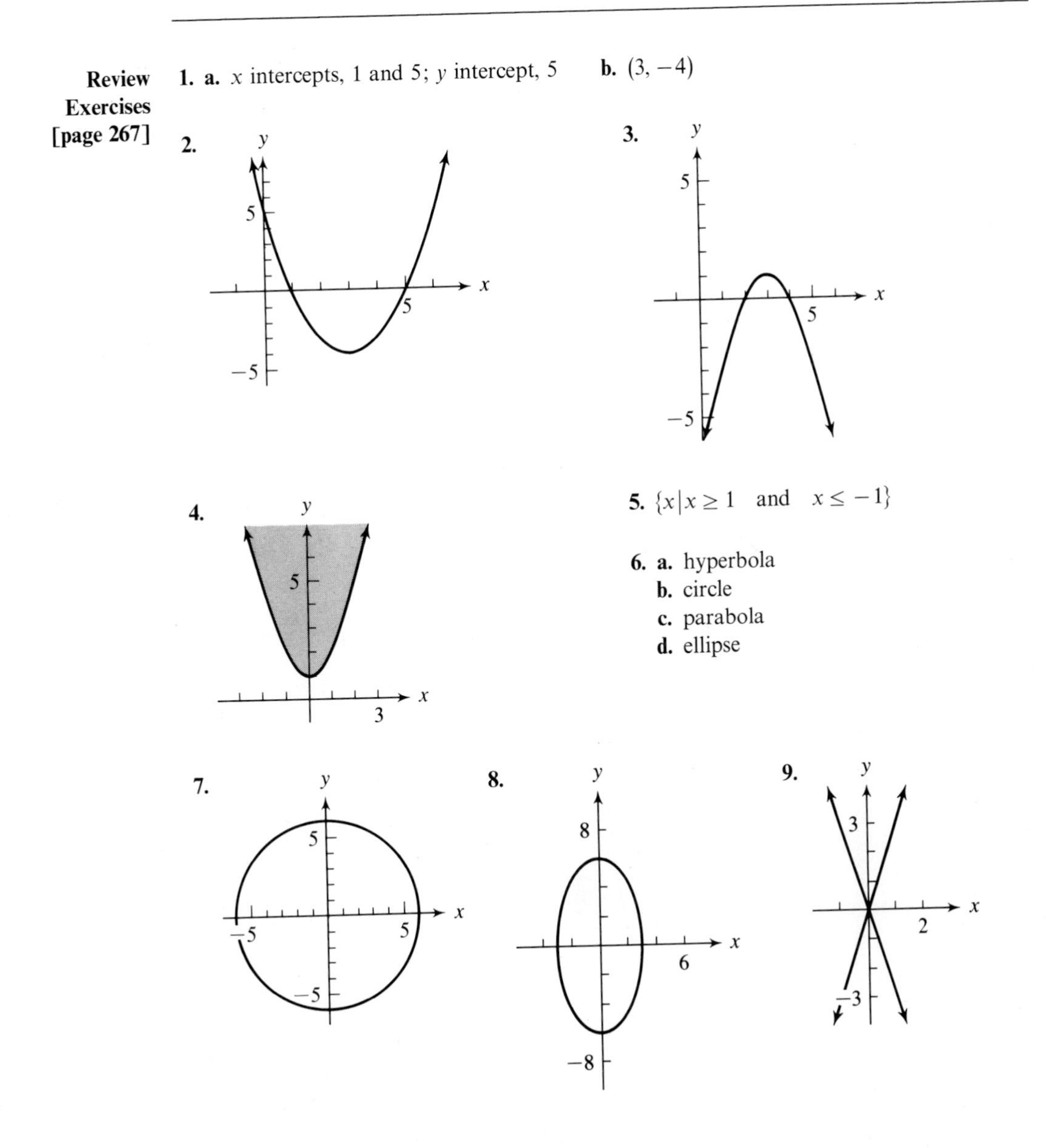

5. $\{x \mid x \geq 1$ and $x \leq -1\}$

6. a. hyperbola
b. circle
c. parabola
d. ellipse

10.

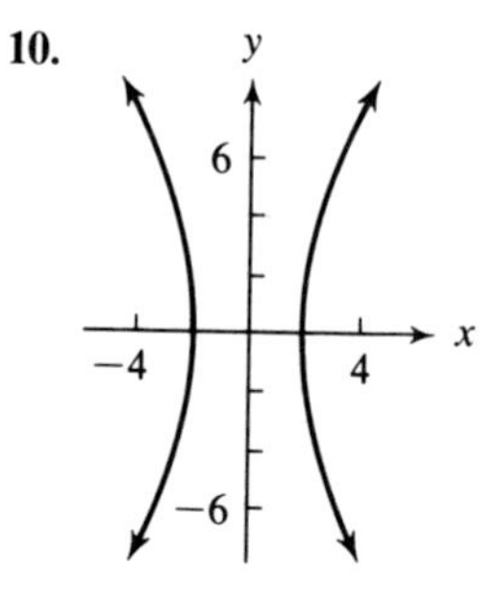

11. 9

12. $\dfrac{12{,}800}{81}$ pounds

13. a. $F^{-1} = \{(7,3), (8,4), (4,8)\}$; F^{-1} is a function

b. $G^{-1} = \{(6,2), (8,3), (8,5)\}$; G^{-1} is not a function

14. $x = y^2 + 9y$; F^{-1} is not a function

15. **16.**

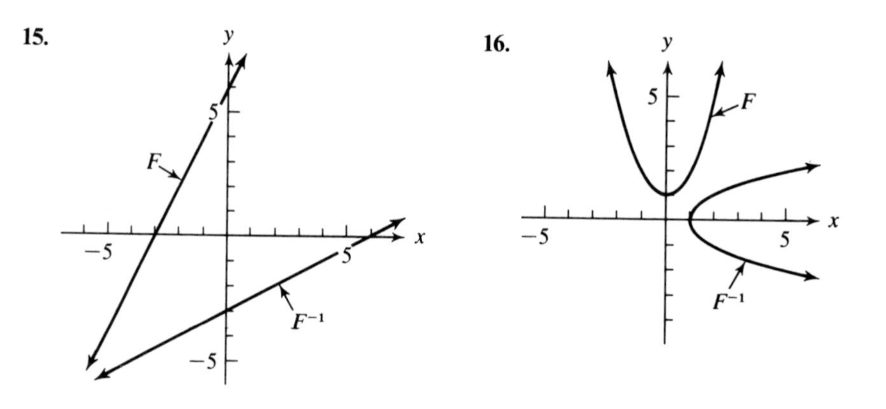

Exercise 9.1 [page 271]

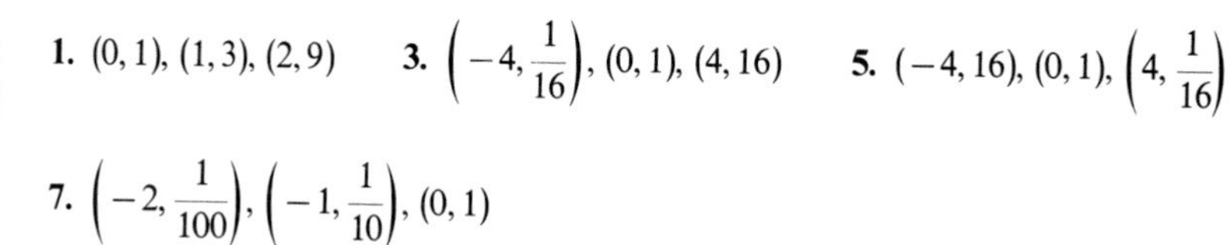

1. (0, 1), (1, 3), (2, 9) **3.** $\left(-4, \frac{1}{16}\right)$, (0, 1), (4, 16) **5.** (−4, 16), (0, 1), $\left(4, \frac{1}{16}\right)$

7. $\left(-2, \frac{1}{100}\right)$, $\left(-1, \frac{1}{10}\right)$, (0, 1)

9.

y

50

x

−3 0 3

11.

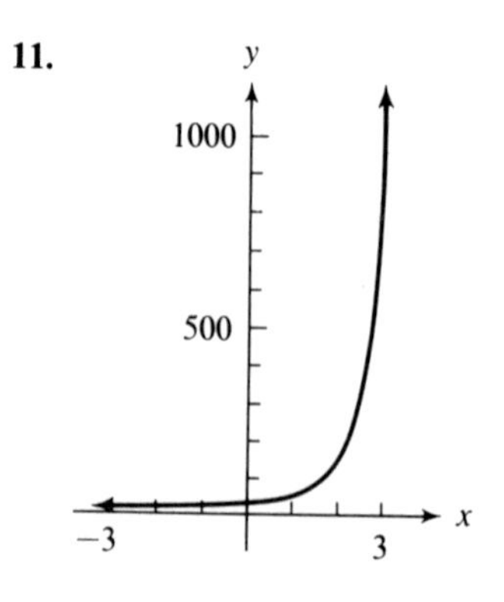

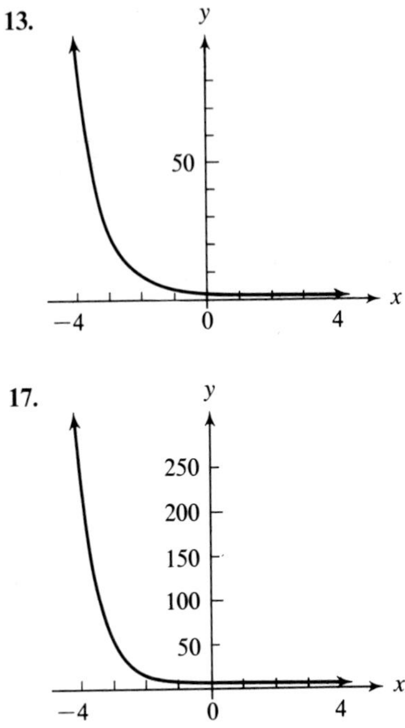

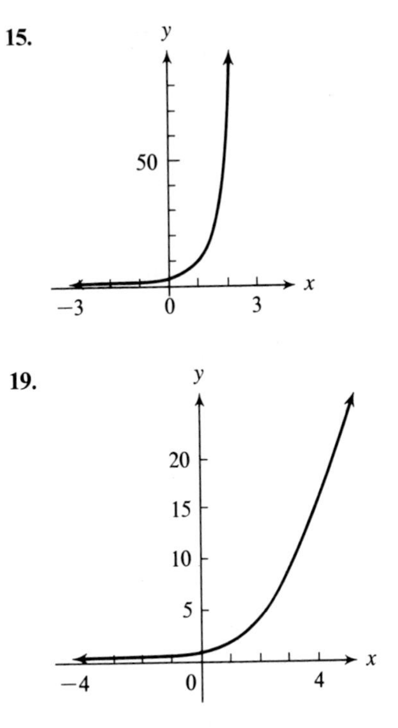

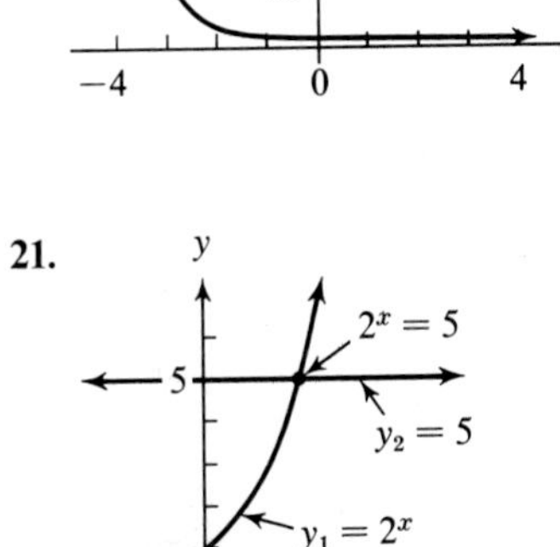

23. It will define an increasing function for all $a > 1$ and a decreasing function for all $0 < a < 1$.

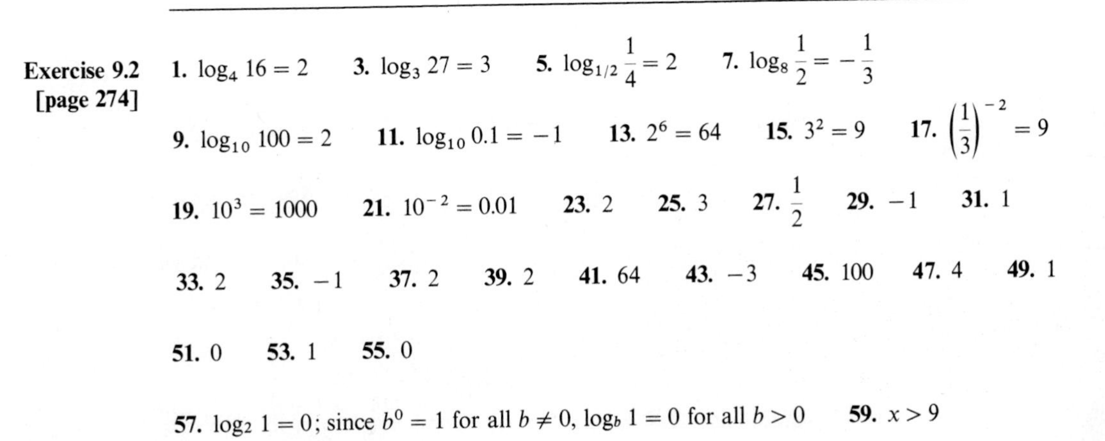

Exercise 9.2 [page 274]

1. $\log_4 16 = 2$ **3.** $\log_3 27 = 3$ **5.** $\log_{1/2} \frac{1}{4} = 2$ **7.** $\log_8 \frac{1}{2} = -\frac{1}{3}$

9. $\log_{10} 100 = 2$ **11.** $\log_{10} 0.1 = -1$ **13.** $2^6 = 64$ **15.** $3^2 = 9$ **17.** $\left(\frac{1}{3}\right)^{-2} = 9$

19. $10^3 = 1000$ **21.** $10^{-2} = 0.01$ **23.** 2 **25.** 3 **27.** $\frac{1}{2}$ **29.** -1 **31.** 1

33. 2 **35.** -1 **37.** 2 **39.** 2 **41.** 64 **43.** -3 **45.** 100 **47.** 4 **49.** 1

51. 0 **53.** 1 **55.** 0

57. $\log_2 1 = 0$; since $b^0 = 1$ for all $b \neq 0$, $\log_b 1 = 0$ for all $b > 0$ **59.** $x > 9$

Exercise 9.3 [page 276]

1. $\log_b 2 + \log_b x$ **3.** $\log_b 3 + \log_b x + \log_b y$ **5.** $\log_b x - \log_b y$

7. $\log_b x + \log_b y - \log_b z$ **9.** $3 \log_b x$ **11.** $\frac{1}{2} \log_b x$ **13.** $\frac{2}{3} \log_b x$

15. $2 \log_b x + 3 \log_b y$ **17.** $\frac{1}{2} \log_b x + \log_b y - 2 \log_b z$

19. $\frac{1}{3} \log_{10} x + \frac{2}{3} \log_{10} y - \frac{1}{3} \log_{10} z$ **21.** $\frac{1}{2} \log_{10} x + \frac{2}{3} \log_{10} y$

23. $\log_{10} 2 + \log_{10} \pi + \frac{1}{2} \log_{10} l - \frac{1}{2} \log_{10} g$ **25.** $\frac{1}{2} \log_{10} (s - a) + \frac{1}{2} \log_{10} (s - b)$

27. $\log_b xy$ **29.** $\log_b \frac{x^2}{y^3}$ **31.** $\log_b \frac{x^3 y}{z^2}$ **33.** $\log_{10} \frac{\sqrt{xy}}{z}$ **35.** $\log_b x^{-2}$ or $\log_b \frac{1}{x^2}$

37. $\{500\}$ **39.** $\{4\}$ **41.** $\{3\}$

43. Show that the left and right members of the equality reduce to the same quantity. For the left side: $\log_b 4 + \log_b 8 = \log_b 2^2 + \log_b 2^3 = 2 \log_b 2 + 3 \log_b 2 = 5 \log_b 2$. For the right side: $\log_b 64 - \log_b 2 = \log_b 2^6 - \log_b 2 = 6 \log_b 2 - \log_b 2 = 5 \log_b 2$.

45. $2 \log_b 6 - \log_b 9 = \log_b 6^2 - \log_b 9 = \log_b 36 - \log_b 9 = \log_b \frac{36}{9} = \log_b 4$

$$= \log_b 2^2 = 2 \log_b 2$$

47. $\frac{1}{2} \log_b 12 - \frac{1}{2} \log_b 3 = \frac{1}{2} (\log_b 12 - \log_b 3) = \frac{1}{2}\left(\log_b \frac{12}{3}\right) = \frac{1}{2} (\log_b 2^2) = \log_b 2 = \frac{1}{3} \log_b 8$

49. If $x_1 = b^{\log_b x_1}$ and $x_2 = b^{\log_b x_2}$, then

$$\frac{x_2}{x_1} = \frac{b^{\log_b x_2}}{b^{\log_b x_1}} = b^{\log_b x_2 - \log_b x_1},$$

and by the definition of a logarithm,

$$\log_b \frac{x_2}{x_1} = \log_b x_2 - \log_b x_1.$$

51. Substitute arbitrary numbers, for example, 1 and 10 for x and y in the given equation.

$$\log_{10} (1 + 10) \stackrel{?}{=} \log_{10} 1 + \log_{10} 10$$
$$\log_{10} 11 \stackrel{?}{=} 0 + 1$$
$$\log_{10} 11 \neq 1$$

Since $\log_{10} 11 \neq 1$, $\log_{10} (x + y) \neq \log_{10} x + \log_{10} y$ for all $x, y > 0$.

Exercise 9.4 [page 282]

1. 2 **3.** 3 **5.** -2 or $8 - 10$ **7.** 0 **9.** -4 or $6 - 10$ **11.** 4 **13.** 0.8280

15. 1.9227 **17.** 2.5011 **19.** $9.9101 - 10$ **21.** $8.9031 - 10$ **23.** 2.3945

25. 4.10 **27.** 36.7 **29.** 0.0642 **31.** 16.0 **33.** 5480 **35.** 0.000718 **37.** 9.10

39. 5000 **41.** 113 **43.** 2 and 3; 2 and 3; 1 and 2 **45.** 49 **47.** 2.97×10^{-1}

49. 2.05×10^{-3}

Exercise 9.5 [page 286]

1. 0.6246 **3.** 0.7937 **5.** 3.1824 **7.** 4.5695 **9.** $9.7095 - 10$ **11.** $7.9218 - 10$

13. 3.225 **15.** 89.38 **17.** 10.52 **19.** 0.05076 **21.** 0.7485 **23.** 0.7495

Exercise 9.6 [page 290]

1. 4.014 **3.** 2.299 **5.** 0.0004613 **7.** 64.34 **9.** 2.010 **11.** 3.435×10^{-10}

13. 0.04582 **15.** 0.2777 **17.** 9.872 **19.** 4.746 **21.** 1.394 **23.** 3.483

25. 57.81 **27.** 2.207 **29.** 1.11 seconds

Exercise 9.7 [page 293]

1. $\dfrac{\log_{10} 7}{\log_{10} 2}$; 2.8076 **3.** $\dfrac{\log_{10} 8}{\log_{10} 3} - 1$; 0.8929 **5.** $\dfrac{\log_{10} 3}{2 \log_{10} 7} + \dfrac{1}{2}$; 0.7823

7. $\sqrt{\dfrac{\log_{10} 15}{\log_{10} 4}}, -\sqrt{\dfrac{\log_{10} 15}{\log_{10} 4}}$; 1.3976, -1.3976 **9.** $-\dfrac{1}{\log_{10} 3}$; -2.0960

11. $1 - \dfrac{\log_{10} 15}{\log_{10} 3}$; -1.4651 **13.** $n = \dfrac{\log_{10} y}{\log_{10} x}$ **15.** $t = \dfrac{\log_{10} y}{k \log_{10} e}$

Certain of the answers below are approximate values. They are consistent with the given data and the use of four-place tables.

17. 4% **19.** 2.5% **21.** \$7396; \$7430 **23.** 7.7 **25.** 2.5×10^{-6}

27. 30 inches of mercury; 16.1 inches of mercury **29.** 2.63

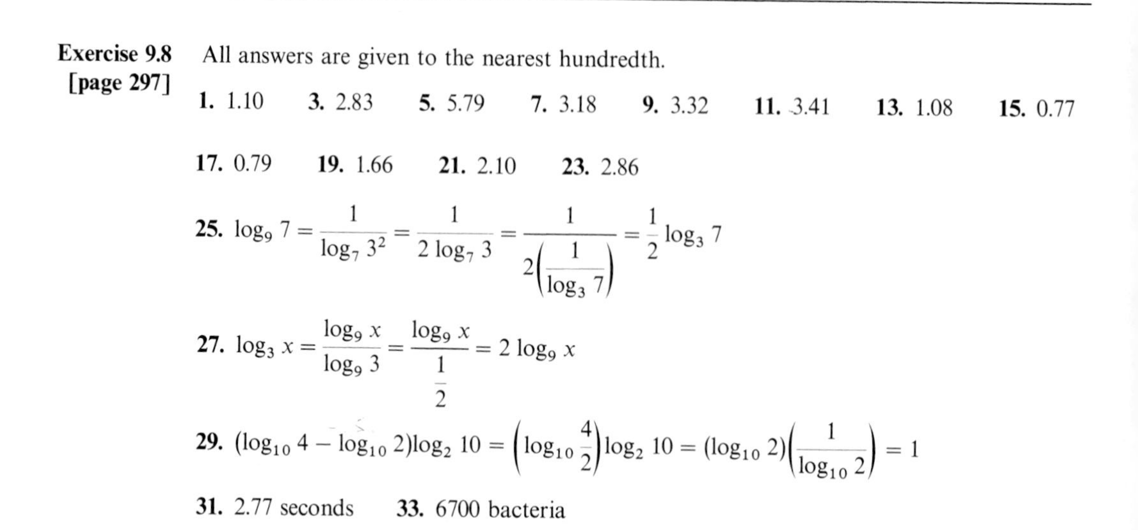

Exercise 9.8 [page 297] All answers are given to the nearest hundredth.

1. 1.10 **3.** 2.83 **5.** 5.79 **7.** 3.18 **9.** 3.32 **11.** 3.41 **13.** 1.08 **15.** 0.77

17. 0.79 **19.** 1.66 **21.** 2.10 **23.** 2.86

25. $\log_9 7 = \dfrac{1}{\log_7 3^2} = \dfrac{1}{2\log_7 3} = \dfrac{1}{2\left(\dfrac{1}{\log_3 7}\right)} = \dfrac{1}{2}\log_3 7$

27. $\log_3 x = \dfrac{\log_9 x}{\log_9 3} = \dfrac{\log_9 x}{\dfrac{1}{2}} = 2\log_9 x$

29. $(\log_{10} 4 - \log_{10} 2)\log_2 10 = \left(\log_{10}\dfrac{4}{2}\right)\log_2 10 = (\log_{10} 2)\left(\dfrac{1}{\log_{10} 2}\right) = 1$

31. 2.77 seconds **33.** 6700 bacteria

Review Exercises [page 300]

1. a. **b.**

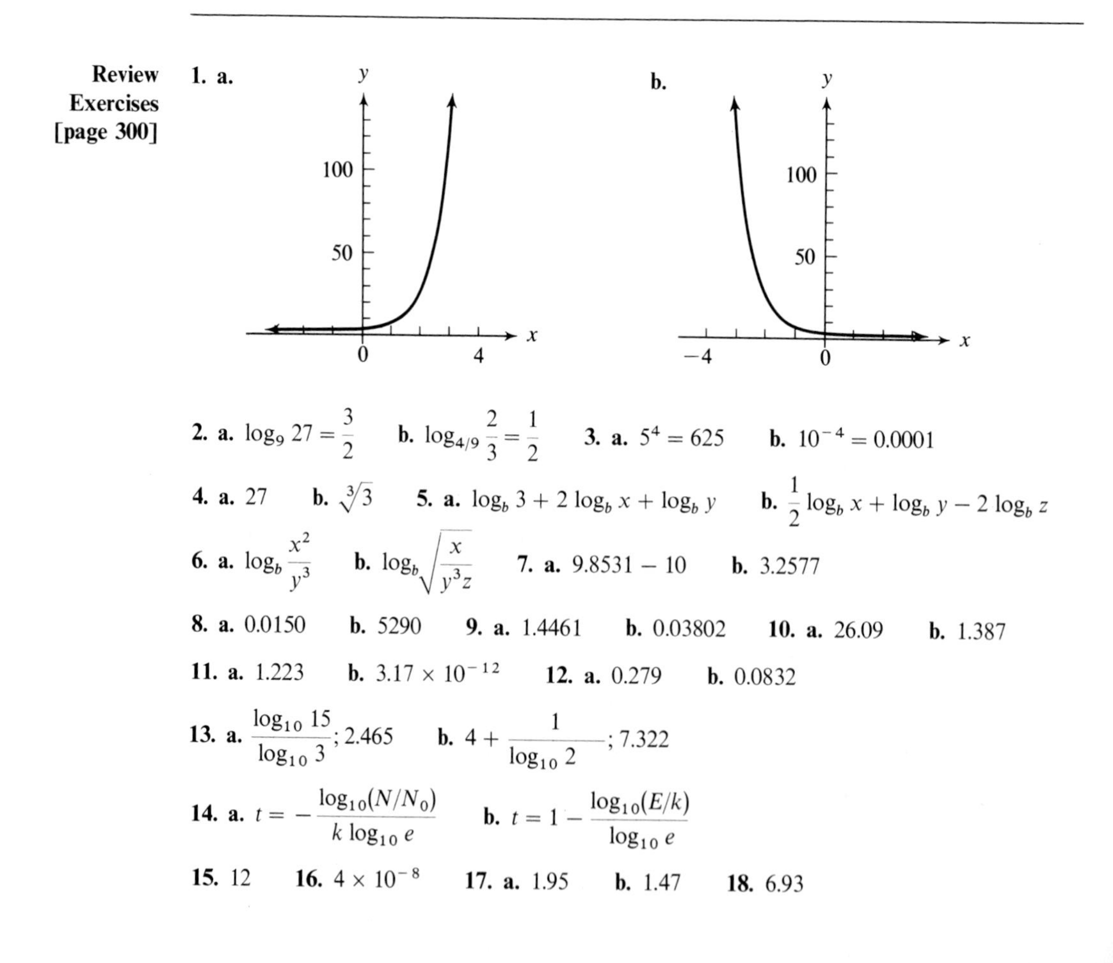

2. a. $\log_9 27 = \dfrac{3}{2}$ **b.** $\log_{4/9}\dfrac{2}{3} = \dfrac{1}{2}$ **3. a.** $5^4 = 625$ **b.** $10^{-4} = 0.0001$

4. a. 27 **b.** $\sqrt[3]{3}$ **5. a.** $\log_b 3 + 2\log_b x + \log_b y$ **b.** $\dfrac{1}{2}\log_b x + \log_b y - 2\log_b z$

6. a. $\log_b \dfrac{x^2}{y^3}$ **b.** $\log_b \sqrt{\dfrac{x}{y^3 z}}$ **7. a.** 9.8531 − 10 **b.** 3.2577

8. a. 0.0150 **b.** 5290 **9. a.** 1.4461 **b.** 0.03802 **10. a.** 26.09 **b.** 1.387

11. a. 1.223 **b.** 3.17×10^{-12} **12. a.** 0.279 **b.** 0.0832

13. a. $\dfrac{\log_{10} 15}{\log_{10} 3}$; 2.465 **b.** $4 + \dfrac{1}{\log_{10} 2}$; 7.322

14. a. $t = -\dfrac{\log_{10}(N/N_0)}{k\log_{10} e}$ **b.** $t = 1 - \dfrac{\log_{10}(E/k)}{\log_{10} e}$

15. 12 **16.** 4×10^{-8} **17. a.** 1.95 **b.** 1.47 **18.** 6.93

Exercise 10.1 [page 308]

1. $\{(3, 2)\}$ 3. $\{(2, 1)\}$

5. $\{(-5, 4)\}$ 7. $\left\{\left(0, \frac{3}{2}\right)\right\}$ 9. $\left\{\left(\frac{2}{3}, -1\right)\right\}$

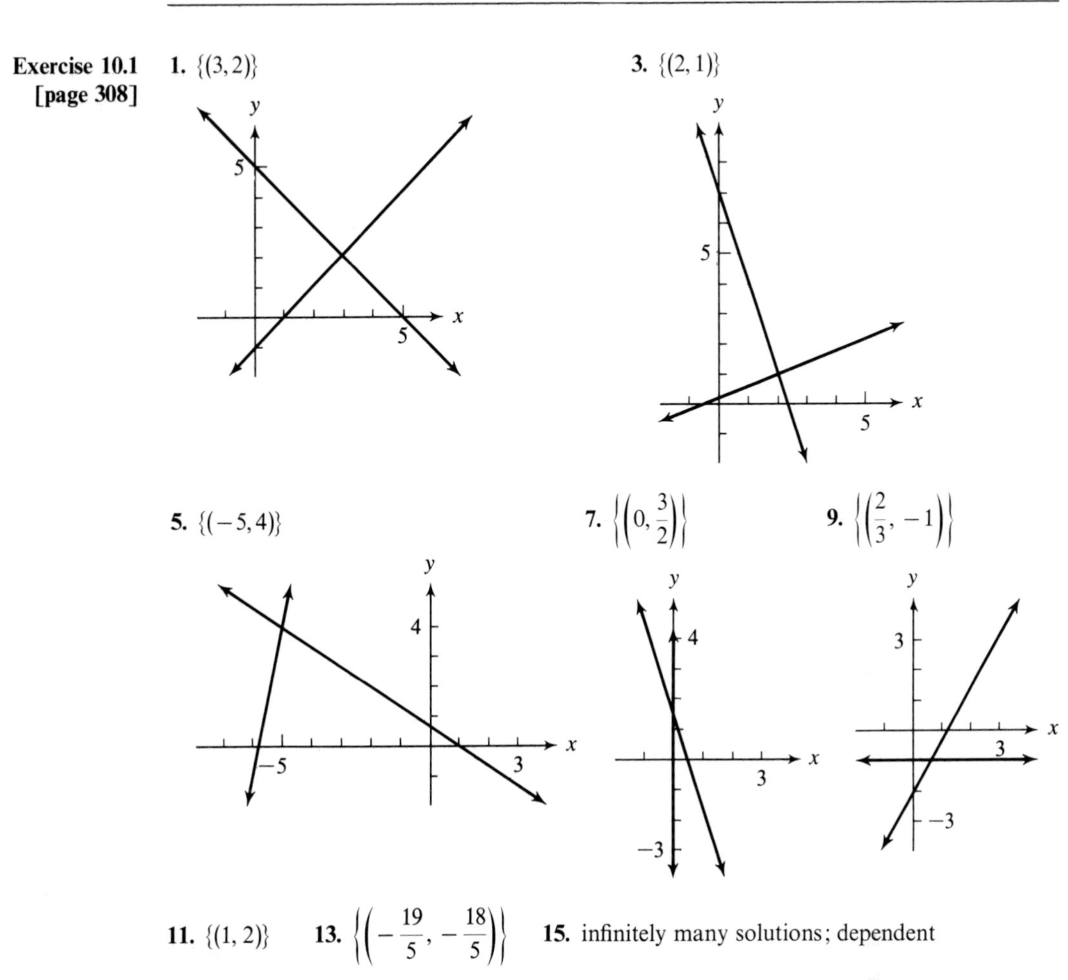

11. $\{(1, 2)\}$ 13. $\left\{\left(-\frac{19}{5}, -\frac{18}{5}\right)\right\}$ 15. infinitely many solutions; dependent

17. $\varnothing$; inconsistent 19. 9; 15 21. 39; 40 23. 46 children; 36 adults

25. $\left\{\left(\frac{1}{5}, \frac{1}{2}\right)\right\}$ 27. $\{(-5, 3)\}$ 29. $\left\{\left(\frac{1}{6}, \frac{1}{2}\right)\right\}$ 31. \$1400 at 6%; \$600 at 8%

33. white, 1957; red, 1961 35. $a = 1; b = -1$ 37. $x = \frac{a+b}{2}; y = \frac{b-a}{2}$

39. no 41. $x = \frac{c_1b_2 - c_2b_1}{a_1b_2 - a_2b_1}; y = \frac{a_1c_2 - a_2c_1}{a_1b_2 - a_2b_1}$ $(a_1b_2 - a_2b_1 \neq 0)$

Exercise 10.2 [page 313]

1. $\{(1, 2, -1)\}$ 3. $\{(2, -2, 0)\}$ 5. $\{(2, 2, 1)\}$ 7. dependent, infinitely many solutions

9. $\{(3, -1, 1)\}$ 11. $\{(4, -2, 2)\}$ 13. 3, 6, 6 15. 60 nickels; 20 dimes; 5 quarters

17. $x^2 + y^2 + 2x + 2y - 23 = 0$ 19. $a = 1; b = 2; c = 3$

ercise 10.3 [page 319]

1. $\{(-1, -4), (5, 20)\}$ **3.** $\{(2, 3), (3, 2)\}$

5. $\{(-3, 4), (4, -3)\}$ **7.** $\{(2, 2), (-2, -2)\}$

9. $\{(i\sqrt{7}, 4), (-i\sqrt{7}, 4)\}$ **11.** $\left\{\left(\frac{20 + i\sqrt{182}}{3}, \frac{10 - i\sqrt{182}}{3}\right)\left(\frac{20 - i\sqrt{182}}{3}, \frac{10 + i\sqrt{182}}{3}\right)\right\}$

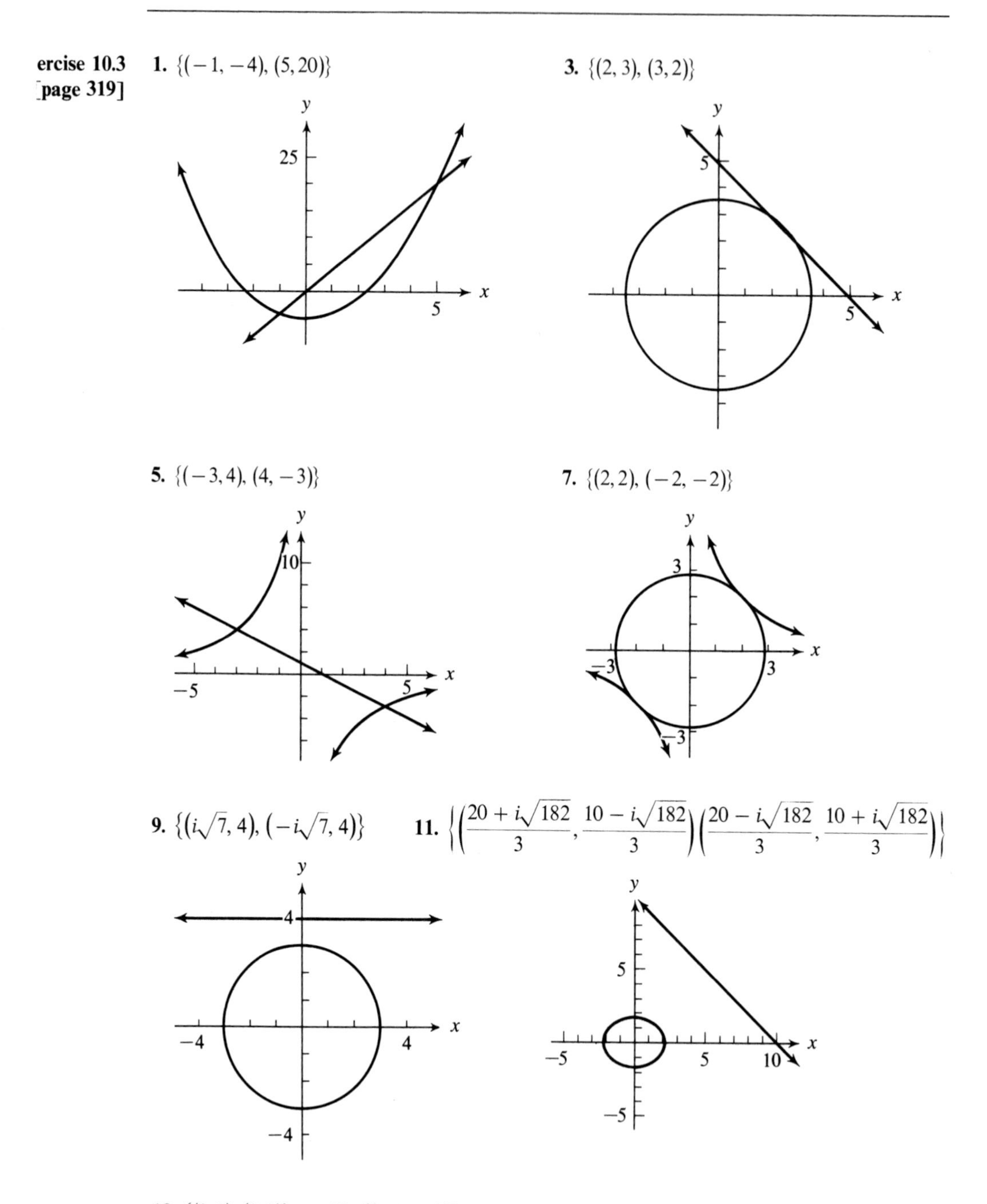

13. $\{(3, 1), (2, 0)\}$ **15.** $\{(-1, -3)\}$ **17.** 2; 3 **19.** length, 12 inches; width, 1 inch

21. length, 7 feet; width, 2 feet **23.** $P = 6$ pounds per square inch; $V = 5$ cubic inches

25. $\{(2, 2), (-2, -2)\}$

Exercise 10.4 [page 323]

1. $\{(1,3), (-1,3), (1,-3), (-1,-3)\}$ **3.** $\{(1,2), (-1,2), (1,-2), (-1,-2)\}$

5. $\{(3, \sqrt{2}), (-3, \sqrt{2}), (3, -\sqrt{2}), (-3, -\sqrt{3})\}$ **7.** $\{(2, 1), (-2, 1), (2, -1), (-2, -1)\}$

9. $\{(\sqrt{3}, 4), (-\sqrt{3}, 4), (\sqrt{3}, -4), (-\sqrt{3}, -4)\}$

11. $\{(2, -2), (-2, 2), (2i\sqrt{2}, i\sqrt{2}), (-2i\sqrt{2}, -i\sqrt{2})\}$

13. $\{(1, -1), (-1, 1), (i, i), (-i, -i)\}$ **15.** $\{(3, 1), (-3, -1), (2\sqrt{7}, -\sqrt{7}), (-2\sqrt{7}, \sqrt{7})\}$

17. $\left\{\left(i, \frac{1}{2}\right), \left(i, -\frac{1}{2}\right), \left(-i, \frac{1}{2}\right), \left(-i, -\frac{1}{2}\right)\right\}$ **19. a.** 1 **b.** 2 **c.** 4

Review Exercises [page 326]

1. $\left\{\left(\frac{1}{2}, \frac{7}{2}\right)\right\}$ **2.** $\{(1,2)\}$ **3.** $\{(12,0)\}$ **4.** $\left\{\left(\frac{22}{35}, \frac{11}{14}\right)\right\}$

5. a. neither dependent nor inconsistent **b.** dependent

6. a. inconsistent **b.** neither dependent nor inconsistent **7.** $\{(2,0,-1)\}$

8. $\{(2,1,-1)\}$ **9.** $\{(2,-5,3)\}$ **10.** $\left\{\left(\frac{4}{17}, -\frac{2}{17}, -\frac{84}{17}\right)\right\}$ **11.** $\{(-2, 1, 3)\}$

12. $\{(2, -1, 0)\}$ **13.** $\left\{\left(3, \frac{\sqrt{3}}{3}\right), \left(3, -\frac{\sqrt{3}}{3}\right)\right.$ **14.** $\{(4i, -2), (-4i, -2)\}$

15. $\{(4, -13), (1, 2)\}$ **16.** $\left\{\left(\frac{6}{7}, -\frac{37}{7}\right), (2, -3)\right\}$

17. $\{(1, \sqrt{5}), (1, -\sqrt{5}), (-1, \sqrt{5}), (-1, -\sqrt{5})\}$ **18.** $\{(2,3), (2,-3), (-2,3), (-2,-3)\}$

Exercise 11.1 [page 331]

1. $-4, -3, -2, -1$ **3.** $-\frac{1}{2}, 1, \frac{7}{2}, 7$ **5.** $2, \frac{3}{2}, \frac{4}{3}, \frac{5}{4}$ **7.** 0, 1, 3, 6

9. $-1, 1, -1, 1$ **11.** $1, 0, -\frac{1}{3}, \frac{1}{2}$ **13.** $1 + 4 + 9 + 16$ **15.** $3 + 4 + 5$

17. $2 + 6 + 12 + 20$ **19.** $-\frac{1}{2} + \frac{1}{4} - \frac{1}{8} + \frac{1}{16}$ **21.** $1 + 3 + 5 + \cdots$

23. $1 + \frac{1}{2} + \frac{1}{4} + \cdots$ **25.** $\sum_{i=1}^{4} i$ **27.** $\sum_{i=1}^{4} x^{2i-1}$ **29.** $\sum_{i=1}^{5} i^2$ **31.** $\sum_{i=1}^{\infty} \frac{i}{i+1}$

33. $\sum_{i=1}^{\infty} \frac{i}{2i-1}$ **35.** $\sum_{i=1}^{\infty} \frac{2^{i-1}}{i}$ **37.** 40 bacteria; 160 bacteria; $10(2)^n$ bacteria

39. 25 grams; $\frac{25}{4}$ grams

Exercise 11.2 [page 335]

1. 11, 15, 19 $\quad s_n = 4n - 1$ **3.** $-9, -13, -17$ $\quad s_n = 3 - 4n$ **5.** $x + 2, x + 3, x + 4$ $\quad s_n = x + n - 1$

7. $x + 5a, x + 7a, x + 9a$ $\quad s_n = x + 2an - a$ **9.** $2x + 7, 2x + 10, 2x + 13$ $\quad s_n = 2x + 3n - 2$ **11.** $3x, 4x, 5x$ $\quad s_n = nx$

13. 31 **15.** $\frac{15}{2}$ **17.** -92 **19.** 2; 3; 41 **21.** twenty-eighth term **23.** 1; 8

25. 16 **27.** 9; 14; 19 **29.** 63 **31.** 806 **33.** -6 **35.** 1938 **37.** 196 bricks

39. 2, 7, 12 **41.** $\frac{14}{15}$

43. The first n odd natural numbers form an arithmetic sequence with $a = 1$ and $d = 2$. Hence, the sum of the first n odd natural numbers is

$$S_n = \frac{n}{2}[2a + (n-1)d] = \frac{n}{2}[2 + (n-1)2] = \frac{n}{2}[2 + 2n - 2] = \frac{n}{2}[2n] = n^2$$

Exercise 11.3 [page 340]

1. 128, 512, 2048 $\quad s_n = 2(4)^{n-1}$ **3.** $\frac{16}{3}, \frac{32}{3}, \frac{64}{3}$ $\quad s_n = \frac{2}{3}(2)^{n-1}$ **5.** $-\frac{1}{2}, \frac{1}{4}, -\frac{1}{8}$ $\quad s_n = 4\left(-\frac{1}{2}\right)^{n-1}$ **7.** $-\frac{x^2}{a^2}, \frac{x^3}{a^3}, -\frac{x^4}{a^4}$ $\quad s_n = \frac{a}{x}\left(-\frac{x}{a}\right)^{n-1}$

9. 1536 **11.** $-243a^{20}$ **13.** 3 **15.** 3, 9 **17.** 18 or -18

19. 16, 8, 4 or $-16, 8, -4$ **21.** 1092 **23.** $\frac{31}{32}$ **25.** $\frac{364}{729}$ **27.** $\frac{3}{4}, \frac{7}{8}, \frac{15}{16}, \frac{31}{32}$; 1

Exercise 11.4 [page 345]

1. 24 **3.** does not exist **5.** $\frac{9}{20}$ **7.** $\frac{8}{49}$ **9.** 2 **11.** $\frac{1}{3}$ **13.** $\frac{31}{99}$

15. $\frac{2408}{999}$ **17.** $\frac{29}{225}$ **19.** 20 centimeters **21.** 30 feet

---

Exercise 11.5 [page 349]

1. $8 \cdot 7 \cdot 6 \cdot 5 \cdot 4 \cdot 3 \cdot 2 \cdot 1$ **3.** $2 \cdot 4 \cdot 3 \cdot 2 \cdot 1$ **5.** $6 \cdot 5 \cdot 4 \cdot 3 \cdot 2 \cdot 1$ **7.** 120

9. 72 **11.** 15 **13.** 28 **15.** 3! **17.** $\frac{6!}{2!}$ **19.** $\frac{8!}{5!}$

21. $n(n-1)(n-2)\cdots 3 \cdot 2 \cdot 1$ **23.** $3n(3n-1)(3n-2)\cdots 3 \cdot 2 \cdot 1$

25. $(n-2)(n-3)(n-4)\cdots 3\cdot 2\cdot 1$ **27.** $x^5+15x^4+90x^3+270x^2+405x+243$

29. $x^4-12x^3+54x^2-108x+81$ **31.** $8x^3-6x^2y+\frac{3}{2}xy^2-\frac{1}{8}y^3$

33. $\frac{1}{64}x^6+\frac{3}{8}x^5+\frac{15}{4}x^4+20x^3+60x^2+96x+64$

35. $x^{20}+20x^{19}y+\frac{20\cdot 19}{2!}x^{18}y^2+\frac{20\cdot 19\cdot 18}{3!}x^{17}y^3$

37. $a^{12}+12a^{11}(-2b)+\frac{12\cdot 11}{2!}a^{10}(-2b)^2+\frac{12\cdot 11\cdot 10}{3!}a^9(-2b)^3$

39. $x^{10}+10x^9(-\sqrt{2})+\frac{10\cdot 9}{2!}x^8(-\sqrt{2})^2+\frac{10\cdot 9\cdot 8}{3!}x^7(-\sqrt{2})^3$

41. $-3003a^{10}b^5$ **43.** $3360x^6y^4$

45. a. $1^{-1}+(-1)(1^{-2})x+1(1^{-3})x^2+(-1)(1^{-4})x^3$ or $1-x+x^2-x^3$

b. $1-x+x^2-x^3$; the results are equal

Exercise 11.6 [page 354]

1. 24 **3.** 362,880 **5.** 2058 **7.** 720 **9.** 1176 **11.** 648 **13.** 448

15. 360 **17.** 17,576 **19.** 676,000 **21.** 60 **23.** 24 **25.** 45,360

27. 6720 **29.** 60 **31.** 34,650

Exercise 11.7 [page 357]

1. 1365 **3.** 7 **5.** 2,598,960 **7.** 15 **9.** 84

11. $\binom{13}{6}\cdot\binom{13}{3}\cdot\binom{13}{4}=350{,}904{,}840$ **13.** 7425

15. $\binom{n}{n-r}=\frac{n!}{(n-r)![n-(n-r)]!}=\frac{n!}{(n-r)!r!}=\frac{n!}{r!(n-r)!}=\binom{n}{r}$ **17.** 7

19. From Section 11.5, we see that the coefficients of the binomial expansion of $(a+b)^4$ are 1, 4, 6, 4, and 1. Since

$$\binom{4}{0}=\frac{4!}{0!4!}=1,\quad \binom{4}{1}=\frac{4!}{1!3!}=4,\quad \binom{4}{2}=\frac{4!}{2!2!}=6,$$

$$\binom{4}{3}=\frac{4!}{3!1!}=4,\quad \binom{4}{4}=\frac{4!}{4!0!}=1,$$

they are equal to the coefficients of the binomial expansion $(a+b)^4$.

Review Exercises [page 360]

1. $1, -\frac{1}{2}, \frac{1}{3}, -\frac{1}{4}$ **2.** $2 + 6 + 12 + 20$ **3.** $4n + 1$ **4. a.** 94 **b.** 138 **5.** 68

6. $5\left(\frac{9}{5}\right)^{n-1}$ **7. a.** $-\frac{81}{8}$ **b.** $-\frac{1261}{216}$ **8.** $\frac{729}{32}$ **9.** $\frac{121}{243}$ **10.** $\frac{1}{2}$ **11.** $\frac{4}{9}$

12. 56 **13.** $x^{10} - 20x^9y + 180x^8y^2 - 960x^7y^3$ **14.** $-15{,}360x^3y^7$

15. a. 125 **b.** 60 **16.** 360 **17.** 1260 **18.** 15 **19.** 24

20. $\binom{13}{4}\cdot\binom{13}{5}\cdot\binom{13}{2}\cdot\binom{13}{2}$

Exercise A.1 [page 365]

1. $x - 2$ **3.** $x + 2$ **5.** $x^3 - x^2 + \frac{-1}{x-2}$ **7.** $2x^2 - 2x + 3 + \frac{-8}{x+1}$

9. $2x^3 + 10x^2 + 50x + 249 + \frac{1251}{x-5}$ **11.** $x^2 + 2x - 3 + \frac{4}{x+2}$

13. $x^5 + x^4 + 2x^3 + 2x^2 + 2x + 1 + \frac{1}{x-1}$ **15.** $x^4 + x^3 + x^2 + x + 1$

17. $x^5 + x^4 + x^3 + x^2 + x + 1$

Exercise A.2 [page 369]

1. 74; -46 **3.** 5; 44

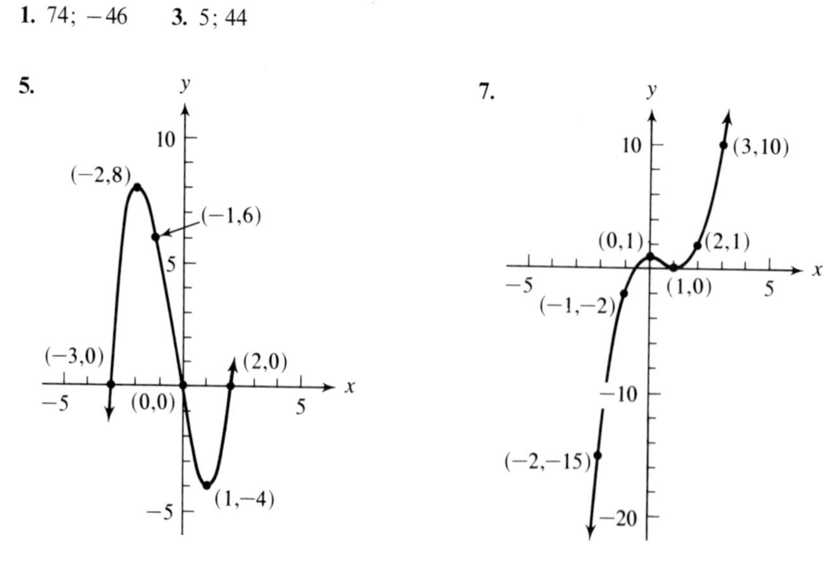

9. 11.

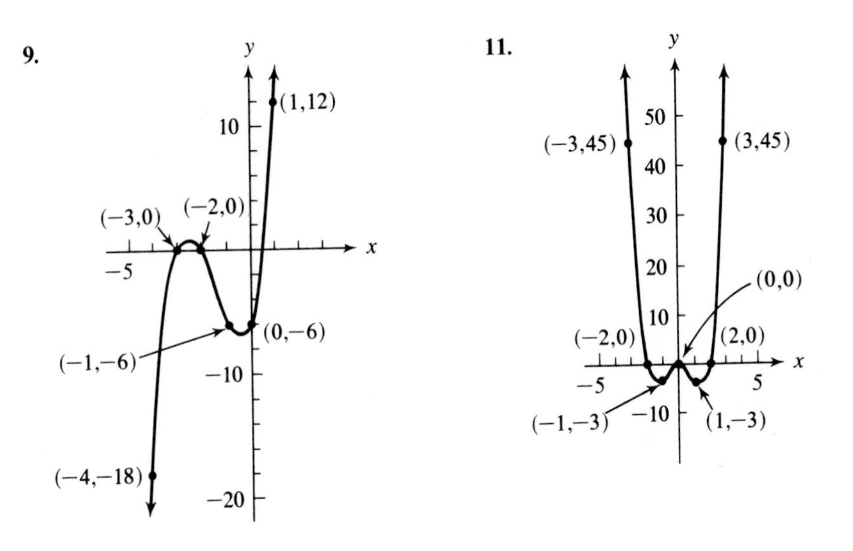

13. no **15.** yes **17.** 1, -1, and -2 **19.** -3, 0, 2, and 4

Review Exercises [page 371]

1. $y^2 + 2y - 4$ **2.** $y^6 + y^5 + y^4 + y^3 + y^2 + y + 1$ **3.** 19; -25 **4.** 4; 8

5. **6.**

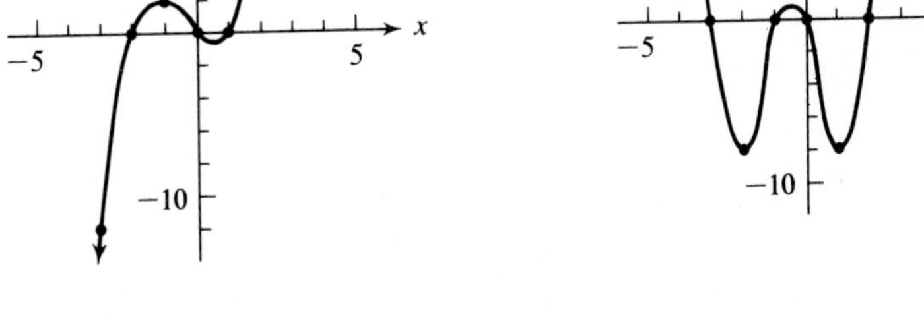

7. no **8.** -1; 3

Exercise B.1 [page 375]

1. $\{(2,3)\}$ **3.** $\{(-2,1)\}$ **5.** $\{(3,-1)\}$ **7.** $\{(1,2,2)\}$ **9.** $\left\{\left(\frac{76}{39}, -\frac{112}{39}, \frac{48}{39}\right)\right\}$

11. $\{(-3,1,-3)\}$

Exercise B.2 [page 378]

1. 1 **3.** -12 **5.** 2 **7.** 5 **9.** $\{(1,1)\}$ **11.** $\{(2,2)\}$ **13.** $\{(6,4)\}$

15. inconsistent, $\varnothing$ **17.** $\{(4,1)\}$ **19.** $\left\{\left(\frac{1}{a+b}, \frac{1}{a+b}\right)\right\}$ **21.** $\begin{vmatrix} a & a \\ b & b \end{vmatrix} = ab - ab = 0$

23. $\begin{vmatrix} a_1 & b_1 \\ a_2 & b_2 \end{vmatrix} = a_1b_2 - a_2b_1$; $-\begin{vmatrix} b_1 & a_1 \\ b_2 & a_2 \end{vmatrix} = -(b_1a_2 - b_2a_1) = -b_1a_2 + b_2a_1 = a_1b_2 - a_2b_1$

25. $\begin{vmatrix} ka & a \\ kb & b \end{vmatrix} = kab - kba = 0$

27. Since $D_x = \begin{vmatrix} c_1 & b_1 \\ c_2 & b_2 \end{vmatrix} = c_1b_2 - c_2b_1 = 0$, then $b_1c_2 = b_2c_1$.

Since $D_y = \begin{vmatrix} a_1 & c_1 \\ a_2 & c_2 \end{vmatrix} = a_1c_2 - a_2c_1 = 0$, then $a_1c_2 = a_2c_1$. Forming the proportion

$$\frac{b_1c_2}{a_1c_2} = \frac{b_2c_1}{a_2c_1},$$

if c_1 and c_2 are not both 0, then

$$\frac{b_1}{a_1} = \frac{b_2}{a_2},$$

from which $a_1b_2 = a_2b_1$ and $a_1b_2 - a_2b_1 = 0$. Since $D = \begin{vmatrix} a_1 & b_1 \\ a_2 & b_2 \end{vmatrix} = a_1b_2 - a_2b_1$, it follows that $D = 0$.

Exercise B.3 [page 382]

1. 3 **3.** 9 **5.** 0 **7.** -1 **9.** -5 **11.** 0 **13.** x^3 **15.** 0 **17.** $-2ab^2$

19. $\{3\}$ **21.** $\left\{2, -\frac{17}{7}\right\}$

23. Expanding about the elements of the third column of the given determinant produces

$$a\begin{vmatrix} y & y \\ z & z \end{vmatrix} - b\begin{vmatrix} x & x \\ z & z \end{vmatrix} + c\begin{vmatrix} x & x \\ y & y \end{vmatrix} = 0 + 0 + 0 = 0.$$

25. For the left side: $\begin{vmatrix} 1 & 2 & 3 \\ 4 & 5 & 6 \\ 0 & 0 & 1 \end{vmatrix} = 1\begin{vmatrix} 1 & 2 \\ 4 & 5 \end{vmatrix} = 5 - 8 = -3.$

For the right side: $-\begin{vmatrix} 4 & 5 & 6 \\ 1 & 2 & 3 \\ 0 & 0 & 1 \end{vmatrix} = -1\begin{vmatrix} 4 & 5 \\ 1 & 2 \end{vmatrix} = -1(8-5) = -1(3) = -3.$

Exercise B.4 [page 385]

1. $\{(1,1,1)\}$ **3.** $\{(1,1,0)\}$ **5.** $\{(1,-2,3)\}$ **7.** $\{(3,-1,-2)\}$

9. no unique solution **11.** $\left\{\left(1,-\frac{1}{3},\frac{1}{2}\right)\right\}$ **13.** $\{(-5,3,2)\}$ **15.** $\left\{\left(\frac{13}{2},0,-11\right)\right\}$

Review Exercises [page 388]

1. $\{(3,-1)\}$ **2.** $\{(-1,0,2)\}$ **3.** -13 **4.** $\{(-7,4)\}$ **5.** -2 **6.** $\{(2,-1,3)\}$

INDEX

COMMON LOGARITHMS

x	0	1	2	3	4	5	6	7	8	9
1.0	.0000	.0043	.0086	.0128	.0170	.0212	.0253	.0294	.0334	.0374
1.1	.0414	.0453	.0492	.0531	.0569	.0607	.0645	.0682	.0719	.0755
1.2	.0792	.0828	.0864	.0899	.0934	.0969	.1004	.1038	.1072	.1106
1.3	.1139	.1173	.1206	.1239	.1271	.1303	.1335	.1367	.1399	.1430
1.4	.1461	.1492	.1523	.1553	.1584	.1614	.1644	.1673	.1703	.1732
1.5	.1761	.1790	.1818	.1847	.1875	.1903	.1931	.1959	.1987	.2014
1.6	.2041	.2068	.2095	.2122	.2148	.2175	.2201	.2227	.2253	.2279
1.7	.2304	.2330	.2355	.2380	.2405	.2430	.2455	.2480	.2504	.2529
1.8	.2553	.2577	.2601	.2625	.2648	.2672	.2695	.2718	.2742	.2765
1.9	.2788	.2810	.2833	.2856	.2878	.2900	.2923	.2945	.2967	.2989
2.0	.3010	.3032	.3054	.3075	.3096	.3118	.3139	.3160	.3181	.3201
2.1	.3222	.3243	.3263	.3284	.3304	.3324	.3345	.3365	.3385	.3404
2.2	.3424	.3444	.3464	.3483	.3502	.3522	.3541	.3560	.3579	.3598
2.3	.3617	.3636	.3655	.3674	.3692	.3711	.3729	.3747	.3766	.3784
2.4	.3802	.3820	.3838	.3856	.3874	.3892	.3909	.3927	.3945	.3962
2.5	.3979	.3997	.4014	.4031	.4048	.4065	.4082	.4099	.4116	.4133
2.6	.4150	.4166	.4183	.4200	.4216	.4232	.4249	.4265	.4281	.4298
2.7	.4314	.4330	.4346	.4362	.4378	.4393	.4409	.4425	.4440	.4456
2.8	.4472	.4487	.4502	.4518	.4533	.4548	.4564	.4579	.4594	.4609
2.9	.4624	.4639	.4654	.4669	.4683	.4698	.4713	.4728	.4742	.4757
3.0	.4771	.4786	.4800	.4814	.4829	.4843	.4857	.4871	.4886	.4900
3.1	.4914	.4928	.4942	.4955	.4969	.4983	.4997	.5011	.5024	.5038
3.2	.5051	.5065	.5079	.5092	.5105	.5119	.5132	.5145	.5159	.5172
3.3	.5185	.5198	.5211	.5224	.5237	.5250	.5263	.5276	.5289	.5302
3.4	.5315	.5328	.5340	.5353	.5366	.5378	.5391	.5403	.5416	.5428
3.5	.5441	.5453	.5465	.5478	.5490	.5502	.5514	.5527	.5539	.5551
3.6	.5563	.5575	.5587	.5599	.5611	.5623	.5635	.5647	.5658	.5670
3.7	.5682	.5694	.5705	.5717	.5729	.5740	.5752	.5763	.5775	.5786
3.8	.5798	.5809	.5821	.5832	.5843	.5855	.5866	.5877	.5888	.5899
3.9	.5911	.5922	.5933	.5944	.5955	.5966	.5977	.5988	.5999	.6010
4.0	.6021	.6031	.6042	.6053	.6064	.6075	.6085	.6096	.6107	.6117
4.1	.6128	.6138	.6149	.6160	.6170	.6180	.6191	.6201	.6212	.6222
4.2	.6232	.6243	.6253	.6263	.6274	.6284	.6294	.6304	.6314	.6325
4.3	.6335	.6345	.6355	.6365	.6375	.6385	.6395	.6405	.6415	.6425
4.4	.6435	.6444	.6454	.6464	.6474	.6484	.6493	.6503	.6513	.6522
4.5	.6532	.6542	.6551	.6561	.6571	.6580	.6590	.6599	.6609	.6618
4.6	.6628	.6637	.6646	.6656	.6665	.6675	.6684	.6693	.6702	.6712
4.7	.6721	.6730	.6739	.6749	.6758	.6767	.6776	.6785	.6794	.6803
4.8	.6812	.6821	.6830	.6839	.6848	.6857	.6866	.6875	.6884	.6893
4.9	.6902	.6911	.6920	.6928	.6937	.6946	.6955	.6964	.6972	.6981
5.0	.6990	.6998	.7007	.7016	.7024	.7033	.7042	.7050	.7059	.7067
5.1	.7076	.7084	.7093	.7101	.7110	.7118	.7126	.7135	.7143	.7152
5.2	.7160	.7168	.7177	.7185	.7193	.7202	.7210	.7218	.7226	.7235
5.3	.7243	.7251	.7259	.7267	.7275	.7284	.7292	.7300	.7308	.7316
5.4	.7324	.7332	.7340	.7348	.7356	.7364	.7372	.7380	.7388	.7396
x	0	1	2	3	4	5	6	7	8	9